Textbook of Drug Design and Discovery

Fourth Edition

Textbook of
Drug Design and Discovery

Fourth Edition

Edited by

Povl Krogsgaard-Larsen
Kristian Strømgaard
Ulf Madsen

GEORGE GREEN LIBRARY OF
SCIENCE AND ENGINEERING

CRC Press
Taylor & Francis Group
Boca Raton   London   New York

CRC Press is an imprint of the
Taylor & Francis Group, an **informa** business

CRC Press
Taylor & Francis Group
6000 Broken Sound Parkway NW, Suite 300
Boca Raton, FL 33487-2742

© 2010 by Taylor and Francis Group, LLC
CRC Press is an imprint of Taylor & Francis Group, an Informa business

---
Library of Congress Cataloging-in-Publication Data

Textbook of drug design and discovery / editors, Povl Krogsgaard-Larsen, Kristian Strømgaard, and
    Ulf Madsen. -- 4th ed.
        p. ; cm.
    Includes bibliographical references and index.
    ISBN 978-1-4200-6322-6 (alk. paper)
    1. Drugs--Design. I. Krogsgaard-Larsen, Povl. II. Strømgaard, Kristian. III. Madsen, Ulf. IV. Title.
    [DNLM: 1. Drug Design. 2. Drug Discovery. QV 744 T3545 2010]

RS420.T49 2010
615'.1--dc22   2009013698
                                        _1006029361_

Visit the Taylor & Francis Web site at
http://www.taylorandfrancis.com

and the CRC Press Web site at
http://www.crcpress.com

Contents

Preface

The areas of medicinal chemistry and drug design are constantly undergoing transformations. The molecular biological revolution and the mapping of the human genome have had a major impact, and these developments have provided new challenges and opportunities for drug research in general and for drug design in particular. The major objectives for medicinal chemists are the transformation of pathophysiological data into a "chemical language" with the aim of designing molecules interacting specifically with the derailed or degenerating processes in the diseased organism.

Potential therapeutic targets are continuously being disclosed, which calls for rapid and effective target validation and for accelerated lead discovery procedures. Consequently, industrial medicinal chemistry laboratories are regularly revising their technology portfolio to meet these demands. Key words in this regard are constructions of diverse compound libraries, fragment-based and high-throughput screening, as well as accelerated ADME and toxicity examinations.

In parallel with this development, biostructure-based drug design and intelligent molecular mimicry are indispensable areas of the medicinal chemistry playing field. Structural biology is playing an increasingly important role, and the borders between biology, biochemistry, and chemistry are broadening and becoming a most fruitful working field for innovative and intuitive scientists in drug design.

Academic medicinal chemistry and drug design departments need to attract the attention of bright students, interested in the creative and fascinating nature of drug design. In order to reach this goal it is of the utmost importance to maintain focus on the integration of the scientific disciplines of chemistry and biology. In relation to industrial "hit-finding" procedures, students should be taught that the conversions of hits into lead structures and further into drug candidates require the integration of a number of related scientific disciplines, such as advanced synthetic chemistry, computational chemistry, biochemistry, structural biology, and molecular pharmacology.

Advances in structural biology have opened up a number of exciting avenues in drug design. Three-dimensional structural information derived from x-ray analyses of enzyme-inhibitor complexes is applied for the design of new types of inhibitors. Similarly, structures of integral membrane proteins and cocrystal structures with ligands are emerging as important tools in receptor-ligand design. Such approaches are the foundation for drug design on a rational basis and are currently an important aspect of student teaching programs in medicinal chemistry.

The use of biologically active natural products has experienced a revival as a starting point for both industrial and academic drug design projects. Not only do natural products often possess novel structural characteristics, but they also frequently exhibit unique biological mechanisms of action, although naturally occurring toxins typically show nonselective pharmacological effects. Thus, by systematic structural modification such toxins can be converted into leads with specific biological functions of key importance in diseases.

This fourth edition of the textbook attempts to cover the diverse aspects of current academic and industrial medicinal chemistry and drug design in an educational context. The book is divided into two parts: Chapters 1 through 10 cover general aspects, methods, and principles for drug design and discovery, while Chapters 11 through 25 cover specific targets and diseases.

Povl Krogsgaard-Larsen
Kristian Strømgaard
Ulf Madsen

Editors

Povl Krogsgaard-Larsen is a professor of medicinal chemistry in the Department of Medicinal Chemistry, Faculty of Pharmaceutical Sciences, at the University of Copenhagen. In 1970, he received his PhD in natural product chemistry. As an associate professor, he established a research program focusing on the conversion of naturally occurring toxins into specific pharmacological tools and therapeutic agents. Key lead structures in this research program were the *Amanita muscaria* constituents, muscimol and ibotenic acid, and the *Areca* nut alkaloid, arecoline, all of which interact nonselectively with GABA, glutamate, and muscarinic receptors, respectively. The redesign of muscimol resulted in a variety of specific GABA agonists, notably THIP and isoguvacine, and specific GABA uptake inhibitors, including nipecotic acid and guvacine. Ibotenic acid was converted into a broad range of subtype-selective glutamate receptor agonists including AMPA, from which the AMPA receptor subgroup was named. Arecoline was redesigned to provide a variety of subtype-selective muscarinic agonists and antagonists. Whereas nipecotic acid was subsequently developed into the antiepileptic agent tiagabine, THIP is currently used in advanced clinical trials.

In 1980, Dr. Krogsgaard-Larsen received his DSc. He has published nearly 440 scientific papers, edited a number of books, and, since 1998, has been European editor of the *Journal of Medicinal Chemistry*. He has been awarded honorary doctorates at the universities of Strasbourg (1992), Uppsala (2000), and Milan (2008), apart from receiving numerous other scientific awards and prizes. He is a member of a number of academies, including the Royal Danish Academy of Sciences and Letters. In 2002, he founded the Drug Research Academy as an academic/industrial research training center. He is currently the chairman of the Carlsberg Foundation and the deputy chairman of the Benzon Foundation.

Kristian Strømgaard is a professor of chemical biology in the Department of Medicinal Chemistry, Faculty of Pharmaceutical Sciences, at the University of Copenhagen. He received his master's degree in chemical research from the Department of Chemistry at the University College London under the supervision of Professor C. Robin Ganellin. In 1999, he received his PhD in medicinal chemistry from the Royal Danish School of Pharmacy (now the Faculty of Pharmaceutical Sciences). Subsequently, Dr. Strømgaard carried out his postdoctoral studies with Professor Koji Nakanishi in the Department of Chemistry at Columbia University. During this period his main focus was on medicinal chemistry studies of neuroactive natural products, with a particular emphasis on polyamine toxins and ginkgolides. In 2002, he returned to the Faculty of Pharmaceutical Sciences as an assistant professor and became an associate professor in 2004 and a professor in 2006. He is currently the head of the chemical biology group in the Department of Medicinal Chemistry that has about 20 people with expertise in applying chemistry and biology in studies of membrane-bound proteins in the central nervous system.

Ulf Madsen is currently an associate dean in the Faculty of Pharmaceutical Sciences at the University of Copenhagen. In 1988, he received his PhD in medicinal chemistry from the Royal Danish School of Pharmacy (now the Faculty of Pharmaceutical Sciences), where he later became an associate professor in the Department of Medicinal Chemistry. Dr. Madsen has been a visiting scientist at the University of Sydney, Australia; at Johan Wolfgang Goethe University, Frankfurt, Germany; and at Syntex Research, California. He has extensive research experience with the design and synthesis of glutamate receptor ligands. This includes structure activity studies and the development of selective ligands for ionotropic as well as metabotropic glutamate receptors, work which has led to a number of important pharmacological tools. Compounds with antagonist activities have shown neuroprotective properties in animal models and are consequently leads for potential therapeutic candidates. Recent projects involving biostructure-based drug designs have resulted in the development of important pharmacological agents with high subtype selectivity. The work generally involves the synthesis of heterocyclic compounds, the use of bioisosteric principles, and molecular pharmacology on native as well as recombinant receptors. The work has led to more than 100 scientific papers. Before becoming the associate dean, Dr. Madsen was the head of the Department of Medicinal Chemistry for nine years.

Contributors

Thomas Balle
Department of Medicinal Chemistry
Faculty of Pharmaceutical Sciences
University of Copenhagen
Copenhagen, Denmark

Benny Bang-Andersen
H. Lundbeck A/S
Valby, Denmark

Fredrik Björkling
Department of Medicinal Chemistry
Faculty of Pharmaceutical Sciences
University of Copenhagen
Copenhagen, Denmark

Klaus P. Bøgesø
H. Lundbeck A/S
Valby, Denmark

Hans Bräuner-Osborne
Department of Medicinal Chemistry
Faculty of Pharmaceutical Sciences
University of Copenhagen
Copenhagen, Denmark

Lennart Bunch
Department of Medicinal Chemistry
Faculty of Pharmaceutical Sciences
University of Copenhagen
Copenhagen, Denmark

Anders Buur
H. Lundbeck A/S
Valby, Denmark

Minying Cai
Department of Chemistry
University of Arizona
Tuscon, Arizona

Guy T. Carter
Wyeth Research
Pearl River, New York

Søren B. Christensen
Department of Medicinal Chemistry
Faculty of Pharmaceutical Sciences
University of Copenhagen
Copenhagen, Denmark

Rasmus P. Clausen
Department of Medicinal Chemistry
Faculty of Pharmaceutical Sciences
University of Copenhagen
Copenhagen, Denmark

Erik De Clercq
Rega Institute for Medical Research
Katholieke Universiteit Leuven
Leuven, Belgium

Robert A. Copeland
Oncology Center of Excellence in Drug
   Discovery
GlaxoSmithKline
Collegeville, Pennsylvania

Bjarke Ebert
H. Lundbeck A/S
Valby, Denmark

Iwan de Esch
Leiden/Amsterdam Center of Drug Research
Faculty of Sciences
VU University Amsterdam
Amsterdam, the Netherlands

Ole Farver
Department of Pharmaceutics and Analytical
   Chemistry
Faculty of Pharmaceutical Sciences
University of Copenhagen
Copenhagen, Denmark

Bente Frølund
Department of Medicinal Chemistry
Faculty of Pharmaceutical Sciences
University of Copenhagen
Copenhagen, Denmark

Ulrik Gether
Department of Pharmacology
The Panum Institute
University of Copenhagen
Copenhagen, Denmark

Richard R. Gontarek
Oncology Center of Excellence in Drug
    Discovery
GlaxoSmithKline
Collegeville, Pennsylvania

Harald S. Hansen
Department of Pharmacology and
    Pharmacotherapy
Faculty of Pharmaceutical Sciences
University of Copenhagen
Copenhagen, Denmark

Helle R. Hansen
Department of Pharmaceutics and Analytical
    Chemistry
Faculty of Pharmaceutical Sciences
University of Copenhagen
Copenhagen, Denmark

Piet Herdewijn
Rega Institute for Medical Research
Katholieke Universiteit Leuven
Leuven, Belgium

Victor J. Hruby
Department of Chemistry
University of Arizona
Tuscon, Arizona

David E. Jane
Department of Physiology and Pharmacology
School of Medical Sciences
University of Bristol
Bristol, United Kingdom

Anders A. Jensen
Department of Medicinal Chemistry
Faculty of Pharmaceutical Sciences
University of Copenhagen
Copenhagen, Denmark

Lars H. Jensen
TopoTarget A/S
Copenhagen, Denmark

Flemming S. Jørgensen
Department of Medicinal Chemistry
Faculty of Pharmaceutical Sciences
University of Copenhagen
Copenhagen, Denmark

Jette S. Kastrup
Department of Medicinal Chemistry
Faculty of Pharmaceutical Sciences
University of Copenhagen
Copenhagen, Denmark

Povl Krogsgaard-Larsen
Department of Medicinal Chemistry
Faculty of Pharmaceutical Sciences
University of Copenhagen
Copenhagen, Denmark

Vinod V. Kulkarni
Department of Chemistry
University of Arizona
Tuscon, Arizona

Rob Leurs
Leiden/Amsterdam Center of Drug Research
Faculty of Sciences
VU University Amsterdam
Amsterdam, the Netherlands

Tommy Liljefors
Department of Medicinal Chemistry
Faculty of Pharmaceutical Sciences
University of Copenhagen
Copenhagen, Denmark

Claus J. Loland
Department of Pharmacology
The Panum Institute
University of Copenhagen
Copenhagen, Denmark

Lusong Luo
Oncology Center of Excellence in Drug
    Discovery
GlaxoSmithKline
Collegeville, Pennsylvania

Ulf Madsen
Department of Medicinal Chemistry
Faculty of Pharmaceutical Sciences
University of Copenhagen
Copenhagen, Denmark

Maria B. Mayo-Martin
Department of Physiology and Pharmacology
School of Medical Sciences
University of Bristol
Bristol, United Kingdom

Niels Mørk
H. Lundbeck A/S
Valby, Denmark

Thomas Mueggler
Institute for Biomedical Engineering
Eidgenössische Technische Hochschule Zürich
University of Zürich
Zürich, Switzerland

Søren-Peter Olesen
Department of Biomedical Sciences
Faculty of Health Sciences
University of Copenhagen
Copenhagen, Denmark

Ingrid Pettersson
Novo Nordisk A/S
Måløv, Denmark

Markus Rudin
Institute for Biomedical Engineering
Eidgenössische Technische Hochschule Zürich
University of Zürich
Zürich, Switzerland

and

Institute for Pharmacology and Toxicology
Zürich, Switzerland

Ulla G. Sidelmann
Novo Nordisk A/S
Bagsværd, Denmark

Kristian Strømgaard
Department of Medicinal Chemistry
Faculty of Pharmaceutical Sciences
University of Copenhagen
Copenhagen, Denmark

Henk Timmerman
Leiden/Amsterdam Center of Drug Research
Faculty of Sciences
VU University Amsterdam
Amsterdam, the Netherlands

Daniel B. Timmermann
NeuroSearch A/S
Ballerup, Denmark

Keith A. Wafford
Eli Lilly
Surrey, United Kingdom

Introduction to Drug Design and Discovery

Povl Krogsgaard-Larsen and Lennart Bunch

CONTENTS

I.1 MEDICINAL CHEMISTRY—AN INTERDISCIPLINARY SCIENCE

Therapeutic agents are chemicals that prevent disease, assist in restoring health to the diseased, or alleviate symptoms associated with disease conditions. Medicinal chemistry is the scientific discipline that makes such drugs available through either discovery or design processes. Throughout history, drugs were primarily discovered by empirical methods, investigating substances or preparations of materials found in the local environment. Over the previous centuries, chemists developed methods and techniques for the isolation and purification of the active principles in medicinal plants. The purification and structure determination of natural products like morphine, hyoscyamine, quinine, and digitalis glycosides represent milestones in the field of drug discovery and the beginning of medicinal chemistry as a fascinating independent field of research (Figure I.1).

In the twentieth century, a very large number of biologically active natural products were structurally modified in order to optimize their pharmacology, and novel drugs were prepared by use of increasingly advanced synthetic methods. Moreover, the rapidly growing understanding of the nature of disease mechanisms, how cells function, and how drugs interact with cellular processes has led to the rational design, synthesis, and pharmacological evaluation of new drug candidates. Most recently, new dimensions and opportunities have emerged from a deeper understanding of cell biology and genetics.

Modern medicinal chemistry draws upon many scientific disciplines, organic and physical chemistry being of fundamental importance. But other disciplines such as biochemistry, molecular biology, pharmacology, neurobiology, toxicology, genetics, cell biology, biophysics, physiology,

1

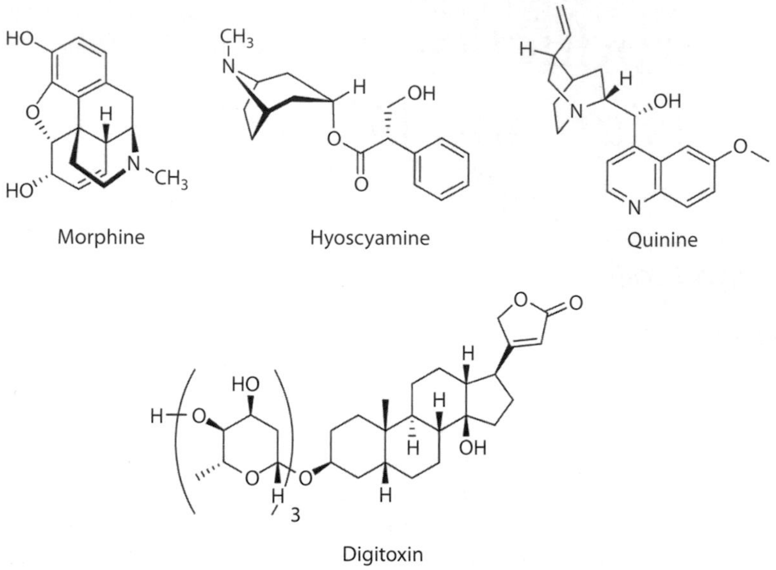

FIGURE I.1 Chemical structures of four naturally occurring classical therapeutic agents.

pathology, and computer technologies play important roles. The key research objective of medicinal chemistry is to investigate relationships between chemical structure and biological effects. When the chemical structure of a particular drug candidate has been optimized to interact with the biological target, the compound further has to fulfill a multifaceted set of criteria before it can be safely administered to patients. Absorption-distribution-metabolism-excretion (ADME) and toxicology studies in animals and humans are time-consuming research tasks, which often call for redesign of the chemical structure of the potential therapeutic agent investigated. It is an iterative process that is bound to end up in an overall compromise.

I.2 DRUG DISCOVERY—A HISTORICAL PERSPECTIVE

In early times, there was no possibility of understanding the biological origin of a disease. Of necessity, progress in combating disease was disjointed and empirical. The use of opium, ephedra, marijuana, alcohol, salicylic acid, digitalis, coca, quinine, and a host of others still in use, long predates the rise of modern medicine. These natural products are surely not biosynthesized by plants for our therapeutic convenience; we believe they have survival value to the plants in dealing with their own ecological challenges.

The presence of biologically active substances in nature, notably in certain plants, was in medieval times interpreted more teleologically. In the early sixteenth century, the German medical doctor and natural scientist, Paracelsus, whose birth given name was Phillip von Hohenheim but later changed to Philippus Theophrastus Aureolus Bombastus von Hohenheim, formulated the "Doctrine of Signatures:"

> Just as woman can be recognized and appraised on the basis of their shape; drugs can easily be identified by appearance. God has created all diseases, and he also has created an agent or a drug for every disease. They can be found everywhere in nature, because nature is the universal pharmacy. God is the highest ranking pharmacist.

The formulation of this doctrine was in perfect agreement with the dominating philosophies at that time, and it had a major impact on the use of natural medicines. Even today, remanences of this doctrine can be observed in countries, where herbal medical preparations are still widely used. Although the "Doctrine of Signatures" evidently is out of the conception of modern medicinal natural product research, the ideas of Paracelsus were the first approach to rational drug discovery.

More than 100 years ago, the mystery of why only certain molecules produced a specific therapeutic response was rationalized by the ideas of Fischer and further elaborated by Langley and Ehrlich that only certain cells contained receptor molecules that served as hosts for the drugs. The resulting combination of drug and receptor created a new super molecule that had properties producing a response of therapeutic value. One extension of this conception was that the drug fits the target specifically and productively like "a key into its corresponding lock." When the fit was successful, a positive pharmacological action (agonistic) followed, analogous to opening the door. On the contrary, a fit which prevented the intrinsic key to be inserted an antagonist action resulted— i.e., the imaginative door could not be opened. Thus, if one had found adventitiously a ligand for a receptor, one could refine its fit by opportunistic or systematic modifications of the drug's chemical structure until the desired function was obtained.

This productive idea hardly changed for the next half century and assisted in the development of many useful drugs. However, a less fortunate corollary of this useful picture was that it led to some limitations of creativity in drug design. The drug and its receptor (whose molecular nature was unknown when the theory was formulated) were each believed to be rigid molecules precrafted to fit one another precisely. Today, we know that receptors are highly flexible transmembranal glycoproteins accessible from the cell surface that often comprise more than one drug compatible region. Further complexities have been uncovered continually. For example, a number of receptors have been shown to consist of clusters of proteins either preassembled or assembled as a consequence of ligand binding. The component macromolecules may be either homo- or heterocomplexes. The challenge of developing specific ligands for systems of this complexity may readily be imagined (Chapter 12).

The opposite extreme to lock and key is the zipper model. In this view, a docking interaction takes place (much as the end of a zipper joins the talon piece) and, if satisfactory complementarity is present, the two molecules progressively wrap around each other and adapt to the steric needs of each other. A consequence of accepting this mutual adaptation is that knowledge of the receptor ground state may not be particularly helpful as it adjusts its conformation to ligand binding. Thus, in many cases one now tries to determine the three-dimensional structure of the receptor–ligand complex. In those cases where x-ray analysis remains elusive, modeling of the interactions involved is appropriate. This is the subject of Chapters 1 through 3.

Earlier, it was also noted that enzymes could be modulated for pharmacological benefit. Enzyme proteins share many characteristics with the glycoprotein components of receptors, although enzymes catalyze biochemical reactions. Receptor ligands interact with the receptor glycoproteins or with the interfaces between the macromolecular subunits of di- or polycomponent receptor complexes and modify the conformation and dynamics of these complexes. Thus, neither receptor agonists nor antagonists directly interfere with chemical reactions and are dissociated from the receptor recognition sites structurally unchanged.

The reaction mechanisms underlying the function of the vast majority of enzymes have been elucidated in detail, and based on such mechanistic information it has been possible to design a variety of mechanism-based enzyme inhibitors, notably k_{cat} inactivators and transition-state analogues, many of which are in therapeutic use (Chapter 11). Until very recently, it was only possible to inhibit enzyme action rather than facilitate it. Actually, diseases frequently result from excessive enzymatic action, making selective inhibition of these enzymes therapeutically useful.

Much later, a number of other classes of receptors have been disclosed, explored, and exploited as therapeutically relevant pharmacological targets. This heterogeneous group of receptors comprises nuclear receptors operated by steroid hormones and other lipophilic biochemical mediators, a broad range of membrane-ion channels (Chapter 13), DNA or RNA (Chapter 23), and a number of other biostructures of known or unknown functions. These aspects will be discussed in different chapters of this book.

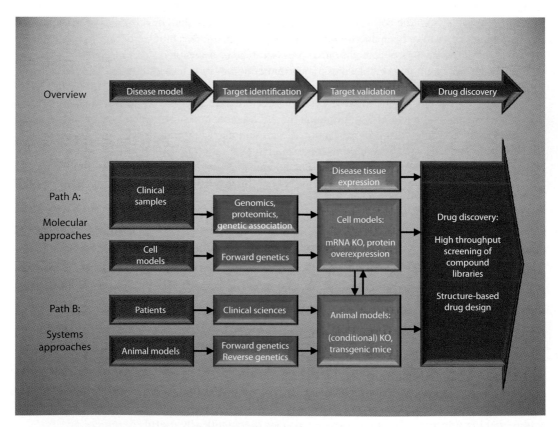

I.3   THERAPEUTIC TARGETS—IDENTIFICATION AND VALIDATION

Target discovery, which comprises identification and validation of disease-modifying targets, is an essential early step in the drug discovery pipeline. A number of approaches to target discovery have been described in recent years. These approaches and models incorporate recent technological advances, such as genomics, proteomics, small interfering RNAs (siRNAs), and mouse gene knock-out models. In this section of the book, these aspects of modern drug discovery will be described in brief (Figure I.2).

The various techniques applied in target identification and validation can be grouped into two broad target discovery strategies: the "molecular" and "systems" approach. In practice, however, both are used in varying proportions within different therapeutic areas. The "systems" approach should not be confused with the recent emergence of "systems biology," which is an attempt to construct models that explain biological responses using the plethora of information being produced from the molecular sciences.

The molecular approach is focused on the cells implicated in the disease and uses clinical samples and cell models. The molecular approach has been driven by the enormous experimental successes of molecular biology, and in particular genomics. In terms of target classes, the molecular approach

FIGURE I.2   Target-based drug discovery. Four step overview (top arrows) and detailed schematic outline. Target-based drug discovery may be divided into molecular- (path A) and system-based (path B) approaches. Each approach is composed of three steps: the provision of disease models/tissues (red), target identification (purple), and target validation (blue). The molecular approach (path A) comprises techniques such as genomics, proteomics, genetic association, and reverse genetics, whereas systems approach (path B) comprises clinical and other in vivo studies to identify targets. Target validation covers conformational experiments in cell and/or animal models. Subsequently, the drug discovery process is commenced.

is more likely to identify intracellular targets, such as regulatory, structural, and metabolic proteins, and has been most extensively deployed in oncology. In recent years, there has been a significant shift toward the molecular approach in an attempt to identify new targets through an understanding of the cellular mechanisms underlying disease phenotypes of interest.

The systems approach is geared toward target discovery through the study of diseases in whole organisms. In general, this information is derived from the clinical sciences and in vivo animal studies in physiology, pathology, and epidemiology. The systems approach has been traditionally the main target discovery strategy and this remains the case for many diseases, including obesity, arteriosclerosis, heart failure, stroke, behavioral disorders, neurodegenerative diseases, and hypertension, in which the relevant phenotype can be only detected at the organismal level. For these historical reasons, the majority of current drugs was identified through this strategy and includes those that act against both disease phenotypes and intracellular/extracellular targets. Interestingly, because many of these drugs are directed against targets which were identified from physiological studies, rather than being directly implicated in the disease mechanism, they would probably not have been identified by the molecular approach. For example, although changes in β_2-adrenoreceptor expression/activity in airway smooth muscle have not been implicated in the mechanism of allergen hyperreactivity that produces airway contraction in asthma; these symptoms are commonly treated with β_2-agonist.

The incidence of many chronic diseases is strongly correlated with age, and such diseases are thought to be influenced by both genomic and environmental factors. The overall contribution of genomic factors is still unknown, although it is believed that many diseases are influenced by the presence of susceptibility genes. With the exception of smoking, the role of environmental factors is controversial, although a number of studies have indicated the importance of infection/inflammation and diet in diseases such as arteriosclerosis, central nervous system (CNS) disease, and cancer.

In undertaking target discovery, one would ideally perform clinical studies and obtain cell/tissue samples using normal and diseased human patients. In reality, this is usually unethical and/or impractical, which means we must rely on cellular and/or animal models. However, such models often suffer from a number of significant problems which make them poor predictors of human disease. In the case of cell models, the central problem lies in simulating the complexity of the in vivo biological interactions, particularly as many of these are unknown. This problem of complexity makes it increasingly difficult to predict the role of a protein as one proceeds from the cellular level to tissue and organism. In addition, the use of immortalized cell lines to overcome the problems of availability prompts the questions regarding their biochemical similarity with primary cells. To overcome the problems of complexity, we often use animal models. However, although these models may reproduce a particular disease phenotype, genomic differences (related to species and strain), and the difficulty of identifying and replicating the long-term environmental influences, imply that the underlying causes could be different.

1.4   THE DRUG DEVELOPMENT PROCESS—AN OUTLINE

The stages through which a drug discovery/development project proceeds from inception to marketing and beyond are illustrated in the following text. From this outline, the complexity of the task of finding new therapeutic agents is evident:

- Identification of target disease, establishment of a multidisciplinary research team, selection of a promising approach, and decision on a sufficient budget. Initiation of chemistry, which normally involves synthesis based on available chemicals or collection of natural product sources. Start of pharmacology, includes suitable screening methods and choice of receptor binding and/or enzymatic assays.
- Confirmation of potential utility of initial class(es) of compounds in animals, focusing on potency, selectivity, and apparent toxicity.

- Analogue syntheses of the most active compounds, planned after careful examination of literature and patents. More elaborated pharmacology in order to elucidate mode of action, efficacy, acute and chronic toxicity, and genotoxicity. Studies of ADME characteristics. Planning of large scale synthesis and initiation of formulation studies. Application for patent protection.

These first project phases, which typically last 4–5 years, are followed by very time- and resource-demanding clinical, regulatory, and marketing phases, which normally last about 10 years:

- Phase I clinical studies, which include safety, dosage, and blood level studies.
- Phase II clinical studies focusing on efficacy and side effects.
- Phase III clinical studies, which involve studies of range of efficacy and long-term and rare side effects.
- Regulatory review.
- Marketing and phase IV clinical studies focusing on long-term safety.
- Very large-scale synthesis.
- Distribution, advertisement, and education of marketing and information personnel.

After these project stages from initiation to successful therapeutic application after approval, the patent protection expires, normally after 17–25 years, and generic competition becomes a reality.

This outline of a drug development project illustrates that, at best, it takes many years to introduce a new therapeutic agent, and it must be kept in mind that most projects are terminated before marketing, even at advanced stages of clinical studies.

I.5   DISCOVERY OF DRUG CANDIDATES

Prehistoric drug discovery started with higher plant and animal substances, and this continues today to be a fruitful source of biologically active molecules frequently belonging to unanticipated structural types. Adding to the long list of classical plant products that are still used in modern medicine, one can list many substances of more recent origin, including antibiotics such as penicillins, cephalosporins, tetracyclines, aminoglycosides, various glycopeptides, and many others (Chapter 25). Anticancer agents of natural origin comprise taxol, camptothecin, vinca alkaloids, doxorubicin, and bleomycin (Chapter 23). Among immunosuppressant agents, cyclosporine and tacrolimus deserve a special mention.

I.5.1   NATURAL PRODUCTS—ROLE IN TARGET IDENTIFICATION

Various biologically active natural products have played a key role in the identification and characterization of receptors, and such receptors are often named after these compounds (Chapter 12).

Morphine is a classical example of a natural product used for receptor characterization. Radiolabeled morphine was shown to bind with high affinity to receptors in the nervous system, and these receptors were, and still are, named opiate receptors. Some three decades ago, the physiological relevance of these receptors was documented by the findings that endogenous peptides, notably enkephalins and endorphins, served as receptor ligands (agonists). Analogues of morphine have been useful tools for the demonstration of heterogeneity of opiate receptors (Chapter 19) (Figure I.3).

The very toxic and convulsive alkaloid, strychnine, has been extensively studied pharmacologically. Using electrophysiological techniques and tritiated strychnine for binding studies, strychnine was shown to be an antagonist for the neuroreceptor mediating the inhibitory effect of glycine, primarily in the spinal cord. This receptor is currently named the strychnine-sensitive glycine receptor or the glycine$_A$ receptor.

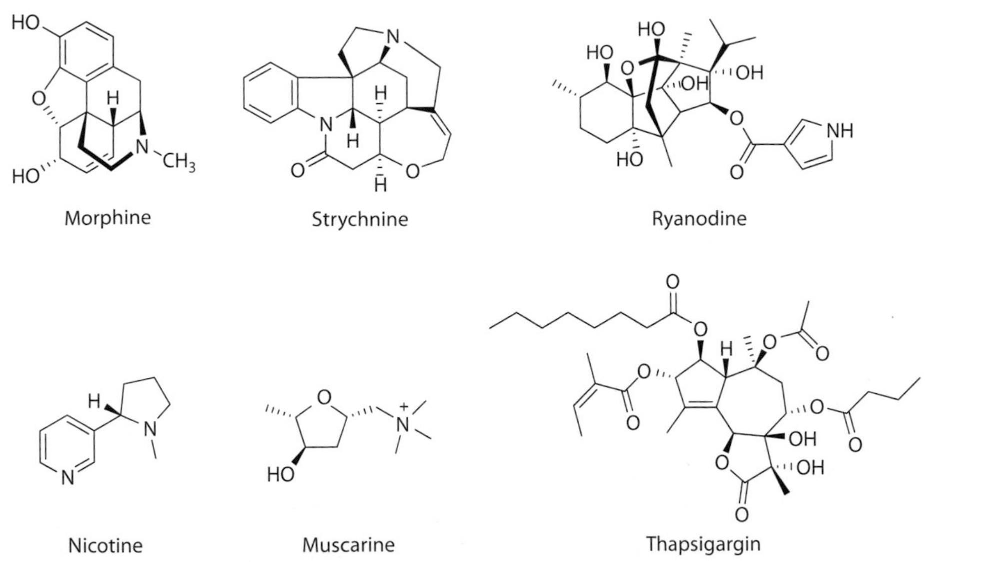

FIGURE I.3 Chemical structures of morphine, strychnine, ryanodine, nicotine, muscarine, and thapsigargin.

Acetylcholine is a key transmitter in the central and the peripheral nervous system. Acetylcholine operates through multiple receptors, and the original demonstration of receptor heterogeneity was achieved using the naturally occurring compounds, nicotine and muscarine. Whereas the ionotropic class of acetylcholine receptors binds nicotine with high affinity and selectivity, muscarine specifically and potently activates the metabotropic class of these receptors. Using molecular biological techniques, a number of subtypes of both nicotinic and muscarinic acetylcholine receptors have been identified and characterized (Chapters 12 and 16).

The ryanodine receptor is named after the insecticidal naturally occurring compound, ryanodine. Extensive studies have disclosed that ryanodine interacts with high affinity and in a calcium-dependent manner with its receptor, which functions as a calcium release channel. There are three genetically distinct isoforms of the ryanodine receptor, which play a role in the skeletal muscle disorder, central core disease.

The sesquiterpene lactone, thapsigargin, which is structurally unrelated to ryanodine, also interacts with an intracellular calcium mechanism. Thapsigargin has become the key pharmacological tool for the characterization of the sarco(endo)plasmic reticulum Ca^{2+} ATPase (SERCA). Thapsigargin effectively inhibits this ATPase, causing a rise in the cytosolic calcium level, which eventually leads to cell death. Although the SERCA pump is essential for all cell types, attempts to target thapsigargin toward prostate cancer cells have been made based on a prodrug (see Chapter 9) approach.

I.5.2 NATURAL PRODUCTS AS LEAD STRUCTURES

Although a number of biologically active natural products have been indispensable as tools for identification and characterization of pharmacological and potential therapeutic targets, these compounds normally do not satisfy the multiple demands on drugs for therapeutic use (Chapter 6). Thus, although morphine is used therapeutically, it is not an ideal drug, and has to some extent been replaced by a number of analogues showing slightly lower side effects and higher degrees of selectivity for subtypes of opiate receptors (Chapter 19). Prominent examples are the μ-selective opiate agonist, fentanyl, and U50,488, which selectively activates the κ-subtype of opiate receptors (Figure I.4).

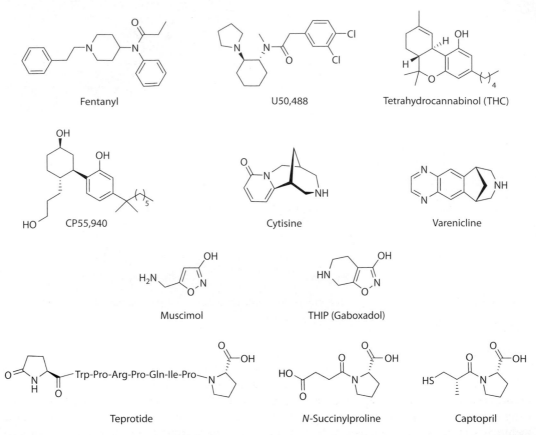

FIGURE I.4 Chemical structures of fentanyl, U50,488, tetrahydrocannabinol (THC), CP55,940, cytisine, varenicline, muscimol, THIP (gaboxadol), teprotide, *N*-succinylproline, and captopril.

The main psychoactive constituent of *Cannabis sativa*, the highly lipophilic tetrahydrocannabinol (THC) has been a useful tool for the identification of the two cannabinoid receptors, CB_1- and CB_2-receptor operated by endocannabinoids. Since different preparations of *C. sativa* have psychoactive effects, health authorities have been reluctant to accept THC and analogues as therapeutic agents for the treatment of pain and other disease-related conditions. This may change with time, as medicinal chemists have synthesized a number of cannabinoid receptor ligands, including the receptor agonist CP55,940, which is markedly less lipophilic than THC (Chapter 19).

The nicotine acetylcholine receptors (nAChRs) have become key targets for therapeutic approaches to treat pain, cognition disorders, depression, schizophrenia, and nicotine dependence. For several reasons, nicotine has limited utility as a therapeutic agent, and a wide variety of nAChR agonists have been synthesized and characterized (Chapter 16). (–)-Cytisine is a naturally occurring toxin acting as a powerful nAChR agonist. Using (–)-cytisine as a lead structure, varenicline was developed as a partial nAChR agonist showing an optimally balanced agonist/antagonist profile for smoking cessation.

Muscimol is another example of a naturally occurring toxin, which has been extensively used as a lead for the design of specific GABA receptor agonists and GABA uptake inhibitors (Chapter 15). Muscimol, which is a 3-isoxazolol bioisostere of GABA, is a constituent of the mushroom *Amanita muscaria*. Muscimol is toxic, it is metabolically unstable, and it interacts with the different GABA synaptic mechanisms and with a broad range of $GABA_A$ receptor subtypes. The cyclic analogue of muscimol, THIP (Gaboxadol) is highly selective for the therapeutically interesting extrasynaptic

GABA$_A$ receptors. Gaboxadol is a clinically active nonopioid analgesic and a nonbenzodiazepine hypnotic, which at present is in clinical trials (see also Chapters 15 and 20).

The angiotensin-converting enzyme (ACE) is a zinc carboxypeptidase centrally involved in the regulation of blood pressure and is an important target for therapeutic intervention. Peptide toxins from the Brazilian pit viper, *Bothrops jararaca* and the synthetic peptide analogue, teprotide, are inhibitors of ACE (Figure I.4), but are not suitable for therapeutic use. Systematic molecular dissection of teprotide led to the nonpeptide ACE inhibitor, *N*-succinylproline, which was converted into the structurally related and much more potent analogue, Captopril that is now marketed as an effective antihypertensive drug.

I.5.3 Basic Principles in Lead Development and Optimization

Potency, efficacy, and selectivity are essential but certainly not the only parameters to fulfill for a pharmacologically active compound to become a therapeutic drug. A large number of additional requirements have to be met, the most important of which have been summarized in the acronym, ADME (Section I.4), which actually should be extended to ADME-Tox (ADME as well as toxicity). Obviously, the drug must reach the site of action in a timely manner and in sufficient concentration to produce the desired therapeutic effect.

After oral administration, the drug must survive the acidic environment of the stomach. In the small intestine, the bulk of absorption takes place. Here, the pH is neutral to slightly acidic. In the gastrointestinal system metabolism can take place. The presence of digestive enzymes creates particular problems for polypeptide drugs, and the gut wall is fairly rich in oxidative enzymes.

Unless the drug acts as a substrate for active energy-requiring uptake mechanisms, which normally facilitate uptake of, for example, amino acids and glucose, it must be significantly unionized to penetrate into the body. Following absorption, the blood rapidly presents the drug to the liver, where Class I metabolic transformations (oxidation, hydrolysis, reduction, etc.) and in some cases Phase II transformations (glucuronidation, sulfatation, etc.) take place. The polar reaction products from these reactions are then typically excreted in the urine or feces.

The rate of absorption of drugs, their degree of metabolic transformation, their distribution in the body, and their rate of excretion are collectively named pharmacokinetics. This is in effect the influence of the body on a drug as a function of time. The interaction of the drug with its receptors, in the broad sense of the word, and the consequences of this interaction as a function of time are pharmacodynamics.

Both of these characteristics are alone governed by the drug's chemical structure. Thus, the medicinal chemist is expected to remedy any shortcomings by structural modifications. In addition to ADME-Tox, a number of other characteristics must also be satisfactory, such as:

- Freedom from mutagenesis
- Freedom from teratogenecity
- Chemical stability
- Synthetic or biological accessibility
- Acceptable cost
- Ability to patent
- Clinical efficacy
- Solubility
- Satisfactory taste
- Ability to formulate satisfactorily for administration
- Freedom from idiosyncratic problems

These challenges emphasize the key importance of scientists trained in interdisciplinary medicinal chemistry in drug discovery and development projects.

I.5.3.1   Bioisosterism

In the broadest sense of the term, bioisosteres are defined as functional groups or molecules that produce a similar biological effect. The tactics of bioisosterism, also named molecular mimicry, has been extensively used by medicinal chemists in the optimization of drug molecules pharmacodynamically or pharmacokinetically.

Bioisosteres have been classified as either classical or nonclassical. In classical bioisosterism, similarities in certain physicochemical properties have enabled investigators to successfully exploit several monovalent isosteres. These can be divided into the following groups: (1) fluorine versus hydrogen replacements; (2) amino-hydroxy interchanges; (3) thiol–hydroxyl interchanges; (4) fluorine, hydroxyl, amino, and methyl group interchanges (Grimm's hydride displacement law, referring to the different number of hydrogen atoms in the isosteric groups to compensate for valence differences).

The nonclassical bioisosteres include all those replacements that are not defined by the classical definition of bioisosteres. These isosteres are capable of maintaining similar biological activity by mimicking the spatial arrangement, electronic properties, or some other physicochemical properties of the molecule or functional group that are of critical importance. The concept of nonclassical bioisosterism, in particular, is often considered to be qualitative and intuitive, but there are numerous examples of effective use of this concept in drug design (Chapters 15 and 16).

The conversion of the muscarinic acetylcholine receptor agonist arecoline, containing a hydrolyzable ester group, into different hydrolysis-resistant heterocyclic bioisosteres is illustrated in Figure I.5. The annulated (I.1) and nonannulated (I.2 and I.3) bicyclic bioisosteres are potent muscarinic agonists. Similarly, compounds I.4 and I.5 interact potently with muscarinic receptors as agonists, whereas I.6, in which the 1,2,4-oxadiazole ester bioisosteric group of I.4 is replaced by an oxazole group, shows reduced muscarinic agonist effects. Thus, the electronic effects associated with these heterocyclic rings appear to be essential for muscarinic activity.

It must be emphasized that a bioisosteric replacement strategy, which has been successful for a particular group of pharmacologically active compounds, cannot necessarily be effectively used in other groups of compounds active at other pharmacological targets.

I.5.3.2   Stereochemistry

Receptors, enzymes, and other pharmacological targets, which by nature are composed of protein constructs, are highly chiral. Thus, it is not surprising that chirality in the drug structures normally plays an important role in pharmacological responses (Chapter 5). In racemic drug candidates, the

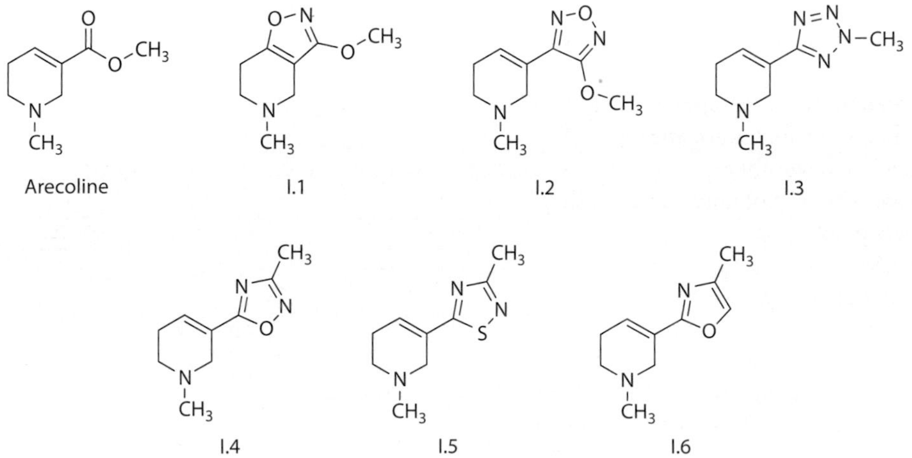

FIGURE I.5   Arecoline and analogues.

desired pharmacological effect typically resides in one enantiomer, whereas the other stereoisomer(s) are pharmacologically inactive or possess different pharmacological effects. Thus, chiral drugs should preferentially be resolved into stereochemically pure isomers prior to pharmacological examination. Since many, especially of older date, synthetically prepared chiral biologically active compounds have been described pharmacologically as racemates, much of the pharmacological literature should be read and interpreted with great care.

Figure I.6 exemplifies the importance of stereochemistry in studies of the relationship between structure and pharmacological activity (SAR studies).

The upper part of Figure I.6 shows the four stereoisomers, which actually are two pairs of enantiomers of two diastereomeric compounds. These 1-piperazino-3-phenylindans were synthesized, resolved, structurally analyzed, and pharmacologically characterized as part of a comprehensive drug research program in the field of central biogenic amine neurotransmission. Whereas one of these stereoisomers turned out to be inactive, two of them were inhibitors of dopamine (DA) and noradrenaline (NE) uptake, and one isomer showed antagonist effects at DA, NE, and serotonin (5-HT) receptors. It is evident that a pharmacological characterization of a synthetic mixture of these compounds would be meaningless.

The 3-isoxazolol amino acid, APPA, is an analogue of the standard agonist, AMPA, for the AMPA subgroup of excitatory glutamate receptors (Chapter 15). APPA was tested pharmacologically as the racemate, which showed the characteristics of a partial agonist at AMPA receptors. Subsequent pharmacological characterization of the pure enantiomers quite surprisingly disclosed that (S)-APPA is a full AMPA receptor agonist, whereas (R)-APPA turned out to be an AMPA antagonist. This observation prompted intensive pharmacological studies, and as a result it was demonstrated that administration of a fixed ratio of an agonist and a competitive antagonist always provides a partial agonist response at an efficacy level dependent on the administered ratio of compounds and their relative potencies as agonist and antagonist. This phenomenon was named "functional partial agonism." An interesting aspect of this pharmacological concept is that administration of an antagonist drug inherently establishes functional partial agonism together with the endogenous agonist at the target receptor.

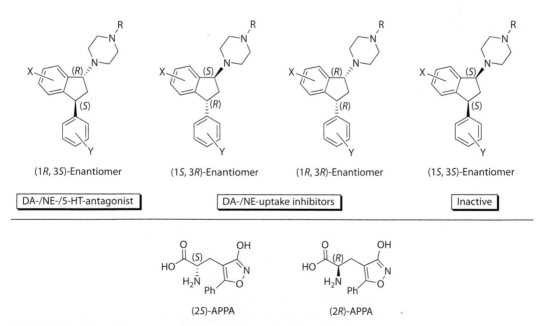

FIGURE I.6 Chemical structures of the four stereoisomers of 1-piperazino-3-phenylindans and the two enantiomers of the phenyl analogue of AMPA (APPA).

I.5.3.3   Membrane Penetration—Including Lipinski Rule of Five

In drug discovery projects, an issue of major importance is the design of drug molecules capable of penetrating different biological membranes effectively and rapidly enough to allow effective concentrations to build up at the therapeutic target. The structure and physiochemical properties of the drug molecule obviously are of decisive importance, and it is possible to establish the following empirical rules:

- Some small and rather water soluble substances pass in and out of cells through water lined transmembrane pores.
- Other polar agents are conducted into or out of cells by membrane associated and energy-consuming proteins. Polar nutrients that the cell requires, such as glucose and many amino acids, fit into this category. More recently, drug resistance by cells has been shown to be mediated in many cases by analogous protein im- and exporters.
- The blood–brain barrier (BBB) normally is not easily permeable by neutral amino acids. However, such compounds with sufficiently small difference between the pK_a values will have a relatively low I/U ratio indicating the ratio between ionized (zwitterionic) and unionized molecules in solution. As an example, THIP (Figure I.4) has pK_a values of 4.4 and 8.5 and a calculated I/U ratio of about 1000. Thus, 0.1% of THIP in solution is unionized, and this fraction permits THIP to penetrate the BBB very easily. Other neutral amino acids typically have I/U ratios around 500,000 and thus very low fractions of unionized molecules in solution, and such compounds normally do not penetrate into the brain after systemic administration.
- Molecules that are partially water soluble and partially lipid soluble can pass through cell membranes by passive diffusion and are driven in the direction of the lowest concentration.
- In cells lining the intestinal tract, it is possible for molecules with these characteristics to pass into the body through the cell membrane alone.
- Finally, it is also possible for molecules with suitable water solubility, small size, and compact shape to pass into the body between cells. This last route is generally not available for passage into the CNS, because the cells are pressed closely together and thus closing off these functions to form the BBB.

Whereas there are no guarantees and many exceptions, the majority of effective oral drugs obey the Lipinski rule of five:

- The substance should have a molecular weight of 500 or less
- It should have fewer than five hydrogen-bond donating functionalities
- It should have fewer than 10 hydrogen-bond accepting functionalities
- The substance should have a calculated log P (clog P) between approximately −1 and +5

The Lipinski rule of five is thus not a rule comprising five paragraphs but simply an empirical rule, where the number five occurs several times. The rule is a helpful guide rather than a law of nature.

I.5.3.4   Structure-Based Drug Design

During the early 1980s, the possibility to rationally design drugs on the basis of structures of therapeutically relevant biomolecules was an unrealized dream for many structural biologists. The first projects were underway in the mid-1980s, and today, even though there are still many obstacles and unsolved problems, structure-based drug design is an integral part of many academic and most industrial drug discovery programs. In Chapter 2, a number of examples of this impressive drug design approach are described.

As structural genomics, bioinformatics, and computational power continue to almost explode with new advances, further successes in structure-based drug design are likely to follow. Each year, new targets are being identified, structures of those targets are being determined at an amazing rate, and our capability to capture a quantitative picture of the interaction between macromolecules and ligands is accelerating.

I.6 INDIVIDUALIZED MEDICINE AND CONCLUDING REMARKS

The mapping of the genome leads us to the identification of targets for therapeutic interventions, as for example chemotherapy, not previously suspected and even allows us to dream of the possibility of correcting genetic defects, enhancing our prospects for a longer and more healthy life, and for devising drugs for specific individuals. Presuming that individual variations in therapeutic response may have a genetic origin, and thus dividing populations into subgroups with similar genetic characteristics, might allow us to prescribe drugs and even dosages within these groups. This form of individual gene typing is already possible, but would be very resource demanding as per days techniques. It is likely that perplexing species differences in response to, for example, chemotherapy, that complicates drug development, may also be understood, when the genome mapping becomes more elaborate.

The new biological capabilities raise many new prospects and problems for drug companies and, in general, for the society, not only scientifically but also morally. Scientific knowledge by itself is morally neutral, but how it is used, is not.

In conclusion, there has never been a more exciting time to take up the study of medicinal chemistry. The technological developments and the amount of information will grow with increasing speed, and scientists may eventually risk to be drowned in this multitude of possibilities. However, the intelligent, intuitive, and skilled medicinal chemist will be able to maneuver in this ocean of multiplicity and to continue the series of brilliant achievements by the pioneers in drug discovery during the past century.

FURTHER READINGS

Anderson, A.C. 2003. The process of structure-based drug design. *Chem. Biol.* 10:787–797.

Bøgesø, K.P. 1998. Drug hunting. The medicinal chemistry of 1-piperazino-3-phenylindans and related compounds. DSc Thesis, The Royal Danish School of Pharmacy, Copenhagen.

Bräuner-Osborne, H., Egebjerg, J., Nielsen, E.Ø., Madsen, U., and Krogsgaard-Larsen, P. 2000. Ligands for glutamate receptors: Design and therapeutic prospects. *J. Med. Chem.* 43:2609–2645.

Ebert, B., Madsen, U., Søby, K.K., and Krogsgaard-Larsen, P. 1996. Functional partial agonism at ionotropic excitatory amino acid receptors. *Neurochem. Int.* 29:309–316.

Lindsay, M.A. 2003. Target discovery. *Nat. Rev. Drug Disc.* 2:831–838.

Lipinsky, C.A., Lombardo, F., Dominy, B.W., and Feeney, P.J. 2001. Experimental and computational approaches to estimate solubility and permeability in drug discovery and development settings. *Adv. Drug Del. Rev.* 46:3–26.

Mitscher, L.A. 2002. Drug design and discovery: An overview. In *Textbook of Drug Design and Discovery*, P. Krogsgaard-Larsen, T. Liljefors, and U. Madsen (eds.), pp. 1–34. London: Taylor & Francis.

Patani, G.A. and LaVoie, E.J. 1996. Bioisosterism: A rational approach in drug design. *Chem. Rev.* 96:3147–3176.

1 Molecular Recognition in Ligand–Protein Binding

Tommy Liljefors

CONTENTS

1.1 INTRODUCTION

Molecular recognition is a basic feature of virtually all biological phenomena. In the case of ligand–protein binding it can be described as the ability of a ligand and a protein (an enzyme or a receptor) to form a "noncovalent" complex. Covalent binding between a ligand and a protein occurs, but is much less common and a discussion of such binding is outside the scope of this chapter. An understanding of the basic principles of molecular recognition is essential for students as well as practitioners of medicinal chemistry. It provides an ability to interpret experimental ligand-binding data and gives an understanding of structure–activity relationships in terms of physical forces acting in the ligand–protein binding process. Such an understanding is a prerequisite for the rational design of new ligands—new potential drug molecules.

The first attempt to understand the basic properties of ligand–protein recognition was formulated in the "lock-and-key" hypothesis by Emil Fischer (1894). A cartoon illustration of this hypothesis is given in Figure 1.1a. The essence of the hypothesis is that the protein (in this case, an enzyme) and the ligand must fit together like a lock and a key in order to initiate a chemical reaction (i.e., enzymatic catalysis). The ligand as well as the protein in this hypothesis is considered to be rigid. Although the lock-and-key hypothesis has been useful for generations of medicinal chemists, it gradually became clear that it is an oversimplification of the properties of ligand–protein recognition. For instance, noncompetitive enzyme inhibition could not be explained by the hypothesis and the fact that some enzymes are highly selective, whereas other enzymes may interact with several structurally different substrates could not be understood. This led Koshland (1958) to introduce the "induced fit theory" (Figure 1.1b) in which the interaction between a ligand and a protein could be described as "a hand in a glove," where the hand and the glove both adjust their shapes in order to provide an optimal fit. Ligands are in general flexible and may change their shape (conformation)

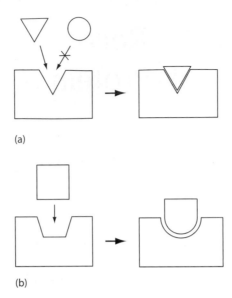

(a)

(b)

FIGURE 1.1   (a) Lock-and-key and (b) induced fit hypotheses.

by rotation about single bonds. In the protein, side-chain conformations as well as the backbone conformation may adjust to optimize the ligand–protein interaction. The rapidly increasing number of crystallographically determined 3D structures of proteins and ligand–protein complexes give strong support to the induced fit hypothesis and this hypothesis is today generally accepted by the scientific community.

Studying an x-ray structure of a ligand–protein complex as visualized by modern computer graphics hardware and software is fascinating and many useful details about ligand–protein interactions can be learned from that. However, a study of a ligand–protein complex alone tells only a part of the story of ligand–protein interactions. For instance, it does not give much information about the strengths of the observed interactions. The purpose of this chapter is to discuss and give examples of ligand–protein interactions in terms of basic physical chemistry. This chapter focuses on the understanding of the type and magnitude of different contributions to the strength of the ligand–protein binding. This may provide answers to questions about what to expect in terms of affinity for a given drug if, for instance, a substituent in this drug is replaced by another one. It also provides a framework for other chapters in this book in which structure–affinity relationships are being discussed.

1.2   DETERMINATION OF THE AFFINITY—THE TOTAL STRENGTH OF THE LIGAND–PROTEIN INTERACTION

It is of utmost importance to understand the affinity of a ligand for a protein and to keep in mind that the affinity is defined by the equilibrium between the unbound ligand and the unbound protein on one side and the ligand–protein complex on the other, as shown in Figure 1.2. Thus, thermodynamics governs the basic physicochemical principles of molecular recognition.

The affinity of the ligand for the protein is given by the free energy difference (ΔG) between the ligand–protein complex (the right-hand side in Figure 1.2) and the "free" (unbound) ligand and the "free" (unbound) protein (the left-hand side). Water plays an important role on both sides of the equilibrium as will be discussed below.

It should be noted that the ligand exists as a mixture of conformations with different shapes on the left-hand side but in a single well-defined conformation on the right-hand side. The protein conformation may also change between the left- and right-hand sides of the equilibrium and the water structure is different on the two sides of the equilibrium.

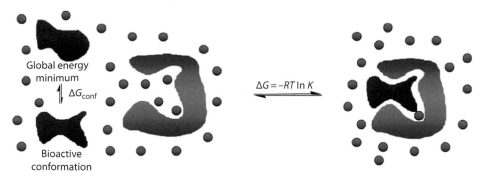

FIGURE 1.2　The equilibrium determining the affinity of a ligand.

The free energy difference is related to the equilibrium constant K by Equation 1.1.

$$\Delta G = -RT \ln K \tag{1.1}$$

where
　R is the gas constant (8.315 J/K/mol)
　T is the temperature in kelvin

A higher affinity implies a larger positive value of K, and, thus, a larger negative value of ΔG. In medicinal chemistry, the affinity of a ligand is most often given as an inhibition constant K_i or by an IC_{50}-value. Since $K = 1/K_i$ the free energy difference in terms of K_i can be written as in Equation 1.2.

$$\Delta G = RT \ln K_i \tag{1.2}$$

ΔG has an enthalpic component (ΔH) as well as an entropic component (ΔS) according to Equation 1.3.

$$\Delta G = \Delta H - T\Delta S \tag{1.3}$$

A higher affinity (a more negative ΔG) corresponds to a smaller value of the inhibition constant K_i (most often given in nM or μM). A K_i of 1 nM corresponds to a ΔG of −53.4 kJ/mol at 310 K and a K_i of 1 μM to −35.6 kJ/mol. Furthermore, using Equation 1.2 it can be calculated that a change in ΔG by 5.9 kJ/mol alters K_i by a factor of 10. An example of the size of this energy in terms of molecular structural change is shown in Figure 1.3. A conformational change of the ethyl group in ethyl benzene from a perpendicular conformation with respect to the phenyl ring (the lowest energy conformation) to a conformation with a coplanar carbon skeleton increases the energy by 6.7 kJ/mol. Thus, even modest changes in the conformation (the shape) of a ligand can result in a significant decrease in the affinity, a fact that should carefully be taken into account in ligand design (for further discussions on ligand conformations, see Section 1.3.2).

IC$_{50}$ expresses the concentration of an inhibitor that displaces 50% of the specific binding of a radioactively labeled ligand in a radioligand experiment. The IC$_{50}$ value can be converted to an inhibition constant K_i by the Cheng–Prusoff equation (Equation 1.4).

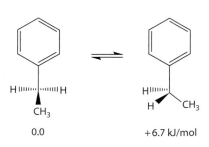

FIGURE 1.3　Conformational energies of ethyl benzene.

$$K_i = \frac{IC_{50}}{\left(1 + \dfrac{[L]}{K_D}\right)} \qquad (1.4)$$

where

[L] is the concentration of the radioligand used in the assay

K_D is the affinity of the radioligand for the receptor

It should be noted that IC_{50} values are dependent on the concentration and the affinity of the radioligand. Care should be taken when comparing IC_{50} values unless the same radioligand and radioligand concentration have been used in all binding experiments. In contrast, K_i is a constant for the ligand with respect to the receptor. It provides a useful measure of the total affinity, but by itself it tells little about the details of the molecular recognition.

1.3   PARTITIONING OF ΔG

The driving forces for ligand–protein recognition are electrostatic interactions (ion–ion, ion–dipole, and dipole–dipole), lipophilicity/hydrophobicity, and shape complementarity. In order to understand the nature and the relative contributions of the different forces it is an useful approximation to partition ΔG into a sum of free energy contributions, as shown in Equation 1.5. Several different partitioning schemes have been proposed in the literature. The partition used here is essentially that suggested by Williams on the basis of studies by Page and Jencks (see Further Readings).

$$\Delta G = \Delta G_{transl+rot} + \Delta G_{conf} + \Delta G_{polar} + \Delta G_{hydrophob} + \Delta G_{vdW} \qquad (1.5)$$

- $\Delta G_{transl+rot}$ accounts for the restrictions of translational movements (movements in x-, y-, and z-directions) and restrictions of rotations (about the x-, y-, and z-axes) of the "whole" molecule from the unbound to the bound state.
- ΔG_{conf} is the difference in the conformational free energies between the unbound and bound states due to conformational restrictions in the ligand–protein complex.
- ΔG_{polar} is the free energy change due to interactions of polar functional groups in the binding cavity of the protein.
- $\Delta G_{hydrophob}$ accounts for the binding free energy due to the hydrophobic effect.
- ΔG_{vdW} gives the difference in free energy due to van der Waals (vdW) interactions in the bound and unbound states.

In the following sections, the different terms in Equation 1.5 and their magnitudes will be discussed in more detail and illustrated in terms of ligand–protein recognition.

1.3.1 $\Delta G_{transl+rot}$—THE FREEZING OF THE OVERALL MOLECULAR MOTION

Outside the binding cavity in the protein, the ligand tumbles freely in the aqueous solution through rotations and translations of the entire molecule. Since the freedom of translation and rotation in the binding cavity becomes severely restricted through formation of the ligand–protein complex, these motions to a large extent become frozen (i.e., three rotational and three translational degrees of freedom are lost). In terms of thermodynamics this leads to a decrease in entropy resulting in a more negative ΔS and consequently a more negative $T\Delta S$. According to Equation 1.3 this gives a more positive ΔG. Thus, the loss of freedom of translation and rotation opposes binding and $\Delta G_{transl+rot}$ is a free energy cost, which must be overcome by the favorable binding forces to make the formation of a ligand–protein complex possible. The magnitude of this free energy cost has been much debated in the literature. Explicit calculations show that it varies only slightly with molecular weight, but an important problem for the estimation of $\Delta G_{transl+rot}$ is that it depends on the "tightness" of the ligand–protein complex. A tighter complex leads to a greater loss of freedom of movement and thus to a more negative $T\Delta S$. Most estimates of $\Delta G_{transl+rot}$ range from 12 kJ/mol for a "loose" complex to 45 kJ/mol for a tightly bound complex. Whatever the exact magnitude of $\Delta G_{transl+rot}$ is in a particular case, it is a very significant energy to overcome by the favorable binding forces. Consider a ligand with an affinity (K_i) of 1 nM corresponding to ΔG of −53.4 kJ/mol at 310 K (Section 1.2). In order to end up with this free energy difference between the bound and unbound states, the favorable binding forces must produce not only 53.4 kJ/mol of ligand–protein binding energy but in addition 12–45 kJ/mol of free energy is required to make the association possible. It should be noted that this free energy cost of ligand–protein association is always present and cannot be reduced by ligand design. However, the exact value of $\Delta G_{transl+rot}$ is only important for predictions of "absolute" ΔG values. To a first approximation it cancels out when comparing the affinities of different ligands to the same receptor.

1.3.2 ΔG_{conf}—Conformational Changes of Ligand and Receptor

The restriction of motions that are accounted for in $\Delta G_{transl+rot}$ described above refer to the "overall" motion of the molecule. However, there is an additional type of motion, which is more or less frozen upon ligand binding. Most ligand molecules are flexible, which means that in the aqueous phase outside the binding cavity in the protein, the ligand undergoes conformational changes by rotation about single bonds. For example, the dihedral angles in hydrocarbon chains changes between gauche and anti conformations resulting in a mixture of ligand conformations, i.e. different ligand shapes. A ligand generally binds to a protein in a single well-defined conformation that positions functional groups used for binding in appropriate locations in space for interactions with their binding partners in the protein. This implies that the motions corresponding to the conformational freedom in aqueous solution are to a large extent frozen in the binding site. As discussed above for $\Delta G_{transl+rot}$, this leads to a decrease in the entropy (conformational entropy) giving a more negative ΔS and $T\Delta S$ and thus a free energy cost for binding. The magnitude of ΔG_{conf} due to $T\Delta S_{conf}$ has been estimated to be 1–6 kJ/mol per restricted internal rotation and depends on the "tightness" of the ligand–protein complex as in the case of $\Delta G_{transl+rot}$ (Section 1.3.1).

A second energy contribution to ΔG_{conf} comes from changes in ligand conformation between aqueous solution and ligand–protein complex. Comparisons of ligand conformations observed in x-ray structures of ligand–protein complexes and ligand conformations in aqueous phase (as calculated by state-of-the-art computational methods) show that a ligand in general does not bind to the protein in its preferred conformation (lowest energy conformation) in aqueous solution. An example of this is shown in Figure 1.4. Palmitic acid prefers the well-known all-anti (zigzag) conformation of the hydrocarbon chain in aqueous solution, but binds to the adipocyte lipid-binding protein with an affinity (K_i) of 77 nM in a significantly folded conformation. The energy required for palmitic acid to adopt the binding conformation has been calculated to be 10.5 kJ/mol. This conformational energy penalty is detrimental to binding and has the effect of increasing the K_i value in comparison to a case in which the ligand binds in its preferred conformation in aqueous solution.

As shown in Section 1.2, a conformational energy penalty of 5.9 kJ/mol corresponds to a decrease in affinity (increase of K_i) by a factor of 10. For each additional 5.9 kJ/mol of conformational energy penalty, the affinity decreases further by a factor of 10. It is consequently of high importance in the design of new ligands using x-ray determined protein structures (see Chapter 2) or pharmacophore models (see Chapter 3) to avoid introducing significant conformational energy penalties in the designed ligands. Calculations of the conformational energy penalties for ligands in a series of x-ray structures of ligand–protein complexes indicate that these energy penalties in general are below 13 kJ/mol. This may be used as a rule of thumb in ligand design. In this context, it is important to

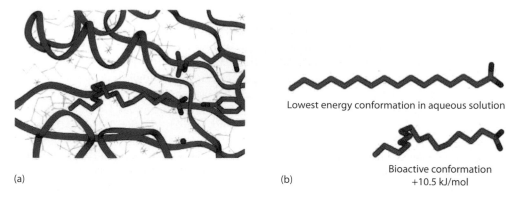

(a)   (b)   Lowest energy conformation in aqueous solution

Bioactive conformation
+10.5 kJ/mol

FIGURE 1.4   (a) Palmitic acid bound to the adipocyte lipid-binding protein (pdb-code 1LIE) and (b) the preferred conformation of palmitic acid in aqueous solution and the conformation bound to the protein.

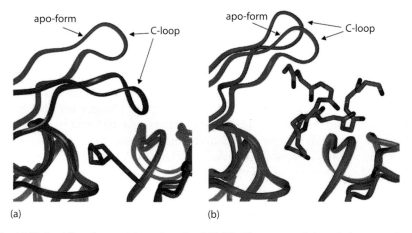

(a)   (b)

FIGURE 1.5   (a) Epibatidine (orange) bound to the AChBP. The orange C-loop belongs to the epibatidine–AChBP complex (pdb-code 2BYQ), the green C-loop belongs to the uncomplexed AChBP (apo-form, pdb-code 2BYN). (b) α-conotoxin ImI (yellow carbons) bound to AChBP (pdb-code 2BYP). The yellow C-loop belongs to the α-conotoxin ImI-AChBP complex, the green C-loop belongs to the uncomplexed AChBP.

note that in calculations of conformational energy penalties, the conformational properties of the unbound ligand "in aqueous phase" must be used as the reference state (see Further Readings).

In terms of ΔG_{conf}, rigid molecules have an advantage relative to more flexible ligands. The binding of a rigid ligand does not result in a loss of conformational entropy and if the ligand has only one possible conformation (or has one strongly preferred conformation corresponding to the binding conformation), the conformational energy penalty is zero. Although highly rigid molecules are ideal as ligands, it is a great challenge to design such ligands. For instance, functional groups taking part in interactions in the binding cavity have to be designed to occupy precisely correct positions in space, as no (or very small) adjustments of their positions are possible in a rigid ligand.

So far only conformational changes in the ligand have been discussed, but conformational changes most often also occur in the protein. Not only may the amino acid side chains adjust their conformations to optimize their interactions with the ligand, but the protein backbone conformation may also change. In some cases, this may result in major movements of, for example, loops or even entire protein domains. Examples of such movements are shown in Figure 1.5. The C-loop in the acetylcholine binding protein (AChBP) and most probably also the corresponding loop in, for example, nicotinic acetylcholine and $GABA_A$ receptors adjusts its position in response to the size of the ligand (Figure 1.5). This has implications for the pharmacological profile of the ligand (see also Chapter 16).

Protein flexibility is a major challenge in structure-based drug design (see Chapter 2) and is currently the focus of much research.

1.3.3 ΔG_{polar}—ELECTROSTATIC INTERACTIONS AND HYDROGEN BONDING

ΔG_{polar} is the free energy change due to interactions between polar functional groups in the ligand and polar amino acid residues and/or C=O and NH backbone groups in the binding cavity of the protein. In addition, indirect ligand–protein interactions via water molecules in the binding cavity are frequently observed. These interactions include ion–ion, ion–dipole, and dipole–dipole interactions, which are well described in books on physical chemistry to which the reader is referred to for details. The attraction between opposite charges or antiparallel dipoles plays an important role in ligand–protein recognition.

The strength of any electrostatic interaction (E_{polar}) is given by Coulomb's Law (Equation 1.6):

$$E_{polar} = \frac{q_i q_j}{\varepsilon r_{ij}} \tag{1.6}$$

where

  q_i and q_j are integer values for ion–ion interactions and partial atomic charges (summed over the participating atoms) for other polar interactions
  ε is the dielectric constant
  r_{ij} is the distance between the charges

In Equation 1.6, it is important to note that the electrostatic energy E_{polar} depends on the dielectric constant (ε), which measures the shielding of the electrostatic interactions by the environment. The dielectric constant of water is 78.4 (25°C). E_{polar} is difficult to quantify in proteins as ε is not uniform throughout the protein but depends on the microenvironment in the protein. A value of about 4 is often used for a lipophilic environment in the interior of a protein.

The relative strength of the different types of electrostatic interactions is ion–ion > ion–dipole > dipole–dipole. Ion–ion interactions do not depend on the relative orientation of the interacting partners, whereas ion–dipole and dipole–dipole interactions are strongly dependent on the relative orientation of the interacting moieties. For instance, the interaction between antiparallel dipoles is attractive, whereas that between parallel dipoles is repulsive.

1.3.3.1 Hydrogen Bonds

A hydrogen bond X–H----Y may be described as an electrostatic attraction between a hydrogen atom bound to an electronegative atom X (in ligand–protein interactions most often nitrogen or oxygen) and an additional electronegative atom Y. The typical hydrogen bond distance is 2.5–3.0 Å (as measured between the heavy atoms X and Y). A hydrogen bond is highly orientation dependent with an optimal X–H----Y angle of 180°. Examples of different types of hydrogen bonds commonly observed in ligand–protein complexes are shown in Figure 1.6.

Figure 1.7 displays the binding of (S)-glutamate to the ligand-binding domain of the ionotropic glutamate receptor iGluR2 featuring a "salt bridge" and a number of other charge-assisted hydrogen bonds between the ligand and the receptor, and also between the ligand and water molecules in the active site.

In order to understand the contribution of hydrogen bonding or other polar interactions to ligand binding, it is crucial to keep in mind that the ligand–protein interaction is an equilibrium process

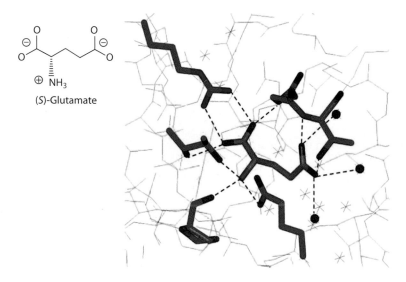

FIGURE 1.6   Examples of different types of hydrogen bonds observed in ligand–protein complexes.

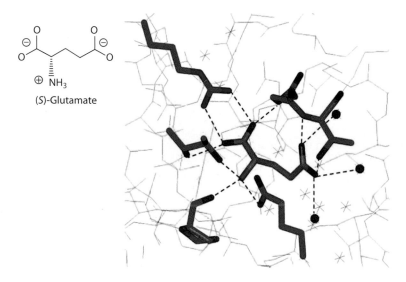

(S)-Glutamate

FIGURE 1.7   The binding of (S)-glutamate to the ligand-binding domain of the ionotropic glutamate receptor iGluR2 (pdb-code 1FTJ).

(Figure 1.2) and that hydrogen bonding is an exchange process. Before formation of the ligand–protein complex (left-hand side in Figure 1.2) the polar functional groups of the ligand, the polar amino acid residues and C=O and NH backbone groups in the protein are engaged in hydrogen bonding with surrounding solvent water molecules. In the ligand–protein complex (right-hand side in Figure 1.2) these hydrogen bonds to the solvent are replaced by hydrogen bonds between the ligand and the protein. The net effect of this hydrogen bond exchange process is the difference in free energy between hydrogen bonding to water and to the protein. As a consequence of this exchange process, a substituent that is hydrogen bonded to water molecules in the aqueous phase, but that is buried in the binding cavity but not hydrogen bonded to the protein (an unpaired hydrogen bond) is strongly unfavorable for binding. It has been shown that the energy cost for an unpaired hydrogen bond is ca. 4 kJ/mol for a neutral substituent and ca. 16 kJ/mol for a charged substituent. This is equivalent to a loss of affinity by a factor of 5 and 500, respectively.

The successful formation of a hydrogen bond in the binding cavity has been estimated to contribute to the binding affinity by 2–6.5 kJ/mol (corresponding to an affinity increase by a factor of

2–13) for a neutral bond and by 10–20 kJ/mol (equivalent to a 50- to 500-fold increase in affinity) for a charge-assisted hydrogen bond or a salt bridge.

1.3.3.2   Polar Interactions Involving Aromatic Ring Systems

Other types of polar interactions often observed in ligand–protein complexes are π–π and cation–π interactions. The exact nature of these interactions is quite complex, but qualitatively they can be easily understood in terms of electrostatics. Figure 1.8 displays the calculated molecular electrostatic potential of benzene. It is obtained by calculating the energies of interaction between a benzene ring and a cation placed in different positions around the aromatic ring. The electrostatic potential in Figure 1.8 is color-coded on the vdW surface. A red color indicates a strong attraction between the cation and the aromatic ring and a blue color indicates a strong repulsion. Thus, a cation, for example an ammonium ion, may favorably interact with the face of the benzene ring, as shown in Figure 1.8. In π–π interactions, the edge of one benzene ring interacts with the face of the other.

Other aromatic rings such as phenol and indole display similar electrostatic potentials as benzene. Thus, the aromatic rings of phenylalanine, tyrosine, and tryptophan side chains may favorably interact with positively charged functional groups of the ligand. It has been estimated that a cation–π interaction may contribute by 8–17 kJ/mol to the overall binding of the ligand, which is equivalent to a 23- to 760-fold increase in affinity. Figure 1.9 shows the binding of nicotine to the

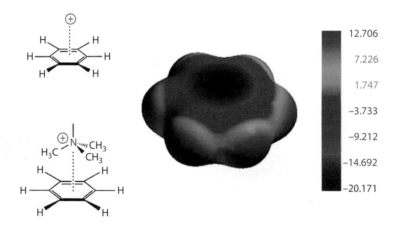

FIGURE 1.8   The molecular electrostatic potential of benzene.

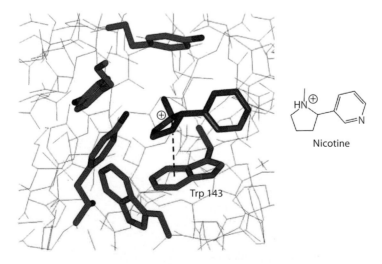

FIGURE 1.9   Nicotine in the binding pocket of the AChBP (pdb-code 1UW6).

AChBP. The binding displays a cation–π interaction between the ammonium group in nicotine and the indole ring system of Trp143.

1.3.4 $\Delta G_{hydrophob}$—THE HYDROPHOBIC EFFECT

The hydrophobic effect is a concept used to describe the tendency of nonpolar compounds to transfer from water to an organic phase, for example, a lipophilic region of a protein. When a lipophilic compound is inserted into water it changes the dynamic network of hydrogen bonds between water molecules in pure liquid. It creates a new interface in which water molecules around the lipophilic compound assume a more ordered arrangement than bulk water. This results in a decrease in entropy. Formation of a ligand–protein complex displaces the ordered water from the ligand and the protein into bulk water, as shown in Figure 1.10. The increased "freedom" of movement of the released water molecules gives an increase in entropy (ΔS) and, according to Equation 1.3, a more negative ΔG—an increase in affinity.

The magnitude of the hydrophobic effect is related to the area of hydrophobic surface that is buried in the binding cavity. Estimates, based on measurements of solvent transfer and ligand binding, range between 0.1 and 0.24 kJ/Å^2 mol. The burial of a methyl group of ca. 25 Å^{-2} is thus expected to result in an affinity increase by a factor of 3–10. In cases of a more perfect fit between a methyl group and the protein, the affinity increase may be even larger.

Analysis of ligand–protein interactions and attempts at ligand design often focus on hydrogen bonding and other electrostatic interactions. However, in many cases even strong hydrogen bond interactions may favorably be replaced by hydrophobic interactions. An example of this is shown in Figure 1.11. The influenza neuraminidase inhibitor **1.1** binds to the protein with an affinity (IC$_{50}$) of

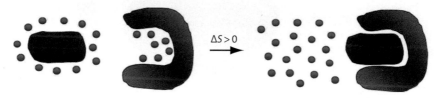

FIGURE 1.10 The hydrophobic effect.

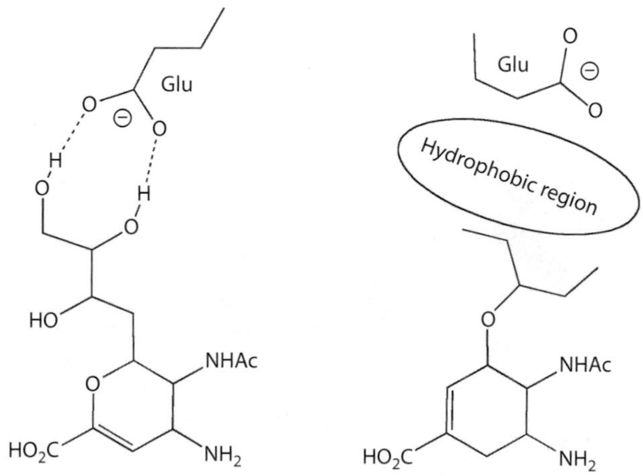

1.1
IC$_{50}$=150 nM

1.2
IC$_{50}$=1 nM

FIGURE 1.11 Influenza neuraminidase inhibitors.

150 nM. According to the ligand–receptor x-ray structure, the binding displays a bidentate charge-assisted hydrogen bond between the terminal glycerol hydroxy groups and a glutamate side chain. Removal of the glycerol side chain as in analogue **1.2** and replacement with an hydrophobic alkyl group increases the affinity to 1 nM (the minor modifications of the six-membered ring are not expected to influence the affinity significantly). An x-ray structure of the protein complex (**1.2**) shows that the glutamate side chain is folded back, opening up a large hydrophobic pocket as indicated in Figure 1.11.

1.3.5 ΔG_{vdW}—Attractive and Repulsive vdW Interactions

Nonpolar interactions between atoms, that is, vdW interactions, may be attractive as well as repulsive as shown by the vdW energy curve in Figure 1.12.

At short atom–atom distances, the vdW interaction is repulsive due to overlap of the electron clouds. The repulsion rises steeply with decreasing atom–atom distance in this region of the energy curve. When a part of the ligand clashes with atoms in the binding site, this steric repulsive vdW interaction is responsible for the often dramatic reduction in affinity that is observed.

At a longer distance, there is a region of attraction between the atoms. This attraction is due to the so-called dispersion forces. These are basically of electrostatic nature and due to interactions between temporary dipoles induced in two adjacent atoms. For a single atom–atom contact, the strength of the interaction is small, ca. 0.2 kJ/mol. However, as the total number of such interactions may be large, the dispersion interaction may in cases of a close fit between ligand and protein be significant. In this context it should be mentioned that the hydrocarbon tails in the core of a bilayer membrane are held together by dispersion forces.

As discussed in Section 1.3.5, a methyl group may be expected to increase the affinity of a compound by a factor of 3–10 due to the hydrophobic effect. Provided that the methyl group can be accommodated in the binding cavity. If it cannot be accommodated, vdW repulsion may instead give a significant decrease in affinity. Thus, the effect on the affinity of introducing a methyl group at different positions in a ligand may be a useful strategy to map out the dimensions of a receptor cavity in lack of experimental information on protein 3D-structure (see the discussion on pharmacophore modeling in Chapter 3). Provided that the introduced methyl group does not change the conformational properties of the parent molecule, the changes in affinity may be interpreted exclusively in terms of hydrophobic interactions and vdW interactions.

An example of effects that may be observed when introducing methyl groups in a ligand is shown in Figure 1.13. As discussed in detail in Chapter 3, flavone (**1.3**) binds to the benzodiazepine site of the GABA$_A$ receptor. The effects on the affinity of methyl groups in different positions of the parent flavone compound (**1.3**) may be interpreted in terms of properties of the binding cavity.

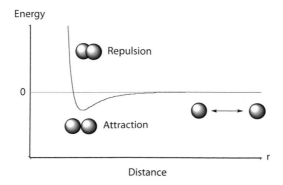

FIGURE 1.12   The vdW energy curve.

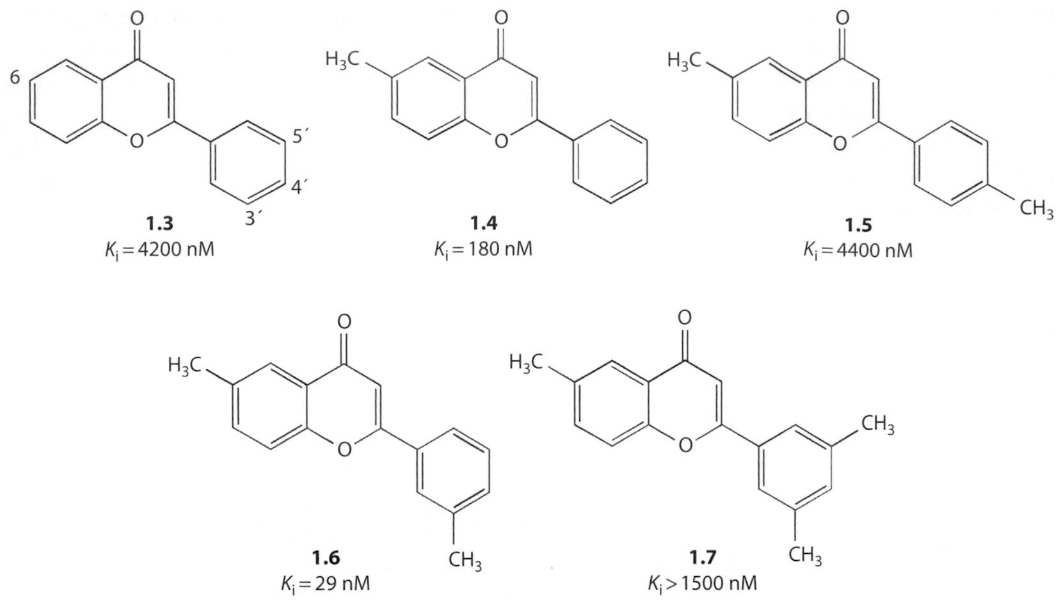

FIGURE 1.13 Affinities of methyl-substituted flavones binding to the benzodiazepine site of $GABA_A$ receptors.

When a methyl group is introduced in the 6-position (**1.4**), the affinity is increased by a factor of 23. This indicates that the methyl group can be very well accommodated in a lipophilic cavity in the binding site and that the affinity increase is due to hydrophobic interactions including dispersion interactions. An introduction of a methyl group to the 4′-position in **1.4** to give **1.5** results in a 24-fold decrease of the affinity. This is most likely due to repulsive vdW interactions between the 4′-methyl group and the receptor giving an indication of the dimensions of the binding site in this region. An introduction of a 3′-methyl group in **1.4** to give **1.6** increases the affinity by a factor of 6. This is a significantly lower affinity increase than shown by the 6-methyl group in **1.4**. Thus, the two receptor regions in the vicinities of the 6- and 3′-positions clearly have different properties. The region in the vicinity of the 3′-position can accommodate a methyl group but the fit between the methyl group and the receptor is not as good as in the case of the 6-methyl compound **1.4**. This is supported by the modest change in affinity when larger substituents are introduced in the 3′-position (see Chapter 3). Finally, the introduction of a 5′-methyl group in **1.6** giving **1.7** strongly decreases the affinity by a factor of more than 52. This is undoubtedly due to strong repulsive vdW interactions with the receptor. This identifies another steric repulsive receptor region adjacent to that identified by compound **1.5**. This example shows that conclusions drawn on the basis of a few compounds may provide valuable information on the properties of the protein binding site. Such information may be fruitfully used in the design of new compounds.

FURTHER READINGS

Ajay, A. and Murcko, M. A. 1995. Computational methods to predict binding free energies in ligand–receptor complexes. *J. Med. Chem.* 38:4953–4967.

Andrews, P. R. 1993. Drug–receptor interactions. In *3D QSAR in Drug Design: Theory, Methods and Applications*, Kubinyi, H. (ed.), ESCOM Science Publishers B.V., Leiden, the Netherlands, pp. 583–618.

Andrews, P. R., Craig, D. J., and Martin, J. L. 1984. Functional group contributions to drug–receptor interactions *J. Med. Chem.* 27:1648–1657.

Boström, J., Norrby, P.-O., and Liljefors, T. 1998. Conformational energy penalties of protein–bound ligands. *J. Comput. Aided Mol. Des.* 12:383–396.

Davies, A. M. and Teague, S. J. 1999. Hydrogen bonding, hydrophobic interactions, and failure of the rigid receptor. *Angew. Chem. Int. Ed.* 38:736–749.

Jencks, W. P. 1981. On the attribution and additivity of binding energies. *Proc. Natl. Acad. Sci. U.S.A.* 78:4046–4050.

Murray, C. W. and Verdonk, M. J. 2002. The consequences of translational and rotational entropy lost by small molecules on binding to proteins. *J. Comput. Aided Mol. Des.* 16:741–753.

Williams, D. H., Cox, J. P. L., Doig, A. J., Gardner, M., Cerhard, U., Kaye, P. T., Lal, A. R., Nicholls, I. A., Salter, C. J., and Mitchellf, R. C. 1991. Toward the semiquantitative estimation of binding constants. Guides for peptide–peptide binding in aqueous solution. *J. Am. Chem. Soc.* 113:7020–7030.

2 Biostructure-Based Drug Design

Flemming S. Jørgensen and Jette S. Kastrup

CONTENTS

2.1 INTRODUCTION

The idea behind biostructure-based drug design is to utilize the information on shape and properties of the binding site of a target molecule (e.g., enzyme or receptor) to design compounds, which possess complementary properties. Thus, biostructure-based drug design requires methods for determination of the three-dimensional (3D) structure of the target molecules as well as knowledge of which molecular interactions are important to obtain the desired binding characteristics.

Examples of ligands (drug molecules) binding to proteins are shown in Figure 2.1. The two ligands have been selected to illustrate different types of molecular interactions between the ligand and the target protein.

The 3D structure of a target protein can be determined experimentally by methods like x-ray crystallography and NMR spectroscopy, or predicted by computational methods like homology modeling (comparative model building). Of 50,000 experimentally determined protein structures 43,000 have been determined by x-ray crystallography (Protein Data Bank, May 2008). An x-ray crystallographic structure determination requires protein crystals, and irradiation with a high-energy x-ray source generates a diffraction pattern by the scattering of x-rays from organized molecules in a continuous arrangement in the crystal. Based on the diffraction pattern, an electron density map of the protein can be derived and subsequently a molecular model reflecting the electron density, the 3D structure, can be determined. Presently, the data collection, data processing, model building, and refinement are highly automated and computerized processes. The present limiting factor for determining the 3D structure of a protein is to get sufficient amounts of pure and stable protein and proper diffracting crystals.

It is important to consider the quality of an x-ray structure before using it for biostructure-based drug design. The resolution is a measure of how detailed the electron density map is and thereby how accurately the positions of the individual atoms can be determined (Figure 2.2A). Structures based on electron densities at 1.2 Å resolution are normally referred to as atomic-resolution structures and, e.g., hydrogen-bonding networks can unambiguously be identified. Generally, a resolution of ca. 2 Å

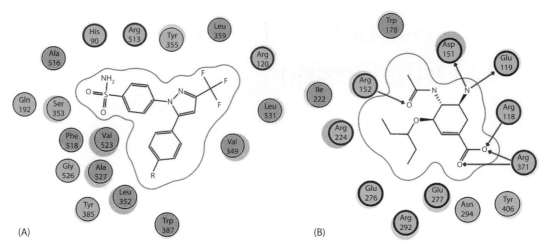

(A)   (B)

FIGURE 2.1 Example of a nonpolar (A) and a polar (B) binding site. In (A), the ligand, which is an analog (R = Br) of the COX-2 inhibitor celecoxid (R = CH$_3$), binds to cyclooxygenase-2 in a pocket primarily formed by nonpolar amino acid residues (pdb-code 1CX2). In (B), the ligand, which corresponds to the active part of the anti-influenza drug oseltamivir (Tamiflu), binds to an influenza virus neuraminidase in a pocket formed by polar residues (pdb-code 2QWK). Green arrows indicate hydrogen bonds. Green and red circles represent nonpolar and polar residues, respectively. The dotted lines illustrate the shape of the binding sites.

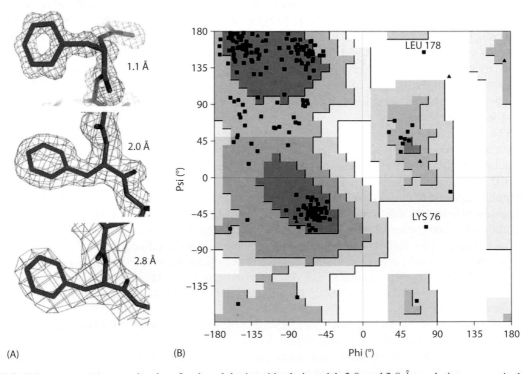

(A)   (B)

FIGURE 2.2 (A) Electron density of a phenylalanine side chain at 1.1, 2.0, and 2.8 Å resolution, respectively. (B) Ramachandran plot of a protein. Two residues (Lys76 and Leu178) adopt unfavorable backbone conformations. The majority of the residues are located in the dark gray regions corresponding to favorable backbone conformations.

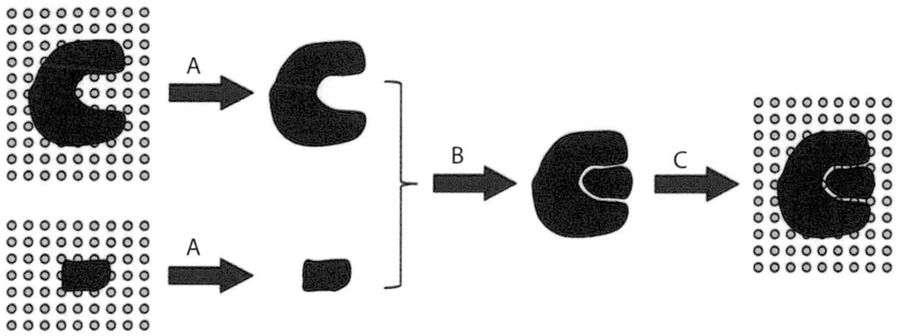

FIGURE 2.3 Simplified binding process. (A) Desolvation of protein and ligand (large unfavorable energy change). (B) The binding process (small favorable energy change). (C) Solvation of the protein–ligand complex (large favorable energy change).

provides an accurate structure, and accordingly such a structure is suitable for biostructure-based drug design. Structures based on 3 Å resolution electron densities should be used with caution.

The stereochemical quality of a protein structure should be carefully evaluated prior to using it for biostructure-based drug design. Planarity of peptide bonds, bond lengths, bond angles, and torsion angles should not deviate from average values. The protein backbone geometry is usually validated by a Ramachandran plot, a plot of the two variable torsion angles (phi and psi) (Figure 2.2B). Special attention should be paid to the part of the protein involved in direct interactions with ligands, e.g., flexible loops and binding site residues.

Binding of a drug molecule to a protein is a complicated process which is often considered as composed of a number of discrete steps in order to simplify the understanding of the process: (1) the protein may contain a cavity where the drug may bind, but prior to binding the water molecules occupying the cavity has to be removed. Similarly, the drug molecule may be surrounded by water molecules that have to be removed before binding. These two desolvation processes are associated with an unfavorable energy change (Figure 2.3A). (2) The next step is the actual binding between the drug molecule and the protein. The drug molecule may change conformation in order to fit into the binding site of the protein, and this conformational change requires energy. The protein may also change conformation, but this is usually ignored. The binding process requires that the drug molecule and protein are complementary not only with respect to shape, but also with respect to molecular properties. Positive parts of the drug molecule will bind to negative parts of the binding cavity and vice versa, and hydrogen-bond donors will bind to hydrogen-bond acceptors and vice versa (Figure 2.3B). (3) Finally, the complex between the drug molecule and the protein has to be surrounded by water molecules (solvated), and this process is associated with a favorable energy change (Figure 2.3C). Thus, the actual binding energy is a relatively small difference (typically on 5–20 kcal/mol) between the processes representing large destabilizing vs. stabilizing forces (for further details see Chapter 1).

2.2 ANTI-INFLUENZA DRUGS

One of the innovative examples on biostructure-based drug design was the discovery of the anti-influenza drug zanamivir (Relenza). Together with the subsequent discovery of the orally active anti-influenza agent oseltamivir (Tamiflu) they now constitute a classical example on a successful drug design story.

Two glycoproteins, namely hemagglutinin and neuraminidase, present on the surface of the virus are important for the replication cycle of the virus. The enzyme neuraminidase, or sialidase, is a glycohydrolase responsible for cleavage of the terminal sialic acid residues from carbohydrate moieties on the surface of the host cells and the influenza virus envelope. This process facilitates the release of newly formed virus from the surface of an infected cell. Inhibition of neuraminidase will leave uncleaved sialic acid residues that may bind to viral hemagglutinin, causing viral aggregation and thereby a reduction of the amount of virus that may infect other cells.

Thus, selective inhibitors of the enzyme neuraminidase are potential drugs against influenza. The determination of a 3D structure of neuraminidase by x-ray crystallography in 1991 yielded a breakthrough in the discovery of neuraminidase inhibitors being sialic acid analogous by biostructure-based methods.

It was shown that neuraminidase forms a homotetramer and that each monomer contains 6 four-stranded antiparallel beta-sheets arranged as blades on a propeller. The active site was identified from complexes with inhibitors of the enzyme and is located in a deep cavity on the surface of the enzyme (Figure 2.4). The active site is primarily formed by charged and polar residues, reflecting that the substrate is also a polar compound.

The mechanism for hydrolysis of glycoconjugates (Figure 2.5A) by neuraminidase yielding sialic acid (Figure 2.5C) is proposed to proceed via a flat oxonium cation transition state (Figure 2.5B).

FIGURE 2.4 3D structure of influenza virus neuraminidase (pdb-code 2QWK). (A) Homotetramer shown as ribbon representation. Four inhibitor molecules bound to the enzyme are shown as space-filling models. (B) Ribbon representation of one monomer colored continuously from blue (N-terminus) to red (C-terminus). An inhibitor is shown as a stick model. (C) Surface representation of monomeric neuraminidase with an inhibitor occupying the active site.

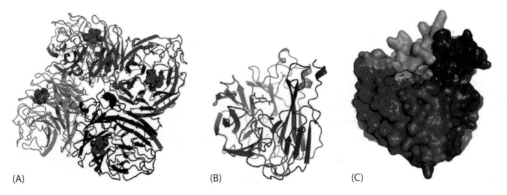

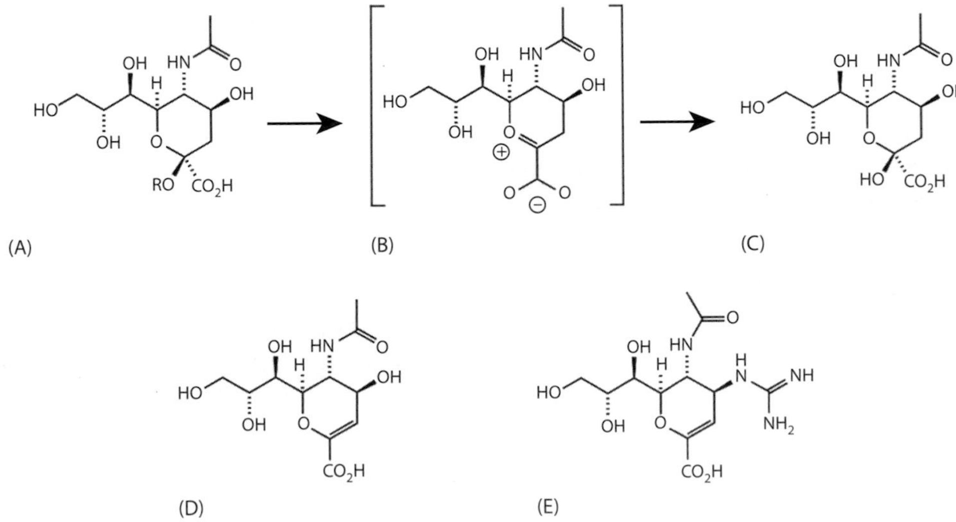

FIGURE 2.5 The biostructure-based design process leading to of zanamivir (Relenza): chemical structure of the glycoconjugates (R = sugar) (A), the proposed transition state (B), sialic acid (C), DANA (D), and zanamivir (E).

Sialic acid is also a weak inhibitor of neuraminidase ($IC_{50} \sim 1\,mM$) and binds to the enzyme in a half-boat conformation. The weak, nonselective inhibitor 2-deoxy-2,3-didehydro-D-*N*-acetyl-neuraminic acid (Neu5Ac2en, DANA, Figure 2.5D) was developed as a sialic acid analog resembling the transition state of the enzymatic process ($IC_{50} = 1$–$10\,\mu M$). Based on the structure of the neuraminidase–sialic acid complex, the complex with DANA and computational studies using the program GRID, which determines favorable binding sites for probes resembling various functional groups, it could be shown that replacement of the hydroxy group at the 4-position on DANA by an amino or a guanidinyl group would enhance binding. It was predicted that a substantial increase in binding could be obtained by introducing an amino group because a salt bridge was formed to a negatively charged side chain in the enzyme. Further replacement of the amino group with the more basic guanidinyl group was as anticipated yielding an even tighter binding inhibitor due to the formation of salt bridges to two negatively charged side chains in the enzyme. Both the predicted increase in affinity and the binding mode were subsequently confirmed experimentally. The 4-guanidinyl analog (Figure 2.5E) was named zanamivir and was the first neuraminidase inhibitor approved for treatment of influenza in humans. It is marketed under the trade name Relenza.

The low oral bioavailability and rapid excretion of zanamivir clearly showed that further improvements were needed in order to obtain a successful anti-influenza drug. Based on the 3D structures of several neuraminidase–inhibitor complexes and computer-based studies, the characteristics of the active site and thereby the properties of the optimal inhibitor were deduced. Replacing the pyranose ring with a benzene ring reduced the affinity, showing that the half-boat conformation of the six-membered ring was important for obtaining a proper orientation of the substituents. Inhibitors with the pyranose ring replaced by a carbocyclic ring system showed promising affinities, and by introducing more lipophilic substituents compounds with significantly better oral bioavailability were obtained.

In the final active drug candidate GS4071 ($IC_{50} \sim 1\,nM$) the carbocyclic ring adopts the half-boat conformation and the polar substituents are all involved in hydrogen bonding to polar residues in the neuraminidase active site (Figure 2.6). The ethyl ester was named oseltamivir and its phosphate salt (a prodrug, see Chapter 9) is marketed under the trade name Tamiflu.

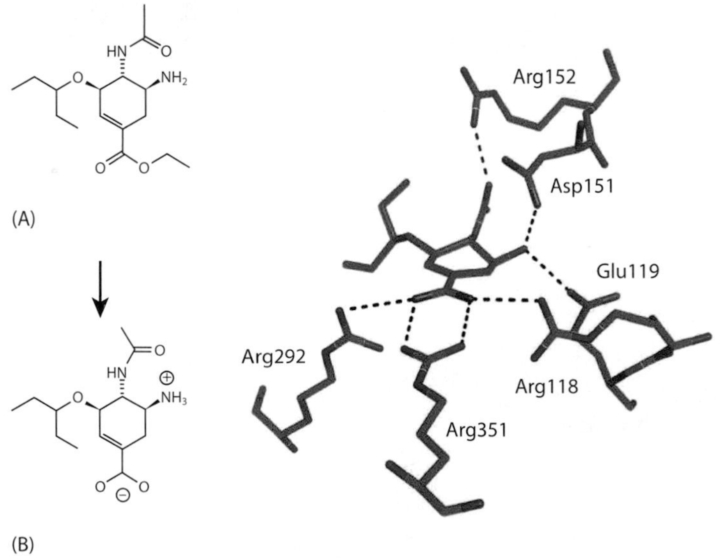

FIGURE 2.6 Left: Chemical structures of oseltamivir (A) (oseltamivir phosphate = Tamiflu) and the active component of Tamiflu (GS4071) (B) formed by enzymatic cleavage. Right: Hydrogen-bonding network between (B) and neuraminidase (pdb-code 2QWK).

2.3   HIV PROTEASE INHIBITORS

When in 1984 it was discovered that the human immunovirus (HIV) caused AIDS it was the start of an intensive drug-hunting process. The DNA in the virus encodes for a number of enzymes, e.g., a reverse transcriptase, an integrase, and a protease. Each of these represented a potential drug target, and drugs have been developed subsequently for each of the enzymes. Presently, cocktails of inhibitors against at least two of these enzymes are used therapeutically. In the following text, we will concentrate on the HIV-1 protease and how biostructure-based design has been applied extensively to this target.

The first 3D structure of an HIV-1 protease in complex with an inhibitor, MVT-101, was reported shortly after it had been shown that inhibition of the HIV-1 protease prevented the virus in producing new virions. MVT-101 binds to the enzyme in an extended conformation and forms a network of hydrogen bonds between the ligand and enzyme (Figure 2.7). Hydrophobic substituents on the inhibitor occupy hydrophobic pockets in the enzyme, and a water molecule mediates contact between the inhibitor and two residues in two flexible beta-sheets, normally referred to as the flaps. Today, close to 300 structures of complexes between HIV-1 protease and ligands have been determined, which makes HIV-1 protease one of the structurally most extensively studied proteins.

The first HIV-1 protease inhibitors like, e.g., indinavir (Crixivan), nelfinavir (Viracept), and saquinavir (Invirase, Fortovase) (Figure 2.8) were derived from the polypeptide sequences cleaved by the protease in the HIV. Accordingly, they were very peptide-like and had poor bioavailability.

Unfortunately, the HIV rapidly developed resistance against these first-generation inhibitors. The shapes of the hydrophobic pockets are sensitive to mutations in the enzyme and accordingly, the virus could easily prevent an inhibitor from binding by mutation of residues forming the hydrophobic pockets.

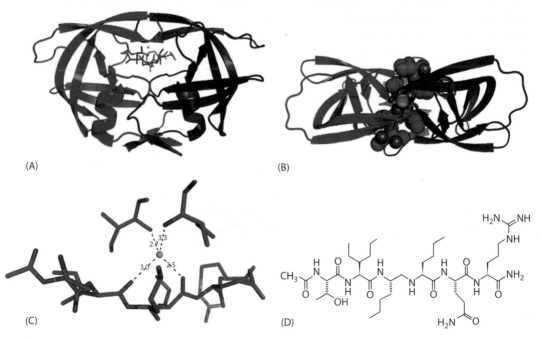

(A)

(B)

(C)

(D)

FIGURE 2.7   3D structure of the HIV-1 protease with bound MVT-101 (pdb-code 4HVP). (A) Side view showing that the active form of the HIV-1 protease is a homodimer (colored red and green, respectively) and that the inhibitor MVT-101 binds between the two monomers. (B) Top view showing the extended form of the inhibitor and that the inhibitor via a structural water molecule (cyan) binds to the flaps. (C) The structural water molecule makes four hydrogen bonds to the inhibitor and to Ile50 in the flaps. (D) Chemical structure of MVT-101.

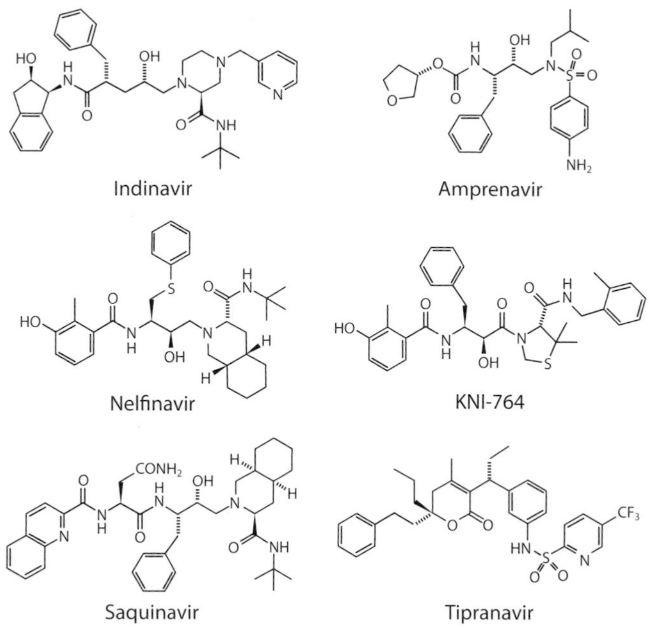

FIGURE 2.8 Examples on HIV-1 protease inhibitors. Indinavir, nelfinavir, and saquinavir are examples of first generation, amprenavir and KNI-764 of second generation, and tipranavir of third-generation HIV-1 protease inhibitors.

In order to circumvent this problem, inhibitors binding not only primarily due to hydrophobic effects but also by hydrogen bonding to the enzyme backbone atoms (amide NH group and amide carbonyl oxygen atom) were designed. These hydrogen-bonding contacts are not sensitive to mutations, since they do not involve the side chains but only the backbone atoms. These second-generation inhibitors like, e.g., amprenavir (Agenerase) and KNI-764 (Figure 2.8) showed significantly different binding characteristics.

Thermodynamic determination of the enthalpy and entropy components to the free energy of binding ($\Delta G = \Delta H - T\Delta S$) can be determined by isothermal titration calorimetry (ITC), and this method has led to a much more detailed understanding of the energetics associated with the process of binding a ligand to a protein. By using ITC to guide the design of new inhibitors it has been possible to optimize the binding characteristics of these new inhibitors.

For the first-generation HIV-1 protease inhibitors, the majority of the free energy of binding is due to an entropy gain associated with filling hydrophobic pockets in the HIV-1 protease with hydrophobic substituents on the inhibitor. The second-generation inhibitors were characterized by both enthalpy and entropy now contributing to the free energy of binding, making the enzyme less likely to develop resistance.

Recently, third-generation HIV-1 protease inhibitors have been developed based on the careful optimization of the structural and energetic contributions to binding. Tipranavir (Aptivus) (Figure 2.8) is an example of an HIV-1 protease inhibitor with unique binding characteristics. It is a highly potent inhibitor ($K_i = 19\,\text{pM}$), which primarily binds to the wild-type HIV-1 protease due to entropy effects. The unusually high binding entropy is most likely caused by release of buried water molecules from the active site of the HIV-1 protease. When binding to the multidrug-resistant HIV mutants tipranavir only looses little in potency, because the reduction in binding entropy is compensated by a gain in binding enthalpy.

In 1994, researchers at the DuPont Merck Pharmaceutical Company reported an important observation. They had realized that in most of the complexes between HIV-1 protease and the peptide-like inhibitors a structural water molecule bridged the ligand and enzyme. They concluded that by

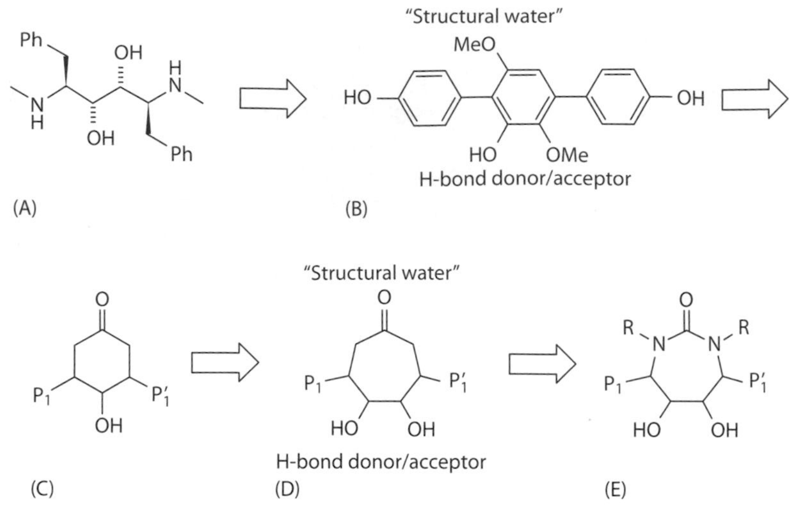

FIGURE 2.9 The DuPont Merck design process leading to the seven-membered urea HIV-1 protease inhibitors. (A) Symmetric dihydroxyethylene inhibitor used to define a pharmacophore. (B) Initial hit from 3D database search. (C) Initial synthetic scaffold. (D) Scaffold modified to accommodate two hydrogen-bond donors/acceptors. (E) Final scaffold optimized for synthetic feasibility and improved hydrogen bonding.

designing an inhibitor that would displace this water molecule, a favorable entropy contribution to the binding energy would be obtained. In addition, selectivity should be gained since this structural water molecule was unique for the viral proteases.

The DuPont Merck scientists defined a pharmacophore (see Chapter 3) from known dihydroxyethylene inhibitors (Figure 2.9A), used this pharmacophore for searching a database and obtained a hit (Figure 2.9B), which subsequently gave the idea to the six-membered cyclic ketone as lead structure (Figure 2.9C). Ring expansion to the seven-membered cyclic ketone (Figure 2.9D) enabled incorporation of two hydrogen-bonding donor/acceptor groups and synthetic reasons led to the

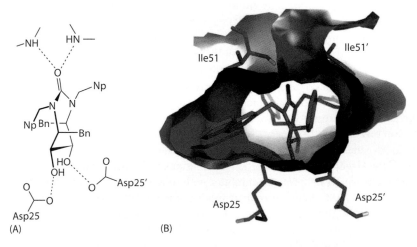

FIGURE 2.10 Schematic representations of the binding of the seven-membered urea HIV-1 protease inhibitor XK-263 to HIV-1 protease (pdb-code 1HVR). In (A), Bn and Np refer to benzyl and 2-naphthyl, respectively. In (B), the two Asp25 and Ile51 residues making hydrogen bonds to the inhibitor are shown as stick models. Heteroatoms are colored red and blue for oxygen and nitrogen, respectively. The surface is colored accordingly.

series of seven-membered cyclic urea HIV-1 protease inhibitors (Figure 2.9E). Using the structural information available from the many peptide-like inhibitors, the nature and stereochemistry of the substituents on the cyclic urea could be designed so they were preorganized for binding to the enzyme (Figure 2.10). By preorganization of a ligand for binding, the conformational energy penalty, often associated with ligand binding, is reduced and the ligand may be more potent.

Although the DuPont Merck cyclic urea inhibitors were based on a brilliant structural idea and potent inhibitors were designed, the inhibitors generally had low bioavailability and the HIV quickly developed resistance against the compounds. Thus, none of these inhibitors are among the drugs used today. Other companies have subsequently adopted the same idea in their design processes. One example is the previously mentioned HIV-1 protease inhibitor tipranavir (Aptivus) where the oxygen atom in the carbonyl group replaces the structural water molecule.

2.4 MEMBRANE PROTEINS

The largest group of biological targets for therapeutic intervention comprises the membrane proteins. These proteins perform various functions in the cell, serving as enzymes, pumps, channels, transporters, and receptors. To emphasize the importance of considering membrane proteins, ca. 40% of all drugs target G-protein-coupled receptors.

The first membrane structure was reported in 1985 of the photosynthetic reaction center from *Rhodopseudomonas viridis*. Today, the number of known membrane protein structures is still limited. Currently, the 3D structures are only known for 160 unique membrane protein structures including proteins of the same type from different organisms (Membrane Proteins of Known 3D Structure Web site, June 2008). The number of membrane protein structures and the change in the number of structures as a function of time are similar to the state of soluble protein structures approximately 25 years ago. The reason for this is primarily that expression, purification, and crystallization of membrane proteins are still nontrivial and require substantial time and resources. At the same time, structure determination of and biostructure-based drug design on membrane proteins represent one of the most challenging areas of modern drug research.

Another strategy taken on membrane proteins is to produce soluble constructs of parts of the proteins, e.g. the extracellular ligand-binding core of ionotropic glutamate receptors (iGluRs). These receptors mediate most fast excitatory synaptic transmission within the central nervous system (see Chapter 15). The glutamate receptors are not only involved in various aspects of normal brain functions but are also implicated in a variety of brain disorders and diseases. Hence, iGluRs are potential targets for biostructure-based drug design.

In 1998, the first structure of a ligand-binding core construct of an iGluR was reported, and presently ca. 90 structures of iGluRs with bound glutamate, agonists, antagonists, or allosteric modulators have been reported. The binding of ligands to the ligand-binding core of iGluRs can be described as a "Venus flytrap" mechanism. In the resting state the ligand-binding core is present in an open form and it is this form that is stabilized by competitive antagonists (Figures 2.11 and 2.12). When glutamate or an agonist binds to the ligand-binding core, a change in conformation occurs, resulting in a closed form of the ligand-binding core. In full-length receptors, this domain closure is thought to lead to the opening of the ion-channel (receptor activation). The presence of more than one conformation of the ligand-binding core of iGluRs clearly stresses the importance of knowing more than one structure of the receptor as fundament for biostructure-based drug design (for further details on glutamate receptor structures, see Sections 1.3.2 and 12.2.2, and Chapter 15).

2.5 FAST-ACTING INSULINS

Biostructure-based drug design is not limited to design of low-molecular weight compounds based on knowledge of the structure of their biological targets. In the following text we are presenting an example on biostructure-based design of macromolecular drug molecules, i.e., insulin analogs.

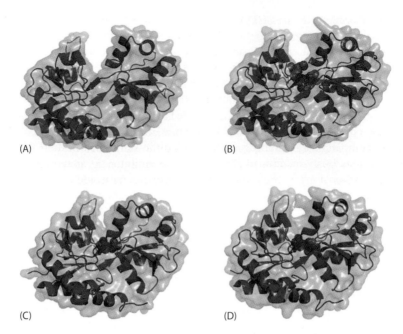

FIGURE 2.11 Structures of the ligand-binding core of the ionotropic glutamate receptor GluR2. (A) The open, unbound form of GluR2 (pdb-code 1FTO). (B) The NeuroSearch compound NS1209 stabilizes GluR2 in the open form (pdb-code 2CMO). (C) The endogenous ligand glutamate introduces domain closure of GluR2 by a "Venus flytrap" mechanism (pdb-code 1FTJ). (D) Various synthetic agonists (here Br-HIBO) also introduce domain closure in GluR2 (pdb-code 1M5C).

This design was made possible only by a detailed insight into the structure of insulin and the intermolecular interactions between the insulin molecules in the crystalline phase.

Insulin is a hormone produced in the pancreas and it is responsible for the regulation of glucose uptake and storage. Insulin is most often associated with diabetes mellitus, which is a disease causing hyperglycemia. Healthy people have a basal level of insulin in the bloodstream, but in response to intake of food or to cover glucose clearance from the blood, peaks of larger insulin concentrations appear throughout the 24 h of a day. Patients with diabetes may have difficulties in maintaining the proper insulin concentrations, basal as well as peak concentrations, and accordingly regulation of their insulin level is essential. Type I diabetes patients need insulin to supplement endogenously produced insulin, whereas Type II diabetes patients often are getting insulin in order to improve glycemic control.

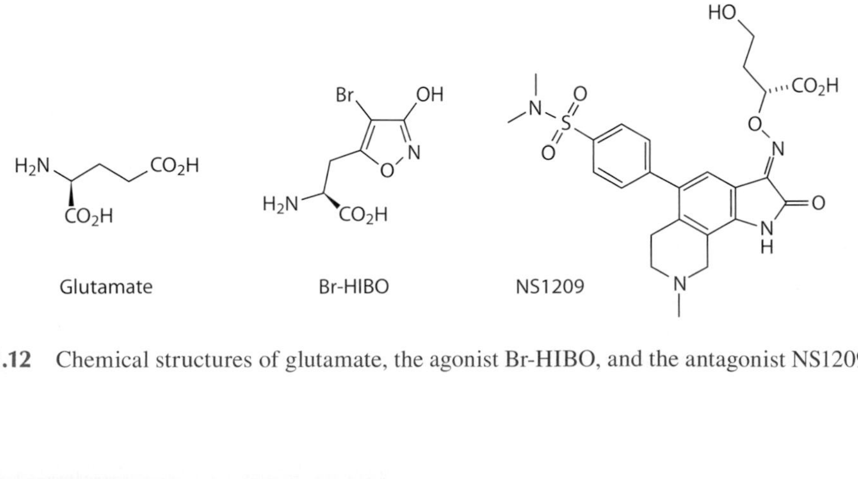

Glutamate   Br-HIBO   NS1209

FIGURE 2.12 Chemical structures of glutamate, the agonist Br-HIBO, and the antagonist NS1209.

Insulin has been available for treatment of diabetes since 1923, and the major form for administration is still by subcutaneous injection. Insulin therapy typically involves multiple doses of different forms of insulin to maintain near-physiological insulin (and thereby glucose) levels. Long-acting insulin maintains the basal insulin level over 24 h with a single administration, and fast- and short-acting insulin analogs, which are instantaneously absorbed, are used to meet the insulin requirements associated with food intake. Thus, the development of insulin formulations with tailored properties, e.g., a prolonged effect or a faster onset, has always had a high priority in insulin research.

With the biosynthesis of recombinant human insulin in the 1980s it became possible to optimize the insulin therapy by designing insulin analogs with optimal pharmacokinetic properties by changing the amino acid sequence of the insulin molecule. A rational design of insulin analogs was only possible because a large number of insulin structures were determined experimentally by NMR spectroscopy and x-ray crystallography. Insulin was one of the first proteins whose 3D structure was determined by x-ray crystallography, and today more than 200 insulin structures are available in the Protein Data Bank.

Insulin exists in the crystalline phase as hexamers, dimers, and monomers. Actually, several forms of hexamers, T6, T3R3, and R6, exist depending on the presence of zinc ions and phenol (Figure 2.13). The presence of several hexameric forms of insulin reflects that even in the crystalline form insulin exerts some kind of conformational flexibility. Both zinc ions and phenol are stabilizing the hexameric form of insulin and are accordingly added to insulin formulations in order to improve their stability. After subcutaneous injection the insulin hexamer dissociates to dimeric insulin and by further dissociation to monomeric insulin, which represents the bioactive form. Thus, from the beginning it was believed that shifting the equilibrium toward the monomeric form would lead to faster-acting insulins, whereas stabilization of the hexameric form would lead to longer-acting insulins.

The insulin molecule consists of two peptide chains, an A-chain of 21 residues and a B-chain of 31 residues. The A- and B-chains are connected by two disulfide bridges linking A7–B7 and A20–B19. With the introduction of recombinant DNA techniques the residues involved in interaction with the insulin receptor or involved in the hexamer vs. dimer stabilization were identified.

From the x-ray structure of hexamer insulin it was evident that the side chain of the HisB10 residue was involved in zinc binding and thereby in stabilizing the hexamer (Figure 2.13). Mutation of the B10 residue from His to Asp yielded an insulin analog being absorbed twice as rapidly as normal insulin. Unfortunately, this analog turned out to be mitogenic and thus not suitable for clinical use.

Based on various structural studies of insulin it could be concluded that the flexibility of the C-terminus is crucial for the binding of insulin to its receptor (Figure 2.14). It also became evident that the B24–B26 residues are stabilizing the dimer by making an intermolecular antiparallel

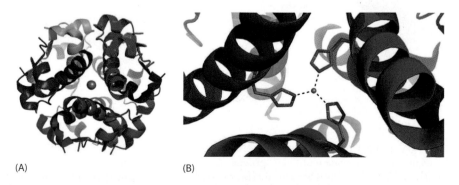

(A)   (B)

FIGURE 2.13   (A) Insulin R6 hexamer (pdb-code 1EV6) showing the threefold symmetry of the three dimers (red–green, cyan–orange, and magenta–blue). (B) Three HisB10 residues coordinate to a zinc ion.

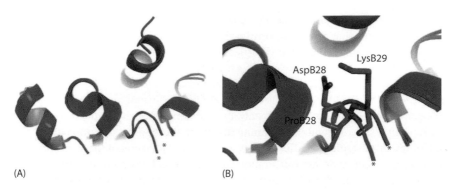

FIGURE 2.14 (A) Human insulin (pdb-code 1ZNJ, green) and insulin aspart (pdb-code 1ZEG, yellow), showing the change in conformation of the C-terminus (marked by an asterisk) by the mutation ProB28Asp. (B) For human insulin ProB28 and for insulin aspart AspB28 and the neighboring LysB29 are shown as stick models. Heteroatoms are colored red and blue for oxygen and nitrogen, respectively.

beta-sheet and at the same time burying the nonpolar side chains. Removal of the B24–B30 residues yields monomeric insulin and eliminates the association of monomers into dimers and hexamers. Furthermore, molecular modeling studies showed that ProB28 was important for dimer formation, but apparently not involved in receptor binding. Mutation of B28 from Pro to Lys or Asp has a profound effect on dimerization. The ProB28Asp mutant is presently marketed as rapid-acting insulin named aspart (Novorapid).

The knowledge of the interactions, which stabilize the hexamer and dimer, has made it possible to engineer the properties of insulin to yield therapeutic insulin analogs with tailored properties. Homology modeling indicated that inversion of the residues B28 and B29 should give faster acting insulins (Figure 2.15). The double mutant, ProB28Lys + LysB29Pro, indeed had a faster onset than normal insulin and it was the first genetically engineered insulin analog to become available for clinical use. Generally, modifications of B29 have a less pronounced effect on dimerization than modification of ProB28. The double mutant is especially interesting because it has the same isoelectric point (pI) as human insulin as their amino acid compositions are identical, and thereby the same solubility. The double mutant called insulin lispro is marketed under the brand name Humalog.

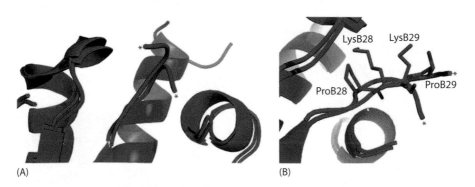

FIGURE 2.15 (A) Human insulin (pdb-code 1ZNJ, green) and insulin lispro (pdb-code 1LPH, yellow), showing the change in conformation of the C-terminus (marked by an asterisk). (B) The ProB28-LysB29 in human insulin and the LysB28-ProB29 in insulin lispro are shown as stick models. Heteroatoms are colored red and blue for oxygen and nitrogen, respectively. The side-chain nitrogen atom in LysB29 could not be observed in the x-ray structure.

Insulin analogs with changes of ProB28, as the above-mentioned insulin analogs aspart and lispro, primarily exist in the monomeric form, which is known to be more vulnerable for unfolding and subsequent fibril formation. To avoid this problem, mutations were limited to the neighboring LysB29 and to residues in the N-terminus of the B-chain. It was already known that changes of the B1–B8 residues primarily affected the stabilization of the dimer form of insulin. The basic LysB29 was exchanged for the acidic, polar, and hydrophilic Glu, and the neutral, polar, and hydrophobic AsnB3 was exchanged for the basic, polar, and hydrophilic Lys. This double mutant has a slightly lower pI of 5.1 compared with human insulin with a pI of 5.5, and accordingly its solubility is also enhanced. The double mutant, AsnB3Lys + LysB29Glu, is especially interesting because higher-order forms dominate relative to the monomeric form, i.e., it is more stable, while still maintaining rapid dissociation to monomeric insulin. This double mutant is normally referred to as insulin glulisine and is marketed as Apidra.

Thus, the design of rapidly absorbed, fast-acting insulin analogs must be characterized as a clear success. The design of the above-mentioned insulin analogs were made possible because the large number of insulin structures provided the researchers with a detailed information about the molecular interactions responsible for receptor binding and the hexamer–dimer–monomer equilibrium.

2.6   NEW METHODS

The experimental and computational methods used in biostructure-based drug designs are constantly evolving. Here, we will only mention a few of the methods that have and are expected to have a great impact on the design process.

The availability of an increasing number of experimentally determined structures of relevant targets is crucial. Crystallization, data collection, and structure determination processes are today partially or fully automated. Centers for high throughput determination of 3D structures have been established in several countries. These developments and initiatives are reflected in an increase in the number of experimentally determined structures deposited at the Protein Data Bank. Many pharmaceutical companies also have in-house groups doing structure determinations, but the number of structures determined here is difficult to estimate.

Computationally, the developments are also considerable. Improved methods for determination of models of structures based on structures of related proteins, i.e., homology modeling or comparative modeling, are being developed. Faster computers allow longer and thereby more realistic simulations of proteins where flexibility and solvation can be considered. Ligand docking programs are being improved by considering not only the flexibility of the ligand but also the flexibility of the protein. The prediction of binding energies have always been a problem in docking methods, but improved quantum mechanics-based or quantum mechanics-derived methods combining speed and accuracy are being developed. Better description and handling of the solvation vs. desolvation processes are also crucial for the correct prediction of binding affinities.

2.7   CONCLUSION

Biostructure-based drug design is being used to efficiently develop new therapeutic candidates and incorporates multiple scientific disciplines, including medicinal chemistry, pharmacology, structural biology, and computer modeling.

The examples described earlier illustrate that biostructure-based drug design in cases where structural information of the targets is available is a powerful method for the design of new or improved drugs.

The DuPont Merck example on biostructure-based design of the cyclic urea HIV-1 protease inhibitors nicely illustrates that in many cases it is possible based on the 3D structures of a target to design ligands to control or regulate a biological system. Unfortunately, the example also illustrates that several other features have to be considered in order for a ligand to become a successful drug.

FURTHER READINGS

Acharya, K.R. and Lloyd, M.D. 2005. The advantages and limitations of protein crystal structures. *Trends Pharmacol. Sci.* 26:10–14.

Brange, J. et al. 1988. Monomeric insulins obtained by protein engineering and their medical implications. *Nature* 333:679–682.

Davis, A.M. et al. 2003. Applications and limitations of x-ray crystallographic data in structure-based ligand and drug design. *Angew. Chem. Int. Ed.* 42:2718–2736.

Hansen, K.B. et al. 2007. Structural aspects of AMPA receptor activation, desensitization and deactivation. *Curr. Oprin. Neurobiol.* 17:281–288.

Hardy, L.W. and Malikayil, A. 2003. The impact of structure-guided drug design on clinical agents. *Curr. Drug Discov.* 3:15–20.

Lam, P.Y.S. et al. 1994. Rational design of potent, bioavailable, nonpeptide cyclic ureas as HIV protease inhibitors. *Science* 263:380–384.

McCusker, E.C. et al. 2007. Heterologous GPCR expression: A bottleneck to obtaining crystal structures. *Biotechnol. Prog.* 23:540–547.

Russell, R.J. et al. 2006. The structure of H5N1 avian influenza neuraminidase suggests new opportunities for drug design. *Nature* 443:45–49.

Thornton, J.M. et al. 2000. From structure to function: Approaches and limitations. *Nat. Struct. Biol.* 7:991–994.

von Itzstein, M. et al. 1993. Rational design of potent sialidase-based inhibitors of influenza virus replication. *Nature* 363:418–423.

3 Ligand-Based Drug Design

Ingrid Pettersson, Thomas Balle,
and Tommy Liljefors

CONTENTS

3.1 INTRODUCTION

The use of computational methods to facilitate the drug discovery process is today well established and plays an important role in modern multidisciplinary drug discovery projects. A wide range of computational methods are used to find new active compounds and to optimize these compounds in order to produce new candidate drug molecules. The methods used depend on the available structural information of the target protein (or more generally the target biomacromolecule). If a three-dimensional (3D) structure of a target enzyme or receptor with a cocrystallized ligand is available from x-ray crystallography, a detailed knowledge of the nature of the ligand-binding site, the ligand-binding mode, and the interactions between the ligand and the receptor/enzyme can be obtained. On this basis, new ligands may computationally be "docked" into the binding site in order to study if they can effectively interact with the receptor. This can be performed by using sophisticated automated flexible docking and scoring computer programs. New and promising compounds identified by such computational experiments may then be synthesized and tested pharmacologically. This procedure is known as "structure-based drug design" and is discussed in Chapter 2.

However, many proteins of high interest as drug targets have so far resisted all attempts of crystallization and 3D-structure determination. This is, for instance, the case for most members of the large and important class of seven-transmembrane (7-TM) G-protein-coupled neurotransmitter receptors (see Chapter 12). In the absence of an experimentally determined 3D structure of the receptor, computational methodologies based on an analysis of the physicochemical and pharmacological properties of known ligands may be used for the design/discovery of new ligands. This computational procedure is called "ligand-based drug design" and is the subject of this chapter. The purpose of this chapter is to introduce and discuss some major methods used for ligand-based drug design and to exemplify the use of these methods by a case study in which the discovery of novel ligands for the benzodiazepine (BZD) site of the γ-aminobutyric acid type A (GABA$_A$) receptor have been successfully accomplished. For details of this case study the reader is referred to the references in the Further Readings section.

The most important and powerful method in ligand-based drug design is "pharmacophore modeling," which is used to develop a pharmacophore model describing the interactions between ligands and the target receptor from the ligand point of view. We will discuss and illustrate how a well-developed pharmacophore model can be used to search databases for new compounds, which fit the model and to optimize identified compounds by a pharmacophore-guided procedure.

A pharmacophore model does not give a quantitative prediction of receptor affinities. The main use of such a model is restricted to the prediction of candidate ligands as active or inactive. Such a classification may be fruitfully used in the selection of new molecules to be synthesized and pharmacologically tested in a drug discovery project. However, a pharmacophore model may additionally be used as a starting point for 3D quantitative structure–activity relationships (3D-QSAR) analysis. The 3D-QSAR combines pharmacophore models, molecular interaction fields and statistical chemometrics methods to give quantitative predictions of receptor affinities and in addition guidelines for ligand optimization. This methodology will be discussed in the last part of this chapter.

3.2 THE BENZODIAZEPINE SITE OF GABA$_A$ RECEPTORS

GABA$_A$ receptors are transmembrane proteins that are assembled from subunits into a pentameric structure forming an ion-channel through which the influx of chloride ions is regulated. In addition to the binding site for the endogenous agonist GABA, the GABA$_A$ receptors have binding sites for compounds that allosterically modify the chloride channel gating of GABA. The most well-known class of such compounds is the BZDs and the binding site for this class of compounds has been named the BZD site. The pharmacological effects of the BZDs (anxiolytic, anticonvulsant, muscle relaxant, and sedative-hypnotic) make them the most important GABA$_A$ receptor-modulating drugs in clinical use.

In addition to the BZD class of compounds, it has been shown that many other classes of compounds bind to the BZD site and a pharmacophore model including several classes of compounds have been developed (see Further Readings). In addition, naturally occurring and synthetic derivatives of flavones have also been shown to bind to the BZD site. The flavone class of compounds is the starting point for the case study described in this chapter.

At present there is no experimentally determined 3D structure for any of the subtypes of the GABA$_A$ receptor. Thus, in the computational part of a drug discovery project dealing with this class of receptors, the use of ligand-based drug design methods is the only alternative. For a more detailed discussion on GABA$_A$ receptors and their ligands, see Chapter 15 and the use of GABA ligands as hypnotics is discussed in Chapter 20.

3.3 PHARMACOPHORE MODELING

Pharmacophores and pharmacophore elements are central concepts in medicinal chemistry. The idea behind these concepts comes from the common observation that variations of some parts of the molecular structure of a compound drastically influence the activity at a target receptor, whereas variations of other parts only cause minor activity changes.

A "pharmacophore element" is traditionally defined as an atom or a group of atoms (a functional group) common for active compounds at the receptor in question and essential for the activity of the compounds. However, the concept of a pharmacophore element may fruitfully be extended to include representations of interactions of ligand functional groups with receptor sites. The "pharmacophore"

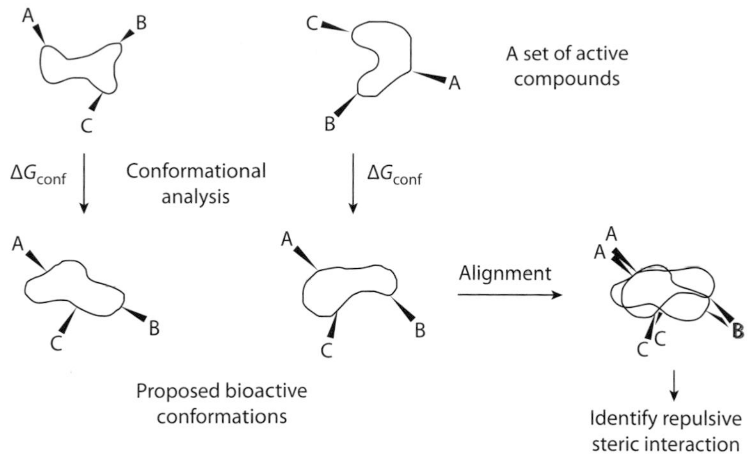

FIGURE 3.1 Basic principles of the development of a 3D-pharmacophore model.

is a collection of pharmacophore elements and the concept of "3D-pharmacophore" may be used when the relative spatial positions of the pharmacophore elements are included in the analysis. Thus, a 3D-pharmacophore consists of a specific 3D-arrangement of pharmacophore elements.

The basic principles of the development of a 3D-pharmacophore model are illustrated in Figure 3.1. On the basis of conformational analysis of a set of active molecules with pharmacophore elements A, B, and C, a low energy conformation of each molecule is selected for which the pharmacophore elements of the molecules overlap in space as shown in the figure. Conformational energies in ligand–protein binding are discussed in more detail in Chapter 1. The selected conformations are the putative bioactive conformations of the molecules and the overlapping pharmacophore elements and their spatial positions make up the 3D-pharmacophore.

The development of a 3D-pharmacophore model requires that a number of active compounds and their affinities for the receptor in question are available. The necessary number of compounds depends mainly on the conformational flexibility of the ligands. In the case of highly flexible molecules, the development of a pharmacophore model generally requires a larger number of compounds compared to the case in which the compounds are less flexible.

In addition to active compounds, it is highly useful that a number of inactive compounds also are available. These may, as will be demonstrated in the following text, fruitfully be used to identify regions of sterically repulsive ligand–receptor interactions and thus provide an estimate of the dimensions of the binding cavity.

3.4 A 3D-PHARMACOPHORE MODEL FOR FLAVONES BINDING TO THE BZD SITE

The development of the pharmacophore model will be first discussed in terms of the different steps in Figure 3.1. The molecular structure and atom numbering of flavone (**3.1**) is shown in Figure 3.2. The carbonyl group, the ether oxygen, and the two phenyl rings are common to all active compounds in the flavone series and are necessary for the activity at the BZD site. These are the basic pharmacophore elements. The identification of the bioactive conformation in the flavone series is straightforward. Flavone (**3.1**) has a very limited conformational flexibility. The only torsional degree of freedom is the rotation about the 2-1′ bond connecting the phenyl ring to the bicyclic system. Conformational analyses of the available compounds in the flavone series and comparisons with the corresponding experimental affinities show that only compounds in which the entire flavone skeleton is planar or close to planar in the global energy minimum are compatible with a

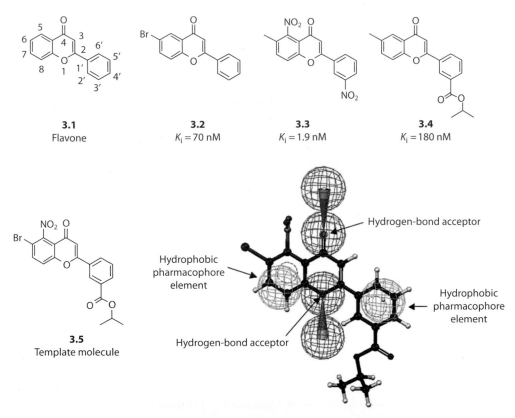

FIGURE 3.2   The structure and atom numbering of flavone (**3.1**), the flavone derivatives (**3.2**–**3.4**), the template structure (**3.5**) and the template structure mapped with four pharmacophore elements (two hydrogen-bond acceptors [green] and two hydrophobic pharmacophore elements [cyan]).

significant affinity. The alignment of pharmacophore elements in Figure 3.1 is in this case trivial as all compounds in the series have the same molecular skeleton.

3.4.1 THE INITIAL PHARMACOPHORE MODEL

In order to develop a computer representation of the pharmacophore model, which also includes information on the available space at important substituent positions (the last step in Figure 3.1), and to extend the simplistic pharmacophore model described earlier to include representations of interactions of pharmacophore elements with receptor sites, three substituted flavones (**3.2**–**3.4**) shown in Figure 3.2 were selected. Since small substituents in the 6-position increases the affinity, compound **3.2** with a bromo substituent in the 6-position was selected. Compound **3.3** with nitro groups in the 5- and 3′-positions was selected due to the favorable effect on the affinity for small substituents in these positions. Finally, compound **3.4** was selected as a representative of compounds in the available series that carry a large substituent in the 3′-position but still display a reasonable receptor affinity. Since all three compounds have the same skeleton, they were for simplicity merged into a single template molecule (**3.5**) displaying all the important features of **3.2**–**3.4**. Alternatively, each of compounds **3.2**–**3.4** could have been analyzed by the computer program CATALYST for common pharmacophore elements.

The template molecule **3.5** was used as input to the widely used computer program CATALYST (Accelrys Software Inc.), which analyzes molecules in terms of pharmacophoric features and displays the pharmacophore elements as spheres. An advantage of this approach is that the pharmacophore model representation produced by CATALYST explicitly includes the physicochemical properties of the pharmacophore elements and that the model directly can be used for database searching as will be described in the following text. The physicochemical properties of the predefined pharmacophore elements in CATALYST are hydrogen-bond acceptor, hydrogen-bond donor, hydrophobic (aliphatic or aromatic), negative or positive charge, negatively or positively ionizable, and ring aromatic. In order to allow for variations in the geometry of the interaction between a ligand and its receptor, distance variation as well as angle variation is taken into account. For instance, a hydrogen-bond acceptor is defined by a distance from the atom, which accepts a hydrogen bond to the site that donates the hydrogen bond with an allowed geometrical variation at both ends. These allowed variations are defined by spheres and the optimal interaction is defined by the center of the spheres.

The resulting initial CATALYST pharmacophore model with the pharmacophore elements of the flavone skeleton is shown in Figure 3.2 (bottom right). The carbonyl oxygen atom and the ether atom are mapped as hydrogen-bond acceptors (green spheres), while the phenyl rings are mapped as hydrophobic pharmacophore elements (cyan spheres). It should be noted that these are both mapped as a single sphere whereas the pharmacophore elements involving the carbonyl group and the ether oxygen are displayed as two spheres in order to take the direction of the modeled hydrogen-bond interaction into account. For example, the sphere centered at the carbonyl oxygen in Figure 3.2 represents the ligand side (the hydrogen bond accepting side) of the hydrogen-bond interaction whereas the outer sphere represents the receptor side (the hydrogen bond donating site) of the hydrogen bond.

3.4.2 RECEPTOR ESSENTIAL VOLUMES AND THE USE OF EXCLUSION SPHERES

Inactive compounds, which fit the initial pharmacophore are, as mentioned earlier, useful to establish the dimensions of the receptor binding site, i.e., to determine parts of the binding cavity where substituents are in steric conflict with the receptor. Figure 3.3 shows four compounds (**3.6–3.9**)

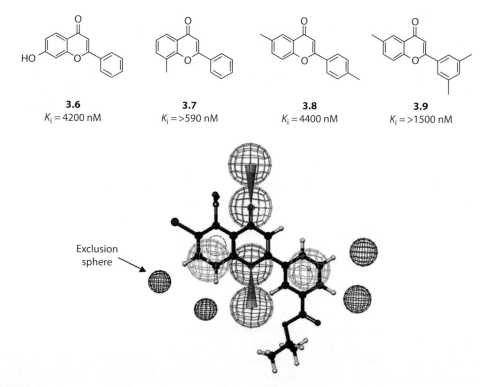

3.6
$K_i = 4200\ nM$

3.7
$K_i = >590\ nM$

3.8
$K_i = 4400\ nM$

3.9
$K_i = >1500\ nM$

Exclusion sphere →

FIGURE 3.3 Compounds **3.6–3.9** and the pharmacophore model including exclusion spheres (black).

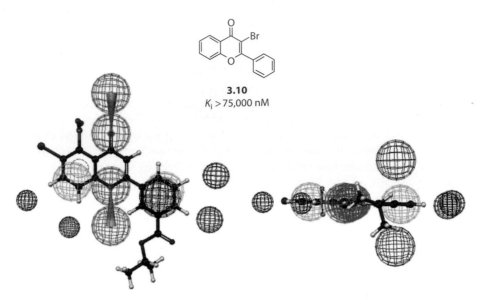

3.10
$K_i > 75,000$ nM

FIGURE 3.4   Compound **3.10** and the pharmacophore with exclusion spheres (black) above and below the phenyl ring.

that all fit the initial pharmacophore model. The affinities of compounds **3.6–3.9** show that even small substituents in the 7-, 8-, 4′-, and 5′-positions of flavone give low affinities, which is most probably due to steric repulsive interactions with the receptor. The ligand–receptor repulsion sites (receptor essential volumes) are represented by black spheres (exclusion spheres) in Figure 3.3.

As mentioned in Section 3.4, only compounds in which the flavone skeleton is planar or close to planar in the global energy minimum are compatible with a significant affinity. This is exemplified by compound **3.10** in Figure 3.4. Due to strong steric repulsion between the bromo substituent and the 6′-hydrogen atom in the phenyl ring, this ring is calculated to be 63° twisted in the preferred conformation of **3.10**. Furthermore, the energy required for the phenyl ring to be coplanar with the bicyclic ring system is as high as 26 kJ/mol. This high conformational energy makes **3.10** inactive at the BZD site. In order to distinguish between compounds that are planar or close to planar and those that are significantly nonplanar, exclusion spheres are positioned above and below the phenyl ring in the pharmacophore model as displayed in Figure 3.4.

3.4.3   EXTENSION OF THE PHARMACOPHORE MODEL

Other classes of compounds binding to the BZD site were examined for the possibility of adding more features to the pharmacophore model. Compound **3.11** (Figure 3.5) displays high affinity for the BZD site. The superimposition of **3.2** and **3.11** shown in Figure 3.5 indicates that both molecules may bind to the BZD site in the same manner. The presence of an NH group in **3.11**, which may be hydrogen bonding to the receptor, makes it of interest to include this as an additional feature in the pharmacophore model. This may be accomplished by extending the template molecule **3.5** to include an NH group in the 6′-position as shown for the updated template molecule **3.12** in Figure 3.6. The updated pharmacophore model including the new pharmacophore element is also shown in Figure 3.6. The final pharmacophore model also includes a "shape" (in light gray), which is the van der Waals (vdW) volume of the template molecule **3.12**. The "shape" gives an estimate of the available space for ligand binding in the receptor binding cavity.

Using the updated pharmacophore model, compound **3.13** was designed and synthesized as a test for the validity of the new pharmacophore feature. The affinity of this compound was found to be 0.9 nM, which makes it the highest affinity compounds in the flavone series with an affinity

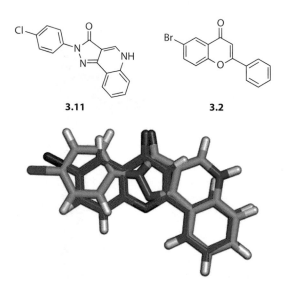

3.11   **3.2**

FIGURE 3.5   Superimposition of compounds **3.11** and **3.2**.

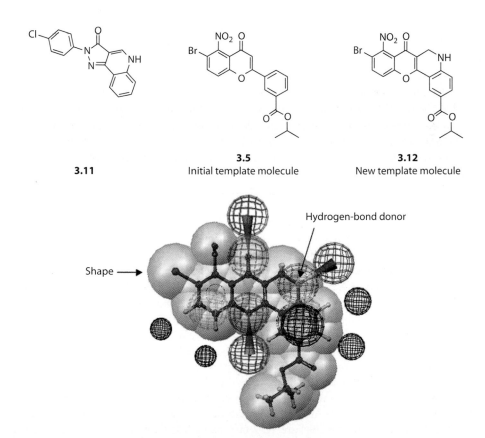

3.11   **3.5**   **3.12**
                    Initial template molecule   New template molecule

FIGURE 3.6   Compound **3.11**, the initial template molecule **3.5**, the new template molecule **3.12**, and the updated pharmacophore model with the new hydrogen-bond donor pharmacophore element (magenta) included.

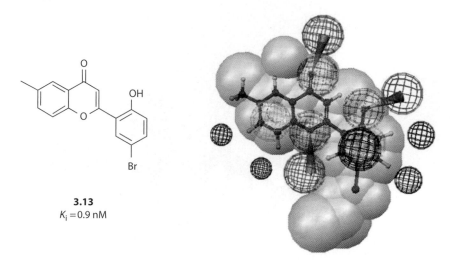

3.13

$K_i = 0.9$ nM

FIGURE 3.7   Compound **3.13** fitted to the pharmacophore model.

4500-fold higher than that of the parent flavone **3.1** as a result of the addition of only three properly placed substituents. Compound **3.13** is shown fitted to the final pharmacophore model in Figure 3.7.

3.5   DATABASE SEARCHING USING THE PHARMACOPHORE MODEL

To identify new classes of compounds, which may bind to the BZD site, the pharmacophore model in Figure 3.6 was used for database searching in CATALYST. Two different compound databases, the Maybridge database (Maybridge Chemical Company) with approximately 47,000 compounds and the Available Chemical Directory (ACD) database (MDL Information Systems, Inc.) with approximately 250, 000 compounds were searched. In order to search a database using a 3D-pharmacophore model, the compounds in the database, which are 2D structures has to be converted to 3D structures and a set of conformations for each compound has to be generated. The conformations can either be generated in advance or for some programs during the search. In CATALYST, a set of conformations for each compound in the database is generated in advance. When a pharmacophore model is used to search a database, compounds fitting all or some of the pharmacophore elements are identified. Such compounds are called "hits." In the present case, it is required that all pharmacophore elements must to some degree be fitted to give a hit. In the search procedure all compounds are given a fit value indicating how well they fit the pharmacophore elements of the model. After the search, a list of hits ranked according to their fit value is available.

As described in Sections 3.4.2 and 3.4.3, exclusion spheres and a vdW shape may be used in a pharmacophore model to represent the dimensions of the receptor binding cavity. This is of high importance for database searching in order to keep the number of hits to a manageable size.

The difference between an exclusion sphere and the vdW shape is that no ligand is allowed to touch an exclusion sphere without a severe penalty, whereas a ligand may be somewhat larger or smaller than the shape without a penalty. In the present case, ligands with vdW volumes 30% smaller or 10% larger than the volume of the shape are allowed.

When the two databases were searched using the pharmacophore model in Figure 3.6, 22 hits in the Maybridge database and 76 hits in the ACD database were obtained. (It should in this context be mentioned that these database searches are very fast, using less than 30 min per database.) Among the 98 hits, five compounds of the highest ranking hits with a significant diversity of the molecular structures were selected and purchased. The highest affinity of these compounds was the 4-quinolone derivative **3.14** ($K_i = 122$ nM) shown in Figure 3.8. As also shown in the figure, this compound is

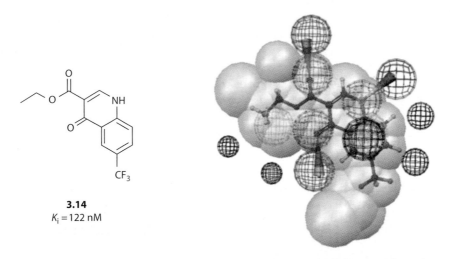

FIGURE 3.8 The most active hit from the database search and its fit to the pharmacophore model.

fitted to the pharmacophore model with its ethyl group corresponding to the fused benzene ring in the flavones and the CF_3 group corresponding to the isopropyl ester in **3.4** (Figure 3.2). The structure of the compound is structurally much different from that of flavones and it is a novel compound in a medicinal chemistry sense as it represents a class of compounds that has not previously been tested for affinity for the BZD site of the $GABA_A$ receptor.

3.5.1 POSTPROCESSING OF DATABASE HITS—AN ESSENTIAL REQUIREMENT

To avoid false positives, i.e., compounds that are ranked high in a database search, but are found to be of low affinity when tested, postprocessing of database hits is, in general, necessary before selection of compounds for synthesis or purchase. As an example, the second best hit in the database search is compound **3.15** fitting to the pharmacophore model with the hydroxyl group as a hydrogen-bond donor as shown in Figure 3.9. Even though the compound displays a good fit to the

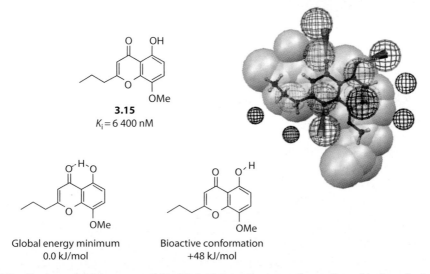

FIGURE 3.9 Compound **3.15**, a top ranking hit in the database search together with the calculated global energy minimum and the bioactive conformations.

pharmacophore, the compound shows only weak affinity ($K_i = 6400\,nM$). According to calculations, an intramolecular hydrogen bond between the hydroxyl group and the carbonyl group is present in the global energy minimum of the compound (Figure 3.9). In order to donate a hydrogen bond from the hydroxyl group as required by the pharmacophore model, the intramolecular hydrogen bond has to be broken giving a high conformational energy penalty (calculated to be $48\,kJ/mol$) and as a consequence of this the affinity is strongly decreased. Such conformational energy penalties may not be taken properly into account by database searching programs. It is also highly advisable to examine hydrogen bond distances to hydrogen bonding pharmacophore elements in the model to remove hits with too long or too short hydrogen bonds.

3.6   PHARMACOPHORE-GUIDED OPTIMIZATION OF COMPOUND 3.14

Examining the fit of compound **3.14** to the pharmacophore model in Figure 3.8, three observations of relevance for optimization of the compound with respect to affinity can be made.

- There is sufficient space at the position of the CF_3 group to replace this group by a larger substituent.
- The ester ethyl group in **3.14** does not completely fill out the cavity in comparison to the bromo substituent in compound **3.2** (compare the fit of the bromo substituent in template molecule **3.12** in Figure 3.6). As this part of the shape is most probably a highly hydrophobic/lipophilic pocket, it is essential for optimal affinity to fill it out as completely as possible. Replacement of the ester ethyl group by a propyl group is an obvious possibility.
- Compound **3.14** has two conformations with respect to rotation around the bond connecting the ester group to the bicyclic ring system (Figure 3.10). A replacement of the ester group by an amide group would stabilize the molecule in the bioactive conformation due to the intermolecular hydrogen bond in the amide compound. This will give a smaller conformational entropy loss for binding and a higher affinity (for more details of entropy effects in ligand binding see Chapter 1).

On the basis of these observations, a small series of compounds were synthesized and tested. The most important compounds and their affinities are shown in Figure 3.11.

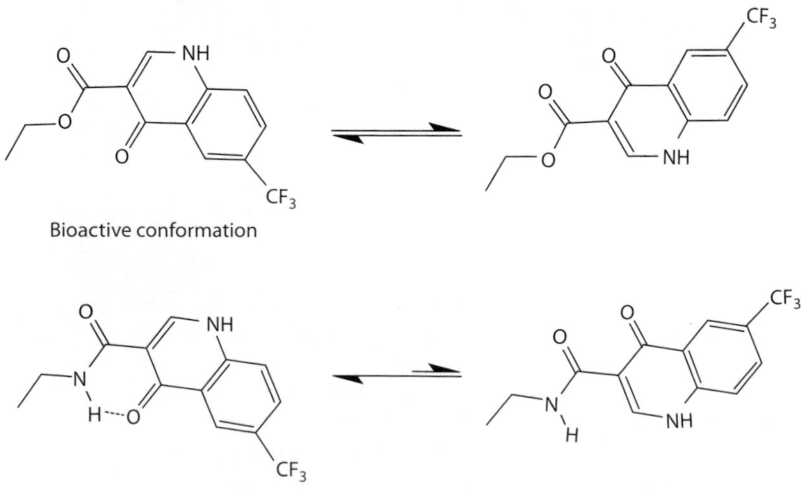

Bioactive conformation

FIGURE 3.10   Conformational equilibria for compound **3.14** and its amide analogue.

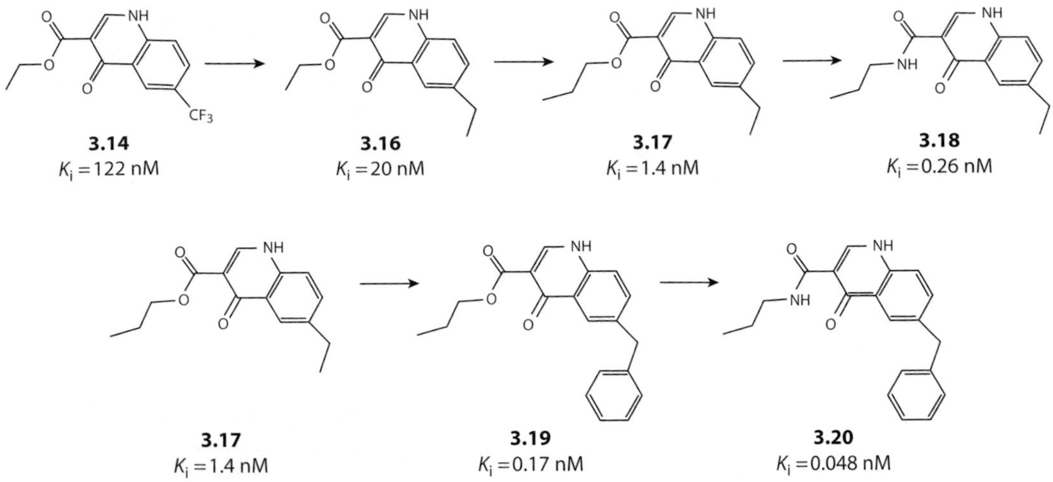

FIGURE 3.11 Pharmacophore-guided optimization of compound **3.14**.

The replacement of the CF_3 group in **3.14** by an ethyl group to give **3.16** increases the affinity from 122 to 20 nM. Replacement of the ester ethyl group in **3.16** by a propyl group to give **3.17** further increases the affinity to 1.4 nM and conversion of the ester group in **3.17** to an amide group gives compound **3.18** with an affinity of 0.26 nM.

By further exploring lipophilic substituents in the 3′-position, compound **3.19** was identified to be a high affinity compound ($K_i = 0.17$ nM). Finally, converting the ester group in **3.19** to an amide group gave the highest affinity compound **3.20** in this series of compounds with $K_i = 0.048$ nM. In comparison with compound **3.14**, compound **3.20** has a higher affinity by a factor of 2500. This increased affinity demonstrates the power of a well-developed pharmacophore model for the optimization of a compound with respect to affinity.

3.7   3D-QSAR ANALYSIS—THE GRID/GOLPE APPROACH

QSAR methods in medicinal chemistry use statistical methods to correlate the variation in molecular properties with the variation in biological activities. The purpose of establishing a QSAR model is to be able to predict activities of new compounds quantitatively. Traditional QSAR methods in general do not explicitly take 3D structures into account, but use substituent parameters to describe the variations in molecular structures/properties. In contrast, 3D-QSAR methods explicitly use 3D molecular structures and use molecular interaction fields to describe the variation of the properties of the molecules.

Several alternative approaches to 3D-QSAR have been developed. In this chapter, we illustrate the 3D-QSAR methodology by the GRID/GOLPE approach.

3.7.1   GRID MOLECULAR INTERACTION FIELDS

Molecular interaction fields describe interaction energies between a molecule and a chemical probe positioned in different locations around the molecule. GRID is a widely used program for the calculation of such fields. The chemical probe may be a methyl group, a water molecule, or any of the more than 60 probes provided by GRID. Interaction energies between the molecule and the probe are calculated by inserting the molecule in a box (Figure 3.12). The probe is then moved through a regular 3D array of grid points at positions around the molecule as shown in the figure. The spacing between the grid points is user-defined but normally 0.25–1 Å. At each grid point the

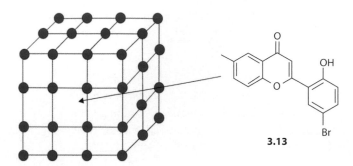

FIGURE 3.12 Setup for calculation of molecular interaction fields.

probe–molecule interaction energy (E_{tot}) between the probe and the molecule is calculated by an empirical force-field as a sum of the vdW energy (E_{vdW}), the electrostatic energy (E_{el}), and the hydrogen bond energy (E_{hbond}) as shown in Equation 3.1

$$E_{tot} = E_{vdW} + E_{el} + E_{hbond} \qquad (3.1)$$

The molecular interaction field for each probe may be visualized by calculating and displaying isoenergy contours at a user-defined energy level. Examples of isoenergy contours are shown in Figure 3.13, displaying the results of GRID analyses for the substituted flavone **3.13**.

Figure 3.13a shows the GRID results using a water probe. As expected, contours are observed around the carbonyl group and the hydroxy group indicating strong hydrogen bond interactions between the water molecule and **3.13**. However, water can donate as well as accept hydrogen bonds but these two modes of interaction cannot be distinguished by the use of a water probe. By choosing an NH+ probe, which can only donate a hydrogen bond, and an O− probe, which can only accept a hydrogen bond, donating and accepting can be distinguished. This is shown in Figure 3.13b, where the blue contour indicates hydrogen bond accepting by the carbonyl group and the red contour hydrogen bond donating by the hydroxy group.

Molecular interaction fields may be used to identify potentially important interactions between a ligand and a protein. The probes are then representing different types of interaction partners in the binding pocket of the protein. However, a more powerful use of molecular interaction fields in ligand-based drug design is the combination of these fields with statistical chemometric methods to develop a 3D-QSAR model for the quantitative prediction of biological activities. The basic idea is

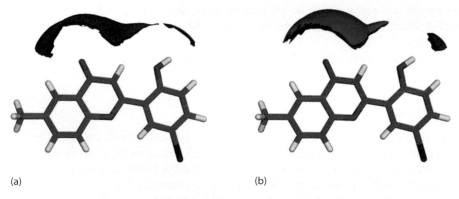

(a)   (b)

FIGURE 3.13 Isoenergy curves for compound **3.13** and (a) a water probe contoured at −21 kJ/mol and (b) an amine cation (NH+) probe contoured at −26 kJ/mol (blue) and a phenolate anion oxygen (O−) probe contoured at −16 kJ/mol (red).

that molecular interaction fields for a series of compounds contain information that can be used for the understanding and prediction of the biological activity of the compounds.

3.7.2 Development of a 3D-QSAR Model for Substituted Flavones

The starting point for the development of a 3D-QSAR model is a series of molecules and their biological activities at a given receptor. To illustrate the methodology, we will use a series of substituted flavones also used in the pharmacophore modeling section earlier.

A crucial first step in a 3D-QSAR analysis is the alignment of the molecules. This is equivalent to the development of a pharmacophore model (see Figure 3.1). For the flavones the alignment has been discussed earlier. Thirty-four substituted flavones were used as a training set for developing the 3D-QSAR model. For each molecule located in a gridbox and aligned according to the pharmacophore model, the interaction energies with two probes, a methyl probe and a water probe, were calculated by GRID. To find a correlation between the biological activity and the calculated molecular interaction fields, the method of partial least squares projections to latent structures (PLS) in GOLPE is used (for more details see Further Readings). In essence, PLS contracts the original description of each molecule (i.e., the molecular interaction fields) into a few descriptive dimensions/variables that are used for the correlation.

It is essential to validate the 3D-QSAR model. This should optimally be done internally as well as externally. For internal validation, also called cross validation, a portion of the training set compounds are left out and a new model of this reduced training set is built. This model is then used to predict the activities of compounds left out. This procedure is repeated a number of times. The results of the predictions of left out compounds are summarized in terms of a predictive correlation coefficient q^2, which should be larger than 0.5 for a high-quality 3D-QSAR model. External validation is performed by predicting the activities of compounds (the test set), which have not been used to build the 3D-QSAR model. The results of this validation may be given as a standard error of prediction (SDEP).

Figure 3.14 displays the results for the series of 34 substituted flavones used as a training set and seven substituted flavones as a test (validation) set. The conventional correlation coefficient r^2 is

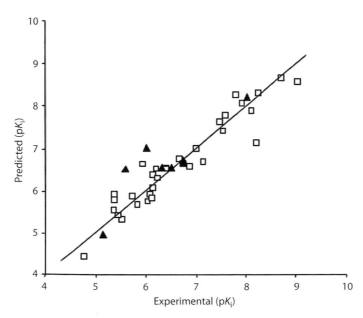

FIGURE 3.14 Experimental and predicted affinities. The training set is shown as unfilled squares and the test set as filled triangles.

such as acetylsalicylic acid, were applied. A general feature, however, has been that small molecules have been used either in a random manner or at best in a case-by-case fashion, therefore Schreiber and others decided to systematize the application of small molecules in studies of proteins. This has led to the conception of chemical genetics.

4.2.2.1 Chemical Genetics

Chemical genetics is a research method that uses small molecules to perturb the function of proteins and does this directly and in real time, rather than indirectly by manipulating their genes. It is used to identify proteins involved in different biological processes, to understand how proteins perform their biological functions, and to identify small molecules that may be of therapeutic value.

The term "chemical genetics" indicates that the approach uses chemistry to generate the small molecules and that it is based on principles that are similar to classical genetic screens. In genetics two kinds of genetic approaches, forward and reverse, are applied, depending on the starting point of the investigation. A classical forward genetic analysis starts with an apparent physical character-istic (phenotype) of interest and ends with the identification of the gene or genes that are responsible for it. In classical reverse genetics, scientists start with a gene of interest and try to find what it does by looking at the phenotype when the gene is mutated. In chemical genetics, small molecules are used to perturb protein function: in forward chemical genetics, a ligand that induces a phenotype of interest is selected, and the protein target of this ligand is identified. In reverse chemical genet-ics, small molecules are screened for effects at the protein of interest, and subsequently a ligand is used to determine the phenotypic consequences of perturbing the function of this protein. Chemical genetics can therefore be regarded as a fruitful and complementary alternative to classical genetics or to the use of RNA-based approaches, such as RNA interference (RNAi) technology.

A now classical example illustrating the power of small molecules in elucidating protein targets is FK506 (or Fujimycin, Figure 4.4), which is a macrolide natural product structurally related to rapamycin (see Chapter 6), and is currently used as an immunosuppressant after organ transplan-tation. The target of FK506 was not known, but using FK506 as molecular bait to fish its binding protein from biological samples, a protein was identified, called FK506 binding protein (FKBP). It

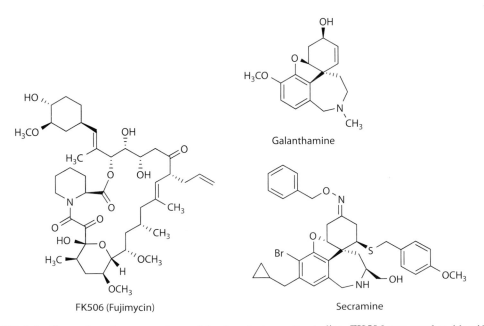

FIGURE 4.4 Examples of compounds used in chemical genetic studies. FK506 was used to identify its protein target, FKBP. Galanthamine was employed as a template for the DOS of structurally diverse analogs, which lead to the identification of secramine, as an inhibitor of vesicular traffic out of the Golgi apparatus.

22222333333333

333

was found that the binding of FK506 to FKBP inhibits immune function by shutting down a specific molecular signaling pathway. Another natural product, galanthamine (Figure 4.4, see Chapter 16), which is an inhibitor of acetylcholinesterase (AChE) isolated from certain species of daffodil and is used for the treatment of mild to moderate Alzheimer's disease, was used as a template for the so-called biomimetic DOS; that is, a range of diverse chemical reactions was applied to a scaffold similar to that of galanthamine. A 2527 compound library of galanthamine-like structures was prepared and underwent phenotypic screening, identifying secramine (Figure 4.4), as an inhibitor of vesicular traffic out of the Golgi apparatus by an unknown mechanism. After an extensive effort it was discovered that secramine inhibits the activation of the Rho GTPase Cdc42, a protein involved in membrane traffic.

4.2.2.2 ChemBank and PubChem

Along the systematization of the application of small molecules to probe protein function, it follows that the results of screening are put into a well-structured format, that is, a format where chemical structures and biological activities are correlated. Two initiatives, ChemBank and PubChem, in this direction have been initiated, which have very similar objectives, to create a public, Web-based informatics environment, with data derived from small molecules and small-molecule screens. The intention of such databases is to guide the chemist in their synthesis of novel compounds or libraries, and assist biologists, who are exploring small molecules that perturb certain biological pathways. These databases contain an increasingly varied set of cell measurements derived from biological objects and cell lines treated with small molecules. Moreover, analysis tools are available and are being developed that allow the relationships between cell states, cell measurements, and small molecules to be determined.

4.3 MODIFYING BIOMACROMOLECULES

Biomacromolecules are oligomeric molecules that are composed of smaller building blocks in nature. The three major classes of biomacromolecules are proteins, nucleic acids, and polysaccharides, built up of amino acids, nucleotides, and carbohydrates, respectively.

The modification of nucleic acids is a very promising and fruitful area of research, where peptide nucleic acid (PNA) and locked nucleic acid (LNA) are prominent examples of modified nucleic acids (Figure 4.5). In PNA, the deoxyribose of the DNA backbone is replaced with N-(2-aminoethyl)-glycine units. This has several advantages, for example, the backbone is linked by peptide bonds,

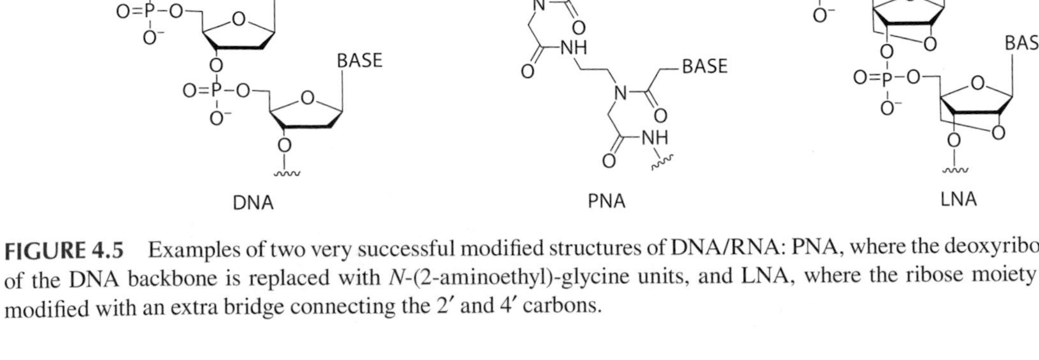

FIGURE 4.5 Examples of two very successful modified structures of DNA/RNA: PNA, where the deoxyribose of the DNA backbone is replaced with N-(2-aminoethyl)-glycine units, and LNA, where the ribose moiety is modified with an extra bridge connecting the 2′ and 4′ carbons.

thus synthesis is easier and also the stability of PNA is increased compared to DNA. Another prominent example of DNA, or rather RNA, modification is LNA, where the ribose moiety of an LNA nucleotide is modified with an extra bridge connecting the 2′ and 4′ carbons (Figure 4.5). The bridge locks the ribose in the 3′-endo structural conformation, which is supposed to be an important bioactive conformation. This locked conformation of LNA enhances base stacking and backbone preorganization, as well as increases the stability of the nucleic acids. Both PNA and LNA are being pursued commercially as potential drug candidates, as well as diagnostic tools. Polysaccharides, such as starch and glycogen, are highly important biomacromolecules, and the term "glycomics" has been introduced analogous to genomics and proteomics, to describe the comprehensive study of sugars in organisms, and glycomics is a subset of studies of sugars in biology in general, termed glycobiology. Carbohydrates also play vital roles as posttranslational modifications (PTMs) of glycoproteins and in the remainder of this chapter focus will be on proteins and methods for modifying proteins.

4.3.1 Protein Engineering

Proteins are the most abundant biomacromolecules in cells, constituting up to 50% of the dry weight of cells. In eukaryotes, proteins are produced in the ribosome, where a messenger RNA (mRNA) carries the code for the primary sequence of the protein, and is read by aminoacylated transfer RNA (aa-tRNA). The code, the genetic code, contains 64 triplet codons, of which 61 codes are for 20 different amino acids, which we call cognate, canonical, or proteinogenic amino acids—that is building blocks for protein biosynthesis. The last three codons, UAG (amber), UAA (ochre), and UGA (opal), are stop codons, also known as nonsense codons. Thus, eukaryotic proteins are generally made up of the 20 proteinogenic amino acids, although in recent years two extra amino acids have been added to this repertoire. The 21st amino acid is selenocysteine, which is found in prokaryotes and eukaryotes and where the sulfur of cysteine is replaced by selenium and the 22nd amino acid is pyrrolysine, where the ε-amino group of lysine is derivatized with β-methylpyrroline (Figure 4.6).

Methods for the residue-specific incorporation of close analogs of natural amino acids have existed for many years, where the depletion of one amino acid and the addition of another, structurally related unnatural amino acid, allows the incorporation of this amino acid. A typical example is the incorporation of selenomethionine (Se-Met), in place of methionine, which is used in structural studies of proteins, as the heavy atom selenium may help in solving the phase problem in x-ray crystallography. Here, we will focus on approaches for the site-specific incorporation of unnatural amino acids into

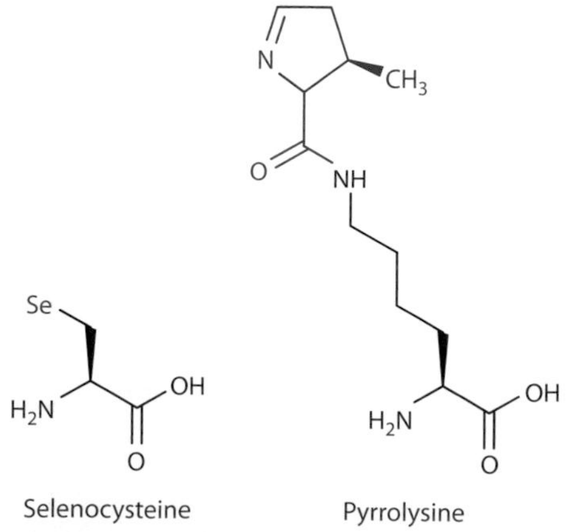

FIGURE 4.6 The 21st and 22nd amino acids, selenocysteine and pyrrolysine, respectively, which are obtained by the conversion of serine and lysine.

proteins; specifically, strategies allowing the site-specific incorporation of unnatural amino acids using the cells own protein synthesis machinery as well as semisynthetic techniques will be discussed.

In general, the use of chemical, rather than conventional genetic methods, to alter protein structure and function offers exciting possibilities. Genetic methods are generally limited to the use of the 20 proteinogenic amino acids, which contain a finite number of functional groups. Nature has increased the diversity by a large number of PTMs (Figure 4.7), which are normally not attainable by genetic methods. Thus, by combining the principles and tools of chemistry with the synthetic strategies and processes of living organisms, it is possible to generate proteins with novel functions. Such proteins can be applied in structural and functional studies of proteins, in ways previously considered unattainable.

The possibilities for generating novel proteins are endless. As previously mentioned, the incorporation of PTMs is a key feature, which allows addressing biological importance of such modifications in great detail. PTMs that can be mimicked are group additions, such as phosphorylation, glycosylation, and lipidation, and the modification of parent amino acids also includes methylation, acetylation, and hydroxylation (Figure 4.7). Another class of modification is those that incorporate biophysical probes or reactive handles, for further derivatization, examples include site-specific labeling with ^{13}C- or ^{15}N-labeled amino acids for biological NMR studies, incorporation of fluorescent amino acids or amino acids containing photolabile groups such as benzophenone (Figure 4.7). Amino acids with reactive groups for selective derivatization are also of great interest; such groups could be azides or alkyne groups (Figure 4.7) to be used in the Huisgen 1,3-dipolar cycloaddition to furnish 1,2,3-triazoles, also known as "click chemistry." Another example is the introduction of ketone functionalities that can be selectively modified, for example, with polyethylene glycol (PEG) linkers. Finally, very subtle changes of proteins, such as the incorporation of D-amino acids, close analogs of encoded amino acids (Figure 4.7), and the modification of the amide backbone can also

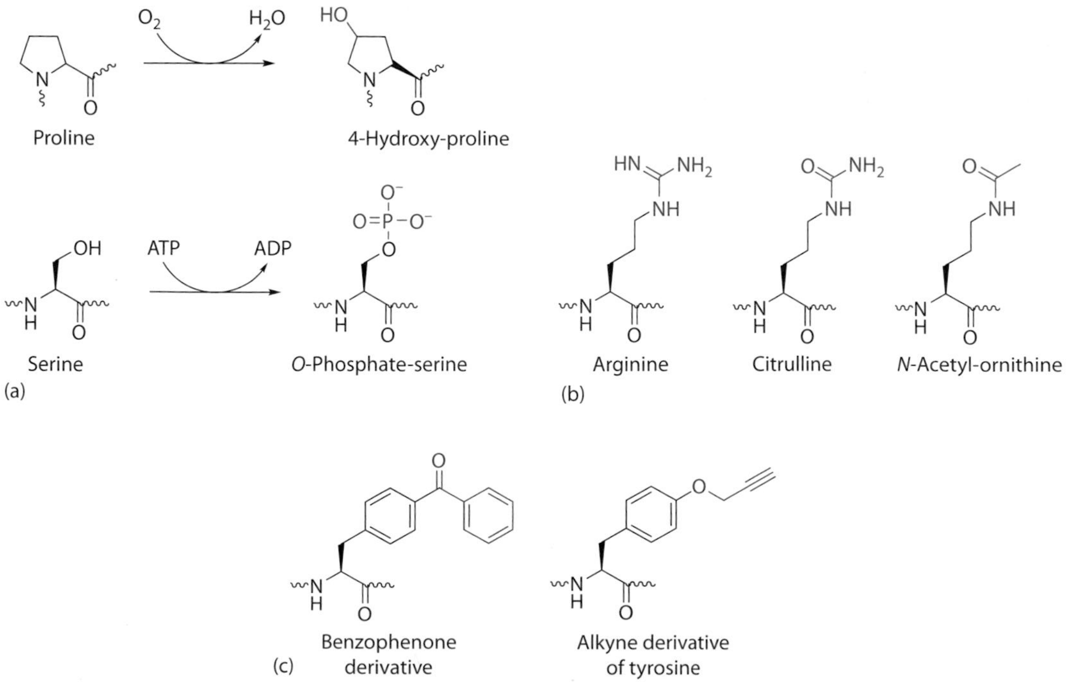

FIGURE 4.7 Modification and incorporation of amino acids that can be achieved by applying chemically based methods. (a) PTMs, such as hydroxylation and phosphorylation. (b) Close analogs of encoded amino acids, arginine, where the subtle modification of the guanidine group is included. (c) Biophysical probes, such as benzophonene, which is a photolabile group and an alkyne derivative that can be used in "click chemistry."

be introduced. This allows very fine-tuned studies of, for example, ligand–receptor interactions and protein function in general and has been described as "protein medicinal chemistry."

A number of technologies have been developed to achieve this objective and it is now possible to generate proteins containing, in principle, any functionality. In the following sections, we will focus on two general methods that allow this: (1) unnatural mutagenesis, which allows the site-specific incorporation of unnatural amino acids into protein and (2) ligation-based strategies, which allows semisynthesis of proteins and thereby the incorporation of a wide range of unnatural functionalities into proteins.

4.3.2 Unnatural Mutagenesis

In 1989, a biosynthetic *in vitro* method that allowed the site-specific incorporation of unnatural amino acids into proteins was introduced based on earlier work on nonsense suppression. The term "nonsense suppression" refers to the use of stop (nonsense) codons and suppressor transfer RNA (tRNA), which recognize stop codons. The method is based on the fact that only one of three stop codons in the genetic code is necessary for the termination of protein synthesis and the two unused stop codons can then be exploited for the introduction of unnatural amino acids.

The primary challenge in this technology is the generation of the modified suppressor tRNA with the unnatural amino acid (Figure 4.8). Once generated, the aa-tRNA is recognized by the mRNA carrying the specific stop codon, whereby the unnatural amino acid is incorporated into the protein at the specific position. Based on this principle, two slightly different methodologies have been developed for the site-specific incorporation of unnatural amino acids: one method applies tRNAs that are chemically aminoacylated with the unnatural amino acid of interest, and the aa-tRNA is subsequently applied in an expression system to generate the protein of interest (Figure 4.8). The other method employs the development of pairs of orthogonal tRNA and aminoacyl-tRNA synthetases (aaRS), where the latter is developed so that it selectively recognizes aminoacylate an unnatural amino acid.

In the chemical aminoacylation of tRNA, a dinucleotide is prepared by chemical synthesis and subsequently aminoacylated with the unnatural amino acid of interest. The aa-tRNA is obtained by the ligation of a truncated tRNA where a dinucleotide at the 3′-terminus is missing with the prepared aminoacylated dinucleotide (Figure 4.8). If an *in vitro* expression system is used, the aa-tRNA is simply added to the media, and when whole cell expression systems are used, the aa-tRNA is injected into the cell. A particular attractive expression system for this methodology is *Xenopus* oocytes, which is generally used for electrophysiological studies of ion channels, receptors, and transporters. The oocyte is coinjected with two RNA species: the modified mRNA encoding for the target protein and the aa-tRNA chemically acylated with an unnatural amino acid. This coinjection results in synthesis and surface expression of the target protein containing the unnatural amino acid (Figure 4.8).

The methodology has been used to incorporate a large number of structurally diverse unnatural amino acids, representing a large variety of functionalities, into proteins. In most cases the unnatural amino acids have been α-amino acids but also non-α-amino acids and most notably α-hydroxy acids have been incorporated with the latter introducing an amide-to-ester mutation in the protein backbone (Figure 4.9). These studies have shown that translation factors and the ribosome are compatible with many types of unnatural amino acids.

In studies of ligand-gated ion channels, such as nicotinic acetylcholine (nACh), γ-aminobutyric acid (GABA), and serotonin (5-HT$_3$) receptors (see Chapters 12 and 14), the technology has proven particularly valuable. These studies were pioneered by Dougherty and Lester, who have explored the molecular details of the cation–π interaction between the quaternary ammonium group of acetylcholine and aromatic residues in the nACh receptor; this was achieved by the site-specific incorporation of fluoro-substituted tyrosine and tryptophan residues, where the fluoro substituent gradually decreases the ability of the aromatic moiety to interact in cation–π interactions. These studies have provided unique details of acetylcholine interaction with subtypes of nACh receptors at the molecular level.

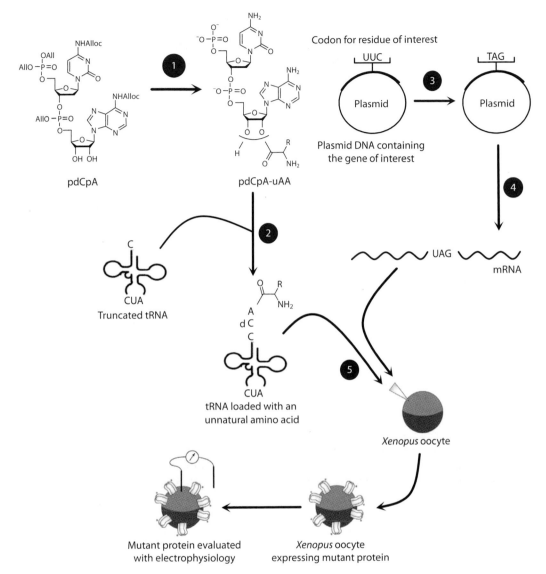

FIGURE 4.8 Example of site-specific incorporation of unnatural amino acids into proteins expressed in *Xenopus* oocytes. ❶ Chemical synthesis and aminoacylation of the dinucleotide, pdCpA; ❷ Ligation of pdCpA-uAA to a truncated tRNA bearing the amber stop anticodon; ❸ Mutation of the codon encoding the residue of interest into TAG amber stop codon by using site-directed mutagenesis; ❹ Generation of mRNA through *in vitro* transcription; ❺ Expression of mutant protein in *Xenopus* oocyte after coinjection of mRNA and tRNA, and the evaluation of the mutant protein with electrophysiology (two-electrode voltage-clamp recordings).

Similarly, the technology has been applied to evaluate the importance of a proline residue for opening and closure of 5-HT$_3$ receptors. It was hypothesized that *cis–trans* isomerism of the proline residue in the ion channel domain was important for the opening and closure of 5-HT$_3$ receptors. This was investigated by the incorporation of proline analogs with either increased or decreased probability of *cis–trans* isomerism and there was a distinct correlation between this ability and the ability of the ion channel to open.

Although the technology has obvious and wide-ranging potential, it also has substantial limitations. Firstly, the generation of aa-tRNAs requires highly skilled persons in both chemistry and molecular biology. Secondly, the amount of protein generated is very low, thus exceptionally

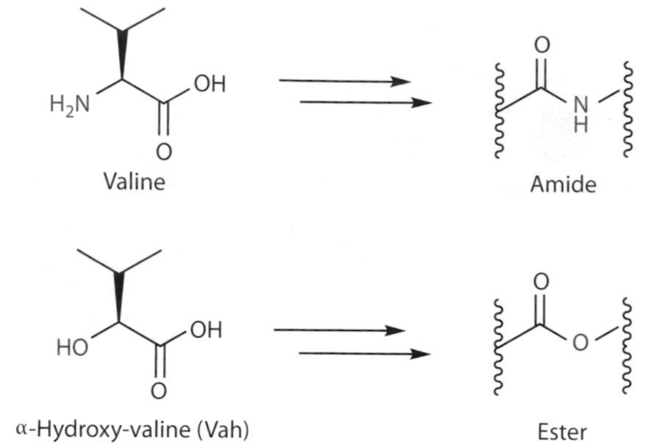

FIGURE 4.9 By using conventional genetic methods, changes in the protein backbone is not possible; however, by applying either unnatural mutagenesis or protein ligation strategies the amide backbone can be changed into, for example, an ester by using α-hydroxy acids rather than amino acids. This can be used to evaluate electronic effects of backbone carbonyl groups.

sensitive detection systems, such as electrophysiology or fluorescence, are required in these studies. Thirdly, the technology is generally limited to *in vitro* systems. Thus, in order to overcome some of these limitations, a modified method for incorporation of unnatural amino acids into proteins *in vivo*, was introduced. In this approach a custom-made pair of tRNA and aaRS is genetically introduced into a cell and the aaRS is engineered so that it only recognizes the unnatural amino acid and efficiently acylates the corresponding tRNA. Subsequently, the unnatural amino acid, which has to be nontoxic and cell permeable, is added to the growth media, taken up by the host organism and incorporated into the protein by the specific tRNA/aaRS pair. This technology, has been successfully applied in both yeast and eukaryotic cell, and allows the generation of proteins with an unnatural amino acid in reasonable yields. The primary challenge of this technology is that specific aaRS have to be generated for each unnatural amino acid, which is done by extensive mutational studies and rounds of positive and negative selections.

The technology has been applied to a number of model proteins, and has been used to specifically incorporate a glycosylated amino acid into myoglobin. In addition, a fully autonomous bacterium, *E. coli*, has been engineered so it could synthesize *p*-amino-phenylalanine, and a specific tRNA/aaRS pair was introduced, which allowed incorporation into myoglobin. The technology also holds commercial prospective, and a company (Ambrx) is developing protein therapeutics based on this technology.

A general limitation of these technologies is that the genetic code only contains three stop codons, which limits the theoretic numbers of different unnatural amino acids, that can be incorporated in a single protein to two. To overcome this limitation, Sisido and colleagues have explored an alternative strategy using extended codons and frameshift suppression. In this approach, an mRNA containing an extended codon consisting of four or five bases is being read by a modified aa-tRNA containing the corresponding extended anticodon. In certain species, some naturally occurring codons are rarely used and the amount of their corresponding tRNA is low. This has been used in the design of four-base codons, which are derived from these rarely used codons, to minimize the competition between the four-base anticodon tRNA and endogenous tRNA. The four-base codon technique has been used to incorporate unnatural amino acids into proteins in *E. coli*. It has also been used to incorporate two different unnatural amino acids into two different sites of a single protein showing that four-base codons are not only orthogonal to their host organism but also to each other.

4.3.3 PEPTIDE/PROTEIN LIGATION

A conceptually different strategy for the modification of proteins is to employ methods based on solid-phase peptide synthesis (SPPS) for the generation of proteins. This would allow the incorporation of principally any amino acid, and thus circumvent the problems of incorporating D-amino acids, which is not feasible by unnatural mutagenesis. SPPS has in a few cases been applied for the synthesis of proteins, although yields are generally rather low. The first example was the synthesis of ribonuclease A (124 residues) by Bruce Merrifield in 1966 and since then a few other proteins have been prepared by this approach, most notably HIV protease (99 residues), which enabled structural characterization of the protein with inhibitors bound.

However, SPPS is generally limited to the preparation of up to 40–60 amino acid peptides, whereas most proteins are considerably larger. Therefore, there has been a considerable interest in developing methods that are not confined to these restrictions and in 1994, a strategy for the preparation of proteins from peptide fragments was introduced, called native chemical ligation (NCL, Figure 4.10). In NCL, two or more unprotected peptide fragments can be ligated together, generating a (native) cysteine residue in the ligation site. The ligation requires a peptide with a C-terminal protein thioester and a peptide with an N-terminal cysteine residue: the thiolate of the N-terminal cysteine attacks the C-terminal thioester to affect transthioesterification, followed by the formation of an amide bond after S → N acyl transfer (Figure 4.10). The reaction takes place in aqueous buffer and generally proceeds in good to excellent yield.

Thus, NCL is a very useful approach for the total chemical synthesis of proteins and has been used for the preparation of numerous proteins, including glycoproteins and proteins with fluorescent labels. An example is the synthesis of an analog of erythropoietin (EPO), which was derivatized with monodisperse polymer moieties in order to improve the duration of action *in vivo*. The 166-residue

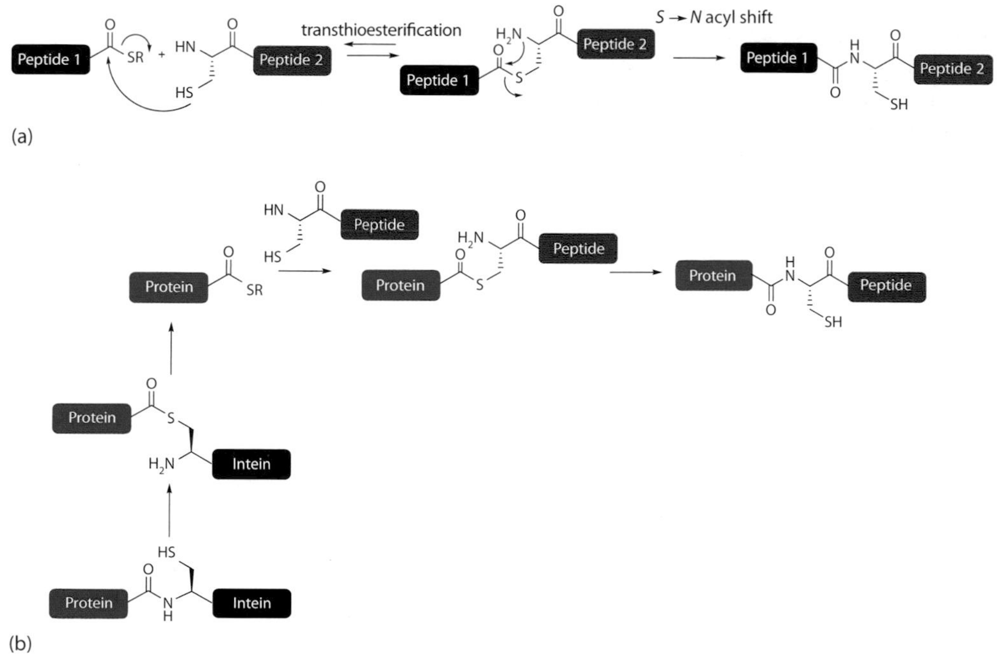

FIGURE 4.10 Principles of NCL and EPL. (a) NCL: a peptide with an N-terminal cysteine and another peptide with a C-terminal thioester can be ligated together. Initially, a reversible transthioesterification takes place and subsequently S → N acyl shift, leading to a cysteine in the ligation site. (b) EPL is applying the same principles, but one of the reactants is a recombinantly expressed protein, which allows the semisynthesis of larger proteins.

protein was prepared by the ligation of four peptide fragments, two of which were modified with the polymer and the EPO analog displayed improved properties *in vivo* compared to EPO.

In 1998, an extension of the NCL principles was introduced, called expressed protein ligation (EPL). The technology applies the same reaction as in NCL, but in contrast to NCL, one of the components is a protein, rather than a peptide (Figure 4.10). The protein is expressed as the so-called intein construct, which allows the formation of a protein thioester, which subsequently can be reacted with a peptide with an N-terminal cysteine in an NCL generating a full-length protein (Figure 4.10). Thus, the EPL methodology combines the advantages of molecular biology with chemical peptide synthesis, and enables the addition of unnatural functionality to a recombinant protein framework.

EPL has been applied in studies of several proteins and here only a few noteworthy examples are provided. Histone complexes are important for the storage of DNA and have flexible N-terminal tails that are heavily modified by PTMs, and is of general importance for epigenetic gene regulation (see Chapter 23). EPL has been applied to prepare full-length, ubiquilated H2B and subsequently used to demonstrate a direct cross talk between PTMs on different histones. The list of proteins prepared by EPL was extended to include integral membrane proteins, specifically the potassium channel KcsA, which is a tetrameric assembly of identical subunits (see also Chapter 13). EPL was used to prepare KscA subunits (122 residues), which were then refolded and reconstituted into lipid membranes. In subsequent studies, unnatural mutations, such as D-alanine and amide-to-ester mutation, in the selectivity filter of the channel have revealed the important information of the function of this important segment of the potassium channel.

4.3.4 CHEMICAL MODIFICATION OF PROTEINS

Besides the two classes of technologies just described, which can be used to alter the very basic structure of proteins, there are a plethora of chemistry-based methods that allows the modification of the parent protein structure.

The endogenous protein structure can be exploited for selective derivatization. The most frequent way of modifying protein structure is by reacting cysteine residues; this can often be successfully carried out with either none or minimal changes to the parent protein. The advantages is that the thiol of cysteine allows for selective modification, relative to the other proteinogenic amino acids and the frequency by which cysteine occurs in proteins is relatively low, thus often allowing the selective modification of specific cysteine residues. Even if a protein contains more than one cysteine residues, these might have different accessibility, which can allow the selective modification of certain residues.

When proteins are being developed as drugs, the pharmacokinetic (PK) and pharmacodynamic (PD) properties of proteins can be improved by the chemical modification of the protein structure. A particularly promising strategy is the introduction of PEG moieties, known as PEGylation, which can help reducing immunogenicity, increasing the circulatory time by reducing renal clearance and also provide water solubility to hydrophobic drugs and proteins. PEGylation is generally performed by the reaction of a reactive derivative of PEG with the target protein, typically with side chains of amino acids such as lysine or cysteine, or by reaction at the C- or N-terminal of the protein or peptide. PEGylated proteins entered the market in the 1990s and today a number of therapeutic proteins are marketed as PEGylated derivatives including PEGylated α-interferons (see also Chapter 24), which are used in the treatment of hepatitis C; the PEGylated α-interferon is injected only once a week, compared to three times a week for conventional α-interferon.

An alternative way of improving protein and peptide therapeutics is by adding lipids to the protein, which can improve half-life. Adding lipids to a protein framework has been achieved by ligation strategies, but in a few cases the differential reactivity of specific residues has been exploited. An example of this is the long-acting insulin analog, insulin detemir (Levemir®), where the N^ε-amino group of a terminal lysine in the B-chain of insulin has been modified with tetradecanoic acid (myristic acid, C_{14} fatty acid chain). This modification increases self-association and binding to albumin, leading to stable insulin supply for up to 24 h. Similarly, liraglutide (Victoza®) is an analog of glucagon-like peptide-1 (GLP-1), where a lysine side chain has been modified; a palmitic acid (hexadecanoic acid, C_{16} fatty acid chain) was added through a glutamate linker. The modification lead to a substantial increase in half-life, due to increased binding to serum albumin, and the modification did not compromise the biological activity.

In some cases, the selective modification of proteins can be achieved by using simple chemical reactions similar to those used in conventional organic synthesis. A general requirement is that such reaction should be compatible with the aqueous (buffer) conditions, in which the protein is present and recently a number of robust and water-compatible reactions have evolved. However, such methods often require the introduction of selective handles, as previously described, in order to be sufficiently selective, but once a reactive handle is incorporated, a wealth of chemical reactions can be performed. The example of "click chemistry," that is, a 1,3-dipolar cycloaddition between an azide and an alkyne providing a 1,2,3-triazole, has already been mentioned. Another prominent example is the Staudinger reaction, which is a phosphine-mediated reduction of an azide to an amine, also known as an aza-Wittig reaction, that has been used particularly in protein glycosylation studies. Interestingly, the Staudinger reaction has recently been applied in the ligation of peptides and proteins.

Finally, enzymes can be used to selectively modify proteins. Enzymes have an inherent advantage that they efficiently add or remove groups to proteins and they are often highly specific for certain sequences (consensus motifs) of amino acids, so modifications are often site-specific. Enzymes are often also highly substrate-specific, that is, kinases add only phosphates groups to serine, theronine, and tyrosine; thus the modification of the enzyme is required if other groups have to be introduced. However, some enzymes, such as glycosyltransferases, which transfers carbohydrates to serines or asparagines, have broader substrate specificity, but in this case, it can be desirable to modify the enzyme to achieve increased reactivity for specific carbohydrates. A particularly powerful method to develop enzymes with desired properties is directed evolution, which basically consist of two steps: (1) the generation of a library of mutants of the enzymes and (2) rounds of screening/selection for the desired properties, which for example can be used to modify substrate specificity of enzymes.

Enzymes are particularly useful to furnish proteins with tailor-made PTMs, which are often essential for the regulation and dynamics of biological activity. For example, most proteins are glycosylated, and controlling glycosylation patterns of proteins is a key challenge. Glycosyltransferases are enzymes that can catalyze the transfer of a monosaccharide to a protein, and using directed evolution was possible to modify the transferases, so monosaccharides of interest could be selectively added to a protein framework. Another example is using transglutaminase (TGase) to obtain selective PEGylation. TGase catalyzes transfer reactions between the γ-carboxamide group of glutamine residues and primary amines, resulting in the formation of γ-amides of glutamic acid and ammonia. Thus, by using an aminoderivative of PEG (PEG–NH$_2$) as substrate for the enzymatic reaction, it is possible to covalently bind the PEG polymer to a therapeutic protein.

4.4 CONCLUDING REMARKS

In this chapter, we have focused on small molecules and how a systematic generation and application of these can be expediently used to probe and discover biology, both in an academic setting and in the initial drug design and development process. We also discussed the application of chemical biology technologies in studies of proteins and how this has opened up new avenues in protein engineering and paved the way for studies of proteins in that has previously not been possible. Similar principles and technologies are applied in studies of nucleic acids and polysaccharides with great benefit for basic research, but likewise in the development of biologicals.

Chemical biology is a scientific discipline that has emerged primarily from chemical sciences to apply chemical tools and principles in studies of biological phenomena. However, chemical biology has evolved in recent years to include scientists from many other disciplines, particularly those that emanate from biological sciences, making chemical biology a truly interdisciplinary playing ground. Like other interdisciplinary sciences, such as nanoscience and synthetic biology, chemical biology will have a growing impact on science in general in the future.

FURTHER READINGS

Dobson, C. M.; Gerrad, J. A.; Pratt, A. J. 2001. *Foundations of Chemical Biology*, Oxford University Press, Oxford.

Morrison, K. L.; Weiss, G. A. 2006. The origins of chemical biology, *Nat. Chem. Biol.* 2: 3–6.

Schreiber, S. L. 2005. Small molecules: The missing link in the central dogma, *Nat. Chem. Biol.* 1: 64–66.

Schreiber, S. L.; Kapoor, T.; Wess, G., Eds. 2007. *Chemical Biology*, Wiley-VCH, Weinheim.

Waldmann, H.; Janning, P. 2004. *Chemical Biology, A Practical Course*, Wiley-VCH, Weinheim.

5 Stereochemistry in Drug Design

Maria B. Mayo-Martin and David E. Jane

CONTENTS

5.1   INTRODUCTION

The proportion of drugs marketed as individual stereoisomers has rocketed over the last decade, a fact that reinforces the importance of the link between chirality and drug design and development. In this chapter, the reasons for this newfound focus in the marketing of chiral drugs will be discussed. First, in this chapter a brief overview of the fundamental principles of stereochemistry will be given. We will then focus on the underlying reasons why individual stereoisomers may have different pharmacological activities. The major challenges in bringing chiral drugs to market are in the production of single stereoisomers on a large scale and in finding methods to assess their purity. Therefore, the final sections of this chapter concentrate on these two important aspects.

5.2   WHAT IS CHIRALITY?

Isomers are compounds with the same molecular formula but with a different arrangements of the atoms. Of the different types of isomers, optical isomers will be the focus of this chapter. A molecule is chiral when it cannot be superimposed upon its mirror image. Hence, a compound and its nonsuperimposable mirror image are two different isomers termed enantiomers. Optical isomerism is a result of this different spatial arrangements of atoms in a molecule. The lack of symmetry can arise from four different substitutions around a tetrahedral carbon atom (stereogenic center), although atoms such as phosphorous may also act as stereogenic centers. For example, lactic acid has a stereogenic center and therefore can exist in two enantiomeric forms. However, propanoic acid possesses a symmetry plane and so is achiral (i.e., the molecule can be superimposed on its mirror image) (Figure 5.1). Enantiomers are identical except for two properties: their optical activity and the way in which they interact with other chiral molecules. The optical activity of an enantiomer is the ability to rotate the plane of polarized light (i.e., light that oscillates in a single plane). A 50:50 mixture of two enantiomers is called a racemic mixture and its optical rotation is zero. The degree of rotation caused by a single enantiomer is measured using a polarimeter. If a molecule rotates plane polarized light anticlockwise it is labeled as *laevorotatory*, abbreviated "l" or (−), or if it is clockwise it is called *dextrorotatory*, "d" or (+).

Specific rotation is an intrinsic property of an optically active molecule that can be used to quantify the amount and purity of a single enantiomer. This value is dependent on the wavelength of light used, the length of the sample tube through which the light is passed, temperature, solvent, and sample concentration. The light source most often used for such determinations is that emitted by a sodium lamp at 589 nm (the so-called D line). In order to compare data, these parameters should be specified when quoting the specific rotation, $[\alpha]_D$. Optical purity (usually expressed as a percentage) can be defined as the ratio of the specific optical rotation of the enantiomeric mixture and the specific optical rotation of the pure enantiomer.

The observed optical rotation (d or l) was the earliest method of distinguishing between enantiomers, but this method gives no indication as to the actual spatial geometry of a molecule

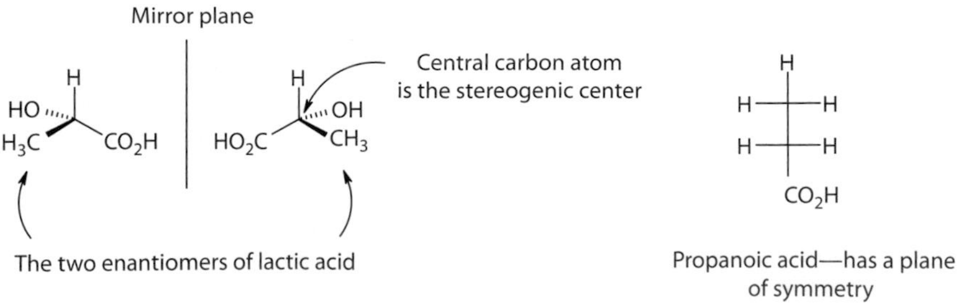

FIGURE 5.1   The two enantiomers of lactic acid are mirror images of each other. However, propanoic acid is achiral as it has a plane of symmetry through the center of the molecule.

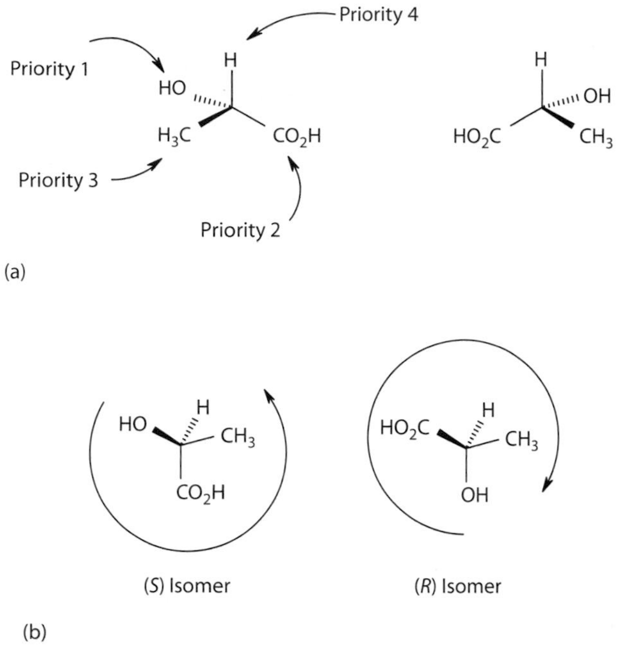

FIGURE 5.2 Procedure for assigning stereogenic centers as possessing either (*R*) or (*S*) configuration. (a) Assign priorities according to the CIP rules. (b) View from opposite the group of lowest priority: Clockwise rotation (13) is (*R*); anticlockwise rotation is (*S*).

i.e., the configuration of atoms or groups about the stereogenic center. This was rectified by the introduction of the Fischer convention, which labeled such centers as having either D or L configuration based on an arbitrary standard, (+)-glyceraldehyde. However, this system has now been superseded by the Cahn–Ingold–Prelog (CIP) system that can be used to unambiguously assign any stereogenic center as possessing either (*R*) or (*S*) stereochemistry. Explanation of the CIP rules can be found in any general organic chemistry textbook. Once the priorities of the substituents have been assigned enantiomers are readily classified as being the (*R*) or (*S*) isomers. Lactic acid is again used as an example to demonstrate this (Figure 5.2).

Molecules such as lactic acid are relatively simple in that they only have one stereogenic center. But what are the implications if multiple stereogenic centers are present? As an example, the drug ephedrine has two stereogenic centers and thus there are four possible isomers (Figure 5.3). Of these, the isomers that are mirror images are enantiomers, while the nonsuperimposable nonmirror images are called diastereomers. It is important to note that diastereomers, unlike enantiomers, will (unless by coincidence) have nonidentical physical and chemical properties such as boiling point, solubility, and spectral properties. The potential applications of these differences are discussed in Sections 5.5.1 and 5.5.2.

As a general rule, the total number of isomers of any given molecule is also given by the rule:

$$\text{Number of isomers} = 2^n, \text{ where } n \text{ is the total number of stereogenic centers}$$

So, as in ephedrine, a compound with two stereogenic centers will have four isomers, three centers leads to eight isomers, and so on. However, there are exceptions to this rule, because some isomers may be meso compounds. These can be described as isomers that contain stereogenic centers but are achiral (and optically inactive) due to the presence of a symmetry plane. Figure 5.4 shows the example of tartaric acid, with two stereogenic centers and three isomers.

The definition of optical purity discussed earlier has been largely superseded by two related terms: enantiomeric excess (ee, or the proportion of the major enantiomer less that of the minor

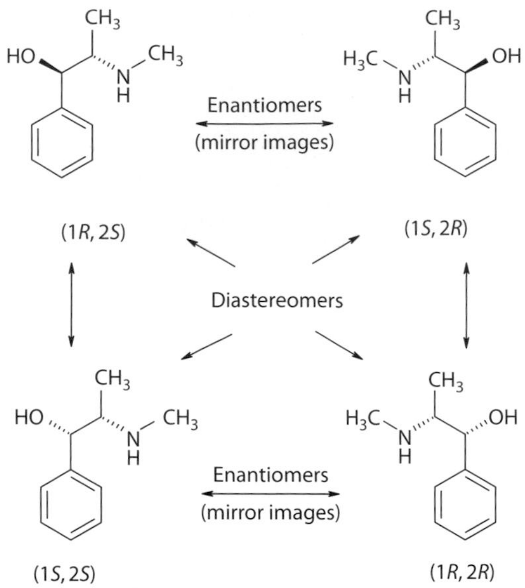

FIGURE 5.3   The relationship between enantiomers and diastereomers. The biologically active forms of ephedrine are those with the $(1R,2S)$- and $(1S,2S)$ configurations, which are diastereomers of each other.

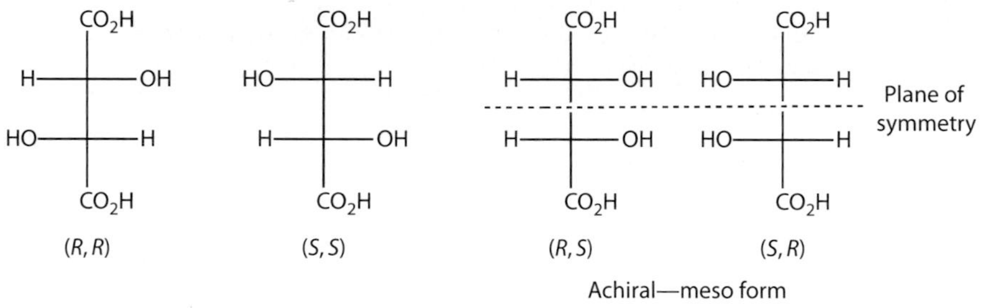

FIGURE 5.4   Tartaric acid has two stereogenic centers but only three stereoisomers.

enantiomer) and diastereomeric excess (de, proportion of the major diastereomer less that of the minor one). Both ee and de are usually expressed as percentages.

If the only difference between enantiomers was their interaction with plane polarized light, then their existence would be little more than academic. However, stereochemistry has important implications in terms of biological activity as described in the following section.

5.3   THE ORIGIN OF STEREOSPECIFICITY IN MOLECULAR RECOGNITION

The lock-and-key hypothesis proposed by Fischer in 1896 was the first attempt to explain the complementarity between a substrate's shape and an enzyme's active site (see also Section 1.1). Koshland's later hypothesis allowed the enzyme to change its shape to accommodate the binding of the substrate (induced fit model). Now we know that both the substrate and the enzyme can change conformation to some extent to ensure optimal binding. Although proposed for enzymes, these models can also be used as the basis to explain drug–receptor interactions.

Receptors (like enzymes) are made up of amino acids, all of which apart from glycine are chiral. The interaction of chiral drugs with receptors would be expected to be enantioselective (i.e., one enantiomer binds with higher affinity than the other). In order to explain the stereoselective action of drugs on receptors, the three-point receptor theory was proposed (Figure 5.5). In this theory only

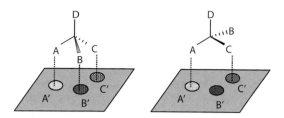

FIGURE 5.5   The three-point receptor theory. Only one enantiomer can form the three correct interactions with points A', B', and C' on the receptor.

one enantiomer has the optimal spatial disposition of the three groups A, B, and C to interact with the complementary sites on the receptor. Interactions (possibly more than three) could be ionic, hydrophobic, steric, or hydrogen bonding. Although the three-point receptor theory is simplistic, it has also been used successfully to understand chromatographic resolution of mixtures of enantiomers on chiral stationary phases (CSPs), which can be thought of as artificial receptors (see Section 5.5.2).

5.4   WHY IS STEREOCHEMISTRY IMPORTANT IN DRUG DESIGN?

Chiral drugs, sold as single enantiomers, either for an economical or regulatory reason, are likely to dominate drug markets in the near future. Pharmaceutical companies see enantiomers as a way of prolonging the patent life of their existing racemic drugs by patenting and then marketing the active enantiomer thereby undercutting competition from generic drug sales. In addition, some companies see this switching from racemate to single active enantiomer as a way into the drug market. However, these are not the only reasons for testing individual enantiomers of chiral drugs. Lessons learned from mistakes made by marketing racemic drugs also play a part, such as the tragic case of thalidomide (Figure 5.8). Racemic thalidomide was developed in the 1950s and was used as a sleeping pill and to treat morning sickness. Unfortunately, the drug had serious side effects as it was found to be teratogenic causing fetal abnormalities. It was later discovered in tests with mice that the (S)-enantiomer possessed the teratogenic activity while the (R)-enantiomer possessed the sedative activity. However, subsequent studies revealed that the enantiomers racemise under physiological conditions. Despite this, thalidomide brought the role of chirality in drug development into the spotlight. Recently, thalidomide has hit the headlines again as the use of the racemate for treatment of leprosy has been approved by the Food and Drug Administration (FDA) but only under the strictest of guidelines. It appears that thalidomide may also have therapeutic utility in the treatment of AIDS-related disorders and tuberculosis.

The FDA strongly urges companies to evaluate both the racemates and the corresponding individual enantiomers as new drugs. Thus, even if a drug is to be sold as a racemate, the individual enantiomers need to be evaluated, which increases the cost and timescale of drug development. Therefore, synthesis of single enantiomeric drugs is becoming a priority.

It should be noted that not only the pharmacodynamic aspects are important in the discussion of the activity of chiral drugs. Pharmacokinetics is also affected as the absorption and clearance of drugs involves interaction with enzymes and transport proteins. Thus, the individual enantiomers of a chiral drug may be metabolized by enzymes at different rates and may be transformed into different chemical entities. As a result of these considerations, it is very important that individual enantiomers of chiral drugs are tested in the clinic.

Ariëns, a pioneer in the field of enantioselective drug actions, has proposed that the active enantiomer of a chiral drug be termed the eutomer while the less active enantiomer should be termed the distomer. The eudismic ratio (ER) is defined as the ratio of the activity of the eutomer to that of the distomer. The presence of the distomer in the racemic drug can have a number of consequences for the biological activity.

5.4.1 The Distomer Is Inactive (High ER)

In this case the distomer is either inactive or displays no undesirable side effects. In the case of the antihypertensive agent (β-blocker) propranolol (Figure 5.6) the (*S*)-enantiomer is 130-fold more potent than the (*R*) enantiomer as a β-adrenoceptor antagonist (i.e., ER = 130). A number of other β-blockers based on this structure show high ERs. These drugs are therefore marketed as racemates as the distomer displays no side effects. Despite this there would have been advantages in marketing the (*S*)-enantiomer if only to extend patent life.

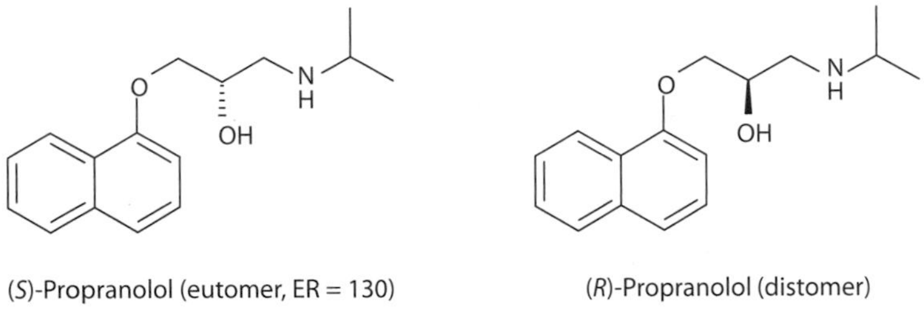

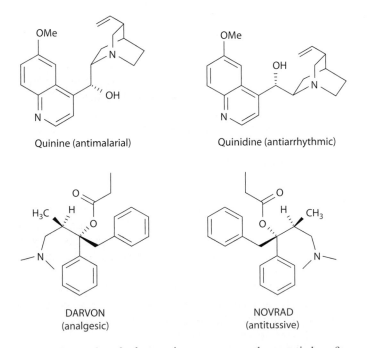

(*S*)-Propranolol (eutomer, ER = 130)   (*R*)-Propranolol (distomer)

FIGURE 5.6 The ER for (*S*)- and (*R*)-propranolol is 130.

5.4.2 Both Enantiomers Have Independent Therapeutic Benefits

In some instances, both enantiomers of a drug may have different therapeutic values. The classical example of this behavior is the diastereomers quinine and quinidine (Figure 5.7). Quinine, which was originally obtained from the bark of cinchona trees was, for centuries, the only treatment for malaria. Quinidine, on the other hand, is used as a class 1A antiarrhythmic agent and acts by increasing action potential duration.

Quinine (antimalarial)   Quinidine (antiarrhythmic)

DARVON   NOVRAD
(analgesic)   (antitussive)

FIGURE 5.7 Examples of drugs where both stereoisomers possess therapeutic benefits.

The drug dextropropoxyphene marketed by Eli Lilly has trade names reflecting the different activities of the enantiomers. Thus the $(2R, 3S)$-enantiomer, DARVON has analgesic properties while the $(2S, 3R)$-enantiomer NOVRAD (Figure 5.7) is an antitussive.

5.4.3   DISTOMER POSSESSES HARMFUL EFFECTS

In some cases, it is known that the distomer produces harmful or undesirable side effects. Thus, dextromethorphan is used as a cough suppressant, while levomethorphan has antitussive properties but it is also an opioid narcotic (Figure 5.8). The harmful teratogenic side effects of the (S)-enantiomer of thalidomide have already been discussed (Section 5.4).

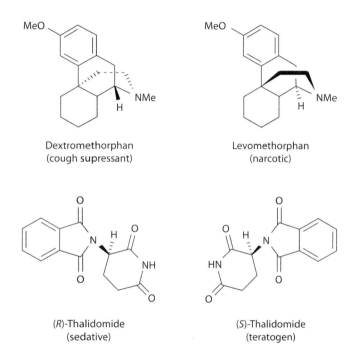

FIGURE 5.8   Examples of drugs where the distomer possesses harmful effects.

5.4.4   THE EUTOMER AND THE DISTOMER HAVE THE OPPOSITE BIOLOGICAL ACTIVITY

It is sometimes observed that the enantiomers of a chiral drug may have opposite biological activity. One example of this is (−)-dobutamine, which is an agonist at α-adrenoceptors while (+)-dobutamine is an antagonist (Figure 5.9). However, (+)-dobutamine is 10-fold more potent than the (−)-isomer as a β1-adrenoceptor agonist and is used to treat cardiogenic shock. The individual enantiomers of the 1,4-dihydropyridine analogue BayK8644 (Figure 5.9) have opposing effects on L-type calcium channels with the (S)-enantiomer being an activator and the (R)-enantiomer being an antagonist.

5.4.5   THE RACEMATE HAS A THERAPEUTIC ADVANTAGE OVER THE INDIVIDUAL ENANTIOMERS

Both enantiomers may contribute to the therapeutic effect though examples of chiral drugs exhibiting this phenomenon are quite rare. We have reported that racemic 3,4-dicarboxyphenylglycine (DCPG, Figure 5.10) displays a greater potency in preventing sound-induced seizures in an experimental model of generalized epilepsy seizures than either enantiomer alone. The (R) enantiomer of DCPG has antagonist activity at the AMPA receptor subtype of ionotropic glutamate receptors while

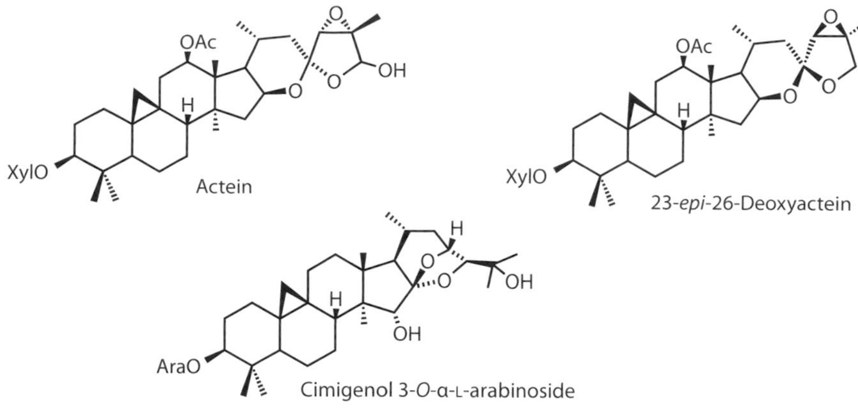

FIGURE 6.2   Some common triterpenoid constituents of black cohosh.

plant preparations marked a turning point for pharmaceutical discovery. Quinine continues to have practical use in the treatment of certain resistant forms of *P. falciperum*.

In Chinese traditional medicine *Artemisia annua* or qinghao, has been employed as an antimalarial agent for many centuries. In this case, the sesquiterpene lactone artemisinin (Figure 6.1), containing a rare peroxide bridge, was isolated from the plant material and shown to possess effective antimalarial activity. Owing to the development of resistance to synthetic antimalarial drugs, artemisinin has become an effective alternative therapy. The yield of artemisinin in *A. annua* is relatively low making its commercial production an expensive process. Considering that the majority of people suffering from malaria live in underdeveloped regions of the world, a practical and cost-effective method for the production of artemisinin is highly desirable. This situation is a prime example of what has become known as the "supply issue" for the practical production of natural products. The issue is focused on the difficulty of producing commercial quantities of complex natural products.

Herbal remedies continue to play a significant role in human medicine. Chemical investigations have identified many of the active principles in many commonly used products. These products are often sold as dietary supplements rather than ethical pharmaceutical products. Because these products are complex mixtures of many natural products there is a need to establish criteria for their standardization. This situation is complicated by the natural variation in secondary metabolites produced by closely related species of medicinal plants. Owing to the possibility that variations in the composition of the products will result in unpredictable potency, the herbal products industry has been developing quality control standards. Black cohosh, for example, which is taken for the relief of menopausal symptoms, has a number of signature triterpenoid constituents including, actein, 23-*epi*-26-deoxyactein, and cimigenol-3-*O*-arabinoside (Figure 6.2). These compounds can be identified by coupled high-performance liquid chromatography–mass spectrometry (HPLC/MS), in comparison with authentic reference compounds. Our ability to quantify the amount of the key constituents in these herbal products will facilitate a better understanding of their efficacy and permit greater confidence in their use in medicine.

6.3   ANTIBIOTICS

The development of antibiotics for the treatment of bacterial infections, which was critical during the war years of the early 1940s, changed the course of drug discovery efforts in the pharmaceutical industry. Following the pioneering experiments of Selman Waksman at Rutgers University on soil actinomycetes, pharmaceutical companies began the systematic evaluation of antibiotics produced by bacteria isolated from the soil. During the succeeding quarter century, often referred to as the

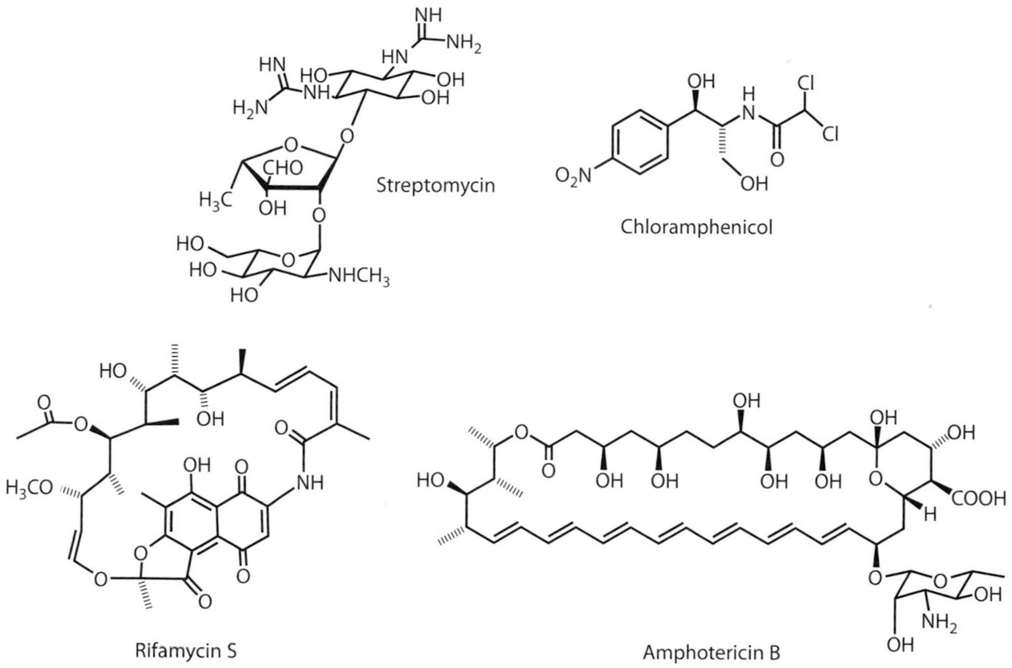

FIGURE 6.3 Additional examples of antibiotics found during the "Golden Age of antibiotics discovery."

"Golden Age" of antibiotic discovery, all of the major classes of life-saving antibiotics were found. The previously illustrated penicillin, vancomycin, tetracycline, and erythromycin represent the progenitors of the most important classes of antibacterial agents still in use. Other examples of these wonderfully complex compounds are illustrated in Figure 6.3. Streptomycin, isolated by Waksman from *Streptomyces griseus*, was the first of the class of aminoglycoside antibiotics to be introduced into therapy. Chloramphenicol, rifamycin, and amphotericin were also discovered during this time and each has a valuable niche in modern chemotherapy.

In addition to antibiotics used for the treatment of microbial diseases, microbial products have been explored for a number of other therapeutic uses. Owing to the relative ease with which new organisms could be isolated from the environment and grown in culture, these provided versatile sources of new chemistry. Beginning in the 1960s these sources were employed for screening against other diseases, such as parasitic and fungal infections, as well as for the ability to differentially kill cancer cells. Notable among the antiparasitic compounds discovered in this way are the milbemycins. These polyketide-derived macrolides, produced by *Streptomyces* species, are exceptionally effective against several types of parasites that infect livestock. Compounds in the milbemycin class, e.g., ivermectin and moxidectin (Figure 6.4) have also found utility against the devastating human disease of river blindness caused by filarial worms, which is endemic to sub-Saharan Africa, and other tropical areas of the world.

Actinomycete-derived antibiotics with efficacy as anticancer agents have also been a major focus of screening programs (see Figure 6.5). Waksman once again discovered the first of these, actinomycin D, from *Streptomyces parvullus*. Today, actinomycin has quite limited use, but it served as a prototype for the discovery of other antitumor or antibiotics. Doxorubicin, which interacts similarly with DNA, was isolated in the 1960s and remains an important component of typical chemotherapy regimes. Another early discovery from the Golden Age that remains in use today for chemotherapy is bleomycin. Bleomycin is a complex glycopeptide antibiotic produced by *S. verticillus* that induces DNA damage through oxidative reactions.

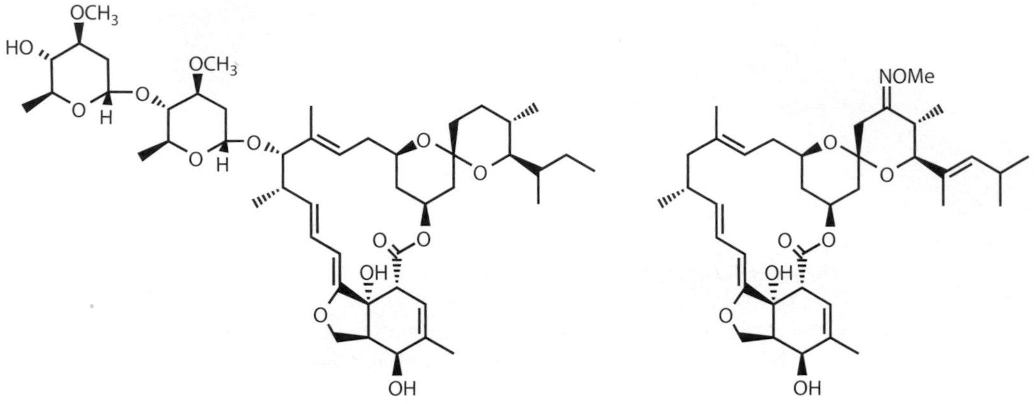

Ivermectin B_{1a} Moxidectin

FIGURE 6.4 Antiparasitic milbemycin analogs.

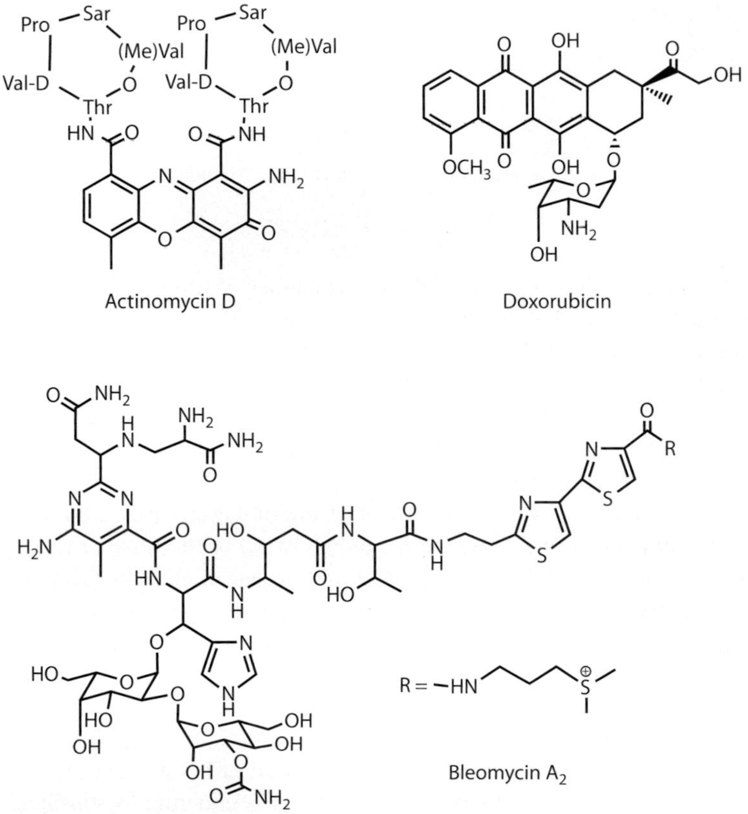

Actinomycin D Doxorubicin

Bleomycin A_2

FIGURE 6.5 Cytotoxic antibiotics.

6.4 SCREENING

6.4.1 General Concepts

In the most general sense screening refers to the process of investigating sources of compounds that exhibit a particular type of property or biological activity. In this chapter, we are exploring ways of using natural products for drug discovery, and therefore the "investigations" are typically linked to a biological assay. A positive response in the assay (a "hit") is determined by the intrinsic potency of a given compound and its concentration in the screening sample. The sources of natural products used in the screening process can be quite diverse, ranging from bacterial products to higher plants and animals, however the processes involved are similar. Once a sample, which is typically an extract of an organism, or a part of an organism (e.g., fermentation broth, fruiting body of a mushroom, leaves, or roots of plants, etc.) has shown a positive response in a given assay the process of "bioassay-guided fractionation" begins. This process is shown as a loop diagram in Figure 6.6. Resolution of the active principle(s) in these materials is a highly experimental process. The ease of resolution is dependent upon such parameters as the concentration of the active compound in an extract, the overall constitution of the extract, in terms of interferences (e.g., tannins, fatty acids and other lipids, and complex carbohydrates), as well as the chemical properties of the compound of interest. Trial and error is the operational mode of these processes and is highly dependent on the preferences and experience of the individual investigator. As indicated in Figure 6.6 the original crude material is initially split into fractions, by a rough process such as differential solubility in solvents with different polarities, or by liquid–liquid partitioning between immiscible phases, usually aqueous versus organic. Subsequent steps are generally of higher resolution often with different forms of chromatography, perhaps using a normal phase high-capacity technique like silica gel chromatography in organic mobile phases first, followed by a reversed phase system with a hydrophobic stationary phase eluted with an aqueous-organic mixture. It is usually the case that a suite of structurally related compounds is isolated in this process, each having some activity in the bioassay of interest. Subtle differences in the potency or selectivity shown by these congeners form the basis for the "natural structure–activity relationship" (SAR) of the series that may be useful in designing improved compounds by synthetic or biosynthetic methods during subsequent optimization of the lead. Once a compound is shown to have the activity of interest and passes a criterion of purity, it can proceed for resolution of its chemical structure and further biological evaluation.

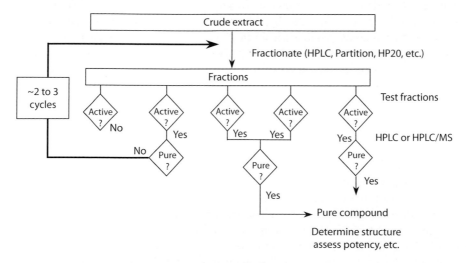

FIGURE 6.6 Bioassay-guided fractionation of natural products.

6.4.2 WHOLE ORGANISM SCREENING

Often referred to as "phenotypic screening," as well, the screening of natural products in whole organisms was historically the original mode, and one that still has a great deal of potential. One of the most robust examples of this approach is screening for antimicrobial activity by inhibiting the growth of selected strains of bacteria on seeded agar plates. A positive response in such an assay is observed as a zone of inhibition of growth around a test sample deposited on the Petri plate. In most cases the diameter of the zone of inhibition is directly related to the concentration of the antibacterial agent in the test sample, so that some degree of quantification of the response is also possible in this simple test. Such assays provide no indication of the mode of action of the active principle, but are generally quite effective tools for following fractionation processes that lead to pure compounds. A crucial extension of this in vitro whole organism assay is another phenotypic test in which the isolated material is administered to rodents that have been challenged with a lethal infection of the target bacterium. The positive end point of the in vivo rodent test is survival beyond that of the control animals. Positive results in the in vivo model of infection provides the critical information that the compound has sufficient drug-like properties to penetrate the normal xenobiotic defenses of the host animal and reach the target population of infecting bacteria.

There are many such whole organism models that have been used for drug screening. Among the simplest of these are those related to infectious diseases, including the aforementioned antibacterial system with various classes of pathogenic agents, as well as those designed for antifungal, antiviral, and antiparasitic agents. In the quest to find new effective agents against cancer, animal models of disease remain a mainstay of the process. Similar models are the norm for advancing the development of drugs in terms of understanding the efficacy, tolerability, metabolism, and long-term effects. These systems are rarely used for high-throughput screening of crude natural products because they require substantial resources for its maintenance. Therefore, most live animal models are used to verify the efficacy of compounds that have been isolated with the aid of a simpler in vitro assay.

As will be discussed in the following section, screening against isolated target biomolecules, such as enzymes or receptors is now favored for high-throughput screening operations. In the case of screening mixtures of natural products, however, the whole organism approach offers tremendous advantages. The discovery of a novel secondary metabolite that confers a positive response in a whole organism screen provides the opportunity to discover a new target, and potentially a new mechanism of action. In current parlance these studies are often referred to as "Chemical Biology" or "Chemical Genetics" (see also Chapter 4). Specifically, in forward chemical genetics a small molecule, in our case a natural product, is employed to probe for their cellular targets. Typical experiments include the creation of affinity binding reagents or affinity matrices that include the small molecule of interest, and these systems are used to fish out target macromolecules from cellular components. Molecular targets for rapamycin and geldanamycin were found by such methods. Once the target macromolecules are verified, additional mechanistic studies are developed to understand the relationship between the binding partners and the disease process.

6.4.3 TARGET-BASED SCREENING

Molecular biology has provided the tools to engineer and produce macromolecular targets of drug action. If it is believed that the inhibition of a particular cell-signaling process will mediate the development of disease, then the isolated enzyme, or receptor that is responsible for the signaling can be used as a target for screening. Alternatively, a selective whole cell screen can be employed that is designed to respond by providing some measurable signal as a result of the interaction with a particular target. Owing to developments in automation for such assay systems, hundreds of thousands of compounds can be conveniently tested for activity in a short period of time. Natural product mixtures may also be tested in these highly automated systems. With mixtures, particularly crude extracts, there is the potential for significant interference with the assay. Such interferences include nonspecific inhibition of targets by ubiquitous classes of natural products like fatty acids, or the presence of a highly potent cytotoxic agent that kills the host cell designed for the specific assay. These issues are not insurmountable, but require diligence in evaluating screening results. Some solutions for these issues are presented in Section 6.4.5.

6.4.4 DEREPLICATION

Dereplication is a term when used by natural products chemists refers to the rapid identification of a compound (or class of compounds). How is this different from the usual process encountered in the isolation of natural products? It is different in that the process refers to the identification of expected (or nuisance) compounds. These nuisance compounds will vary depending upon the particular assay system that is followed, and therefore it takes some experience with a given bioassay to identify the classes of interfering compounds. Before the advent of high-throughput HPLC/MS systems, specific tests were developed to identify the nuisances. In the early days of antibiotic discovery paper chromatography was used, as were thin layer chromatography with specific detection, and liquid chromatography with diode-array detection, and so on. This process has been greatly expedited by the use of HPLC/MS such as the system diagrammed in Figure 6.7. The system is based upon the separation of a mixture by reverse-phase HPLC, the continuous recording of UV/visible absorption spectra and mass spectra throughout the chromatogram, and the correlation of these data with biological activity. Once the active wells in the bioassay plate are related to a retention time, the optical and mass spectral data are correlated and used for querying suitable databases that provide matches of known compounds. Once an active compound is identified in this way, one can simply rely on the chromatographic and spectral data for dereplication of future samples.

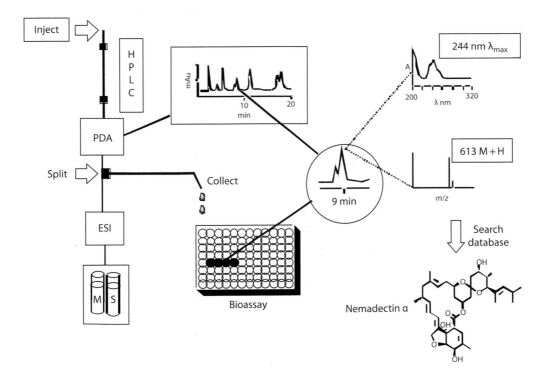

FIGURE 6.7 Schematic of HPLC/UV/MS system for dereplication of natural products.

6.4.5 SCREENING IMPROVEMENTS

In order to enhance the quality of screening data generated from highly valuable natural products, the nature of the extracts can be improved. There are a number of approaches to this problem, but the common goal has been to simplify the materials that are tested. When the offending materials are well understood, such as the presence of tannins in plant extracts, pretreatment of the crude extracts may suffice to remove the tannins from the screening samples. A broader approach that is applicable to samples derived from a variety of sources is diagrammed in Figure 6.8. This HPLC-based method seeks to remove offending nonspecific materials, as well as simplifying the actual screening samples. As diagrammed in Figure 6.8, the natural products are concentrated, often in the form of a solvent extract, and then subjected to a separation by reversed phase HPLC. The compounds are separated by employing gradient elution and only the material contained in the shaded area is retained for testing. This area contains components with polarities consistent with drug-like properties, having eliminated the early-eluting highly polar compounds such as saccharides and amino acids and the highly retained lipophilic materials that elute at the end of the run. The area is divided into 10 fractions that are concentrated and plated for use in high-throughput screening.

Fractionated screening samples derived from chromatographic separations as indicated in Figure 6.8 offer several advantages over crude extracts. The major one is the ability to independently evaluate diverse components produced by a particular source organism. In cases where one component in an extract is toxic, its adverse effect on the test organism in a phenotypic assay may obscure the positive response of a second component. Actinomycetes are particularly notable in their ability to produce numerous families of compounds. For example, it has been well documented that *Streptomyces* species that produce the milbemycin class of macrolides, such as the previously mentioned nemadectin, generally also produce oligomycins. The oligomycin macrolides are known

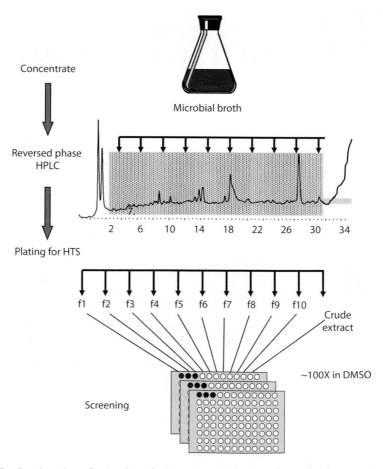

FIGURE 6.8 Prefractionation of natural products to generate improved samples for screening.

respiration inhibitors that are highly toxic to eukaryotic cells. Therefore, if one were screening in a rodent model for antiparasitic activity, a crude extract containing both the milbemycin and oligomycin would likely only show the toxicity. On the contrary, if the components were resolved chromatographically prior to screening the cryptic antiparasitic effect of the milbemycin would also be observed.

Apart from unmasking activities, there are benefits that accrue from enhancing the concentration of minor components present in a complex mixture. This is particularly true if in the preparation of fractionated screening samples the effort is made to normalize the concentration of the samples. The benefits of obtaining the maximum positive responses from these samples, often representing precious material collected under unique conditions, argue for expending the extra effort required.

In the course of resolving "hits" from natural product screening through bioassay-guided fractionation, as shown in Figure 6.6, it was emphasized that this is an empirical and highly experimental process, each new extract requiring an individual strategy for the isolation of its biologically active principles. If one has prefractionated the extract prior to initial screening, and the activity falls into a neat cluster of fractions, then one has valuable information on how to begin the purification process. This information will facilitate the resolution of the hit and lead to greater efficiency of the entire isolation and purification process.

6.5 OPTIMIZATION OF NATURAL PRODUCT LEADS

Nature has preserved the ability of a given organism to make these fascinating secondary metabolites, although their inherent biological roles remain obscure. As scientists seek to co-opt these metabolites as medicinal agents, attempts are typically made to enhance their pharmaceutical effectiveness. Such enhancements may be to improve the spectrum of activity against a range of targets, as in the case of antibiotics where broad-spectrum activity in inhibiting the growth of both gram-positive and gram-negative bacteria is important. In other cases it may be crucial to enhance the specificity to a narrower range of targets, such as the ability to selectively inhibit a particular kinase reaction in a signaling cascade. Furthermore, it may be necessary to improve the drug-like properties of a natural product lead. Here improvements in solubility, chemical stability in biological matrices or metabolic stability may be crucial. These and a host of other reasons drive the process to make structural modifications of the core natural product, which may be effected by chemical or biosynthetic means.

6.5.1 SEMISYNTHESIS

Semisynthesis refers to the process of performing synthetic chemical transformations starting with a natural product, for the purpose of enhancing the pharmaceutical performance of the natural product. This approach has been most effectively used with complex microbial products, owing to the ready availability of the starting material through fermentation of highly productive variants of the parent organism. The challenge in these experiments is twofold: one, to achieve adequate selectivity in the chemical process and two, the subsequent purification of reaction mixtures. A good example of the successful application of semisynthesis is in the case of the rifamycin antibiotics, shown in Figure 6.9. Rifamycin B is the originally isolated natural product derived from *Nocardia mediterranei*. Rifamycins have potent activity against gram-positive bacteria and are of greatest importance to inhibit the growth of the tuberculosis causing organism *Mycobacterium tuberculosis*. Rifamycin B was only modestly effective when administered to infected animals and this led to investigation of derivatives in search of improved potency. Greater potency was achieved through substitutions on the aromatic portion as exemplified by rifamide, rifampicin, and rifabutin. The latter two compounds continue to be important drugs for the treatment of reemerging epidemics of tuberculosis.

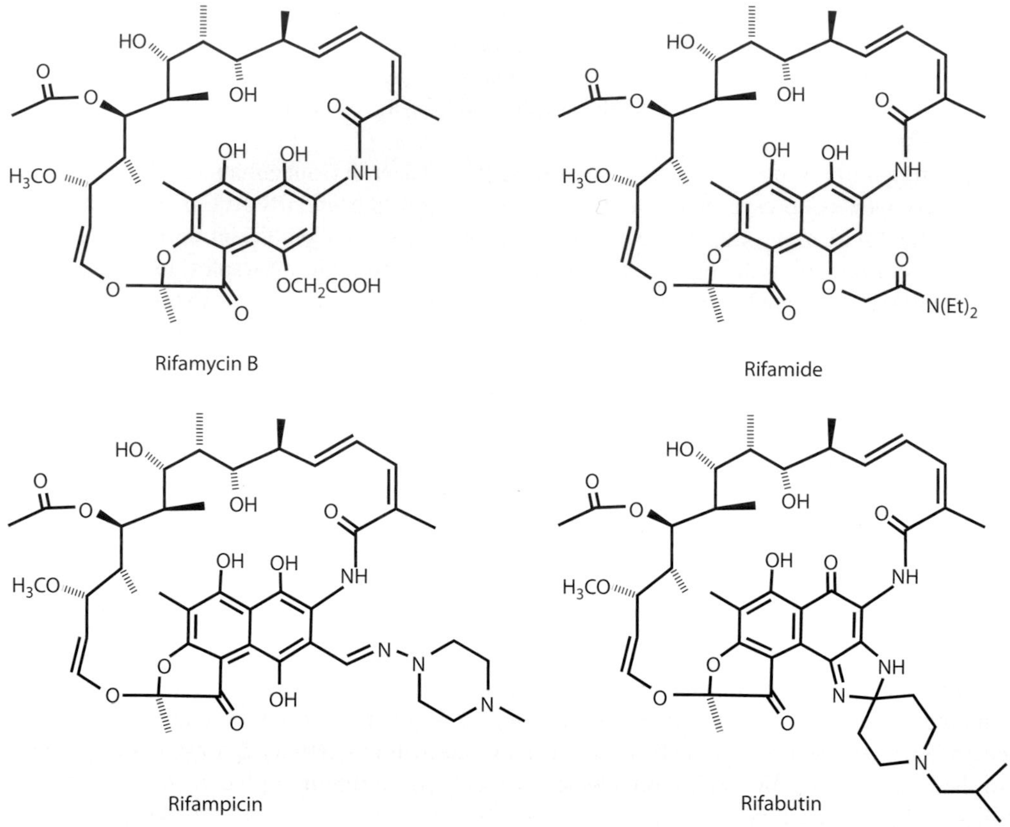

FIGURE 6.9   Rifamycin B and semisynthetic analogs.

6.5.2   Improvements in Natural Products through Total Synthesis

Although the total synthesis of natural products has been the forte of many prominent academic laboratories, only a few totally synthetic analogs of natural products have been introduced into commerce. The continued development of efficient and selective synthetic methods could provide alternative supply routes for simpler natural products in the future. Regardless of the issue of practical scalability, total synthesis enables the production and testing of analogs that often illuminate key features of the structure that are critical for biological activity. Paul Wender's research on the bryostatins, potent cytotoxic principles isolated from marine invertebrates, illustrates some of the key insights that can be revealed through total synthesis.

6.5.3   Biosynthetic Modifications

Genetic engineering of biosynthetic pathways to create specific modifications in chemical structure of secondary metabolites is now a practical reality in bacterial systems. In the simplest cases, a single enzymatic function is eliminated by inactivating the respective gene, resulting in an altered product. One such example is shown in Figure 6.10 from the work of scientists at Biotica Technologies, Cambridge, U.K., where the oxidation state of the aromatic ring in the ansamycin antibiotic macbecin is reduced from quinone to phenol. This was accomplished by inactivation of the gene *macM* that was found through genetic analysis to code for the specific enzyme responsible for addition of the *para* oxygen to the phenol ring that is further oxidized to the quinone.

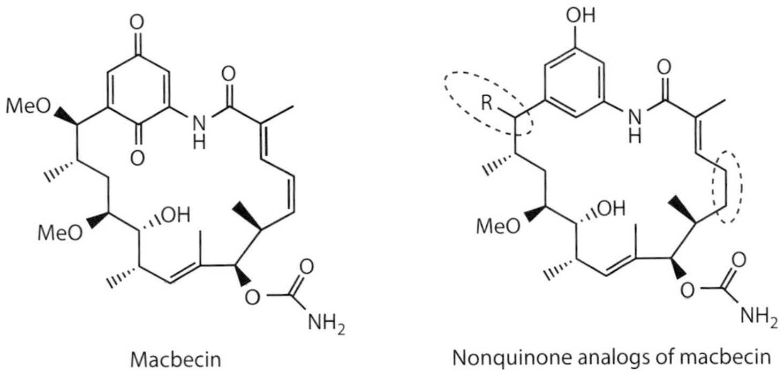

Macbecin   Nonquinone analogs of macbecin

FIGURE 6.10   Macbecin and nonquinone analogs derived from the knock out of a key oxidative function.

The macbecins are promising HSP-90 inhibitors with potential in cancer chemotherapy whose off-target effects have been linked to the reactive quinone moiety. The new products lack this reactive unit and are expected to have reduced side effects. This precise alteration in the structure was made possible by the identification of the functions associated with the key genes found in the macbecin biosynthetic gene cluster.

6.5.3.1   Mutasynthesis

Another technique that relies upon the knockout of an enzymatic function is known as mutasynthesis. In mutasynthesis a key step in a biosynthetic sequence is knocked out such that no product is made without the addition of a suitable precursor. In the past, these processes were done by random mutagenesis followed by screening of the resultant mutants for the desired phenotype. Today, it is a straightforward process to obtain the fully annotated genetic map of a biosynthetic pathway and to specifically design experiments to knock out the targeted function. One such example is shown in Figure 6.11 for the microbial product rapamycin. This work was pioneered by Peter Leadlay at the University of Cambridge, England who mapped the biosynthetic gene cluster for rapamycin. As illustrated, knock out of the gene rapL results in the organism's inability to make pipecolic acid, which is the usual amino acid incorporated into the rapamycin macrocycle. Supplementing the fermentation medium of the knockout strain with alternative cyclic amino acids, such as substituted proline analogs, results in efficient incorporation of these units yielding selectively modified rapamycin analogs.

6.5.3.2   Polyketide Synthase (PKS) Engineering

In the case of polyketide-derived compounds, the biosynthetic modules that are responsible for the iterative addition of two-carbon units to the nascent chain can be exchanged within sequences to alter both the substitution and oxidation state of the resultant unit. The most studied case of this biosynthetic class is erythromycin. The polyketide assembly of this macrolide antibiotic is illustrated in Figure 6.12, where three multifunctional enzymes DEBS 1, 2, and 3, encoded by *eryAI*, *II*, and *III* genes, assemble a starter unit propionate residue with six propionate extender units to produce the substituted poyketide chain. The chain is indicated in Figure 6.12, growing as the successive condensations add the propionate units. The resulting keto groups are reduced to alcohols by keto-reductase functions (KR, modules 1, 2, 5, and 6), not reduced at all as in module 3 (note the lack of a reductive loop), or fully reduced to the bare methylene by ketone reduction, enolization, and hydrogenation (KR, ER, DH, and module 4). The chain is terminated and cyclized through the action of the final active site in DEBS 3, the thioesterase (TE). Additional tailoring enzymes further modify 6-deoxyerythronolide B, by oxidation, glycosylation, and methylation processes to yield the fully functionalized erythromycin A.

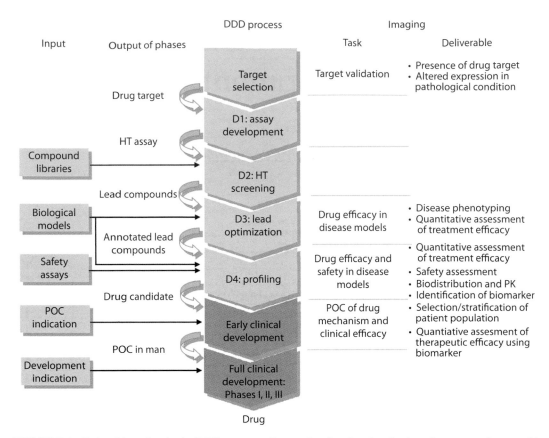

FIGURE 7.1   Role of imaging in the DDD process. Conventional and molecular imaging approaches provide information for validation of the drug target, for the evaluation of treatment efficacy in models of human disease during lead optimization and profiling. A critical aspect in view of future clinical development is the identification and characterization of biomarkers with prognostic value for clinical outcome. These biomarkers are used in clinical proof-of-concept studies. *Abbreviations used*: drug discovery and development (DDD), high throughput (HT), pharmacokinetics (PK), proof-of-concept (POC). The different gray levels of the individual phases in the DDD chain indicate the translation from preclinical to clinical phases.

been obtained by using positron emission tomography (PET). Today, progress in imaging technologies has greatly enhanced the scope of imaging in DDD: so-called molecular imaging methods enable annotation of tissue structure with cellular and molecular information in animals and in humans. These methods are based on the design of target-specific exogenous imaging agents, which selectively target the process of interest or on assays originally developed for imaging of cells and cell networks, frequently involving genetic engineering to produce detectable imaging signals (reporter gene assays).

7.2   MULTIMODAL IMAGING TECHNIQUES FOR DDD

In the subsequent sections we will discuss the role of conventional and molecular imaging in the context of DDD. Before digging into imaging applications we will briefly discuss three important modalities, magnetic resonance imaging (MRI), fluorescence imaging, and PET. Figure 7.2 schematically describes the source of signals, the physical principle leading to image generation, the influence of the (biological) environment on the signals detected, the principle of spatial encoding, and the spatial resolution provided for the respective modality.

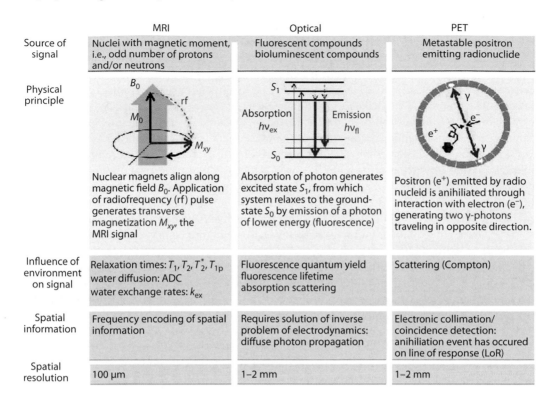

	MRI	Optical	PET
Source of signal	Nuclei with magnetic moment, i.e., odd number of protons and/or neutrons	Fluorescent compounds bioluminescent compounds	Metastable positron emitting radionuclide
Physical principle	Nuclear magnets align along magnetic field B_0. Application of radiofrequency (rf) pulse generates transverse magnetization M_{xy}, the MRI signal	Absorption of photon generates excited state S_1, from which system relaxes to the ground-state S_0 by emission of a photon of lower energy (fluorescence)	Positron (e^+) emitted by radio nucleid is anihiliated through interaction with electron (e^-), generating two γ-photons traveling in opposite direction.
Influence of environment on signal	Relaxation times: T_1, T_2, T_2^*, T_{1p} water diffusion: ADC water exchange rates: k_{ex}	Fluorescence quantum yield fluorescence lifetime absorption scattering	Scattering (Compton)
Spatial information	Frequency encoding of spatial information	Requires solution of inverse problem of electrodynamics: diffuse photon propagation	Electronic collimation/ coincidence detection: anihiliation event has occured on line of response (LoR)
Spatial resolution	100 µm	1–2 mm	1–2 mm

FIGURE 7.2 Features of imaging modalities MRI, fluorescence imaging, and PET. MR images represent the weighted distribution of tissue protons, predominantly those of water and adipose tissue. The signal is weighted by parameters such as the relaxation times, diffusion properties, or proton exchange rate, which depend on the local environment. Spatial encoding is achieved by applying magnetic field gradients: as the resonance frequency depends on the local magnetic field, the spatial information is directly encoded in frequency information. Spatial resolution in animal imaging is of the order of 100 µm. Fluorescence imaging measures the distribution of fluorescent molecules (dyes, quantum dots, or fluorescent proteins). As light is heavily scattered by tissue, photon propagates as a diffusion wave. In order to determine the localization and intensity of a fluorescent source the inverse problem has to be solved by iterative procedures. Spatial resolution is of the order of 1 mm. PET measures the distribution of radionuclides that decay by emitting a positron. As antimatter particles, positrons are captured by electrons after traveling through tissue for a short distance (positron range), the annihilation process generating two γ-photons traveling in opposite direction. The detector system consists of a ring of scintillation crystals. Coincidence detection of the two photons allows determining a line of response (LOR), on which the annihilation process (not the radionuclide decay) has occurred. Measuring a sufficient number of LORs allows reconstructing the PET image.

7.2.1 Conventional Imaging

Classical imaging yields morphological and physiological information. Intrinsic contrast in images is governed by the interaction of radiation with tissue and hence on the biophysical properties of tissue. For x-ray-based techniques such as computer tomography (CT) the incident x-ray beam is attenuated upon passage through matter due to scattering at electrons. MRI maps the distribution of protons (water) in tissue. The signal intensity is governed by a multitude of parameters: proton density, spin relaxation properties that describe interaction of the protons interrogated with their environment, macro- and microscopic motion in a magnetic field and chemical exchange reactions that alter the local environment of protons. This multiparameter dependence explains the high soft-tissue contrast provided by MRI methods. In ultrasound, the pressure wave is reflected by tissue interfaces and the contrast arises from differences in tissue elasticity, compressibility and backscattering.

Contrast-to-noise ratio (CNR) and spatial resolution are critical parameters in determining the quality of structural images. It is important to realize that these two parameters are related; increasing the spatial resolution decreases CNR assuming all other parameters being constant. CNR might be improved by administration of contrast-enhancing agents, which are compounds comprising electron-dense atoms (e.g., iodine, barium) for x-ray, paramagnetic, or superparamagnetic agents that enhance relaxation rates for MRI (e.g., gadolinium chelates, iron oxide nanoparticles), or microbubbles with different compressibility than adjacent tissue. Dynamic measurement of changes in CNR following the administration of an exogenous contrast agent allows monitoring of physiological processes, such as tissue perfusion, vascular permeability, or the function of excretory organs.

Structural and physiological imaging has been primarily used as a diagnostic tool to identify pathologies, for monitoring disease progression, and for the evaluation of the response to therapeutic interventions. An important aspect for applications in DDD is quantitative analysis of biomedical imaging data based on morphometric or densitometric (intensity) measures. Morphometric readouts are, for example, the volume of a tissue structure, i.e., organ, tumor mass, or an ischemic region in the heart or brain, a cross-sectional area such as the lumen in a major vessel, or a distance such as the thickness of articular cartilage. Preferentially, such measures should be obtained in an automated fashion with minimal operator interaction. Morphometric analysis involves image segmentation, which is relatively straightforward for image data sets displaying high CNR for the structure of interest. However, due to limited CNR, operator interaction is still required in many cases slowing down the analysis procedure and bearing the risk of introducing operator bias. Quantitative structural information may also be obtained from the relative signal intensities or more accurately from absolute values of image contrast parameters such as MRI relaxation rates. For example, it has been shown that alterations of the so-called transverse relaxation rate (R_2) and water diffusion coefficients in ischemic brain tissue following cerebral infarction are indicative of the severity of the ischemic insult (Figure 7.5) normalization of these values following cytoprotective therapy reflects drug efficacy.

Quantitative analysis of dynamic changes in image intensity is applied to derive physiological or functional information. The signal intensity is monitored in response to a pharmacological or physiological challenge or to the passage of an exogenous contrast agent. Derivation of physiological information from dynamic MRI data sets involves the development of physiological models. For instance, tissue perfusion can be assessed using the tracer dilution method involving determination of the local tissue blood volume (from the integral of the tracer concentration–time curve) and the tracer mean transit time (MTT). Relative tissue blood flow (TBF) is then obtained from the ratio TBF = TBV/MTT. In order to derive absolute perfusion rates the arterial input function for the tracer must be known.

7.2.2 Molecular Imaging Approaches

Morphological and physiological aberrations are the result of abnormal molecular processes in tissue and it is a reasonable hypothesis that quantitative mapping of these molecular events in vivo should increase both the sensitivity and the specificity of diagnostic tools. Complementing structural and functional information, "molecular" imaging methods provide readouts on levels of gene transcription and translation products, on critical molecules involved in signal transduction, and/or protein–protein interactions. Furthermore, the fate of labeled cells can be studied in the intact organism. As molecular events occur at low frequency, highly sensitive imaging modalities are required, in particular, nuclear imaging approaches such as single photon emission computer tomography (SPECT) or PET, and optical imaging such as fluorescence and bioluminescence imaging.

7.2.2.1 Target-Specific Imaging Assays

Target-specific imaging probes: Derivation of specific information on molecular processes requires target-specific reporter systems. Their generic design combines a targeting moiety, for example, a receptor ligand, antibody, or oligonucleotide with a signal generating entity, for example, a radionuclide, a fluorescent molecule, or an MRI contrast agent. It is important that the labeling procedure preserved the pharmacophore of the receptor ligand, i.e., the part of the molecule that is critical for its interaction with the target. The development of such specific reporter constructs has to tackle the same issues encountered in the development of a therapeutic: (1) the probe has to be target-specific with minimal cross-reactivity to other receptor systems; (2) it must have good pharmacokinetic (PK) properties, i.e., the probe should be able to penetrate tissue barriers to reach the target and the exposure time has to be long enough to allow for the target-specific interaction, and the nonreacting fraction of the label should be cleared rapidly from the system to maximize signal-to-background ratio (SBR); (3) the reporter construct should be biocompatible, not posing any safety issues; and (4) in addition as the concentration of targets in tissue is generally low, the signal generated by the reporter group should be amplified when possible. This can be achieved by increasing the payload, by ligand trapping, or enzymatic probe processing. So-called target-activatable probes constitute an elegant strategy to improve SBR. These are reporter moieties that change their biophysical properties upon interaction with their target. Fluorescent molecules or MRI contrast agent qualify for such designs, as these fluorescence and magnetic dipole interactions can be modulated by the local environment. Enzyme activatable probes fall into this category. A critical issue with exogenous probes is their delivery to the target, in particular when the target is located intracellular. Cellular uptake of low molecular weight probes is optimized by derivatization following strategies such as Lipinski's rule of fives. For larger probes, target delivery may be achieved by exploiting cellular uptake mechanisms such as transporter systems or receptor-mediated endocytosis or by conjugation of cell-penetrating peptides to the reporter moiety.

Reporter gene assays: Complementary to the use of exogenous probes, reporter assays generated by the biological system itself might be used. Reporter gene assays are established tools in molecular and cell biology. They involve the generation of a genetically modified cell (or animal) that expresses a reporter molecule under the control of the promoter of the gene of interest, or use a strategy, in which the target protein drives the expression a reporter gene. A remarkable number of reporter gene assays suited for application in whole animals meanwhile suitable for MRI, PET, fluorescence imaging have been developed the best known being fluorescent proteins such as green fluorescent protein (GFP) or bioluminescent proteins such as luciferases. It is obvious that these systems are largely limited to animal studies (except for gene therapy approaches or potentially cell tracking in humans).

7.2.2.2 Molecular Imaging Strategies, Imaging Targets

Imaging approaches may target different aspects that relate to the efficacy of drugs: its biodistribution, the interaction of the drug with its therapeutic target, the initiation of the signaling cascade, or the response of the biological system in terms of morphological, physiological, or metabolic changes (Figure 7.3). In the following list, these different aspects will be addressed.

a. *Drug biodistribution*: Unfavorable PK properties are an important reason for failure of drugs during development. In view of this fact, detailed knowledge on the drug's biodistribution is of key importance in the development of novel therapeutics. During the preclinical phase this information is commonly obtained from quantitative whole-body autoradiographic studies, which measures the distribution of radiolabeled drug molecules. More recently, matrix-assisted laser desorption and ionization mass spectrometric (MALDI-MS) imaging has been introduced for PK studies in tissue samples. Molecular identification is based on the mass determination; hence, MALDI-MS imaging does not require labeling with radioisotopes. In addition, using the mass filter parent molecule and metabolites can be distinguished, in contrast to radiolabel-based techniques. Both autoradiographic and MALDI-MS imaging are ex vivo techniques and will not be discussed further.

In vivo drug biodistribution studies almost exclusively use PET. As drug labeling should not affect its PK and pharmacodynamic properties, its molecular structure must be unaffected by the introduction of a reporter group. The only possibility to achieve this is isotopic substitution by a radionuclide. Moreover, introduction of a radioactive

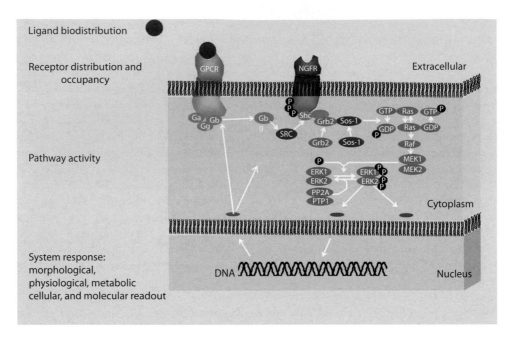

FIGURE 7.3 Imaging targets relevant for DDD. Currently available imaging techniques allow visualization and quantification of the drug's mechanism of action. Labeling of the drug molecule itself (or of a competitive receptor ligand) reveals information on its biodistribution and receptor interaction. The expression level of a receptor can be visualized using specific reporter ligands or following a reporter gene strategy. Activation of the signaling cascade is visualized by targeting individual pathway molecules (e.g., caspases for studying apoptosis) or by measuring protein–protein interaction (see text). Finally the result of the therapeutic intervention such as morphological, physiological, metabolic, cellular, or molecular changes can be monitored.

reporter nuclide yields the sensitivity that is required to detect small amounts of the drug ligand in tissue.

b. *Expression of the molecular target*: A critical step in early drug discovery is target validation, i.e., demonstration of the presence of a drug target in the tissue of interest. The common strategy to visualize and quantify the presence of a drug target, such as a membrane
receptor or an enzyme, uses target-specific imaging probes, the vast majority using either
radionuclides or fluorescent dyes as reporter moiety. There are numerous examples of such
studies (Section 7.3.1). It is important to realize that unless one uses target-activatable
probes, it cannot be discriminated whether the signal observed arises from the reporter
fraction that is bound to its target or just from free or unspecifically bound molecules. Thus,
it is important to wait until the unbound probe is cleared from circulation. Alternatively,
reporter gene assays can be used to visualize target expression.

c. *Imaging pathway activities*: Two strategies can be pursued to study pathway activities,
either by monitoring critical molecules in the signal transduction cascade or by visualizing protein–protein interactions. The first approach uses the concept outlined in the
previous paragraph. For example, a reporter gene assay has been developed to visualize
the activity of caspases-3, a critical player in cellular apoptosis. Signal propagation relies
on protein–protein interactions. A number of assays have been developed to study these
key processes in cellular systems; some of them have been translated for applications
in intact animals such as the two-hybrid assay or the protein fragment complementation
assay. As an example, a split luciferase assay has been developed to study the interaction
of the two proteins FRB and FKBP12, which is induced by the administration of the
macrolide rapamycin.

d. *Monitoring cell migration*: Monitoring the trafficking and fate of labeled cells in vivo has a wide range of applications in DDD. Tumor cells have been transfected to express fluorescent or bioluminescent proteins for the noninvasive assessment of tumor growth or metastasis load in murine tumor models. Inflammatory processes have been studied using MRI by monitoring the infiltration of monocytes and lymphocytes labeled with superparamagnetic iron oxide nanoparticles into inflamed tissue. Cell therapy is becoming an increasingly important therapeutic strategy requiring tools to visualize the location, migration, and viability of stem or progenitor cells. The fate of such cells has been monitored using either MRI and bioluminescence imaging in models of cerebral and cardiac ischemia or in brain tumor models. While phagocytotic cells such as monocytes can be efficiently labeled in situ, other cell types have to be harvested and labeled in vitro. Alternatively, genetically engineered cells expressing a reporter gene have been applied. While for most molecular imaging approaches the use of MRI is limited by its low intrinsic sensitivity, cells tolerated relatively high amounts of superparamagnetic iron oxide; therefore low amount of cells, in favorable cases even single cells, can be detected with unsurpassed spatial resolution.

7.3   SELECTED IMAGING APPLICATIONS IN DDD

7.3.1   D1: TARGET VALIDATION

We will illustrate the role of imaging for target validation with several examples from oncology. Peptide receptors are frequently over-expressed in tumors and therefore constitute attractive targets both for therapeutic interventions and for diagnostic imaging. Extensively studied examples are the membrane-bound somatostatin receptors (SSTRs) that are highly expressed in neuroendocrine tumors. A number of imaging probes based on octreotide, a metabolically stabilized analogue of the endogenous ligand somatostatin (SST-14 or SST28), have been developed, using either radionuclides or fluorescent groups as reporters. Feasibility of visualizing SSTR expression has been demonstrated both in patients suffering from neuroendocrine tumors and in animal models of the human disease. The imaging probes specifically accumulated at the tumor site(s). One of the ligands (^{111}In-pentetreotide, Octreoscan®) has been approved for clinical use as diagnostic SPECT probe. There are several other radiopeptides that are currently being evaluated as tumor-specific imaging agents such as bombesin or cholecystokinin, for example, for detection of prostate cancer. Alternatively fluorescent probes have been developed for experimental studies in animals (Figure 7.4).

Overexpression of the Her-2/neu tyrosine kinase receptor is observed in approximately 25% of human breast cancers and is associated with poor prognosis. Inhibition of Her-2/neu signaling, for example, using a specific antibody (trastumazab) therefore constitutes a therapeutic strategy in these cancer patients. Correspondingly, an imaging probe suited for demonstrating Her-2/neu overexpression would be essential for the selection of patients amenable to therapy. A PET imaging probe based on Her-2 antibody fragments has been used in murine tumor models to study the effects of inhibition of heat shock protein 90, a molecular chaperon, on Her-2/neu levels. A significant reduction of Her-2/neu levels could be observed noninvasively within hours after therapy onset, which later translated into reduced tumor growth.

A third example relates to tumor angiogenesis, a critical step in the formation of a neoplastic lesion. A high degree of neovascularization is commonly associated with rapid tumor proliferation and thus malignancy. Inhibition of angiogenesis is considered an attractive strategy in tumor therapy, in particular in combination with other therapeutic strategies. Critical proangiogenic factors induced are vascular endothelial growth factor (VEGF) and its receptor (VEGF-R). When selecting patients for treatment with VEGF-R inhibitors, the demonstration of high levels of the receptor in tumor tissue would be relevant; this has prompted the development of target-specific assays suitable for in vivo imaging. As VEGF-R is expressed at the endoluminal side of the endothelial cell layer it

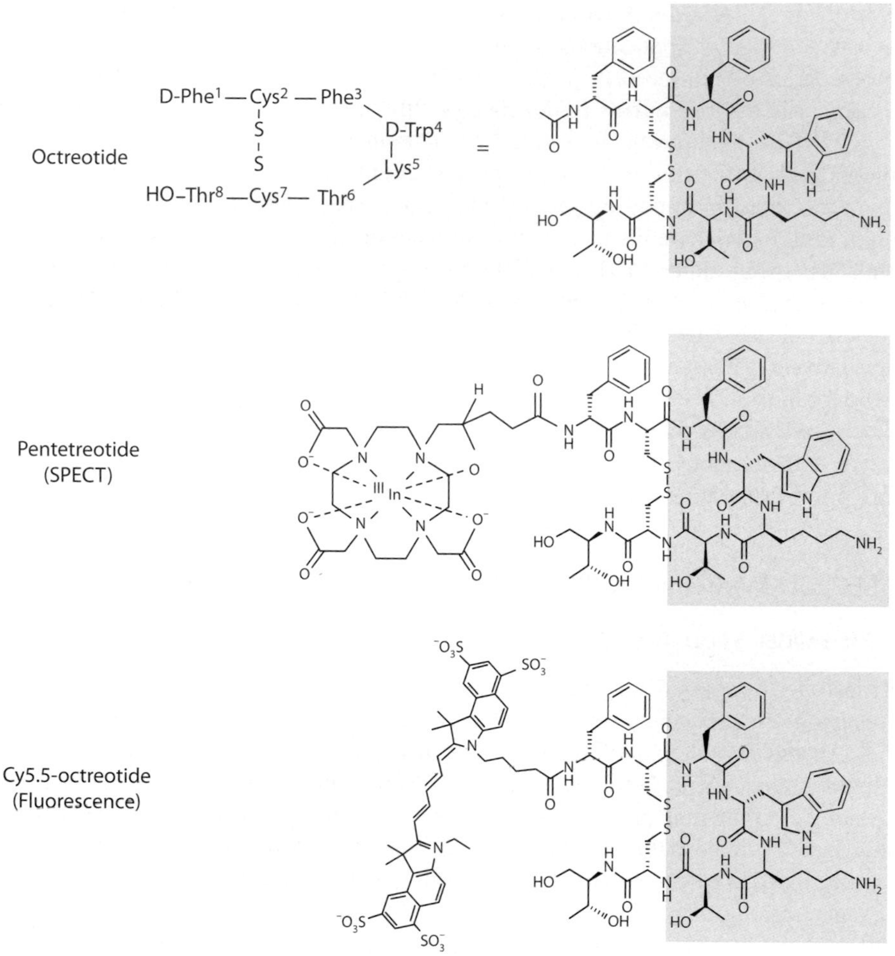

FIGURE 7.4 Structure of octreotide and related imaging ligands for measuring the distribution of SSTRs. Octreotide is a somatostatin analogue that is stabilized against degradation by peptidases through cyclization (disulfide bridge). Structure–activity relationship revealed that the amino acids phenylalanine-tryptophane-lysine-threonine (Phe-Trp-Lys-Thr) are essential for the interaction with the somatostatin receptor 2 (SSTR2). This pharmacophor (shaded area) must be retained when designing a targeted imaging ligand. The SPECT ligand pentetreotide comprises a chelated indium-111 reporter group (g-photon emitter radionuclide) while for the optical probe the fluorescent indocyanine dye Cy5.5 has been used. Both reporter groups are linked to the terminal ᴅ-phenylalanine, distant from the pharmacophor group.

can easily be reached also by macromolecular probes. VEGF-R expression has been demonstrated using radiolabeled probes that are based on the endogenous ligand VEGF or on VEGF-R targeting antibodies.

The demonstration of altered expression levels of a potential drug target in a pathological condition is an important, yet not sufficient, step in the target-validation process.

7.3.2 D3/D4: Lead Optimization and Drug Profiling

By far, most imaging studies in DDD have been carried out in the lead optimization phase. Imaging has been used for phenotyping disease models and to assess the efficacy of drug candidates using structural, functional, metabolic, and more recently also molecular readouts. We will discuss one example to illustrate the potential and also problems associated with imaging approaches in preclinical DDD: the evaluation of anti-ischemic therapy in models of focal cerebral ischemia.

Stroke is a leading cause of death in industrialized nations and the only clinically approved treatment is thrombolysis using recombinant tissue plasminogen activator (rtPA) applicable in approximately 5% of stroke patients. Hence, there is a high medical need to develop novel pharmacological therapies.

Focal cerebral ischemia is caused by transient or permanent occlusion of a major cerebral artery. Cessation of local perfusion initiates a cascade of detrimental effects within minutes such as energy failure due to the shutdown of aerobic ATP synthesis that affects all energy-dependent processes of the cell including membrane pumps required to maintain ion homeostasis, and intracellular signaling cascades via ATP-dependent protein kinases. Failure of membrane pumps causes intracellular Ca^{2+} to accumulate prompting a cascade of deleterious downstream events for the cell ultimately leading to cell death. Excessive levels of excitatory neurotransmitters such as glutamate are another major reason for tissue exhaustion and damage as the inactivation of glutamate via glial and neuronal uptake is an energy-dependent process. Elevated glutamate levels cause opening of N-methyl-D-aspartate (NMDA) receptors enhancing Ca^{2+} influx with the consequences already discussed. Glutamate interaction with metabotropic glutamate receptors (mGluR) activates second-messenger-mediated signaling. In addition to these acute effects, delayed infarct growth due to recruitment of penumbral regions has been observed at time points beyond 48 h after infarction, associated with apoptosis and neuroinflammation. Therapeutic strategies in stroke target the individual processes of the pathophysiological cascade.

Long-term tissue survival can only be achieved through restoration of blood supply to the ischemic lesion, which has to occur within a short time frame of a few hours. In the acute phase, protective effects can be achieved by reducing the energy demand in affected brain areas that show some residual perfusion (ischemic penumbra). This can be achieved by administration of Ca^{2+} channel blockers, which reduce Ca^{2+} influx through voltage-gated channels, or by inhibition of receptor-operated channels such as the NMDA receptor. Both strategies have been rather successful in animal models of focal cerebral ischemia. A number of other therapeutic targets have been investigated such as glycine receptors, prevention of excitotoxicity using antagonists of the α-amino-3-hydroxy-5-methyl-isoxazole-4-propionate (AMPA) receptor, free-radical scavengers, inhibitors of death protease, or anti-inflammatory treatment. Yet, all these compounds failed upon translation into the clinics, mostly due to lack of efficacy. More recently, tissue repair strategies using neuronal stem or progenitor cells have been proposed, which revealed beneficial effects in a rat stroke model.

Today, imaging techniques, and in particular MRI-based methods, enable the visualization of individual aspects of the pathophysiological cascade both in humans and in animals from the initial vascular occlusion to the infiltration of inflammatory cells during the postacute phase. The same techniques have been applied to evaluate the efficacy of anti-ischemic drugs.

For example, it has been shown that in models of global ischemia pretreatment with cytoprotective Ca^{2+} inhibitors significantly delayed ATP depletion and tissue acidosis. Yet, most of the preclinical stroke studies evaluating drug efficacy using MRI are based on morphometric readouts, i.e., they use infarct volumes as efficacy biomarker. The underlying assumption is that reduction of the structural damage, i.e., reduction of the infarct volume, will necessarily translate into an improved behavioral or correspondingly clinical outcome. The classical MRI method for the assessment of cerebral infarct volume is based on R_2 contrast: formation of a vasogenic edema leads to significantly reduced R_2 values (and correspondingly increased $T_2 = 1/R_2$ values) providing a good demarcation between ischemic and intact tissue (Figure 7.5). A considerable number of drug candidates have been evaluated using this approach. It has to be kept in mind that cerebral tissue displaying a decreased R_2-value is already irreversibly damaged; hence the method indicates an endpoint. Earlier indicators that provide a significant contrast for tissue that is still salvageable are the apparent water diffusion coefficients (ADC) in brain parenchyma and local cerebral blood flow (CBF) rates. It has been demonstrated, at least in animal models of global and focal ischemia, that ADC changes are

in man. This is in line with the Critical Path Initiative of the U.S. Food and Drug Administration (FDA) that aims at "translating basic scientific discoveries more rapidly into new and better medical treatment by creating new tools to find answers about how the safety and effectiveness of new medical products can be demonstrated in faster time frames, with more certainty, at lower cost and with better information." The agency is convinced that DDD programs will benefit from the availability of biomarkers, imaging methods being considered key enabling technologies. The FDA sees an important role for biomarkers in providing proof-of-principle of a therapeutic intervention, for stratification patient populations, and for the evaluation of therapy response or eventual side effects. This enthusiasm has to be viewed in relation to the fact that currently only very few biomarkers are accepted by the FDA as efficacy readouts (see below), and the development of novel ones seems to be a major undertaking. Aspects such as specific versus generic biomarkers, validation, standardization, biomarker profile versus individual markers, and quantification have to be addressed.

As an example, we will discuss two biomarkers for studying the efficacy of tumor therapy that are in a fairly advanced stage of development. The measurement of glucose utilization rates via [^{18}F]-2-fluoro-2-deoxyglucose (FDG) PET has evolved as a sensitive diagnostic tool for the characterization of primary tumors and for the detection of metastases. FDG is taken up by cells via the glucose transporter and phosphorylated by hexokinase to yield FDG-6-phosphate (FDG-6P) and trapped in the cell (Figure 7.7). The PET activity, hence, reflects glucose transporter and hexokinase activity and thus glycolytic activity of the tissue of interest. There is evidence from several clinical

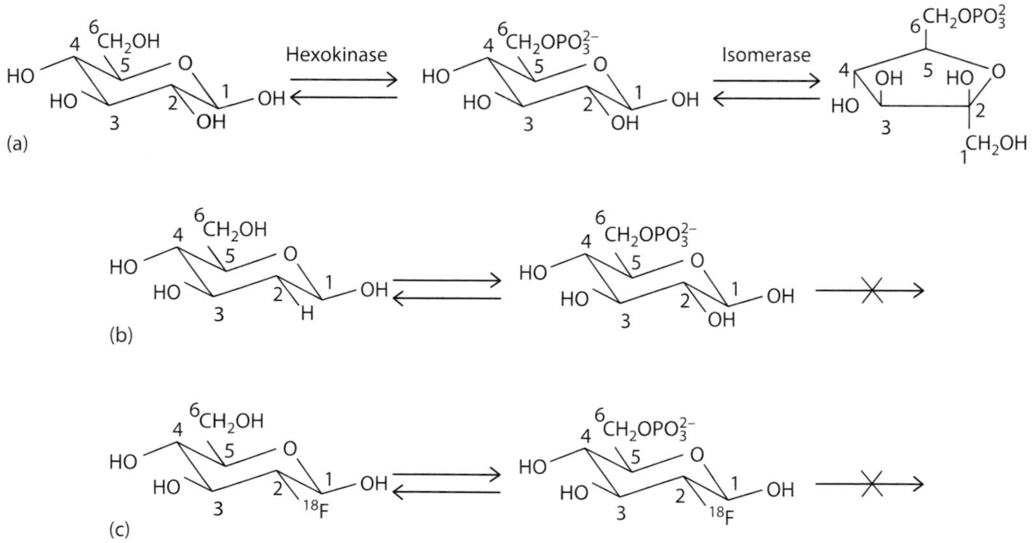

FIGURE 7.7 Measurement of tissue glucose utilization using PET and 2-fluoro-2-deoxyglucose. The primary energy substrate glucose is taken up from the circulation via glucose transporters and is phosphorylated within the cells by hexokinase to yield glucose-6-phosphate. The next step in the metabolic cascade is the isomerization to fructose-6-phosphate, which after further processing enters the citric acid cycle. By using 2-deoxy-glucose (2DG) as substrate it could be demonstrated that uptake and phosphorylation are identical as for glucose. Yet, isomerization is not possible due to the lack of the hydroxyl group in the 2-position, and 2DG-6-phosphate is trapped in the cell. Hence, the accumulation of 2DG-6-phosphate, which is measured using autoradiographic techniques, is considered a surrogate for glucose utilization by tissue. In the PET version of the 2DG assay the hydrogen at the 2-position is replaced by a fluorine-18 radionuclide (with half-life 110 min) yielding [^{18}F]-2-fluoro-2-deoxyglucose (FDG). As glucose and FDG are different compounds, the rate constants for enzymatic processing of the two substrates differ. This is accounted for by correcting glucose utilization rates derived from FDG measurements using so-called lumped constants.

drug trials that changes in glycolytic rate precede effects on the tumor volume. For example, in patients suffering from gastrointestinal stromal tumors treated with imatinib glucose utilization was significantly reduced within 24 h after treatment onset, while there was no effect on tumor volume for several months, indicating that glucose utilization rate predicts therapy response.

A common technique to assess tumor angiogenesis is the so-called dynamic contrast-enhanced (DCE) MRI methods, which exploits the fact that newly formed immature vessels are characterized by increased vascular permeability. DCE-MRI measures the leakage of low molecular weight contrast agents such as GdDTPA (Magnevist®) or GdDOTA (Dotarem®) into the extracellular space. This method is currently evaluated as biomarker for evaluating the efficacy of antiangiogenic therapy. Inhibition of VEGF receptor signaling should be reflected by decreased vascular permeability and potentially also reduced tumor blood volume, as demonstrated for the VEGF tyrosine kinase inhibitor vatalanib for several tumor models in mice. Drug effects could be detected within 48 h following onset of treatment. Clinical studies in patients with liver metastases yielded corresponding results indicating that, in fact, vascular permeability measures may serve as a biomarker of efficacy. The method has been used for translational studies with a number of compounds.

Currently used imaging biomarkers are based on structural (e.g., response evaluation criteria for solid tumors [RECIST] for tumors, infarct volume for stroke, and lesion load for MS) or physiological and metabolic readouts (e.g., DCE-MRI and glucose utilization rates for tumors). A next generation of (molecular) imaging approaches will provide specific mechanistic information tightly linked to the therapeutic pharmacological principle; a term coined in this regard is theranostics, the merger of therapeutic and diagnostics.

7.4 DOES IMAGING ADD VALUE TO THE DRUG DISCOVERY PROCESS?

What is the added value of using resource-intense imaging methods in DDD? The obvious expectation is that the use of noninvasive analytical techniques facilitates the translation of the therapeutic concept from preclinical to clinical development and thus might contribute to shortening of DDD times. Convincing as a concept, there is little evidence today that the use of imaging has greatly impacted development. This is likely to change in the future as all major stakeholders have recognized the importance of speeding up DDD as outline in the FDA Critical Path Initiative. Biomarkers and eventually surrogate markers will allow assessing treatment efficacy and safety aspects significantly earlier than classical clinical endpoint measures. Noninvasive imaging will be a key enabling technology in that context as illustrated in the previous section. It is important to realize that the purpose of translational imaging applications in DDD are not large-scale multicenter phase III trials, which pose high demands on standardization of imaging protocols and potentially would be very expensive. Translation imaging studies serve the purpose of establishing pharmacological proof-of-concept in a selected patient population in a small, well-controlled clinical study—critical information for decision makers before entering large-scale clinical evaluation.

Apart from these translational aspects imaging readouts have turned out to provide essential information to the drug developer. The possibility to derive morphological, physiological, metabolic, cellular, and molecular information in a noninvasive manner from an intact organism is highly relevant, in particular when studying chronic degenerative diseases that require longitudinal evaluation. It is therefore not surprising that the majority of imaging applications refers to disease phenotyping; in view of the increasing number of genetically engineered mouse lines available, this application will certainly become even more important. The information gathered during disease phenotyping should be used to stratify treatment groups: prior to therapy administration, patients or animals can be classified into "homogenous" treatment groups, which should translate into data with better statistical relevance.

Imaging is inherently quantitative although translation from primary imaging parameters to biomedical information is not straightforward and constitutes a major challenge for the imaging

FURTHER READINGS

Beckmann N (Ed.) (2006) *In Vivo MR Techniques in Drug Discovery and Development.* Taylor & Francis (New York).

Phelps ME (Ed.) (2004) *PET: Molecular Imaging and Its Biological Applications.* Springer (New York).

Rudin M (2005) *Molecular Imaging—Principles and Application in Biomedical Research.* Imperial College Press (London).

Rudin M (Ed.) (2005) *Imaging in Drug Discovery and Early Clinical Development*, Birkhäuser Verlag (Basel).

Rudin M (2005) Imaging readouts as biomarkers or surrogate parameters for the assessment of therapeutic interventions, *Eur. Radiol.*, 17, 2441–2457.

Webb A (2003) *Introduction to Biomedical Imaging.* John Wiley & Sons Inc. (Hoboken, NJ).

8 Peptides and Peptidomimetics

Minying Cai, Vinod V. Kulkarni, and Victor J. Hruby

CONTENTS

8.1 INTRODUCTION TO PEPTIDES AND PEPTIDOMIMETICS

Since the 1950s with the discovery of the peptide hormones and neurotransmitters, oxytocin and vasopressin, and the peptide antibiotics such as penicillin and valinomycin there has been a growing interest in the use of peptides as drugs for the treatment of diseases. The field is now extremely vast with thousands of papers published each year, with over 50 peptide- and peptidomimetic-based drugs, and with literally hundreds of other peptides in various phases of development as drugs. In addition, it is increasingly realized that the modulation and control of most biological processes, especially in more advanced forms of life such as multicellular animals, are controlled or modulated by peptide–protein, peptide–nucleic acid, peptide–lipid, protein–protein, protein–nucleic acid, and protein–lipid interactions (in the latter protein–X interactions, it is often a continuous or discontinuous polypeptide segment of the protein that is involved in the interaction, signaling, modulation, or other physiological effects). The interest on how to effectively utilize peptides and peptidomimetics to control their effects on health and disease has become a central theme of modern biology and medicine.

We have chosen to emphasize those aspects of peptide and peptidomimetic chemistry that are most important to know and to understand if one is to utilize bioactive peptides as the basis for drug design and discovery. For this purpose we have chosen peptide hormones and neurotransmitters as our major focus. These peptides are known to be critical molecules in virtually all bodily functions, in all aspects of human behavior, and in many aspects of disease—both peripheral diseases such as cardiovascular disease and diabetes, and in most diseases of the central nervous system (CNS) from anxiety and depression to sexual function, addiction, and obesity.

Therefore, in this chapter, we will examine the intrinsic properties of peptides, including structural, conformational, topographical, and dynamic properties of peptides that can be controlled and modified to obtain insights into the structural and conformational properties that can lead to peptidomimetics. In this regard, we should begin by carefully defining our terms. A "peptide" is a biopolymer made up of amide (peptide bond) linked α-amino acids. A "peptidomimetic" is a derivative of a peptide that possesses modifications of a common peptide structure including peptides containing β- or γ-amino acids; peptoid structures; amide bond replacements in their structure; side chain–modified structures such as β-alkyl or aryl-substituted α-amino acids; unusual side chain cyclizations including lactam bridges, lactone bridges, alkane, or alkene bridges, and other side-chain-to-side-chain bridges; side-chain-to-backbone and backbone-to-backbone cyclizations. Nonpeptide peptidomimetics whose design is based on peptides will not be discussed here, nor will so-called peptide mimetics that are discovered as part of some screening processes of chemical libraries. In any case, these compounds often do not truly mimic the peptide structure related to its pharmacophore and often when examined carefully have different biological activity profiles.

8.2 GENERAL FEATURES OF PEPTIDE STRUCTURE

Peptides are short biopolymers naturally synthesized from α-amino acids, linked by an amide bond, also called a peptide bond. The features of a peptide structure or backbone, encompasses critical information about the backbone torsional angles ϕ, ψ, and ω (Figure 8.1) were first investigated extensively by Ramachandran and coworkers, and χ angles were carefully examined later.

The backbone torsional angles dictate the conformational space that is occupied by the peptide and are the most decisive feature of a bioactive peptide that exhibits affinity and biological activity toward receptors such as G protein-coupled receptors (GPCRs), proteases, or other proteins. In addition, side chain torsional angles (χ) play crucial roles in protein–protein folding and peptide ligand–receptor interactions.

Synthetically designed α-helices, β-sheets, β-turns, and modifications of them, are important considerations for the design of a potent ligand/peptide. Mimicking or replacing atoms or bonds can introduce conformational constraints on the peptide structure, which can greatly affect the biological activity. Tuning such modifications forms the central theme for designing potent ligands for research and development and for designing peptide-based drugs.

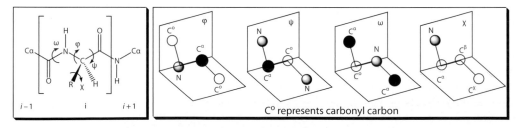

FIGURE 8.1 Definitions of peptide backbone (ϕ, ψ, and ω) and Chi (χ) conformations.

8.3 DESIGN CONSIDERATIONS

The paradigm that underlies biological active peptides and peptidomimetics drug design can broadly be considered as ligand-based drug design and receptor/acceptor biological activity-based drug design (Figure 8.2).

For GPCRs, cytokine receptors, and the like, endogenous ligands have affinity for the extracellular region of the receptor, binding to a specific region of the receptor. For GPCRs, a ligand whose binding causes a measurable increase in basal activity of second messengers is defined as an agonist. A ligand that interacts with the receptor in the same binding pocket and causes no

Ligand-based design

Ac-Ser-Tyr-Ser-Met-Glu-His-Phe-Ar-Trp-Gly-Lys-Pro-Val-NH₂ (α-MSH)

Based on ligand similarity
• Truncation
• Amino acid scan
• Scaffold, fragment,
  chemotype, pharmacophore
• Conformational constraint
• β-Turn peptidomimetics
• 3D structure of ligand (NMR, x-ray)
• Computational-aided drug design

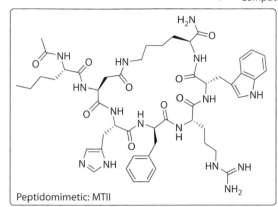

Peptidomimetic: MTII

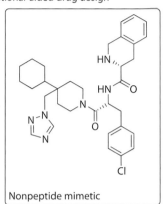

Nonpeptide mimetic

Receptor/acceptor-based design

Based on the similarity of receptor/acceptor
• Receptor class (e.g., GPCR); overall sequence homology (phylogenetic tree)
• Multiple site directed mutagenesis
• Domain shift or chimeric receptors
• 3D structure of binding site
• Cell functional response (signaling pathway)

FIGURE 8.2 Design of biologically active peptides and peptidomimetics.

change in the basal activity of the second messenger functions is defined as a neutral competitive antagonist. A ligand that interacts with the receptor in the same binding pocket and causes the biological activity below the basal level is defined as inverse-agonist. Agonists and antagonists are of vital significance in drug design for targeting specific diseases. For example, a potent and selective agonist toward human melanocortin receptor 4 (hMC4R) may be important for the treatment of obesity. Development of antagonists may be critical in the areas of addiction and tolerance, which are generally induced by prolonged use of opiates. Thus, designing of agonists, antagonists, and inverse-agonist is of utmost importance depending on the targets. Such design can be based on two broad categories; (1) ligand-based; and (2) receptor/acceptor-based design (Figure 8.2). Although design of agonists and antagonists share some common strategies, at some point they must diverge due to their different biological effects. We will discuss here some common approaches toward the design of biological active peptides.

8.3.1 Ligand-Based Design

The ligand is a natural peptide whose structure/sequence is known (a natural endogenous peptide hormone or neurotransmitter) or discovered by combinatorial chemistry, biology, or through proteomic strategies/identification.

8.3.1.1   Identifying Pharmacophore Elements

Identifying the pharmacophore elements plays a crucial role to develop a potent and optimized design of peptides. (1) "Truncating" one amino acid at a time from the amino and/or carboxy termini can lead to a minimum active fragment or sequence responsible for biological activity. (2) "Amino acid scans" can reveal the binding affinities and activities of a primary ligand. These scans help to deduce the importance of the side chain groups required for biological activity that is responsible for molecular recognition and signal transduction. Amino acid scans in peptide or protein design include the following. (a) *Alanine scans*: For a given peptide each amino acid is replaced by an L-alanine to evaluate the relative significance of the side chain group in the binding and biological activity of the peptide. (b) *D-Amino acid scans*: Each L-amino acid is replaced by its corresponding D-amino acid. Advantages of D-amino acid offer the stability of certain reverse (hairpin) turns or destabilization of α-helices, thus providing insights about the chirality of the amino acids in the interested peptide sequence. "D-alanine scans" offer the dual-advantage of the aforementioned two scans thus disclosing the crucial role of chirality and side chain group at the same time. (c) *Proline scans*: Replacing a given amino acid by proline, or other *N*-alkylated amino acids provide specific insights into the importance of backbone conformations. Many GPCRs recognize β-turn structural type of ligands and a judicious replacement by proline can, for example, induce turns and give rise to conformers that may be crucial toward design of a ligand. (d) *Bulky amino acid scans*: Bulky amino acids and aromatic amino acids can play important roles in peptide ligand binding with receptors/acceptors because of their size or hydrophobicity. Often bulky amino acids will cause hindrance of binding of the peptide to the receptor/acceptor binding site, or they can show vast differences in binding to one receptor subtype over another to enhance selectivity for a particular receptor subtype. (3) *Cyclic scans*: As discussed earlier in the global restriction section, side-chain-to-side-chain, side-chain-to-backbone, C-terminal-to-N-terminal, backbone-to-backbone, and other combinations are employed to stabilize or favor segments of the peptide to adopt a global conformation. Varying types of cyclized moieties and ring sizes are adapted to explore functional and biological changes. Cyclization of peptides is often used to bias the peptide α-helix, β-turn, and other hairpin type conformations that may be crucial for the biological activity. (4) *Other scans*: Amide bond replacement scans and aza scans (aza peptides adopt β-turn conformations) are additional methods used for the design of peptides or peptidomimetics.

8.3.1.2   Conformation of the Pharmacophore

Most peptides are highly flexible in a conformational manner in aqueous solution, but upon interacting with another biologically relevant molecule they adapt a preferred conformation. Thus, the reduction of conformational freedom may eventually lead to insights regarding the receptor/acceptor-bound conformation, and can also result in selective interaction of a ligand with a receptor. Conformationally constrained peptides can provide crucial information about biologically active conformations. A major goal of using conformational constraints is to determine which peptide conformation is required for binding to the receptor. Conformational constraint of flexible bioactive peptides can significantly improve potency, selectivity, stability, and bioavailability compared with endogenous peptides. The determination of a biologically active conformation of peptide is a tedious process. However, general strategies have been developed and tested in many laboratories. In many cases, the relevant conformation would be one from the major low energy secondary structures such as α-helix, β-sheet, a reverse (β-turns or γ-turns) or extended structures.

Turns are important conformational motifs of peptides and proteins, besides α-helix and β-sheets. Reverse turns contain a diverse group of structures with well-defined three-dimensional (3D) orientation of amino acid side chains. Turns represent the most important subgroup. A β-turn is formed from four amino acids and is stabilized by a hydrogen bond between the carbonyl group of the first amino acid residue and the amino group of the fourth amino acid residue (Figure 8.3).

The correlation of biological activity with peptide conformation provides useful information about the best fit to the corresponding finding region of the receptor. Once the primary structure of

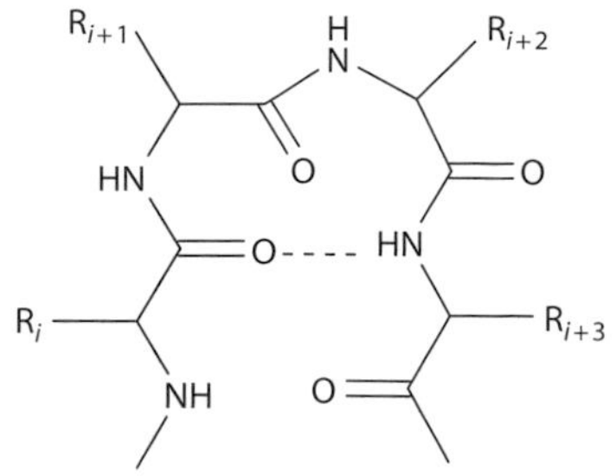

FIGURE 8.3 General structure of a β-turn.

a biologically active peptide has been identified, the next step is to determine the key amino acid residues that are involved in the interaction with a receptor/acceptor via alanine scan, D-amino acid scan, etc., and evaluation of the truncated sequences of the biologically active peptide as described. Once the minimal peptide sequence and key amino acid residues have been determined, the next step involves further modification of peptide conformation. There are two general strategies to constrain the peptide conformations. The flexibility of a peptide chain can be restricted either by global or local constraints.

(1) *Local conformational constraints.* Local conformational constraints can often provide important insights into the structural basis of agonist, antagonist, and inverse-agonist biological activity. The most informative local conformational constraints are those that constrain the backbone ϕ, ψ, and ω torsional angles. Local constraints can be achieved by introduction of unnatural α- or β-substituted amino acids, or modification of amino acid side chains, and/or modification of the peptide backbone.

(2) *The modification of amino acid side chains.* If the conformational flexibility of the side chain groups of key pharmacophore are restricted to varying degrees in a bioactive peptide, important insights into their biologically active 3D topography can be obtained. Usually, the side chain conformation can be controlled in several ways. One general approach is to introduce an alkyl group at the β-position or on the 3′ and/or 5′ position of the aromatic ring of an aromatic amino acid residue. These kinds of modifications can constrain χ^1 and χ^2 angles; on the other hand, they generally do not perturb the backbone conformation drastically, and still allow the peptides to have some degree of flexibility. In a similar manner, substitution on the aromatic ring of an aromatic amino acid in the 3′ and/or 5′ positions will limit the conformational flexibility of a peptide to varying degrees depending on the nature of substitution. Furthermore, the introduction of alkyl groups, halogens, or other functional groups can enhance the lipophilicity or other chemical properties and thus help the peptide bind to receptors and/or cross membrane barriers. Incorporation of these highly constrained amino acids into peptides and studies of such peptidomimetics have provided a valuable approach to probe the stereochemical requirements of binding pharmacophore for recognition of receptors, and sometimes such changes alone can lead to completely different biological activities. Some examples from the Hruby research group are given in Figure 8.4 as illustration of the many possibilities. Among natural amino acids, proline is unique with a constrained cyclic system and substituted versions of this amino acid can be used as a semirigid template in design of conformationally constrained peptidomimetics.

(3) *The modification of the peptide backbone.* Another strategy in the design of peptide drugs is the peptide backbone modifications, which generally refer to the isosteric or isoelectronic exchange of NHCO units in the peptide chain or introduction of additional groups. Some of the most frequent modifications to the peptide backbone are listed in Figure 8.5.

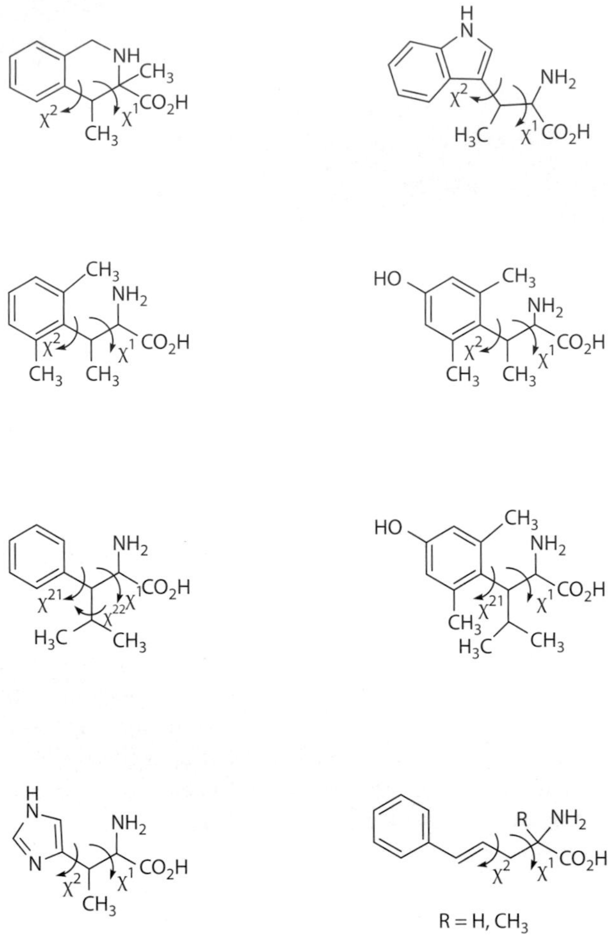

FIGURE 8.4 χ-Constrained amino acids synthesized in the Hruby group. (From Kazmierski, W.M. et al., *J. Org. Chem.* 22, 231, 1994; Boteju, L. et al., *Tetrahedron* 50, 2391, 1994; Xiang, L. et al., *Tetrahedron*, 6, 83, 1995; Qian, X. et al., *Tetrahedron*, 51, 1033, 1995; Liao, S. and Hruby, V.J., *Tetrahedron Lett.*, 37, 1563, 1993; Han, Y. et al., *Tetrahedron Lett.*, 38, 5135, 1997; Wang, S. et al., *Tetrahedron Lett.*, 41, 1307, 2000; Qiu, W. et al., *Tetrahedron*, 56, 2577, 2000.)

The modification to the peptide backbone can also serve to introduce local backbone constraints. For example, N-alkylation restricts the ϕ torsional angle but eliminates the hydrogen bonding capability of the amide bond. *N*-Methyl amino acids have been incorporated into bioactive mimetics of opioid peptides, bradykinin, thyrotropin releasing hormone (THR), angiotensin II, and cholecystokinin (CCK), and many others. α-Methyl (and α-alkyl or α-anyl) substituted amino acids often can induce or stabilize particular turn and helical structures.

(4) *Globally constrained peptide conformation.* Cyclization of a peptide is another general approach to constrain the conformation by limiting the flexibility of the peptide. In this approach, the amino acid side chain groups and backbone moieties that are unimportant in biological activity are chosen as the sites to construct a cyclic structure. The cyclization can be formed between side chains through different types of bonds, such as disulfides, lactams, and thioethers (Figure 8.6). Other kinds of cyclic constraint are also possible between side chains and C- or N-termini or between side chain and backbone nitrogens (Figure. 8.7). By making an appropriate covalent bond between two groups one can stabilize a particular conformer and get a relatively rigid structure that adopts a particular secondary structure.

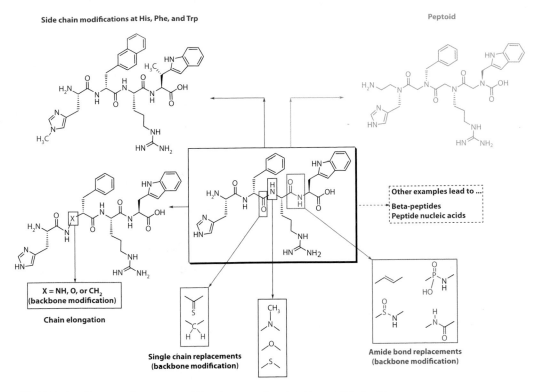

FIGURE 8.5 The most frequent modifications employed to the peptide backbone.

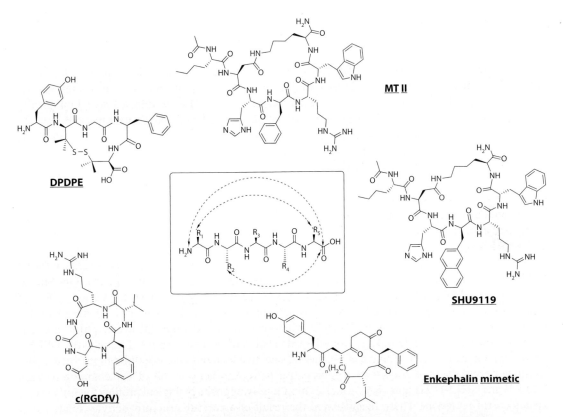

FIGURE 8.6 Examples of converting biologically active linear peptides to potent cyclic peptides.

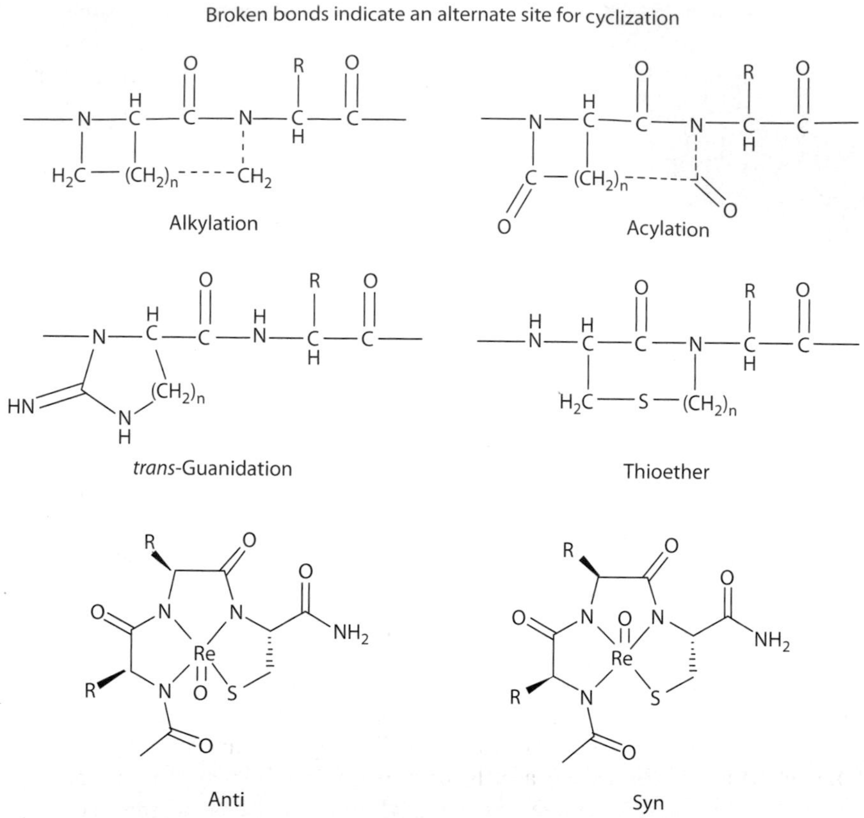

Broken bonds indicate an alternate site for cyclization

Alkylation

Acylation

trans-Guanidation

Thioether

Anti

Syn

FIGURE 8.7 Other cyclic constrains.

By restricting the flexibility the number of conformations of linear peptides can be reduced. To reduce these dynamic degrees of freedom, cyclization is an excellent protocol that has revolutionized the discipline of peptide chemistry. Such global modifications within a linear biologically active peptide offer essential advantages such as: (a) increasing agonist and antagonist potency; (b) reducing proteolytic degradation; (c) increasing receptor selectivity; (d) enhancing bioavailability; and (e) providing conformational insight for receptor/acceptor binding and drug design.

Global constraints can be established by introducing a covalent bond between any two given positions along the peptide chain. The classic examples include, lactam bridges, disulfide bonds, or by introduction of spacers like a succinyl [-CO-(CH₂)₂-CO-] moiety. These linkages or bridges can be broadly categorized into four distinct ways: (1) side chain to side chain; (2) side chain to N- or C-terminus; (3) N- to C-terminal; and (4) backbone residue to backbone residue (for instance, N or C$_\alpha$ of the backbone).

A classic and natural example is oxytocin, which has a disulfide bridge that is necessary for its full biological activity. Among synthetic constrained peptides, DPDPE (H-Tyr-c[D-Pen-Gly-Phe-D-Pen]-OH) and MT-II (Ac-Nle-c[Asp-His-D-Phe-Arg-Trp-Lys]-NH₂) are well-known examples of side chain to side chain cyclized bioactive superagonists toward the δ-opioid and melanocortin receptors-respectively (Figure 8.6). DPDPE, which is derived from the linear enkephlin pharmacophore, bears the disulfide linkage from the side chains of D-penicillamine. Ligands like DPDPE are of high interest for development of potent and selective ligands that can exhibit high efficacy and facilitate development of drugs toward neuropathic pain. Another example of side chain cyclization is MT-II and SHU9119 (Figure 8.6), which exhibit totally opposite biological profiles toward certain melanocortin receptors. These ligands were the result of a careful structure–activity relationship

(SAR) studies of the linear ligands, starting from the natural α-melanocyte stimulating hormone (α-MSH), which led first to a potent and stable linear peptide NDP-α-MSH and then to the cyclized ligands (Figure 8.6, MT-II and SHU9119). Melanocortin receptors are involved in many critical physiological actions such as pigmentation, feeding behavior, and sexual behavior. As shown in Figure 8.6, MT-II and SHU9119, bear the lactam ring using the side chain groups of lysine and aspartic acid. SHU9119 was derived from superagonist MT-II, by the replacement of the amino acid D-phenylalanine to D-2′-naphthylalanine. This introduction of four carbons (a local modification) on the side chain of the phenyl ring led to SHU9119, a superpotent antagonist at the melanocortin 3 and 4 receptors, which demonstrates the drastic change in biological action that a small change in structure can induce. Cyclic RGD analogues are a class of $\alpha_v\beta^3$ antagonists and are of particular interest in human tumor metastasis and in angiogenesis. Kessler's group has demonstrated an excellent example of N- to C-terminal cyclized bioactive peptides. Incorporation of D-amino acids to induce a β-turn structure within the cyclized moiety has resulted in very potent $\alpha_v\beta^3$ antagonists.

Examples of side chain to the C-termini are well demonstrated by Schiller's group in exploring opiate receptor selectivity by altering the conformational restriction on the modified enkephalin sequences. The series of cyclic enkephalins that were synthesized exhibited the subtle variation in conformational restriction thus disclosing the conformational space exhibited by the opiate receptors. Many other constraining moieties exist including alkylation, *trans*-guanidation, acylations, thioether bridges, and metal complexed peptides, are not discussed here (Figure 8.7).

8.3.1.3 β-Turn Peptidomimetics and Other Peptide Mimetics

β-Turn structures are an important class of peptides and are recognized by many GPCR's for their affinity and potency. Besides the aforementioned strategies of globally restricting dynamics of a peptide, enhancing β-turns also provides an excellent opportunity for design of novel bicyclic conformations. A β-turn can be viewed as a 10-membered ring formed from four amino acids that are stabilized by a hydrogen bond between the carbonyl group of the first amino acid residue and the amine-hydrogen of the fourth amino acid residue. When such interactions are stabilized by inducing greater constraints on the second and third amino residues can lead to novel templates for β-turn peptidomimetics.

The development of peptide mimetics can be a critical drug discovery strategy. The evolution of peptide ligands to small molecule mimetics has been a major goal in the field, with several notable successes. Peptides are ideal drug leads, but often their stability to proteolytic enzymes and their bioavailability need to be enhanced. This can be done by incorporating various computational tools for molecular design, proprietary scaffolds, conformational constraints, conformation–activity analysis, and lead optimization strategies to design mimetics that will retain the desired biological properties of the peptide lead, but are metabolically stable, have appropriate diversity, and can be tailored to have desired drug-like pharmaceutical properties. The peptide mimetic design strategies are summarized in Figure 8.8. The interaction of a peptide with a receptor/acceptor may occur via a direct binding of a linear sequence of the peptide in one of the conformations accessible to a particular peptide. Linear peptides are the mode of recognition of many peptide ligands for receptors/acceptors. In other cases, turn structures or cyclic structures are important, as in MTII, where the -His-D-Phe-Arg-Trp- peptide sequence is oriented in a turn motif. In larger peptides or proteins, the folding of the peptide may bring groups that are distant in a sequence into close spatial arrangements, and thus the "binding motif" is a 3D arrangement of the peptide side chain groups involved in recognition. In other cases, the recognition may be along a face of a β sheet or α helical sequence. The linear, folded, and 3D presentations can all be translated from peptides to peptide mimetics.

8.3.1.4 Applying Computational and Biophysical Studies for Design

Rational design should be and is currently a central theme for drug discovery. Experimental and theoretical developments in peptide chemistry and biophysical methods, in conjunction with new biological methods have offered a wide variety of new design, synthesis, and analysis methods for

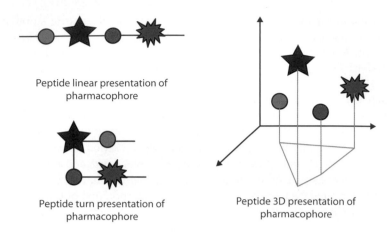

Peptide linear presentation of
pharmacophore

Peptide turn presentation of
pharmacophore

Peptide 3D presentation of
pharmacophore

FIGURE 8.8 Peptide structures to mimic in peptide mimetic design.

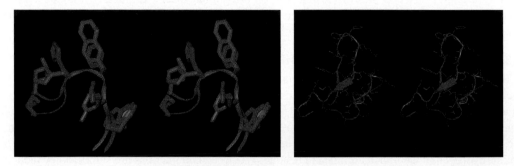

FIGURE 8.9 NMR structures of superimposed MTII (green) and SHU9119 (yellow) (left) and NMR structure of AGRP (right).

peptides with powerful biological properties. The use of computational methods in conjunction with molecular modeling is a powerful tool in design considerations, with the caveat that for novel structures, current force fields might not give an accurate picture of the structure or energetics of the system. As an example of melanocortin system, along with the nuclear magnetic resonance (NMR) structure of Agouti-related protein (AGRP, an endogenous antagonist of hMC3R and hMC4R) and MTII (superagonist of hMC4R), the 3D pharmacophore of human melanotropin has been partially deciphered (Figure 8.9). The primary structure of both MTII and AGRP are quite different, but from their NMR it can be found that they both share the same β-turn like structure. These accomplishments combined with the computational chemistry can be very crucial for designing novel selective agonist and antagonist compounds.

There is one computational approach for mapping molecular interaction space without knowledge of the 3D structure, known as proteochemometrics. It is an extension of traditional ligand-based 3D QSAR approaches. It exploits affinity data for a series of diverse organic ligands binding to different receptor subtypes, providing proteochemometric analysis, and insights into ligand recognition.

8.3.2 DESIGN BASED ON THE BIOLOGICAL FUNCTION

The paradigm that underlies the chemogenomics approaches to ligand design can be stated as "similar receptor/acceptor binds similar ligand." This implies that for a novel receptor/acceptor, the information obtained from a known ligand for a related receptor/acceptor, can serve as a starting point for ligand design. Proteins that belong to the same target family or class (e.g., the family of GPCRs) can be considered as similar. Although all GPCRs have seven transmembrane helices and translate an extracellular signal into an intracellular response mediated by G proteins, there is a great diversity of ligands for GPCRs. Moreover, there is currently insufficient reliable 3D structural information available regarding the GPCRs. Thus, insights into ligand–receptor interactions have to rely on molecular recognition experiments. Multiple site-directed mutagenesis have been used to identify the ligand-binding site as well as the biological function site. Recently, domain shift or chimeric receptors also have been used to provide information about the binding domain for a ligand.

8.4 MODELING AND DOCKING

In this chapter, we have emphasized that a peptide or endogenous ligand in a biological system upon interaction with an enzyme or GPCR leads to specific biological functions. Thus, knowledge about such receptors or enzymes and the like in terms of their binding site for a ligand, and an understanding of the bound ligand–receptor complex is of utmost importance. A critical approach to understand receptor structure is its orientation in 3D space. X-ray crystal structures provide 3D conformation, but may be misleading in terms of function. Incorporation of these x-ray coordinates in to a computer-aided examination of a specific function in 3D space is being pursued. This allows to further explore the region/site or the surrounding 3D space occupied by the key amino acids of the protein (where the potent ligand has an affinity) so as to better understand the biological actions.

Peptide analogues that are sufficiently constrained, whose conformational and dynamic properties are known from multidimensional NMR and other spectroscopic methods, in combination with modern molecular mechanics methods and docking experiments can greatly aid peptide ligand development. Exploring the accessible conformational space on the docking site of the receptor/ acceptor serves as the starting point for further design of analogues to develop conformation–biological activity relationships. Such molecular modeling studies provide a useful tool toward the generation of potent peptide or peptidomimetics ligand structures. Docking for the enzyme-based proteins are widely available because the x-ray structures are available in many cases. Unfortunately, the same is not true for GPCRs and many other proteins, especially integral membrane proteins. In these cases, homology modeling with the few known structures in these classes has been somewhat successful, but further developments are still required.

8.5 CONCLUSIONS

The design of peptides and peptidomimetics is often the first step of drug discovery. To obtain a useful peptide or peptide mimetics drug requires a multidisciplinary approach including chemical biology, biophysics, and biological approaches. It is necessary to be aware of all available information and design methods in order to begin such a drug discovery program. Several chemical, biophysical, and bioinformatics approaches have been discussed and need to be evaluated in conjunction with the results from the best possible and appropriate *in vitro* and *in vivo* assays. This multidisciplinary approach will be essential for the development of rational approaches to peptide-based drug design in the future.

FURTHER READINGS

Boteju, L.W., Wegner, K., Qing, X., and Hruby, V.J. (1994) Assymetric synthesis of unusual amino acids: Synthesis of optically pure isomers of N-indole-(2-mesitylenesulphonyl)-β-methyltryptophan. *Tetrahedron* 50:2391.
Cowell, S.M., Balse-Srinivasan, P.M., Ahn, J.M., and Hruby, V.J. (2003) Design and synthesis of peptide antagonists and inverse agonists for G protein coupled receptors. In *Methods in Enzymology*, eds. R. Iyengar and J.D. Hildesbrandt. New York: Academic Press.

Han, Y., Liao, S., Qiu, W., Cai, C., and Hruby, V.J. (1997) Total assymetric syntheses of highly constrained amino acids β-isopropyl-2′,6′-dimethyl-tyrosines. *Tetrahedron Lett.* 38:5135.

Hruby, V.J. (2002) Design peptide receptor agonists and antagonists. *Nat. Rev. Drug Discov.* 1:847–858.

Hruby, V.J., Al-Obeidi, F., and Kazmierski, W.M. (1990) Emerging approaches in the molecular design of receptor selective peptide ligands: Conformational, topographical and dynamic considerations. *Biochem. J.* 268:249–262.

Kazmierski, W.M., Urbanczyk-Lipkowski, Z., and Hruby, V.J. (1994) New amino acids for the topographical control of peptide conformation: Synthesis of all isomers of α, β-dimethylphenylalanine and α, β-dimethyl-1, 2, 3, 4 tetrahydroisoquinoline-3-carboxylic acid of high optical purity. *J. Org. Chem.* 59:1789.

Liao, S. and Hruby, V.J. (1996) Assymetric synthesis of optically pure β-isopropylphenylalanine: A new β-branched unusual amino acid. *Tetrahedron Lett.* 37:1563.

Marshall, G.R. (1993) A hierarchical approach to peptidomimetic design. *Tetrahedron* 49:3547–3558.

Olson, G.L., Bolin, D.R., and Bonner, M.D. et al. (1992) Concepts and progress in the development of peptide mimetics. *J. Med. Chem.* 36:3039–3049.

Qian, X., Russel, K.C., Boteju, L.W., and Hruby, V.J. (1995) Stereoselective total synthesis of topographically constrained designer amino acids: 2′, 6′-dimethyl-β-methyltyrosines. *Tetrahedron* 51:1033.

Qiu, W., Soloshonok, V.A., Cai, C., and Hruby, V.J. (2000) Convenient, large-scale assymmetry synthesis of enantiomerically pure transcinnamylglycine and α-alanine. *Tetrahedron* 56:2577.

Sawyer, T.K. (1997) Peptidomimetic and non-peptide drug discovery: Impact of structure-based design. In *Structural Based Drug Design*, ed. P. Veerapandian, pp. 559–634. New York: Marcel Dekker, Inc.

Wang, S., Tang, X.-J., and Hruby, V.J. (2000) First stereoselective synthesis of an optically pure β-substituted histidine: (2S, 3S) β-methylhistidine. *Tetrahedron Lett.* 41:1307.

Wiley, R.A. and Rich, D.H. (1993) Peptidomimetics derived from natural products. *Med. Res. Rev.* 13:327–384.

Xiang, L., Wu, H., and Hruby, V.J. (1995) Stereoselective synthesis of all individual isomers of β-methyl-2′, 6′-dimethylphenylalanine. *Tetrahedron* 6:83.

9 Prodrugs: Design and Development

Anders Buur and Niels Mørk

CONTENTS

9.1 INTRODUCTION

Although it is well recognized that a clinically useful drug must possess balanced characteristics in terms of efficacy and safety as well as drugability, traditional drug discovery and design efforts have not focused sufficiently on the latter. Consequently, challenges concerning oral bioavailability, pharmacokinetics, and pharmaceutical technical properties have only been dealt with during the development phase, which in many cases has proven to be difficult.

The inappropriateness of this old paradigm to drug design has become evident to the pharmaceutical industry. The implementation of high-throughput screening and combinatorial chemistry have led to the identification of numerous lead compounds with excellent pharmacological properties in terms of affinity and specificity but at the same time inadequate pharmaceutical, biopharmaceutical, and absorption, distribution, metabolism, and excretion (ADME) properties such as low aqueous solubility or high metabolic liability, thus resulting in a significant increase in the development time and the cost or even the closure of projects.

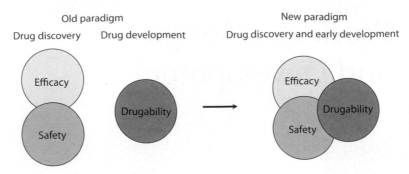

FIGURE 9.1 Shift in drug discovery/drug development paradigm.

In the new paradigm, an integrated approach to drug design is introduced that takes into account optimization of lead compounds based upon both pharmacological (affinity and specificity), ADME (clearance, metabolic enzymes, and protein binding) and biopharmaceutical (lipophilicity, cell membrane permeability), as well as pharmaceutical (solubility, physical and chemical stability) properties. This is illustrated in Figure 9.1.

However, with the integrated approach it may still prove to be a Herculean task to compress all the desired characteristics of efficacy, safety, and drugability into a single molecule. One potential solution to this challenge is the utilization of the prodrug principle as part of drug design and discovery.

9.2 THE PRODRUG PRINCIPLE

The term "prodrug" or "proagent" was introduced by A. Albert in 1958 to describe compounds that undergo biotransformation prior to eliciting a pharmacological effect.

As shown in Figure 9.2, the formation of a prodrug provides a transient change of physicochemical and biological properties thereby altering or eliminating undesirable properties of the parent drug molecule. This prodrug to drug biotransformation may take place before absorption, during absorption, immediately after absorption, or at a specific site of action.

The conversion of prodrugs to the active parent drug molecule can take place through various reactions such as hydrolytic cleavage either as a spontaneous or enzyme-mediated reaction.

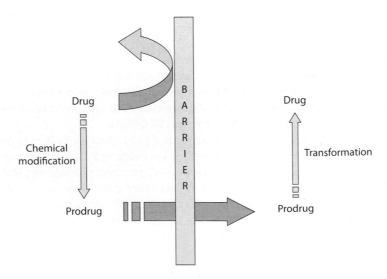

FIGURE 9.2 The prodrug principle.

The prodrug approach has been successfully utilized to overcome significant barriers for many drugs aiming

1. To improve the oral bioavailability by increasing aqueous solubility, by increasing biomembrane permeability, and/or by improving (metabolic) stability
2. To increase the duration of pharmacological action
3. To decrease toxicity or adverse reactions
4. To obtain drug targeting using site-specific biotransformation or site specific transporters

The prodrug approach has been used over a span of more than three decades mainly as a means to address drugability problems with drugs already introduced to the market. The aim has primarily been to develop improved versions of such drugs or as a rescue for compounds with drugability problems in late-stage development; and several drugs have been developed and are used in clinical practice as prodrugs (e.g., pivampicillin, enalapril, and vanaciclovir). In modern drug design, prodrug strategies are considered part of the drug discovery effort aiming to expand the chemical space of drugable molecules and, thus, it can be expected that many future drugs will appear as prodrugs.

The purpose of this chapter is to introduce the reader to the prodrug principle and discuss important elements in prodrug strategies within drug discovery and development. Finally, we will provide examples of achievements offered by prodrugs.

9.3 DESIGN OF PRODRUGS—CHEMICAL CONSIDERATIONS

Prodrugs are most often taken into consideration after identification of a pharmacological active lead compound or structure. Consequently, the first step in prodrug design is identification of functional groups such as hydroxyl, carboxyl, carbonyl, amide, NH-acidic, and/or amino groups in the active compound that are available for chemical derivatization. Second, potentially bioreversible derivatives such as esters, *N*-acyl, *N*-hydroxymethyl, or *N*-acyloxyalkyl derivatives, *N*-Mannich bases, enaminones, and lactones may be synthesized and subjected to further testing. To this end, the most important requirement for a prodrug is its ability to adequately regenerate the active drug *in vivo*. In addition to this, it must be chemically stable in the bulk form and together with common excipients used in drug formulation leading to an acceptable shelf-life and, finally, the toxicity of the promoiety and the prodrug itself must be acceptable.

The necessary conversion or activation of the prodrug in the body to the active drug molecule can take place by both enzymatic- and nonenzymatic-mediated reactions, which is discussed in the following text.

9.3.1 Prodrugs Transformed by Enzymatic Reactions

Prodrugs can be designed to target specific enzymes in the body. This is based upon knowledge and considerations on enzyme activity, specificity, tissue distribution, and abundance and can be utilized to obtain the desired ability of a prodrug to overcome one or more barriers to development of the active drug molecule.

The most common prodrugs are those requiring hydrolytic cleavage mediated by enzymatic catalysis. Drugs containing hydroxyl, carboxyl, or amino functional groups can be converted into prodrug esters or amides from which the active forms are readily regenerated by hydrolytic enzymes such as esterases (see Section 9.4.1.2), amidases, peptidases, or phosphatases (see Section 9.4.1.1) (Figure 9.3).

Less often prodrugs are designed to undergo reductive or oxidative processes mediated by enzymes such as cytochrome P450, monoamine oxidases, azoreductases (see Section 9.4.3), or nitroreductases. A novel prodrug principle (HepDirect™) has recently been introduced for highly site-specific delivery of phosphate drugs to the liver. These prodrugs consist of cyclic 1,3-propanyl esters

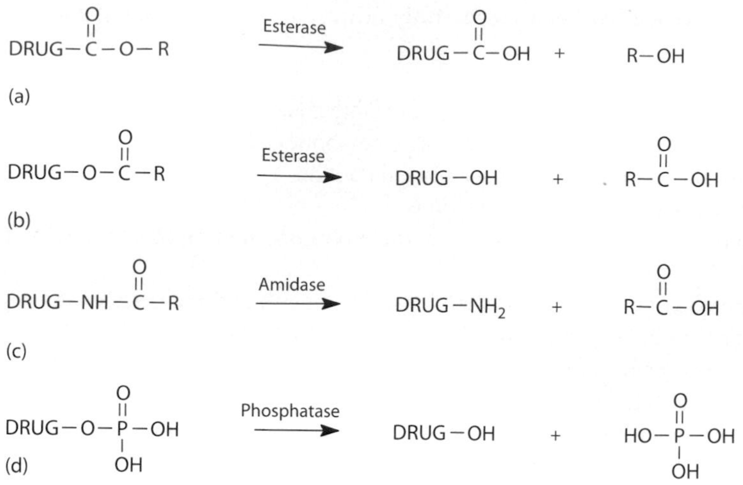

FIGURE 9.3 Schematic examples of prodrugs designed to undergo enzymatic-mediated hydrolysis.

of phosphates designed to undergo enzyme-catalyzed oxidative conversion into active phosphates specifically mediated by CYP3A4 enzymes present in the liver.

A challenge in the design of prodrugs susceptible to enzymatic conversion is the limited availability of predictive *in vitro* and *in vivo* models for selection and optimization of the prodrug candidate. *In vitro* assays in which reaction kinetics is studied in the presence of human or animal material such as blood, serum, plasma, and intestinal or liver tissue can provide qualitative information on drug conversion whereas to a much lesser extent they offer quantitative predictions on the rate and extent to which biotransformation takes place *in vivo*. *In vivo* models using experimental animals such as mice, rats, dogs, or monkeys can indeed provide *in vivo* relevant information but suffers from the only sparse knowledge available on species differences in terms of enzyme abundance, distribution, activity, and specificity compared to the human *in vivo* situation. Thus, there is a risk that prodrugs dependent on bioconversion mediated by enzymes may show high interindividual variability due to variability in enzyme levels and activity between individuals.

9.3.2 PRODRUGS TRANSFORMED BY SPONTANEOUS REACTIONS

As an alternative to the enzymatic-mediated bioconversion, prodrugs may be designed to undergo spontaneous (or chemical) transformation dictated by the physicochemical environment such as the pH in various parts of the human body.

Some important examples of prodrugs biotransformed by nonenzymatic-mediated reactions are hydroxymethyl derivatives of NH-acidic drug molecules such as 5-fluorouracil, phenytoin, *N*-Mannich bases derived from drugs containing amino or amide functions (e.g., tetracycline, carbamazepine) and ring-opened derivatives of cyclic drugs (e.g., pilocarpine Figure 9.5). The rate of spontaneous transformation of *N*-hydroxymethyl prodrugs of NH-acidic drugs can be predicted from the acidity (pK_a) of the parent drug compound. A linear relationship between the logarithm to the half-life (log $t_{1/2}$, pH 7.4, 37°C) of *N*-hydroxymethyl derivatives and the pK_a of the parent NH-acidic drug has been established allowing easy evaluation to whether a *N*-hydroxymethyl derivative may consist an attractive prodrug principle for a given drug molecule:

$$\log t_{1/2} = 0.77 \times pK_a - 8.34$$

Evaluation of *N*-Mannich bases as potential prodrugs for NH-acidic drugs offers considerable room for the design of appropriate characteristics with respect to lipophilicity, aqueous solubility, and biological half-life ($t_{1/2}$). This is achieved by proper selection of pK_a and substituent pattern (R_1 and R_2 in Figure 9.4b) of the attached amino moiety as the rate of biotransformation is controlled by both the pK_a and the sterical hindrance caused by substituents connected to the amino moiety.

Nonenzymatic catalyzed bioconversion of prodrugs has the advantage that physicochemical properties such as pH in the bloodstream is predictable and show low interindividual variability leading to predictable and reliable conversion of the prodrug *in vivo*.

On the other hand, it should be noted that such prodrugs possess an inherent drawback in that incorporation of a chemically labile function may lead to formulation stability problems and, consequently, to reduced shelf-life. An elegant solution to this problem is the double-prodrug concept where an enzymatic-mediated transformation is combined with a subsequent spontaneous reaction releasing the active parent drug.

An example of this is the protection by acylation of very labile *N*-hydroxyalkyl derivatives of NH-acidic drugs with pK_a values below approximately 11 ($t_{1/2} < 1$ min at 37°C and pH 7.4) as illustrated in Figure 9.6. After enzymatic cleavage of the ester function the highly unstable *N*-hydroxyalkyl intermediate is rapidly converted into the parent HN-acidic drug molecule.

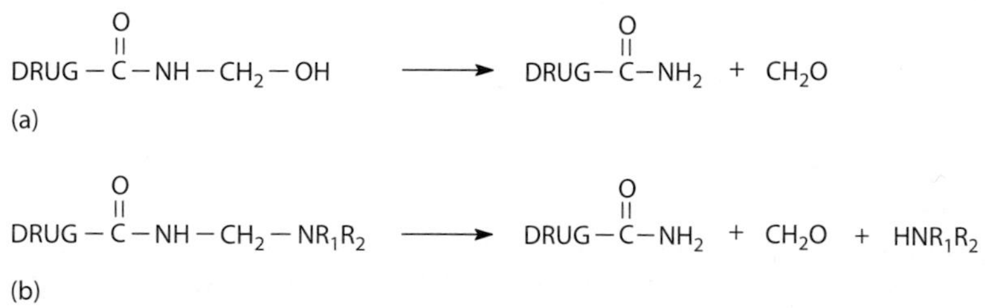

(a)

(b)

FIGURE 9.4 Schematic examples of prodrugs designed to undergo spontaneous nonenzymatic transformation.

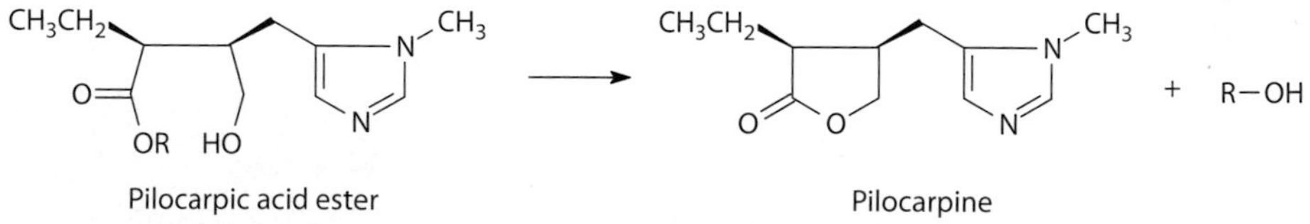

Pilocarpic acid ester

Pilocarpine

FIGURE 9.5 Pilocarpic acid ester prodrug cyclization into pilocarpine.

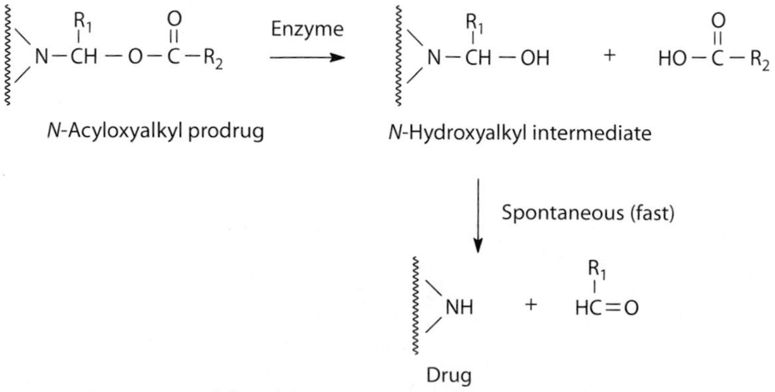

N-Acyloxyalkyl prodrug

N-Hydroxyalkyl intermediate

Spontaneous (fast)

Drug

FIGURE 9.6 The double-prodrug principle.

9.4 DESIGN OF PRODRUGS—APPLICATION OF THE PRODRUG PRINCIPLE

The prodrug principle offers an opportunity for optimization of drug therapy for a variety of reasons. The intention of this section is to illustrate important features of the principle by providing selected examples of achievements by prodrug design and development.

9.4.1 DESIGN OF PRODRUGS WITH IMPROVED BIOAVAILABILITY

The oral bioavailability (F) of a compound is determined by a number of parameters such as aqueous solubility (determines the amount of drug available at the site of absorption), permeability (the ability of the molecule to permeate biological membranes), and stability in the gastrointestinal tract.

The relationship between lipophilicity—as measured by the distribution coefficient (D) between n-octanol and aqueous buffer at physiological pH—and the biomembrane permeability characteristics—as measured by the permeability coefficient (P_{app}) across, e.g., Caco-2 cell monolayers, i.e., the absorption (A) and the oral bioavailability (F), respectively, is shown in Figure 9.7.

It appears that the biomembrane permeability (P_{app}) increases with increasing lipophilicity until a certain point where the permeability is no longer controlled by the rate at which a given molecule permeates the cell layer and from where no further significant improvement can be induced. However, the mass transport that can be obtained (i.e., the flux) and thereby the oral absorption (A) and the bioavailability (F) tend to level off (illustrated by the shaded area in Figure 9.7) as the increase in lipophilicity continues. This is due to the fact that an increase in lipophilicity often dictates a significant decrease in aqueous solubility and thereby limiting the amount of drug available for absorption (i.e., limiting the concentration gradient that can be obtained) and, consequently, decreasing oral absorption and bioavailability. In addition to this, increased lipophilicity, in general, promotes the affinity to proteins such as enzymes and, thus, decreases the metabolic stability of the compounds and thereby increases systemic clearance leading to decreased oral bioavailability.

Thus, increasing the lipophilicity of drug compounds is advantageous in terms of oral bioavailability as long as it is not at the "expense" of aqueous solubility or other important parameters for drug absorption.

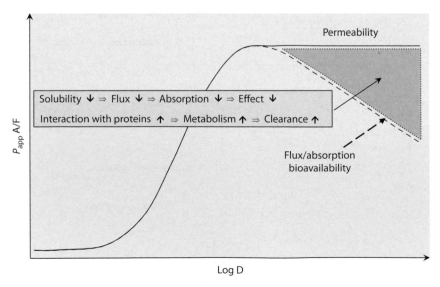

FIGURE 9.7 Relationship between lipophilicity (D) and (1) biomembrane permeability (P_{app}), (2) absorption (A), and (3) oral bioavailability (F).

This is in agreement with Fick's first law of diffusion that describes the rate of diffusion across a (bio)membrane:

$$dQ/dt = P_{app} \times A \times (C - C_o)$$

where

dQ/dt is the mass flux across the membrane
A is the membrane surface area
$C - C_o$ determines the initial concentration gradient between the apical and basolateral side of the epithelium
P_{app} is the permeability coefficient

From this relationship, it follows that drug transport across the epithelium is governed by both aqueous solubility (generating a high initial concentration gradient) and permeability characteristics of the compound.

Thus, optimization of the flux and thereby the absorption potential is a balance between high aqueous solubility and adequate permeability characteristics. This has been successfully obtained by using the prodrug approach.

9.4.1.1 Improved Aqueous Solubility

For compounds suffering from dissolution rate or solubility limited absorption due to low aqueous solubility and/or high therapeutic dose introduction of a hydrophilic moiety may prove beneficial for the oral bioavailability. The most appealing type of prodrugs to overcome solubility problems are the phosphate esters because they are generally chemically stable and at the same time enzymatically labile and, furthermore, because the solubility increase obtained by phosphate esters can be of several orders of magnitude.

The protease inhibitor amprenavir suffers from a combination of low aqueous solubility (0.04 mg/mL) and high dose requirements (1200 mg BID corresponding to eight capsules of the marketed product). Thus, Vertex Pharmaceuticals together with GlaxoSmithKline developed the phosphate ester prodrug fosamprenavir (Figure 9.8) with the aim of increasing the oral bioavailability of the parent drug. Fosamprenavir is rapidly and extensively hydrolyzed by alkaline phosphatases at the brush border of the gastrointestinal tract during or after absorption to yield the parent active drug, amprenavir. Only minimal amounts of intact fosamprenavir reach the systemic circulation. The development of fosamprenavir has reduced the "pill burden" for HIV patients considerably and the prodrug can be administered without any food or water restrictions.

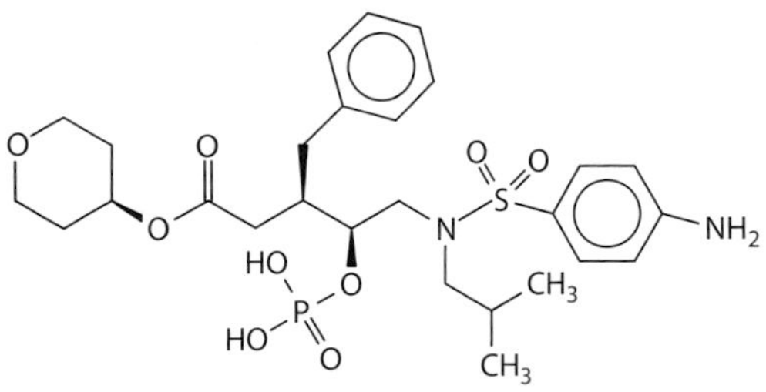

FIGURE 9.8 Fosamprenavir.

9.4.1.2 Improved Biomembrane Permeability (Passive Diffusion)

Due to the large surface area of the gastrointestinal epithelium, transcellular (passive) diffusion is a very important absorption principle for drug molecules. Thus, the ability of a compound to permeate this epithelium is of vital importance for the absorption of the drug. Transcellular diffusion of a compound is governed by its ability to interact with lipophilic cell membranes, and, thus, its lipophilicity (Figure 9.7). This has been utilized in many prodrug strategies adding lipophilic promoieties to hydrophilic drug molecules or lead compounds. A few examples are given in the following text.

Pivampicillin, talampicillin, and bacampicillin are the pivaloyloxymethyl, the phthalidyl, and the ethoxycarbonyloxyethyl esters, respectively, of the parent penicillin ampicillin (Figure 9.9). These esters offer a significant increase in lipophilicity in comparison with the parent ampicillin by blocking the carboxylic function and adding lipophilic groups to the molecule.

As it appears from Figure 9.10, the prodrug strategy for pivmapicillin involves a double ester in which an initial enzymatic-mediated process is followed by a fast and spontaneous reaction. This principle has been successfully used to improve the oral bioavailability of ampicillin from about 30%–40% to essentially complete absorption. Thus, by employing the prodrug principle, unabsorbed drug that may negatively interact with bacteria in the gastrointestinal tract will be significantly limited for the benefits of patients receiving the drug.

A major problem in the use of the β-adrenergic antagonist timolol (Figure 9.11) in the treatment of glaucoma is the high incidence of cardiovascular and respiratory side effects. This is due to absorption of the topically applied drug into the systemic circulation, which again can be attributed to insufficient corneal penetration of the drug. The use of more lipophilic prodrugs has been investigated to improve the corneal penetration of timolol and it was found that O-butyryl timolol, which is 50 times more lipophilic than timolol (Figure 9.11), was absorbed into the aqueous humor

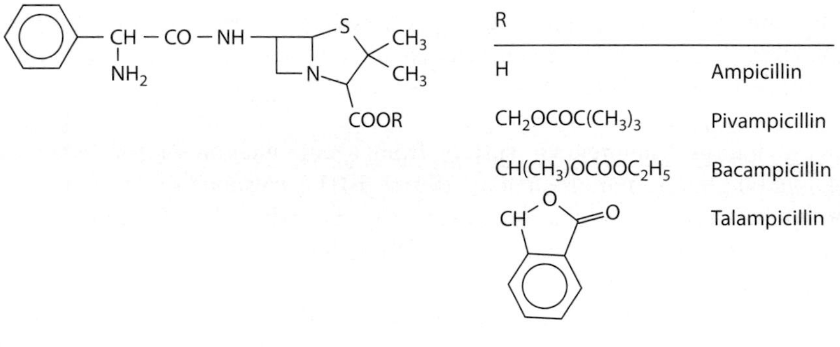

FIGURE 9.9 Ampicillin, pivampicillin, talampicillin, and bacampicillin.

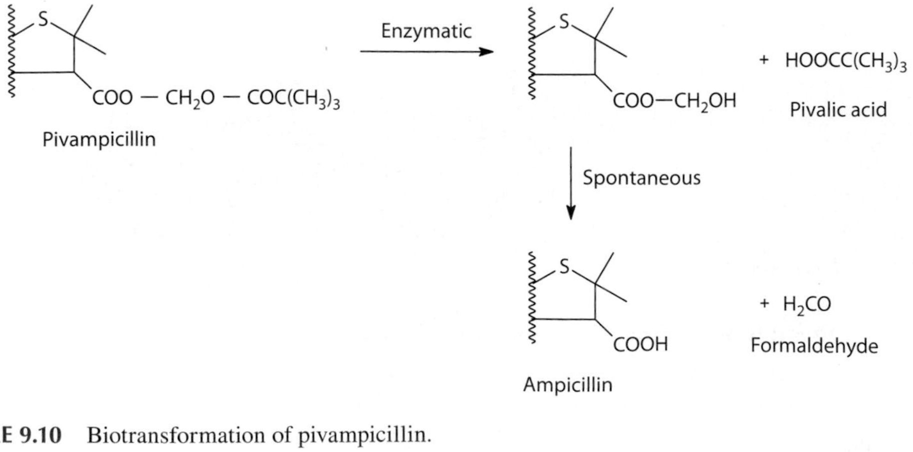

FIGURE 9.10 Biotransformation of pivampicillin.

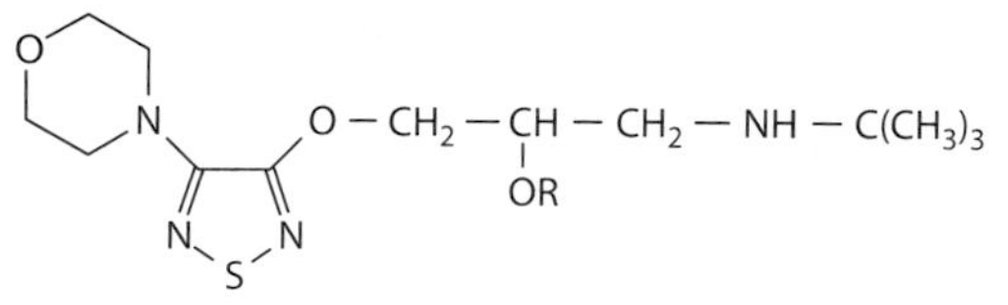

FIGURE 9.11   Timolol and its *O*-alkyl esters.

R	Log D	
—H	−0.04	Timolol
—CO(CH$_2$)$_2$CH$_3$	2.08	*O*-Butyryl
—COCH(C$_2$H$_5$)$_2$	3.26	*O*-2-Ethylbutyryl
—COCH$_2$C(CH$_3$)$_3$	3.09	*O*-3,3-Dimethylbutyryl
—COcC$_3$H$_5$	1.74	*O*-Cyclopropanoyl

four to five times better than timolol, while yielding similar plasma timolol concentrations. In addition to this, it has been shown that the *O*-butyryl timolol prodrug offers extended duration of action in comparison with the parent timolol in an experimental animal model. This is most likely due to the more lipophilic prodrug being held in the ocular tissue compartment, which may act as a depot from which the drug is being slowly released. It should, however, be noted that although the *O*-butyryl timolol ester and other alkyl esters indeed increase corneal absorption they at the same time suffer from chemical instability in aqueous solution. However, the chemical stability could be substantially improved by introducing sterically hindered esters such as the 2-ethylbutyryl, 3,3-dimethylbutyryl, and cyclopropanoyl derivatives.

9.4.1.3   Improved Transporter-Mediated Permeability

Various active transport mechanisms for amino acids, small peptides, monocarboxylic acids, monosaccharides, and nucleosides exist in the human body and it is generally recognized that such mechanisms play a major role in the absorption of many drug molecules. This has been used in prodrug design attempting to provide chemical bioreversible modifications mimicking natural substrates for various active transporters with the aim of improving intestinal absorption of various drugs.

The oral absorption of the antiviral compound gangciclovir (Figure 9.12) is less than 10%. Valgangciclovir, the corresponding L-valine ester of the compound, however, shows a marked increase in oral bioavailability of about 60%. This is mainly ascribed to the transport of the prodrug across the gastrointestinal epithelium, which is mediated by a small peptide active transport mechanism for which gangciclovir is not a substrate.

Another example of facilitating active absorption processes by the use of prodrugs is levodopa. Levodopa (or L-dopa) is a prodrug of the neurotransmitter dopamine and utilizes the L-aromatic amino acid transporter for permeation of the intestine and the blood–brain barrier. The active compound dopamine is regenerated by decarboxylation mediated by dopamine decarboxylase enzymes (Figure 9.13).

9.4.1.4   Increased Stability in the Gastrointestinal Tract

The poor absorption of carbenicillin (Figure 9.14) after oral administration can to a large extent be attributed to fast acid-catalyzed degradation in the gastrointestinal tract. More acid resistant prodrugs such as carindacillin and carfecillin (Figure 9.14) have been shown to significantly increase the bioavailability of cabenicillin.

9.4.2 Design of Prodrugs for Prolonged Drug Action

The utility of prodrugs as a means to achieve prolonged pharmacological action of a drug has been effective mainly in terms of intramuscular (IM) injection preparations and other local administrations. For oral administration, the prodrug principle has not been investigated extensively with the aim to prolong drug action and it is questionable if this approach will offer any benefit to well-established formulation principles such as matrix systems, osmotic systems, and pellet systems.

9.4.2.1 Intramuscular Depot Injections

Prodrugs have been used to prolong the pharmacological effect of drugs by providing a sustained release of the prodrug form to the systemic circulation. The principle has provided long-term (1–4 weeks) delivery of neuroleptic drugs such as haloperidole, flupentixole (Figure 9.16), and fluphenazine by IM preparations of highly lipophilic prodrug derivatives such as decanoate and other long-chain fatty acid esters formulated in oily vehicles. This has offered a significant increase in patient compliance and, thus, treatment of patients suffering from psychiatric disorders. The sustained release over weeks of neuroleptic drugs from IM preparations was initially attributed to the high affinity of the lipophilic prodrugs to the vehicle leading to slow release from the vehicle. Although this explains the prolonged drug presence in the bloodstream to a certain extent, partitioning of the lipophilic prodrug derivatives into the lymphatic system is of equal importance. In the lymph nodes, a slow biotransformation to the parent drug takes place, followed by the release of parent drug to the systemic circulation that will all together play a significant role for the overall release profile.

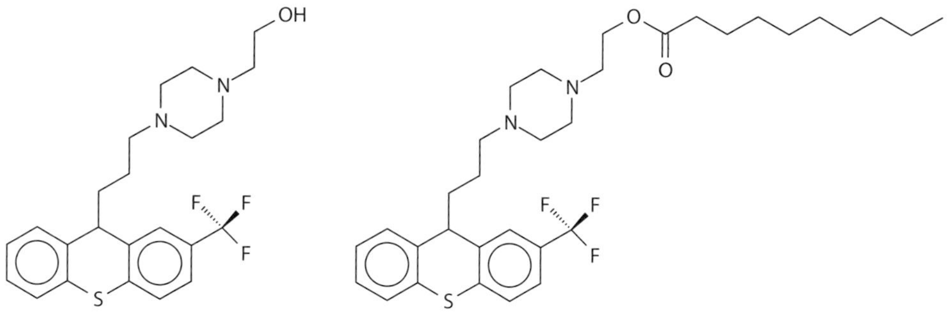

FIGURE 9.16   Flupentixole and its decanoate ester.

9.4.2.2 Macromolecular Prodrugs

Conjugation of small drug molecules to high molecular weight promoieties such as polyethylene glycols (PEGs), polysaccharides such as dextrans or other polymers may be used to obtain prolonged drug action. The high molecular weight conjugates may prevent rapid clearance of the active drug and provide a long term circulating depot from which active drug is liberated at a rate dependent on the nature of the drug-high molecular weight promoiety linkage. This principle has been primarily investigated for anticancer drugs such as paclitaxel.

It should be noted that introducing a high molecular compound, for example, in the form of a macromolecular prodrug to the body may give rise to an immunological response. This is of major concern for drug development and should be investigated thoroughly.

9.4.3 Design of Produgs for Drug Targeting

Drug targeting by site-specific bioactivation was achieved using intelligent prodrug design of the blockbuster drug omeprazole, which is widely used in the treatment of gastric ulcers. Omeprazole specifically inhibits the enzyme gastric H^+-K^+-ATPase that is responsible for the gastric acid production and is located in the secretory membranes of parietal cells. Omeprazole itself does not inhibit this enzyme but is biotransformed within the acid compartments of the parietal cells into the active inhibitor cyclic sulfonamide, which reacts with cysteine thiol groups of the enzyme thereby inactivating it (Figure 9.17).

Since omeprazole is only converted to the active molecule at acidic conditions, bioactivation almost exclusively takes place at the low pH at the parietal cells where high concentrations of the active species are generated.

Another example of site-specific delivery using the prodrug approach is olsalazine, which is being used in the treatment of ulcerative colitis. The active drug 5-aminosalicylic acid (mesalazine) is delivered from olsalazine to the colon due to specific bioactivation mediated by azo-reductases produced by anaerobic colonic bacteria. Olsalazine acts as a so-called twin prodrug in that it is cleaved to yield two molecules of the parent drug (Figure 9.18).

Drug targeting to the brain has been investigated using uptake transporters in the blood–brain barrier as targets for prodrugs. For example, glucose and mannose derivatives of levodopa and dopamine have shown increased *in vivo* activity in animal models compared to parent compounds. This has been attributed to enhanced drug transport into the brain by prodrugs utilizing the glucose transporter in the blood–brain barrier. Other brain uptake systems such as the large amino acid transporter have been the target for prodrugs.

Finally, it should be mentioned that the use of monoclonal antibodies as potential promoieties for anticancer agents may be of interest since such antibodies offer high specificity toward its target. The principle of using antibodies as promoieties will theoretically offer selective drug attachment to cancer cell antigens followed by release of the active drug leading to a higher drug accumulation at the site of action and thereby increased therapeutic effectiveness.

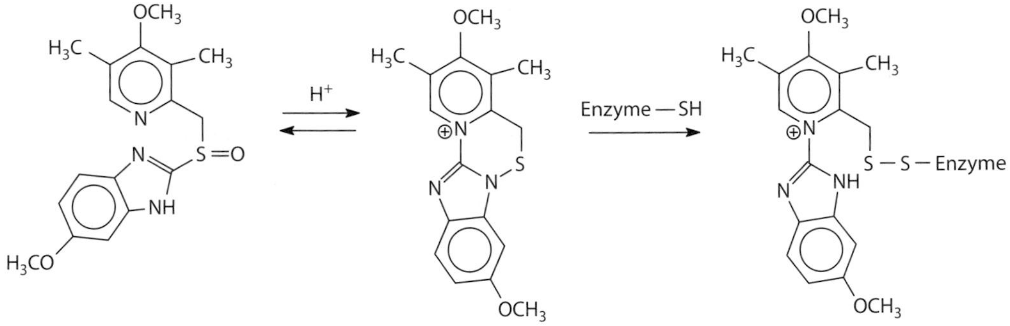

FIGURE 9.17   Omeprazole—bioconversion and mechanism of action.

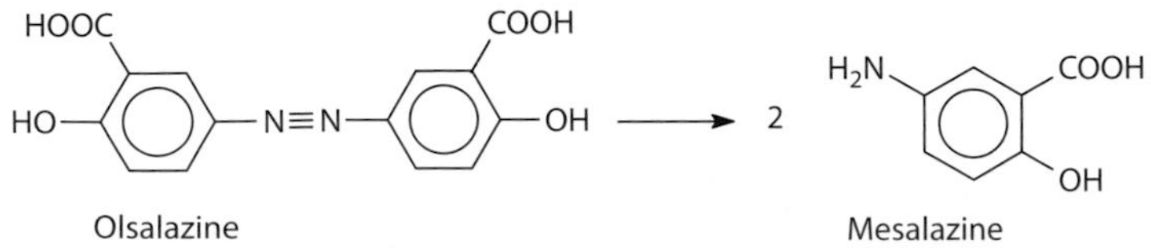

FIGURE 9.18   Bioactivation olsalazine—a twin prodrug.

9.5 DEVELOPMENT OF PRODRUGS

In the past, prodrug design and development was focused on solving problems with existing drugs approved for marketing. Therefore, bridging data to existing documentation and clinical experience with the active drug was widely accepted by regulatory bodies, thus, resulting in simplified and shortened development for such prodrugs. This has, however, changed and as prodrug design is becoming an integral part of drug discovery it is in principle no different from development of a *per se* pharmacologically active molecule. Thus, from a development and regulatory perspective, a prodrug is regarded as a novel chemical entity (NCE) having one—and preferably only one—major pharmacological active metabolite i.e., the active drug. Consequently, there are no specific guidelines or regulatory requirements for developing a prodrug. However, the fact that prodrugs are designed to undergo extensive and often fast metabolism *in vivo* does increase the complexity and the pitfalls during development.

9.5.1 Safety Assessment of Prodrugs

The safety assessment for a prodrug is generally more complicated compared to a *per se* pharmacologically active drug molecule. Thus, sufficient exposure in experimental preclinical safety studies in animal models must be obtained and documented not only in terms of the prodrug itself but also for the parent drug and other relevant metabolites. This is required for the establishment of relevant predictions, safety ratios, and therapeutic indices in humans and may prove very difficult for a prodrug. For example, selection of appropriate preclinical species such as mice, rats, dogs, and monkeys for risk assessment is critical as it is not only important to select the proper species in terms of pharmacological and toxicological relevance compared to humans. It is of equal importance to ensure that the species used in preclinical safety assessment of a prodrug adequately biotransform the prodrug into the active species to a similar extent and at the right site in order to fully explore and evaluate the risk/benefit of the prodrug. In addition, a development program for prodrugs must also include safety assessment of the promoiety or promoieties that are released in the body when the prodrug undergoes biotransformation. Although, a promoiety is normally designed to be inactive, potential toxicological effects must be properly addressed and documented during development.

One example of a potential problematic promoiety is formaldehyde, which is released during the biotransformation of various prodrugs, most importantly those utilizing the double-prodrug concept (Figure 9.6). Formation of formaldehyde from a prodrug could be considered a problem, but the amount released in the body from prodrugs is low compared to the release of formaldehyde from normal metabolic processing of endogenous compounds. Thus, the release of formaldehyde from prodrugs is not considered to pose a safety risk.

Another potentially more critical promoiety that is also widely used in prodrug design is pivalate or pivalic acid (Figure 9.10). This has been associated with changes in carnitine homeostasis through reaction with cellular coenzyme A and subsequent formation of pivaloyl-CoA, which can lead to depletion of carnitine from the body. The potential negative effects has gained attention from regulatory bodies can be evaluated based upon the intended daily dose of the prodrug and the duration of treatment. In most cases, however, exposure to the pivalic acid is regarded to have no or little toxicological impact, which may be outweighed by the clinical advantages offered by the prodrug. However, extended treatment with high dose prodrugs such as pivampicilin (Figure 9.9) may in extreme situations lead to clinical important carnitine deficiency. Thus, careful consideration of factors such as daily dose and duration of treatment should be given prior to a decision of designing and developing a pivalate-based prodrug and included in the overall benefit/risk evaluation of the development program.

9.6 CONCLUDING REMARKS

Prodrug design and development has proven to be a valuable complementary tool in modern drug design. In an integrated approach to drug design including optimization of compound structures in terms of efficacy, safety, and drugability, the prodrug principle offers opportunities for a temporarily change and improvement of physicochemical, pharmaceutical, biopharmaceutical, and ADME properties of a pharmacologically active molecule. The prodrug principle constitutes a tool for building additional properties into interesting pharmacologic active molecules, which are of vital importance for the development of clinically useful drugs. Thus, integration of prodrug considerations into the drug design paradigm may increase the opportunity for discovery and development of novel innovative drugs and treatment concepts to the benefit of patients.

FURTHER READINGS

Bundgaard, H. (Ed.). 1985. *Design of Prodrugs*. Elsevier, Amsterdam.

Bundgaard, H. 1989. Themed issue. Prodrugs for improved drug delivery. *Adv. Drug Deliv. Rev.* 3(1), 1–154.

Sloan, K.B. (Ed.). 1982. *Drugs and the Pharmaceutical Science Vol. 53. Prodrugs: Topical and Ocular Drug Delivery*. Taylor & Francis, Boca Raton, FL.

Stella, V.J. 1996. Themed issue. Prodrugs. *Adv. Drug Deliv. Rev.* 19(2), 111–330.

Stella, Valentino J., Borchardt, Ronald T., Hageman, Micahel J., Oliyai, Reza, Maag, Hans, and Tilley, Jefferson W. (Eds.). 2007. *Prodrugs: Challenges and Rewards Part 1 and 2*. 1st edition, Springer, New York.

10 Metals in Medicine: Inorganic Medicinal Chemistry

Helle R. Hansen and Ole Farver

CONTENTS

10.1 INTRODUCTION

Although biology is generally associated with carbon chemistry, most of the chemical elements from hydrogen to bismuth bear potential in drug design. Given the enormous variety and range in reactivity of inorganic compounds, the application of inorganic chemistry in improving human health opens new vistas in the field of research, and bioinorganic chemistry plays an increasingly important role in modern medicinal chemistry.

Inorganic compounds have been applied in medicine for thousands of years. However, serendipity has played a major role throughout time. In modern rational drug design and development it is of utmost importance to understand the reaction mechanism of the inorganic compounds, including the identification of the target centers. And, indeed, research in this area is now slowly gaining momentum.

10.1.1 Essential and Nonessential Elements

Essential metals (Table 10.1) are commonly found as natural constituents in proteins where they perform a wide spectrum of specific functions associated with biological processes. Metalloproteins with catalytic properties, metalloenzymes, implement chemical transformations of certain substrate molecules, and almost half of all enzymes in the human organism depend on the presence of one or more metal ions. Obviously, these metal ions are key pharmaceutical targets for drugs.

Pharmaceuticals may control metabolism of essential elements in two ways:

1. Supply of specific drugs with target properties may enable delivery or removal of elements to/from specific sites.
2. The natural physiological pathways may be blocked by the drug.

Nonessential (or toxic) elements can be useful in medicine since they either show insignificant toxic effect in a certain limited concentration range (left-hand side of Figure 10.1) or because their pharmaceutical benefit overrules their toxic effect. Such compounds are potentially of pharmacological interest, e.g., in killing certain cell types, as cancer cells or microorganisms while only expelling minor harm to the host. The examples include drugs based on platinum, gold, antimony, and bismuth complexes, which are described in more detail in Section 10.6. Their toxicity depends on concentration, oxidation state, and ligand and thus should be administered

TABLE 10.1
Constitution of the Human Organism (Adult 70 kg)

Element	Mass (g)	Recommended Daily Dose (mg)	Element	Mass (g)	Recommended Daily Dose (mg)
Oxygen	45,500		Copper	0.11	3–5
Carbon	12,600		Aluminum*	0.1	
Hydrogen	7,000		Lead*	0.08	
Nitrogen	2,100		Antimony	0.07	
Calcium	1,050	800–1200	Cadmium*	0.03	
Phosphorous	700	800–1200	Tin*	0.03	
Sulfur	175	10	Iodine	0.03	0.15
Potassium	140	2000–5500	Manganese	0.03	2–5
Chlorine	105	3200	Vanadium*	0.02	
Sodium	105	1100–3300	Selenium	0.02	0.05–0.07
Magnesium	35	300–400	Barium*	0.02	
Iron	4.2	10–20	Arsenic*	0.01	
Zinc	2.3	15	Boron*	0.01	
Silicon	1.4		Nickel*	0.01	
Rubidium*	1.1		Chromium	0.005	0.05–0.2
Fluorine	0.8	1.5–4.0	Cobalt	0.003	0.2
Bromium*	0.2		Molybdenum	<0.005	ca. 0.1
Strontium*	0.14				

Note: Elements marked with * are either nonessential, or their function is unknown.

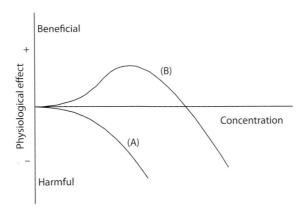

FIGURE 10.1   Dose/effect diagram. Response to (A) nonessential or toxic and (B) essential elements. Note that at sufficiently small concentrations, toxic compounds are tolerated, while essential elements become toxic at elevated concentrations.

with utmost care. This underlines the major point: "Specific biological activity of inorganic compounds can be achieved by proper design."

10.2   CLASSIFICATION OF INORGANIC PHARMACEUTICALS

Inorganic drugs may be divided into three different categories (Table 10.2):

1. *Active complexes.* Here, the entire complex, metal ion, and ligands, determines the action. Many coordination compounds act as neurotoxins by blocking acetylcholine receptors. Cisplatin and other uncharged Pt(II) and Ru(II) complexes are active as antitumor drugs (Section 10.6.4) and Bi(III) complexes are used in the treatment of gastrointestinal diseases (Section 10.6.7). Insoluble salts of certain heavy metals can be applied as x-ray contrast compounds, such as $BaSO_4$. The rare earth gadolinium is used in NMR diagnostics. Tin coordinated to protophorphyrin inhibits heme oxygenase degradation of a certain iron-heme product to bilirubin. The latter compound is the most frequent cause of neonatal jaundice.

2. *Active elements.* Here, the metal ion is decisive for the action of the drug while the anion or ligand only serves to keep the metal ion in solution or simply as a counter ion. The effect of lithium in the manic-depressive psychosis is well known although not fully understood. Interestingly, the lithium ion is counteracting both phases of the cyclic course of this disease (Section 10.6.1). The cariostatic effect of fluoride is well established although the mechanism is still unclear. Silver(I) and mercury(II) are potential antibacterial agents, and silver sulfadiazine is used clinically (Section 10.6.5). Technetium (the ^{99m}Tc isotope) is applied in radiodiagnostics, administered as a $Tc^{VII}(CNR)_6^+$ complex. The significance of gold in treating rheumatoid arthritis and of antimony in treatment of "leishmaniasis" will be discussed further in Section 10.6.

3. *Active ligands.* Many ligands can be delivered to or from a metal ion in the organism. The classic iron coordination compound, nitroprusside $[Fe(CN)_5NO]^{2-}$, releases NO, which functions as an hypotensive agent causing smooth muscle relaxation. Some selected chelates are presented in Section 10.5.

TABLE 10.2

Classification of Inorganic Drugs

Active complexes

- Cr, Co, Rh (neuromuscular blockers)
- Pt, Ru (anticancer drugs)
- Gd (NMR probes)
- Co (vitamin B12)
- Al, Zr (antiperspirant)
- Ba (x-ray contrast)
- Sn (jaundice)
- Bi (antiulcer and antibacterial agents)

Active metals

- Li (manic-depressive psychosis)
- F (tooth paste)
- Ag, Hg (antimicrobial compounds)
- ^{99m}Tc, ^{111}In (radiodiagnostics)
- Au (rheumatoid arthritis)

Active ligands

Delivered by a metal ion

- Ca, Mg, Al (antacid compounds)
- Fe (antihypertensive)
- Ti, Au (anticancer)

Delivered to a metal ion

- Bleomycin (Fe)
- Penicillamine (Cu)
- Desferrioxamine (Fe, Al)
- Bisphosphonates (Ca)

Source: Sadler, P.J., *Adv. Inorg. Chem.*, 36, 1, 1991.

10.3   THE HUMAN BODY AND BIOINORGANIC CHEMISTRY

Since the human body functions by uptake, accumulation, transport, and storage of chemical compounds (cf. Table 10.1) the arbitrary differentiation between "organic" and "inorganic" matter is irrelevant and is only a historical relic. The double helix structure of DNA, for instance, could not be stabilized without the presence of mono- and divalent cations that compensate for the electrostatic repulsion between negatively charged phosphate groups. Electric nerve impulses as well as more complex trigger mechanisms are initiated by rapid bursts of ions across membranes, particularly Na^+, K^+, and Ca^{2+} ions. Degradation of organic molecules requires acid and base catalysis, which at physiological pH could not take place without the presence of either Lewis acids like zinc(II) or Lewis bases that could be inorganic anions. Electron transfer is essential for all energy conversion processes, and here redox active transition metals like iron and copper become indispensable.

It is evident that fundamental biological processes proceed in reactions that often involve inorganic substances in central roles. Thus, inorganic chemistry holds a huge potential for developing new pharmaceuticals that requires a thorough knowledge of interactions between metal ions and organic molecules; a field also known as coordination chemistry. Stability and kinetics of metal ions complexes will therefore be a central subject in this chapter describing inorganic drugs.

10.4 COORDINATION CHEMISTRY

All metal ions are Lewis acids since they can coordinate to free electron pairs (i.e., Lewis bases). The outcome of this reaction is called a coordination compound or a complex between the central metal ion (Lewis acid) and the electron donor (Lewis base). A complex is thus composed of ions or molecules that may exist individually in solution, but in combination they produce the coordination compound. The ions or molecules coordinated to the central metal ion are called ligands and make up the coordination sphere. The number of points at which ligands are attached to the metal ion is called the coordination number. The different categories of ligands are shown in Figure 10.2.

10.4.1 CHELATE EFFECT

Ligands (Lewis bases) with several binding sites are called chelates and form particularly stabile complexes: metal–chelates. The stability of a coordination compound increases with the number of binding centers on the ligands (chelate effect). Amino acids, peptides, and proteins contain many metal binding groups that make them excellent chelates. In proteins, besides peptide NH and C=O groups many side chains may serve as complex agents for metal ions. These include thiolate in cysteine, the imidazole ring of histidine, carboxylates of glutamic acid and aspartic acid, and the amino side chain of lysine.

The rationale behind the chelate effect is quite straightforward. As soon as a metal ion coordinates to one group in a multidentate ligand, the probability for coordination of other potential donor groups is enhanced. A favorable entropic factor further adds to the stability since chelation is accompanied by release of nonchelating ligands like water from the coordination sphere.

A closely related effect is termed the macrocyclic effect, which relates to the notion that a complex with a cyclic polydentate ligand has greater thermodynamic stability when compared with a similar noncyclic ligand. As a consequence, macrocyclic complexes occur widespread in nature, and are found in, e.g., crown ethers, cryptands (alkali metals), cytochromes (iron), chlorophyll (magnesium), and coenzyme-B_{12} (cobalt).

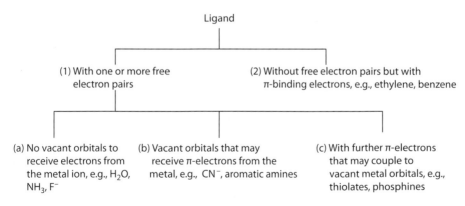

FIGURE 10.2 Classification of the different ligand types.

10.4.2 HARD AND SOFT ACIDS AND BASES (HSAB PRINCIPLE)

Metal ions can be divided into two categories:

a. "Class a" metals coordinate to bases that bind strongly to the proton ("hard bases"); i.e., bases in the ordinary sense of the word.
b. "Class b" metals bind preferentially to large polarizable or unsaturated bases ("soft bases") that usually show insignificant basicity toward the proton.

TABLE 10.3
Classification of Lewis Acids

Hard	Soft
H^+, Li^+, Na^+, K^+	Cu^+, Ag^+, Au^+, Tl^+, Hg_2^{2+}
Mg^{2+}, Ca^{2+}, Sr^{2+}, Mn^{2+}	Pd^{2+}, Cd^{2+}, Pt^{2+}, Hg^{2+}
Al^{3+}, La^{3+}, Gd^{3+}, Cr^{3+}, Co^{3+}, Fe^{3+}, As^{3+}	Tl^{3+}, Au^{3+}
Si^{4+}, Ti^{4+}, Os^{4+}	

Borderline
Fe^{2+}, Co^{2+}, Ni^{2+}, Cu^{2+}, Zn^{2+}, Sn^{2+}, Pb^{2+}, Sb^{3+}, Bi^{3+}, Ru^{2+}, Os^{2+}, NO^+

For "class a" metal ions, the order of complex stability is as follows, whereas for "class b" metal ions the order is virtually the opposite.

$$F > Cl > Br > I$$
$$O >> S > Se > Te$$
$$N >> P > As > Sb > Bi$$

In Table 10.3, metal ions that exhibit some importance in the bioinorganic chemistry are classified.

The general feature of a "class a" metal ions includes a small ionic radius, high positive charge, and are called "hard (Lewis) acids." "Class b" metal ions are in contrast associated with low oxidation state, large ionic radius, and are called "soft (Lewis) acids." This leads to a useful corollary, which is as simple and useful:

"Hard acids prefer to coordinate to hard bases, while soft acids prefer soft bases"

The stability order for soft acid complexes with Lewis bases is as follows:

$$S \sim C > I > Br > Cl > N > O > F$$

For hard acids, the division is even sharper since only complexes with oxygen or fluorine donor atoms will exist in aqueous solution. The HSAB (hard and soft acid and base) principle will be widely applied in the following text.

10.4.3 KINETICS. INERT AND LABILE COMPLEXES

Any complex formation takes place in a substitution reaction by replacement of one ligand by another. Thus, any substitution reaction is fundamentally a Lewis acid–base reaction.

The rate of a substitution reaction is primarily determined by the ratio between charge and size (charge density) of the metal ion, but when transition metals are involved the d-electron structure should also be taken into account. The term "labile" will be used for very reactive complexes while less reactive ones are called inert. Care should be taken not to confuse the term "labile" (kinetic) with the thermodynamic designation, stable.

Knowledge of the kinetic properties of complexes will obviously be decisive in the design of drugs. If a pharmaceutical in the form of an organic molecule is transferred to a target site by means of a metal ion, the complex should not be highly inert. On the other hand, complex formation between Pt(II) and DNA bases should be sufficiently inert in order to have adequate time to affect the division of tumor cells.

The electronic structure of transition metal ion complexes determines their reactivity due to the particular occupancy of the d-orbitals. Following the crystal field theory, the five d-orbitals split in the presence of the electrostatic field provided by the ligands (crystal field).

The doubly degenerate energy levels are denominated e while the triply degenerate levels are called t_2. The two high-energy d-orbitals of an octahedral complex are thus type e-orbitals while the

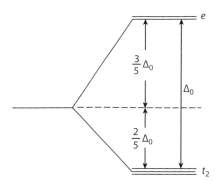

FIGURE 10.3 Crystal splitting of d-orbitals. The diagram shows the splitting of a set of d-orbitals in a metal ion complex having an octahedral symmetry. The energy difference between e and t_2 orbitals is designated Δ_0.

three lower-lying orbitals are of the t_2 type (cf. Figure 10.3). The energy difference, Δ_0, is called the ligand field splitting. In a tetrahedral coordination compound the arrangement is the opposite.

By preferentially filling up the lower-lying t_2-orbitals the d electrons will stabilize the system relative to an average arrangement of the electrons among all available orbitals. The gain in binding energy obtained by distributing the charges in a nonsymmetrical way is called crystal field stabilization energy (CFSE). The e-orbitals clearly have higher energy than the t_2-orbitals. We now assign an energy of $-2/5 \times \Delta_0$ to the three t_2-orbitals and $+3/5 \times \Delta_0$ to the e-orbitals, and can calculate the stabilization energies for complexes with any number of d electrons. For example, a d^5 high-spin octahedral complex will acquire a CFSE of $(-3 \times 2/5 + 2 \times 3/5) \times \Delta_0$ equal to 0. In a low-spin d^5 complex, the energy will be lowered by $-5 \times 2/5 \times \Delta_0$ or $-2 \times \Delta_0$. Thus, the latter will be considerably less reactive than the former.

The d^8 configuration deserves special attention since this system leads to very stable and inert square planar compounds. Platinum(II) complexes belong to this group and will be discussed in detail in Section 10.6.4. Cu(II) (d^9) and Zn(II) (d^{10}) coordination compounds are found frequently in enzyme systems where their large reactivity is fully utilized.

10.4.4 REDOX REACTIONS

A reduction–oxidation (redox) reaction is a process in which changes in oxidation states or oxidation numbers take place. Many transition metals exist in several stable oxidation states, which render them particularly interesting also in biological redox chemistry. Redox reactions play a central role in biochemistry; pertinent examples are photosynthesis and respiration where cascades of electron transfer reactions are coupled to synthesis of high-energy molecules like ATP and similar compounds. However, one of the expenses for living under oxygen rich conditions is the danger of unwanted radical formations. Oxygen easily gets reduced to hydrogen peroxide, and in the presence of reducing metal ions like Fe^{2+} or Cu^+ further reactions may take place like the Fenton reaction, generating hydroxyl radicals:

$$Fe^{2+} + HO_2^- \leftrightarrow FeO^+ + \cdot OH$$

Fortunately, the organism possesses a number of effective chelates, proteins like albumin, transferrin, and the like that, to a certain limit, will sequester redox-active iron- and copper ions.

10.5 CHELATE THERAPY

Heavy metals pose health hazards and toxication can be treated by using antagonists (chelate therapy), which involves complex binding (sequestration) and transport of acutely poisonous elements by means of polydentate ligands (Table 10.4). Obviously, selectivity plays a vital role and thus

TABLE 10.4
Chelating Ligands Toward Toxic Metal Ions

Ligand	Commercial or Trivial Name	Preferred Metal Ions
(a) 2,3-Dimercapto-1-propanol	Dimercaprol, BAL	Hg^{2+}, As^{3+}, Sb^{3+}, Ni^{2+}
(b) D-β,β-Dimethylcysteine	D-Penicillamine, PEN	Cu^{2+}, Hg^{2+}
(c) Ethylenediaminetetraacetic acid	EDTA	Ca^{2+}, Pb^{2+}
(d) Desferrioxamine	DFO, desferral	Fe^{3+}, Al^{3+}

constitutes a fundamental challenge in bioinorganic chemistry, and development of chelating pharmaceuticals that specifically sequester the undesired (heavy) metal ions becomes imperative. The most successful ligands demonstrate selectivity by (1) exclusive fitting to ions of definite size and charge; (2) comprising donor atoms that prefer Lewis acids of certain hardness or softness (Section 10.4.2). Further, the chelates must (3) form thermodynamically stable and kinetically inert coordination compounds (Sections 10.4.1 through 10.4.3); and finally (4) be able to excrete the undesired metal ion rapidly and effectively.

10.5.1 SELECTED CHELATES

10.5.1.1 BAL

The first example of chelate therapy was performed during World War II when BAL (2,3-dimercapto-1-propanol; British Anti Lewisite [Figure 10.4A]) was applied as antagonist against arsenic containing poison gas. BAL, being a very soft Lewis base, preferentially coordinates to soft heavy metal ions. Thus, aside from arsenic, As(III), the chelate will be highly efficient in treatment of mercury toxication. Today, however, BAL is exclusively utilized in connection with acute gold poisoning in patients undergoing treatment with gold containing pharmaceuticals (Section 10.6.5). An advantage of BAL is its lipophilic character, which facilitates transport into the cells. However, the drug itself is toxic and must be administered with great care.

10.5.1.2 D-Penicillamine

The structure of D-penicillamine (D-PEN) is shown in Figure 10.4B where the three different donor groups should be noticed: Two hard donor atoms (amine-N and carboxylate-O) together with a soft thiolate (−SH) group making the chelate a universal drug for both soft and hard Lewis acids although with limited ion selectivity. D-PEN is water soluble and, in contrast to BAL, not inherently toxic. But the L-isomer is a vitamin-B_6 antagonist and thus harmful to the organism. D-PEN has found wide application and may in most cases replace BAL, often applied simultaneously with EDTA as in the treatment of lead poisoning. It is also effective in sequestering gold and mercury. The administration of D-PEN to patients suffering from Wilson's disease is of particular interest (Section 10.6.5).

10.5.1.3 EDTA

Ethylenediaminetetraacetic acid (EDTA) and its analogues are all excellent chelates sequestering most metal ions, but for the same reason they are also not selective. EDTA coordinates preferentially to hard metal ions (Section 10.4.2) and due to the large chelate effect, quite stable complexes are formed. Since EDTA is inadequately absorbed only from the gastrointestinal tract, it is usually administered by intravenous injections. But due to the low degree of selectivity the hazard of eliminating essential metal ions is high. Adding the drug as Na_2H_2EDTA, the serum

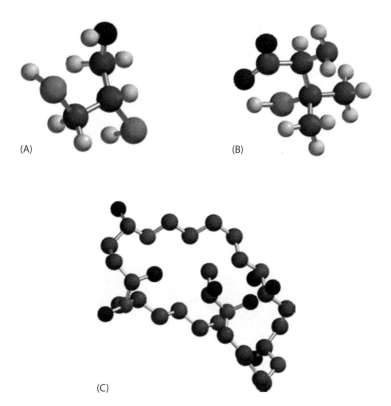

FIGURE 10.4   Selected ligands from Table 10.4. Structures of BAL (A); D-PEN (B); DFO (C); carbon atoms are black, oxygen red, nitrogen blue, and sulfur green. Hydrogen atoms (light gray) are only shown in the two former structures.

concentration of calcium ions will be lowered, often with severe muscle spasms as a result. Instead, the CaH_2EDTA salt is recommended, and in general the diet should be supplemented with essential metal ions during EDTA treatment.

10.5.1.4   Desferrioxamine

The siderophores are naturally occurring small molecule chelates secreted by many microorganisms. One important example of a siderophore is desferrioxamine (DFO, desferral) (Figure 10.4C). A large amount of (hard) oxygen donor atoms in an octahedral geometry render it highly specific toward iron(III), with a stability constant of not less than 10^{31} M^{-1}. The chelate is used in the treatment of acute iron poisoning and in certain cases of anemia where iron is accumulated in liver and heart. Binding constants for the corresponding Fe(II) complexes are much smaller due to larger ionic radius and smaller charge, and release of iron can be induced simply by reduction of Fe(III). Since the coordination chemistry of Al^{3+} is similar to that of Fe^{3+}, DFO can also be used in cases of aluminum poisoning. Incidentally, chelate formation also lies behind the body's strategy of producing fever in cases of infections. The higher temperature reduces the bacteria's ability to synthesize particular iron-chelating ligands.

10.6   INORGANIC CHEMISTRY AND PHARMACEUTICALS

So far, the pharmaceutical industry has mainly developed drugs based upon organic chemistry and on natural products. However, the remaining part of the periodic table offers an ever-increasing number of diagnostics and genuine pharmaceuticals. In the following text we shall give a brief review of some pertinent examples that illustrate the aforementioned principles.

FIGURE 10.5 Crown ether with potassium. The potassium ion (orange) is squeezed in between the two ring systems and is coordinated to eight oxygen atoms (red). Hydrogen atoms are not shown.

10.6.1 ALKALI METALS

The most important biological role of sodium and potassium is to stabilize cell membranes and enzymes by electrostatic effects and osmosis. Besides, these ions transmit electrical signals by diffusion through a certain concentration gradient.

Crown ethers (Figure 10.5) form stable complexes with alkali metals and their biological importance is binding heteroatoms on the inside of the macromolecule while the surface is more lipophilic. As a result, pharmacologically active natural products of this kind function as antibiotics since they may transfer alkali metals in and out of the cells and thereby perturb the natural metal ion balance. The examples include valinomycin (Figure 10.6) and nonactin. Another efficient method for controlled cation transport through lipid double layers involves incorporation of ionic channels in membranes.

Lithium salts play a particular role in treatment of manic-depressive psychosis. The effective plasma concentration is 1 mM while 2 mM exhibits toxic side effects, and already 3 mM is a lethal dose. The Li^+ ion has approximately the same radius as the Mg^{2+} ion, and both metal ions demonstrate high affinity phosphate binding. Lithium ions inhibit the enzymatic function of inositol monophosphatase thereby preventing release of phosphate from the active site. Inositol phosphatases are magnesium-dependent, and structural studies have shown that Li^+ may bind to one of the catalytic Mg(II) sites. Inositol phosphates are responsible for mobilizing calcium ions, and Li^+ will therefore influence the calcium ion level in cells, which makes it imperative to monitor the calcium concentration carefully in the patients during lithium treatment.

10.6.2 ALKALINE EARTH METALS

The calcium ion is engaged in a series of fundamental physiological processes from skeleton stabilization, cell division, and blood coagulation to muscle contraction and immune responses.

FIGURE 10.6 Potassium complex with valinomycin. Potassium (orange) is coordinated to six oxygen atoms (red).

FIGURE 10.7 The metal ion center of parvalbumin. The calcium ion (green) is shown five-coordinated to peptide carbonyl and carboxylate oxygen atoms (red). Nitrogen atoms are blue. A water molecule may further bind to the calcium ion. The coordinates are taken from the Protein Data Bank (1CDP).

The concentration of Ca^{2+} inside the cell is extremely small, about 10^{-7} M, while the extracellular concentration is in the mM range. In order to sustain this huge concentration gradient very effective, specific calcium pumps are required. Calcium is taken up in the small intestine bound to the active form of vitamin-D, in a yield of 50%.

Proteins with many acidic groups are particularly effective ligands for calcium ions, and a well-known example is parvalbumin (Figure 10.7), a protein located in smooth muscles. This protein is related to the extended family of structurally flexible calmodulins, which are calcium-binding proteins that can bind to and regulate a multitude of different protein targets, thereby affecting many different cellular functions.

10.6.3 IRON AND COBALT

Iron plays the leading role in all biological processes wherein oxygen turnover takes place. Iron(II) coordinates to a certain type of porphyrin and forms a complex labeled heme (Figure 10.8). Vertebrates utilize two such heme proteins for reversible O_2 transport and storage: hemoglobin in red blood cells and myoglobin in muscle tissue. Anemia results from insufficient dioxygen supply usually due to a low hemoglobin blood level.

A significant heme enzyme is cytochrome P-450, a dioxygen activating metalloporphyrin that catalyzes a series of important biological oxidation processes. Enzymatic monooxygenation reactions, e.g., the conversion of vitamin-D or transformation of drugs like morphine are indicative of such processes. Unwanted transformations like epoxidation of benzene to produce carcinogenic derivatives or oxidation of nitrosamines to form reactive radicals are examples of a toxicological function of cytochrome P-450.

For every 30 seconds, a child in Africa dies from malaria and the commonly used antimalarial therapy has become increasingly ineffective due to chloroquine resistance. However, a new series of drugs based on tervalent metal ion coordination compounds, like the ethylenediamine-bis[propylbenzylimino]Fe(III) complex, exhibit highly selective activity, ironically, particularly against chloroquine-resistant parasites. Heme, released from hemoglobin in the parasite, is very toxic to eukaryotic cells due to lysing of the membranes. In order to prevent this destructive action, the parasite polymerizes heme, but the aforementioned imino complexes inhibit this protective process thereby destroying the parasite.

Transport and storage of iron have been studied assiduously. As Fe(II) easily becomes oxidized to Fe(III) the products formed are in general highly insoluble at pH 7, unless Fe(III) becomes sequestered to some chelate like the siderophores (Section 10.5.4; cf. Figure 10.4C). The salmonella

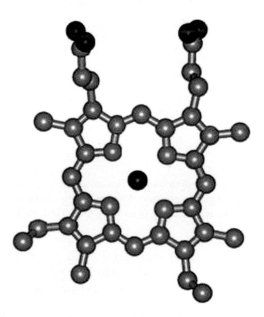

FIGURE 10.8 Heme. The heme prosthetic group consists of an iron(II) ion (magenta) complexed in a square planar geometry to four pyrrole nitrogen atoms (blue) of a substituted porphyrin ligand.

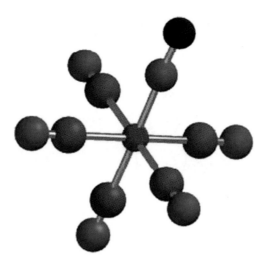

FIGURE 10.9   Nitroprusside. The "nitroprusside" anion $[Fe(CN)_5NO]^{2-}$. N is blue while O is red.

bacteria produce a siderophore that binds iron(III) with a stability constant of not less than 10^{50} M^{-1}. Another siderophore, desferrioxamine produced by the *Streptomyces* fungus, is used in order to prevent iron poisoning in connection with blood transfusion. Pathogenic microorganisms rely on a constant supply of iron, and therefore the availability of iron to bacteria invading the organism plays an important role in many diseases, like cholera and tuberculosis where a decrease in iron content in the blood is invariably observed. Effective iron scavenging chelates will thus act as potent antibiotics and naturally occurring iron complexing agents are, therefore, of great interest in medicine both as antibiotics and as drug delivery agents. An example is bleomycin, an antitumor agent that is isolated from the *Streptomyces* fungus.

"Sodium nitroprusside," $Na_2[Fe(CN)_5NO]$ (Figure 10.9) is an active hypotensive agent used in the treatment of heart infarct and in the control of blood pressure during heart surgery. The release of NO causes relaxation of the muscles surrounding the blood vessels, probably by the coordination of nitric oxide to an iron porphyrin receptor within the guanylate cyclase enzyme, which converts guanine triphosphate to cyclic guanine monophosphate. NO is also synthesized in the human body in a process where an iron containing (heme) enzyme catalyzes oxidation of the amino acid, arginine to nitric oxide.

The role of cobalt as essential trace element is confined to only one function, namely, as the redox active metal ion in coenzyme-B_{12} (Figure 10.10), which contains a Co–C (adenosyl) bond. As early as in the 1920s, it was well established that pernicious anemia (a state of anemia due to vitamin B_{12} deficiency) could be cured with injections of extracts from liver samples, and trace element analysis demonstrated later that the extracts contained cobalt. One of the axial cobalt ligands can be an alkyl residue: C_nH_{2n+1}. This is the only known example of a naturally occurring beneficial metal–carbon bond. The rate of alkylation may be accelerated enzymatically up to 10^{10} times.

10.6.4   PLATINUM AND RUTHENIUM

Platinum anticancer agents have been used extensively during the last 35 years in chemotherapy. The drugs target and interact with DNA in cells and thus prevent cell division. The cytotoxic effect is most serious on rapidly dividing cells, i.e., in addition to cancer cells there are also the cells of normal bone marrow, gut, skin epithelium, and mucosa. The lack of high selectivity of the Pt drugs is one of its main problems and thus constitutes a major challenge in developing new pharmaceuticals.

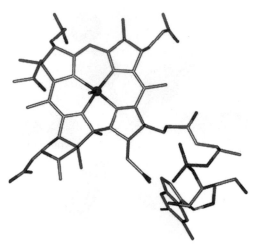

FIGURE 10.10 Coenzyme B_{12}. The prosthetic group in vitamin B_{12}. Besides coordinating to four corrin nitrogen atoms (blue), the cobalt(III) ion is also bound to an axial ligand, nitrogen from a benzimidazole group (not shown). The vacant sixth position is a binding site for substrates and is indicated with a stick pointing upward. A phosphate group (orange) is seen in the lower right corner. The coordinates are taken from the Protein Data Bank (1CB7).

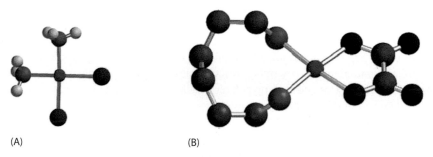

(A)   (B)

FIGURE 10.11 Platinum anticancer agents. Cisplatin (A). Oxaliplatin (B).

However, the drugs are partly selective toward cancer cells as they are more effective against proliferating cell than resting cells, where no DNA replication may occur over long periods of time.

Despite extensive research, only three platinum drugs have succeeded in reaching the market in the Western World while a fourth (satraplatin) is expected to be released in 2008. The relevant drugs are shortly introduced in the following text. For details concerning platinum interactions with proteins, biochemical pathways, and cancer type specific activities we refer to Chapter 23.

Cisplatin (*cis*-diamminedichloroplatinum(II); *cis*-DDP; Figure 10.11A), the first platinum anticancer agent to be used clinically, was discovered serendipitously in 1970 as an inhibitor of *E. coli* cell division. The two ammine ligands represent nonleaving groups and the chloride ions constitute exchangeable groups, which can be replaced by other ligands (nucleophiles). The second and third generation analogues to *cis*-DDP have been developed by simple substitution of the ammine ligands or chloride leaving groups of cisplatin. In "carboplatin" [*cis*-diammine-1,1-cyclobutanedicarboxylatoplatinum(II)], the two chloride ions have been substituted with less labile carboxylato groups. Carboplatin shows less toxic side effects than cisplatin and has now replaced the latter in many clinical situations.

In the third generation analogues of cisplatin, "oxaliplatin" [(diaminocyclohexane) oxalato-platinum(II); cf. Figure 10.11B] and "satraplatin" [bis(aceto)amminedichloro-(cyclohexylamine) platinum(IV)] all ligands have been replaced by more bulky ligands. They are effective in cells that show resistance to cisplatin. The unique feature of satraplatin is that platinum is present in the +4 oxidation state (compared to the more cytotoxic +2 oxidation state in the remaining platinum anti-cancer drugs). As a consequence it will undergo fewer reactions en route that make the drug suitable as an oral agent, which will potentially improve the quality of life millions of cancer patients.

Presumably, platinum agents mainly cross the cell membranes as uncharged molecules by passive diffusion, but studies indicate that active transport via Cu transporters are also involved. Once inside the cell, the compounds hydrolyze, i.e., *cis*- and carboplatin form the same positively charged di-aqua species $[Pt(NH_3)_2(H_2O)_2]^{2+}$. This ligand exchange is essential as the aquated and positively charged molecules are very reactive toward nucleophilic centers in biomolecules. Thus, the aqua-platinum complex favors binding to the N7 atoms of the imidazole rings of guanosine (G) (Figure 10.12) and adenosine (A). Three different types of purine base adducts can be formed in DNA, all involving coordination to G: monadducts d(Gp), intrastrand crosslinks (1,2-d(GpG), 1,2-d(GpA), 1,3-d(GpXpGp)), and interstrand crosslinks like d(GpG). Platinum crosslinks cause bending of the double helix of DNA and thus induce changes in the secondary structure of DNA. Oxaliplatin, with its bulky and hydrophobic ligand causes a larger bend than *cis*- and carboplatin. In almost all cases of pt–DNA binding, the metal alone cannot be held responsible for binding and stability, but hydrogen bonding between the ligand and DNA is an additional factor, thus making the pt drugs "active complexes."

It is generally assumed that the cytotoxicity of platinum compounds is due to the ability of the cross-links to block DNA replication and/or prevent transcription, as polymerase enzymes cannot pass the lesions. The 1,2-intrastrand crosslink may be responsible for the cytotoxicity of cisplatin. In the *trans*-isomer, this crosslink cannot be established in line with the smaller cytotoxicity of this complex. Additionally, high mobility group (HMG) proteins recognize and bind to DNA at the 1,2-intrastrand crosslink. The HMG binding prevents the NER system (nucleotide excision repair system, which normally removes impaired DNA sequences) in removing the platinum adducts.

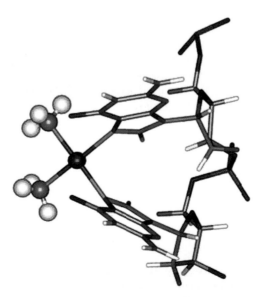

FIGURE 10.12 Platinum complex with DNA. Platinum (blue) is seen coordinated to two neighboring guanosine molecules in the DNA string. Nitrogen is blue, oxygen red, and phosphorus orange. The coordinates are taken from the Protein Data Bank (1A84).

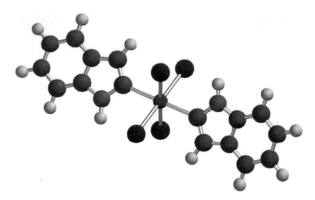

FIGURE 10.13 A Ru containing anticancer drug, KP1019. Ruthenium (green) is coordinated to two indazole molecules and four chloride ions (orange).

It should be noted that most platinum containing agents bind to proteins rather than DNA and that the degree of platinum cytotoxicity cannot be explained by inhibition of DNA synthesis alone. Thus, other mechanisms such as direct binding and damage of proteins or other biomolecules may also be of significance in triggering apoptosis or necrosis. Currently, research is focused on the development of platinum (and ruthenium) anticancer drugs with other targets than DNA and on combination therapies.

In the last few years, the search for effective anticancer compounds based on other metal centers than Pt has been intensified and particularly anticancer drugs based on ruthenium are making progress in clinical trials. Two ruthenium-based anticancer drugs, NAMI-A (Imidazolium *trans*-[tetrachloro(DMSO)(imidazole)ruthenate(III)] and KP1019 (Indazolium *trans*-[tetrachlorobis (1H-indazole)ruthenate-(III); cf. Figure 10.13], are scheduled to enter phase 2 trials in the near future.

Both NAMI-A and KP1019 are prodrugs that have Ru present in the +3 oxidation state and are activated by reduction. The "activation by reduction" mechanism may contribute to the lower toxicity of Ru(III) compounds compared with Pt-based anticancer compounds. However, the lower toxicity may also be due to its ability to mimic iron in the binding to biological molecules such as albumin and transferrin. Binding of ruthenium to transferrin allows for effective uptake of Ru-based drugs into cancer cells, as transferrin receptors are generally overexpressed in rapidly dividing cells.

The first Ru-based anticancer drugs were designed to target DNA, similarly to cisplatin. However, NAMI-A and KP1019 belong to a newer generation of Ru-based drugs that do not target primary tumors and DNA, but instead target metastases cells and proteins. A recent approach is design of Ru-drugs with multiple modes of activity by combining Ru with an active targeted ligand. Some examples are the Ru-SERMs complexes (selective estrogen receptor modulators) that target both hormone-dependent and -independent breast tumors and RAPTA complexes designed to inhibit Glutathione-*S*-Transferases (GST), a cytosolic detoxification enzyme associated with drug resistance.

10.6.5 COPPER, SILVER, AND GOLD

Copper and iron constitute the most important redox active transition metals in bioinorganic chemistry, and they seem to complement each other. Both copper and iron proteins are involved in oxygen transport and charge transfer. But while the iron containing proteins and enzymes always are found intracellularly, copper proteins and enzymes mainly operate outside the cells.

In humans most copper is found in the brain, the heart, and the liver. The high metabolic rate of these organs requires relative large concentrations of copper containing enzymes, some of which are presented in the following text. Not surprisingly, copper deficiency leads to brain diseases and anemia.

Copper is also found in many oxygenating enzymes, i.e., proteins that catalyze the incorporation of oxygen into organic substrates. An important example is dopamine-β-hydroxylase found in the brain where it catalyzes insertion of oxygen into the β-carbon of the dopamine (a neurotransmitter in the brain) side chain to produce norepinephrine. Another member of this class of proteins, peptidyl-α-amidase, catalyzes the conversion of C-terminal glycine extended peptides to their bioactive amidated forms, and hence is responsible for the biosynthesis of essential neuropeptide hormones like vasopressin and oxytocin.

The human variant of the antioxidant enzyme, superoxide dismutase, contains both copper and zinc (Figure 10.14). The toxic superoxide anion, O_2^-, is sometimes deliberately produced by organisms for particular objectives. Thus, some phagocytes, which are part of the immune system in higher organisms, produce large quantities of superoxide together with peroxide and hypochlorite by means of oxidases in order to kill invading microorganisms. In unfortunate cases this protection system may fail giving rise to certain autoimmune diseases like rheumatoid arthritis. Under these circumstances, the superoxide dismutase enzyme is administered as an anti-inflammatory pharmaceutical. The same therapy is consistently applied during open heart surgery in order to protect the tissue against oxidative attack by the superoxide radical.

The process of aging and neurodegenerative disorders like Parkinson's (PD) and Alzheimers' disease (AD) have been also linked to O_2^- production. AD is characterized by deposition of the amyloid-β peptide accompanied by neuronal loss. Although it is generally accepted that AD is associated with accumulation of Cu in the brain, very opposing treatment strategies exist. The traditional therapy is the treatment with chelates in order to remove excess Cu from the brain, whereas a newer strategy is to introduce Cu(II)-complexes in order to increase the number of free Cu ions in the brain. The rationale of the new strategy is that Cu in the brain is incorporated into plaques, which consequently results in intracellular depletion of free available Cu.

Free radical formation caused by metal ions like Cu(II) results in a continuous production of cytotoxic species leading to loss of dopaminergic neurons associated to PD. Possibly, antioxidant systems that are supposed to preserve life become dysregulated by abnormal metal ion interaction, which eventually lead to neurodegeneration. The traditional treatment of patients suffering from PD is pharmacotherapy by supply of the neurotransmitter, dopamine. However, metal ions are a common denominator in the pathogenesis of neurodegenerative processes in the brain and therefore relocate metal–protein interaction to an important role in neuroscience.

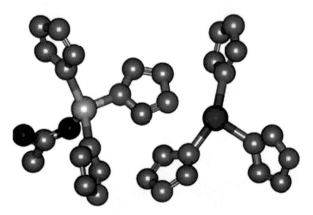

FIGURE 10.14 The active Cu–Zn site in superoxide dismutase. Zinc (gray) is coordinated to an aspartate and three histidines, one of which bridges the two metal ions (the coordinate bond to Cu is not shown here). The copper ion (blue) has one vacant position for substrate (O_2^-) binding. The coordinates are taken from the Protein Data Bank (1YAI).

Copper is a potent poison for any cell and thus proteins of the metallothionein type exist, which will transport excessive copper ions out of the cells. Due to the delicate balance between excess and deficiency of copper, a tight control of uptake and excretion of this metal is needed. Excess copper leads to copper, accumulation in liver and brain, which untreated leads to severe damage of these organs and results in early death (Wilson's disease). Therapy with powerful copper chelates like D-penicillamine (Section 10.5.1.2) can keep the copper concentration on a suitable level. Deficiency in copper, is just as serious since it leads to grave mental and physical illnesses (e.g., Menkes' disease) that involve a hereditary dysfunction in copper metabolism.

Many silver(I) compounds can be used as effective antibacterial drugs, like silver sulfadiazine, which is used clinically in ointments as an antimicrobial agent in instances of severe burns. Silver nitrate has also been applied in dilute solutions in cases of eye infections due to its antiseptic property.

Gold has been applied in certain contexts during history. The ancient Chinese, several thousand years ago produced an elixir containing colloidal gold, which should ensure eternal life. The benefit of this treatment, however, has never been fully documented. Nevertheless, gold(I) compounds are currently the only class of drugs known to halt the progression of rheumatoid arthritis. Initially, gold compounds like gold sulfide and gold thiomalate were painfully administered as intramuscular injections. Later it was discovered that the triethylphosphinegold(I) tetra-*O*-acetylthioglucose (auranofin, Figure 10.15) was equally effective and could be administered orally.

As an extremely soft metal ion Au(I) shows a large affinity toward soft bases like sulfur (thiolates) and phosphorous (phosphines) while the affinity toward oxygen and nitrogen containing ligands is small. The Au(I) coordination in auranofin is shown in Figure 10.15. It is interesting to note that the copper level is directly related to the severity of rheumatoid arthritis, which has led to proposals that antiarthritic drugs like D-PEN and auranofin operate by affecting the center of coordination for copper ions, like the one found in human serum albumin. As demonstrated by NMR studies, aura-nofin coordinates to cysteine-34 in albumin and induces a conformational change in the protein. This affects the copper binding center (imidazole from a histidine group) whereby the copper homeostasis becomes perturbed. It has been suggested that the damage of the joints due to tissue inflammation is the result of lipid oxidation caused by free radicals such as O_2^-. This notion provides a link from gold to copper. Yet, in another hypothesis gold(I) complexes are suggested to inhibit formation of undesired antibodies in the collagen region.

Gold-based pharmaceuticals unfortunately possess unpleasant side effects that include allergic reactions as well as gastrointestinal and renal problems. These side effects may be linked to the

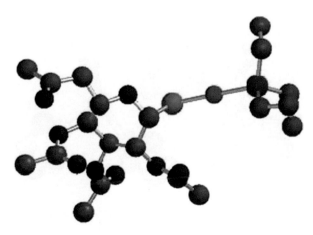

FIGURE 10.15 Auranofin. Gold(I) (green) is coordinated linearly to a phosphine group (orange) and a thiolate sulfur (small green).

production of strongly oxidizing gold(III) metabolites and a better understanding of the mechanism of gold preparations is indeed needed in order to produce more effective and less toxic gold-based drugs.

10.6.6 ZINC

Zinc is involved in a large number of biological processes and today more than 200 proteins containing Zn^{2+} are known. Among these, many essential enzymes are found that catalyze the transformation or degradation of proteins, nucleic acids, lipids, and the like. Besides, the zinc ion stabilizes many different proteins like insulin. Obviously, zinc deficiency will lead to severe pathological effects.

Carbonic anhydrase is a zinc containing enzyme that catalyzes the hydrolysis of CO_2:

$$CO_2 + H_2O \leftrightarrow H^+ + HCO_3^-$$

and is of fundamental significance in respiration. The catalytic process occurs 10^7 times faster in the presence of the zinc enzyme compared with the uncatalyzed reaction. Certain antiepileptic pharmaceuticals like acetazil amide coordinate directly to zinc(II) in the active center of the carbonic anhydrase enzyme and thus obstructs the catalytic transformation of carbon dioxide. With accumulation of CO_2 in the blood stream pH drops, and it has been suggested that this perturbs the gamma-aminobutyric acid (GABA, an inhibitory neurotransmitter) concentration in brain cells, either by increasing the GABA synthesis or by blocking the process of degradation (see Chapter 15).

10.6.7 ANTIMONY AND BISMUTH

Antimony and bismuth have been applied in medicine for centuries due to their antiparasitic and antibacterial properties. Sb(III) and Bi(III) are borderline metal ions and exhibit a high affinity for oxygen, nitrogen, and sulfur. Unlike Sb, where the +5 oxidation state is favored, the +3 oxidation state is the most common and stable form of Bi.

Millions of people suffering from "leishmaniasis" worldwide, particularly in the developing countries, are subject to intravenous treatment with drug formulations of pentavalent antimony, Sb(V) complexes with polyhydroxy carbohydrate ligands. Very limited resources have been invested by the developed world in optimizing these drugs. Thus, at the moment, patients infected with the parasites *Leishmania* spp. undergo long treatments with extremely high "nontarget" doses of Sb, which often result in severe side effects. Although, alternative antileishmania drugs are available in the market, or are in the development phase, antimony-based drugs have remained the main treatment worldwide since the 1930s (except for certain domains where resistance has curbed their use).

Sodium stibogluconate (Pentostam) and meglumine antimonate (Glucantime) are the two drugs in current use and typically administered intravenously. The carbohydrate ligands in Pentostam and Glucantime, gluconic acid, and *N*-methyl-D-glucamine, respectively, increase the general solubility of antimony and may serve to deliver Sb to the macrophages, where the protozoa that cause "leishmaniasis" undergo division.

Pentostam and Glucantime [both based on Sb(V)] are believed to be only prodrugs and Sb(III) accounts for the active form of antimony at the target site. Antimony(III) possesses higher antileishmania activity than Sb(V), but due to its higher toxicity it has no direct therapeutic use. Trypanothione, the most abundant low molecular weight thiol-containing ligand in the parasites (while nonexisting in human) may act as the reductant in the parasites. Once Sb(III) is present in the parasite it can interfere with several enzymes and proteins, which eventually destroys the parasite.

Although bismuth preparations have historically been applied in the treatment of a number of diseases, including syphilis and hypertension, their application during the last decades has been

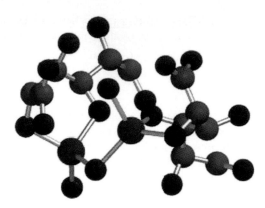

FIGURE 10.16   Bismuth antiulcer complex. Suggested structure of a binuclear bismuth(III) complex with citrate. The Bi atoms are orange while oxygen is red and carbon black. The complex may form larger clusters with bridging citrate anions.

restricted to gastrointestinal therapy. Since the 1970s, two Bi(III) compounds have been most commonly used worldwide; bismuth subsalicylate (BSS) for the prevention and treatment of diarrhea and dyspepsia, and colloidal bismuth subcitrate (CBS; Figure 10.16) for the treatment of peptic ulcers. Most ulcers are associated with the bacterium, *Helicobacter pylori*. In the 1990s, a new Bi(III)-containing drug was developed, ranitidine bismuth citrate (RBC), which combines the antisecretory action of ranitidine with the bactericidal properties of bismuth. Although the use of bismuth containing drugs for years was declining, they are now again becoming increasingly popular as combination pharmaceuticals due to developed antibiotics resistance by *H. pylori*.

All bismuth(III) drugs are chelates of complicated polymeric nature. In BSS, salicylate ligands coordinate to bismuth atoms via chelation, with extraordinary variations in binding modes. In CBS, Bi(III) is aggregated through citrate bridges and H-bonds to form complex polymers. The actual structure determination of bismuth-based drugs is complicated by the ability of the drugs to change composition with pH and concentration.

The exact reaction mechanism of Bi(III) drugs is not fully understood, but the therapeutic activity may result from mucosa-protective properties and from degradation of *H. pylori*. There is no doubt that Bi(III) possesses antimicrobial activity itself and not only its organic ligand as was once believed. The major target for Bi(III) appears to be proteins and enzymes. Pathogenic microorganisms such as *H. pylori* produce large amounts of enzymes, and Bi(III) inhibits several of these enzymes, including alcohol dehydrogenase (ADH) and urease. The inhibition of ADH is likely due to the displacement of Zn(II) by Bi(III) at the active site since Bi(III) has higher affinity for thiolate groups than Zn(II). Urease has long been regarded as a potential target for bismuth-based drugs, as the enzyme catalyzes the degradation of urea into carbon dioxide and ammonia, which helps to neutralize the acidic environment in order for the bacteria to survive the hostile environment. Presumably, Bi(III) binds to exposed thiolates of the enzyme and blocks the entrance to the active cavity of the enzyme.

10.7   CONCLUDING REMARKS

As this chapter has demonstrated, inorganic chemistry plays an important role in biology and an increasingly important role in modern drug development. In many cases, metabolism of essential metal ions may be controlled by means of organic pharmaceuticals since an intimate synergism exists between the function of inorganic elements and organic compounds of the body. Metal ions control some of the fundamental biochemical processes such as DNA and RNA replication, and many enzymes hold a key position as pharmaceutical targets. Emphasis should be put on kinetics rather than stability of inorganic compounds, since the most interesting biological processes, more often than not, take place rapidly and far from equilibrium. As a result, a new and exciting field of research with a variety of challenges has opened in inorganic medicinal chemistry.

FURTHER READINGS

Bertini, I., Gray, H.B., Stiefel, E.I., and Valentine, J.S. (2007) *Biological Inorganic Chemistry*. Sausalito, CA: University Science Books.

Cotton, F.A., Wilkinson, G., and Gaus, P.L. (1995) *Basic Inorganic Chemistry*. New York: John Wiley & Sons.

Guo, Z. and Sadler P.J. (2000) Medicinal inorganic chemistry. *Adv. Inorg. Chem.* **49**, 183–306.

Kaim, W. and Schwederski, B. (1994) *Bioinorganic Chemistry: Inorganic Elements in the Chemistry of Life*. New York: John Wiley & Sons.

Lippard, S.J. and Berg, J.M. (1994) *Principles of Bioinorganic Chemistry*. Mill Valley, CA: University Science Books.

Sadler, P.J. (1991) Inorganic chemistry and drug design. *Adv. Inorg. Chem.* **36**, 1–48.

Williams, R.J.P. and Frausto da Silva, J.J.R. (2003) *The Natural Selection of the Chemical Elements*. New York: Oxford University Press.

11 Enzyme Inhibitors: Biostructure-Based and Mechanism-Based Designs

Robert A. Copeland, Richard R. Gontarek, and Lusong Luo

CONTENTS

11.1 INTRODUCTION

Inhibition of disease-associated enzyme targets by small molecular weight drugs is a well-established modality for pharmacologic intervention in human disease. Indeed, a recent survey of the FDA Orange Book showed that more than 300 marketed drugs work through enzyme inhibition. Among orally dosed drugs in clinical use, nearly half of them function by inhibition of specific enzyme targets. Likewise, much of current preclinical drug discovery efforts in biotechnology and pharmaceutical companies—as well as those same efforts in government and academic laboratories—is focused on the identification and optimization of small molecules that function by inhibition of specific enzyme targets. The reasons for the popularity of enzymes as targets for drug discovery have been reviewed a number of times recently (see for example, Copeland (2005)). In brief, enzymes make good drug targets for two significant reasons. First, the catalytic activity of specific enzymes is often critical to the pathophysiology of disease, such that inhibition of catalysis is disease modifying. Second, the binding pockets for natural ligands of enzymes, that play a crucial role in catalytic activity, are often uniquely well-suited for interactions with small molecule drugs. Thus, the very nature of the chemistry of enzyme catalysis makes these proteins highly vulnerable to inactivation by small molecule inhibitors that have the physicochemical characteristics of oral drugs.

Enzyme catalysis involves the conversion of a natural ligand (the substrate) into a different chemical species (the product), most often through a process of chemical bond breaking and formation steps. The chemical transformation of substrate to product almost always involves the formation of a sequential series of intermediate chemical species along the reaction pathway. Paramount in this reaction pathway is the formation of a short-lived, high-energy species referred to as the transition state. To facilitate this sequential process of intermediate species formation, the ligand-binding pocket(s) of enzymes must undergo specific conformational changes that induce strains at correct locations and align molecular orbitals to augment the chemical reactivity of the appropriate functionalities on the substrate molecule(s), at defined moments during the reaction cycle. The bases of mechanistic enzymology include understanding the chemical nature of the various intermediate species formed, and their interactions with those elements of the enzyme-binding pocket that facilitate chemical transformations. When these studies are coupled with structural biology methods, such as x-ray crystallography and multidimensional nuclear magnetic resonance (NMR) spectroscopy, a rich understanding of the structure–activity relationships (SAR) that attend enzyme catalysis can be obtained. What is germane to the present discussion is that this structural and mechanistic understanding can be exploited to discover and design small molecule inhibitors—mimicking key structural features of reaction intermediates—that form high-affinity interactions with specific conformational states of the ligand-binding pocket of the target enzyme. In this chapter, we describe the application of mechanistic and structural enzymology to drug discovery efforts with an emphasis on the evolution of structural changes that attend catalysis and the exploitation of these various conformational forms for high-affinity inhibitor development.

11.2 MODES OF INHIBITOR INTERACTION WITH ENZYMES

The simplest enzyme-catalyzed reaction one can envisage is that of a single substrate (S) being converted by the enzyme (E) to a single product (P). This reaction can be summarized by the following equation:

$$E + S \underset{k_2}{\overset{k_1}{\rightleftharpoons}} ES \xrightarrow{k_{cat}} E + P \qquad (11.1)$$

As summarized by Equation 11.1, enzyme and substrate combine to form a reversible initial encounter complex (ES) that is governed by a forward rate constant for association (k_1) and a reverse rate constant for dissociation (k_2). The equilibrium dissociation constant for the ES complex is given the symbol K_S and is mathematically equivalent to the ratio of the rate constants k_2/k_1. Subsequent to initial complex formation, a series of chemical steps ensue that are collectively quantified by the cumulative rate constant k_{cat}. Thus, k_{cat} is not a microscopic rate constant, but rather summarizes all of the intermediate states that must be formed during the chemical transformation of substrate to product (see Section 11.3 for more details on the individual intermediate steps that may contribute to k_{cat}).

Three modes of inhibitor interaction with an enzyme target can be defined, based on their effects on the catalytic steps summarized in Equation 11.1. Competitive inhibitors bind to the free enzyme in a manner that blocks the binding of substrate so that they increase the apparent value of K_S, but have no effect on the apparent value of k_{cat}. Noncompetitive inhibitors can bind to both the free enzyme and to the ES complex (or intermediate species that follow the formation of the ES complex). Such inhibitors can have some effect on the value of K_S but show the greatest effect on k_{cat}, as they inhibit by blocking catalytic steps subsequent to substrate binding. Finally, uncompetitive inhibitors have no affinity for the free enzyme and only bind subsequent to the formation of the ES complex. These inhibitors decrease the apparent value of K_S (i.e., increasing the apparent affinity of the enzyme for substrate) and also decrease the apparent value of k_{cat} (i.e., diminishing the ability of the enzyme to catalyze chemical steps subsequent to substrate binding). Among drugs in current clinical use one finds multiple examples of each of these three modalities of enzyme inhibition.

11.3 PROTEIN DYNAMICS IN ENZYME CATALYSIS AND INHIBITOR INTERACTIONS

The catalytic pathway summarized in Equation 11.1 is a gross oversimplification of even the simplest of enzymatic reactions. At minimum, this reaction requires the formation of two additional forms of enzyme–ligand binary complex, these being the enzyme-transition state complex and the enzyme–product complex. In practice, one often finds that additional intermediate states are accessed during the catalytic cycle of an enzyme. Thus, one can say that the catalytic cycle of an enzyme is a sequential series of protein–ligand complexes, each representing a unique chemical form of the ligand with attendant changes in the protein conformation of the ligand-binding pocket. Each conformational state of the ligand-binding pocket that is accessed during this catalytic cycle is a potential target for small molecule drug interactions. Hence, one can think of the ligand-binding pocket of an enzyme not as a single target for drug intervention, but rather a collection of targets that evolve and interconvert over the time course of catalytic turnover.

To illustrate these concepts, let us consider the reaction cycle of an aspartyl protease. The aspartyl proteases constitute a family of protein/peptide hydrolyzing enzymes that use a pair of aspartic acid residues within the enzyme active site to facilitate peptide bond cleavage. Figure 11.1 provides

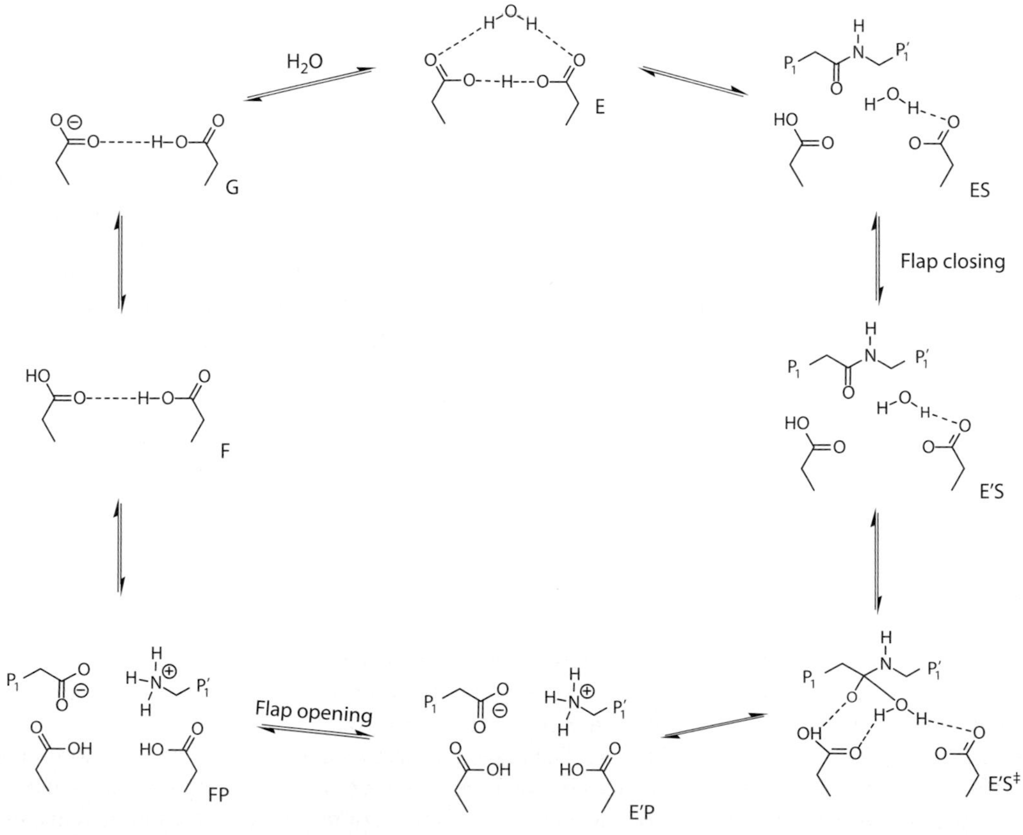

FIGURE 11.1 The reaction pathway for an aspartyl protease. (Adapted and modified from Copeland, R.A., *Evaluation of Enzyme Inhibitors in Drug Discovery: A Guide for Medicinal Chemists and Pharmacologists*, Wiley, Hoboken, NJ, 2005.)

a schematic representation of the canonical reaction cycle of an aspartyl protease, illustrating the changes in active site structure that attend catalysis. Before substrate binding the enzyme is in a resting form (E) in which the two active site aspartic acid residues are bridged by a water molecule. One of the aspartates is present in the protonated acid form while the other is present as the conjugate base form, and the two residues share the acid proton through a strong hydrogen bond. Initial substrate binding causes disruption of the hydrogen bonding interactions and displacement of the water molecule in species ES. After initial substrate binding, a flexible loop of the protein, referred to as the "flap" closes down over the active site to occlude the active site groups from bulk solvent. The substrate-bound enzyme in this altered conformation is referred to as form E′S in Figure 11.1. Subsequently, the water of the enzyme active site attacks the carbonyl carbon of the scissile peptide bond to form a dioxy, tetrahedral carbon center on the substrate. This constitutes the enzyme-bound transition state of the reaction and is symbolized as E′S‡ in the figure. Bond rupture then occurs, leading to a species with both active site aspartates protonated and with both the anionic and cationic peptide products bound (form E′P). After that, the flap retracts from the active site to generate a new conformational state of the active site, referred to as FP. With the flap out of the way, the product peptides can now dissociate from the enzyme, forming enzyme state F. Deprotonation of one of the active site aspartates occurs next to form state G. Finally, addition of a water molecule returns the enzyme back to the original conformational state E, thus completing the catalytic cycle.

It is clear from Figure 11.1 that protein dynamics is an important component of the catalytic cycles of enzyme reactions. As stated earlier, the importance of this concept to drug discovery is that each intermediate state accessed along the reaction pathway provides unique opportunities for inhibitor interactions. For example, in the case of the aspartyl proteases, there are three distinct ligand-free conformational forms of the enzyme (states E, F, and G); small molecule inhibitors are known that preferentially bind to each of these individual states.

The protein structures that are represented along the reaction pathway of an enzyme reflect a collection of conformational microstates that interconvert among themselves through rotational and vibrational excursions. Thus, any particular conformation of the enzyme can be represented as a manifold of conformational substates (microstates), each stabilized to different degrees by specific interactions with ligands. Isomerization of the enzyme from one structure to another therefore involves the potential energy stabilization of certain microstates at the expense of others.

Similarly, high-affinity inhibitor interactions often involve isomerization of the enzyme from an initial structure resembling the unliganded enzyme to a new conformation in which a particular microstate(s) is highly stabilized by interactions with the inhibitor. Kinetically, this requires two distinct steps in overall binding of high-affinity inhibitors. As illustrated in Figure 11.2A, the first step involves the formation of a reversible encounter complex between the enzyme and inhibitor (EI), which often displays only modest affinity. Forward binding of the inhibitor to the enzyme is dictated by the association rate constant k_3 and dissociation of the initial EI complex is dictated by the dissociation rate constant k_4. Once the EI complex is formed, enzyme isomerization can occur to form the much higher affinity complex E*I. The forward conversion of EI to E*I is dictated by the isomerization rate constant k_5 and the reverse conversion of E*I back to EI is dictated by rate constant k_6. All three enzyme forms (E, EI, and E*I) can also be represented as potential energy diagrams, as illustrated in Figure 11.2B. It should be noted that the affinity of the enzyme–inhibitor complex is related to the potential energy stabilization of the system, which is reflected in the depth of the potential well in the energy diagram. The deeper the potential energy well for the inhibited form, the more energy that is required to escape this well and thus access the other conformational microstates required for continued catalysis. The potential energy stabilization of the inhibitor-bound form is mediated by productive interactions between the inhibitory ligand and the binding pocket of the protein, in the form of hydrogen bonds, electrostatic interactions, hydrophobic interactions, van der Waal forces, and the like. These concepts can also be conceptualized in terms of an induced fit between the binding pocket and the inhibitor, as illustrated in cartoon form in Figure 11.2C.

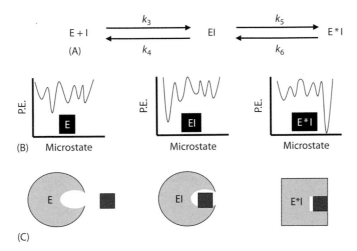

FIGURE 11.2 A two-step inhibitor-binding mechanism involving initial binding of the inhibitor to the enzyme in one conformation and a subsequent isomerization of the enzyme to a new conformation. (A) Reaction sequence illustrating the forward and reverse kinetic steps of binding and enzyme isomerization. (B) Potential energy diagrams representing the three conformational states of the enzyme: E, EI, and E*I. (C) Cartoon representation of the inhibitor binding and enzyme isomerization steps in this mechanism.

The state E*I then represents a state of high-affinity interactions between the enzyme and the inhibitor. As long as the inhibitor is bound to the enzyme, either in the form of EI or E*I, the biological activity of the enzyme is blocked. Dissociation of the inhibitor from the enzyme can occur for any reversible inhibition process; once the enzyme is free of inhibitor, catalytic activity is restored. In the case of tight-binding inhibitors that induce enzyme isomerization, the overall rate constant for inhibitor dissociation, k_{off}, must take into account reversal of the isomerization step, reisomerization via k_5, and dissociation of the inhibitor from EI via k_4. Mathematically, the value of k_{off} is given by

$$k_{off} = \frac{k_4 k_6}{\left(k_3 + k_5 + k_6\right)} \tag{11.2}$$

For this two-step binding mechanism, it is almost always the case that the reverse isomerization step, mediated by rate constant k_6, is by far the slowest step in overall inhibitor dissociation. Thus, the lower the value of k_6, the longer the duration of potent inhibition by the drug. There are a large and growing number of examples of highly efficacious drugs that demonstrate tight-binding interactions with their target enzyme through a two-step enzyme isomerization mechanism as described here. In some cases, the slowness of the reverse isomerization step leads to prolonged duration of inhibition that may translate into an extended duration of pharmacodynamic activity *in vivo*; this concept is considered further in Section 11.6.

11.4 MECHANISM-BASED INHIBITOR DESIGN

Enzymes are designed by nature to catalyze a specific chemical reaction. As described earlier, every enzyme accesses a sequential series of intermediate states along the reaction pathway, thus providing unique opportunities for inhibitor interaction. Consequently, enzyme inhibitors can be effectively designed based on an understanding of the mechanistic and structural details of the catalyzed reaction pathway. The majority of known enzyme inhibitors are structurally related to natural ligands of the enzymatic reaction; a recent survey suggested that more than 60% of marketed drugs that target enzymes are either analogs of substrates or enzyme cofactors, or they undergo catalyzed structural conversion within the active site of an enzyme. Substrate, cofactor, and product mimicry, however, is not the only method for the design of high-affinity, selective enzyme inhibitors. Advances in transition state theory during the past three decades have helped to establish an alternative approach for mechanism-based design: intermediate-state-based design (sometimes also referred to as transition-state-based design). In this latter approach, inhibitors that mimic the steric and electronic features of high-energy reaction intermediate states are designed to capitalize on the specific interaction of active site residues with the reaction intermediate. In the next two sections, cases for substrate structure-based design and intermediate-state-based design will be discussed to exemplify inhibitor design strategies that have led to successfully marketed products or clinical candidates.

11.4.1 Substrate Structure-Based Design

11.4.1.1 Nucleoside and Nucleotide Inhibitors of HIV Reverse Transcriptase

HIV reverse transcriptase (RT) is one of two main targets for antiacquired immunodeficiency syndrome (AIDS) therapy (the second target being the HIV protease; *vide infra*). The RT enzyme catalyzes the synthesis of double stranded proviral DNA from single stranded genomic HIV RNA. Drugs targeting HIV RT can be divided into two categories: (i) nucleoside and nucleotide RT inhibitors (NRTIs), which are competitive with respect to the natural deoxynucleotide triphosphates (dNTPs) and serve as alternative substrates for catalysis (resulting in chain termination); (ii) nonnucleoside RT inhibitors (NNRTIs), which are allosteric, noncompetitive inhibitors that bind at a site distal to the RT active site. NRTIs were the first class of chemotherapeutic agents to be utilized in the clinic to treat AIDS patients and offer excellent examples of inhibitor design based on substrate mimicry. The first NRTI, Zidovudine (AZT) was approved by the FDA in 1987 (Figure 11.3).

This molecule is a thymidine analog with an azido group in place of the hydroxyl group at the 3′ position of the ribose. Since the advent of AZT-based therapy, a number of NRTIs have joined the

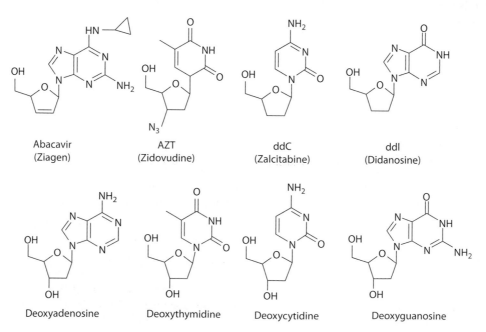

Abacavir (Ziagen)	AZT (Zidovudine)	ddC (Zalcitabine)	ddI (Didanosine)
Deoxyadenosine	Deoxythymidine	Deoxycytidine	Deoxyguanosine

FIGURE 11.3 Representative FDA-approved nucleoside/nucleotide RT inhibitors (top panel) that closely mimic the natural deoxynucleotides (bottom panel).

anti-AIDS treatment armamentarium. Most of these NRTIs are nucleoside analogs with the exception of tenofovir disoproxil fumarate (TDF), which is a nucleotide analog of adenosine phosphate. The NRTIs are administered as unphosphorylated prodrugs. Upon entering the host cell, these prodrugs are recognized by cellular kinases and further converted to the tri-phosphorylated form. The tri-phosphorylated NRTIs then bind to the active site of RT and are catalytically incorporated into the growing DNA chain. The incorporated NRTIs block the further extension of the chain since the NRTIs lack the 3′ hydroxyl group on their ribose or pseudoribose moiety and thus cannot form the 3′–5′ phosphodiester bond needed for DNA extension. NRTIs are one of the major classes of inhibitors used in all combination therapies for the treatment of HIV-infected patients. However, the clinical successes of these agents are limited by viral resistances to NRTIs, arising through mutations in the coding region of RT. These mutations confer viral resistances through improved discrimination of a nucleotide analog relative to the natural substrate, or by increased phosphorolytic cleavage of an analog-blocked primer. To overcome these acquired resistances, the design of the next generation of NRTIs has been mainly focused on two fronts: (i) nucleoside analogs possessing a 3′ hydroxyl group that can induce delayed polymerization arrest; (ii) nucleotide analogs that are designed to be incorporated into the viral genome during replication. These nucleotide analogs can introduce mutations into the HIV genome through mispairing and blockade of the replication process.

11.4.1.2 Human Steroid 5α-Reductase Inhibitors

The human enzyme steroid 5α-reductase is responsible for the conversion of testosterone (T) to the more potent androgen, dihydrotesterone (DHT). It has been shown that abnormally high 5α-reductase activity in humans leads to excessively high DHT levels in peripheral tissues. Inhibition of 5α-reductase thus offers a potential treatment for DHT-associated diseases, such as benign prostate hyperplasia, prostate cancer, acne, and androgenic alopecia. In humans, there are two types of steroid 5α-reductase: type I and type II. The type I 5α-reductase is mainly expressed in the sebaceous glands of skin and the liver, while the type II enzyme is most abundant in the prostate, seminal vesicles, liver, and epididymis. The first 5α-reductase inhibitor approved for clinical application in the United States was finasteride; it is currently employed in the treatment of benign prostatic hyperplasia (BPH) in men. This compound is approximately 100-fold more potent toward the type II than the type I isozyme of 5α-reductase. In humans, finasteride decreases prostatic DHT levels by 70%–90%, resulting in reduced prostate size. The detailed biochemical characterization of finasteride inhibition suggested that finasteride is a mechanism-based inhibitor. It is proposed that by closely mimicking the substrate (testosterone), finasteride is accepted as an alternate substrate and forms an NADP-dyhydrofinasteride adduct at the enzyme active site (Figure 11.4). This covalent NADP-dyhydrofinasteride adduct represents a bisubstrate analog with extremely high affinity ($K_i \leq 1 \times 10^{-13}$ M) to the type II 5α-reductase. Interestingly, finasteride is also a mechanism-based inhibitor of the human type I 5α-reductase. However, the NADP-dyhydrofinasteride adduct formation rate at the type I 5α-reductase active site is reduced by more than 100-fold compared to that for the type II isozyme. This difference in NADP-dyhydrofinasteride adducts formation rate accounts for the isozyme selectivity of finasteride both *in vitro* and *in vivo*. Knowledge of the mechanism of inhibition of 5α-reductase by 4-azasteroids (represented by finasteride) and of the SAR for dual 5α-reductase inhibition, led to the discovery of a potent, dual inhibitor of 5α-reductase, known as dutasteride. Dutasteride, is equipotent versus type I and type II 5α-reductase and demonstrates exceptional *in vivo* potency. This compound has also been approved for clinical use in the treatment of BPH.

11.4.2 Intermediate State-Based Design

11.4.2.1 Inhibitors of Hydroxymethylglutaryl-CoA Reductase (HMG-CoA Reductase)

The biosynthetic pathway for cholesterol involves more than 25 different enzymes. The enzyme 3-hydroxy-3-methylglutaryl coenzyme A (HMG-CoA) reductase catalyzes the conversion from

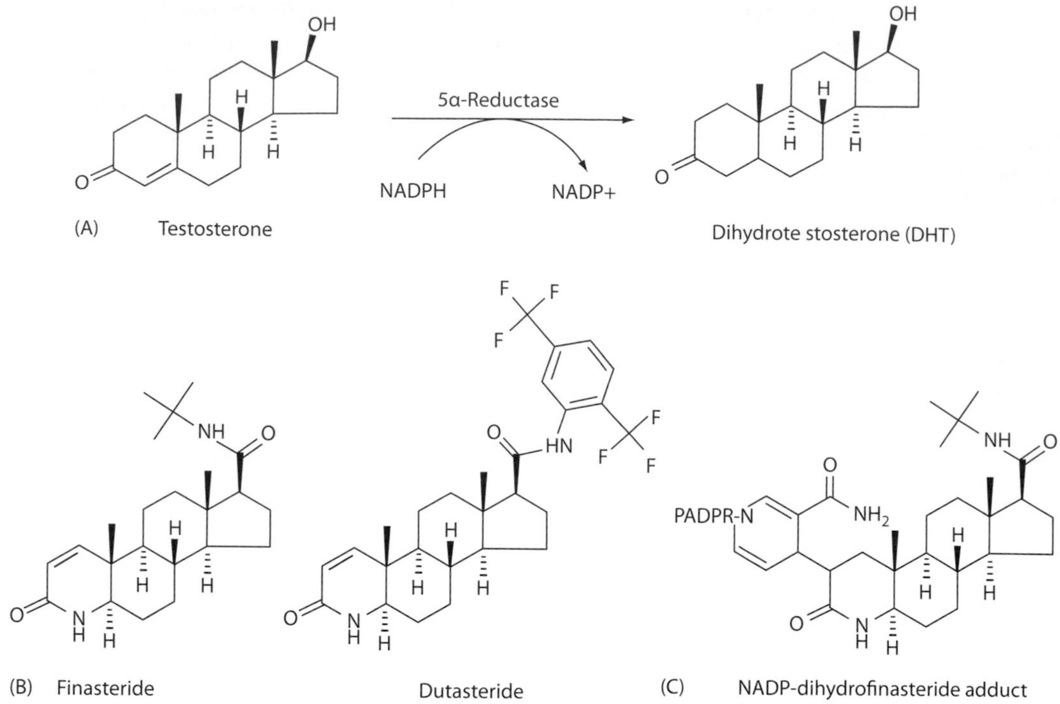

FIGURE 11.4 (A) 5α-Reductase catalyzed conversion of testosterone (T) to dihydrotesterone (DHT); (B) chemical structures for finasteride and dutasteride; and (C) the proposed structure of the NADP-dihydrofinasteride adduct. PADPR, phosphoadenosine diphosphoribose.

HMG-CoA to mevalonate, the rate-limiting step of the entire pathway. Inhibition of HMG-CoA reductase provides a very attractive opportunity to inhibit cholesterol biosynthesis because no buildup of potentially toxic precursors occurs upon inhibition. In 1976, Japanese microbiologist Akira Endo isolated a series compounds including ML236B (compactin) from *Penicillium citrinum* with powerful inhibitory effect on HMG-CoA reductase. Since then, seven HMG-CoA reductase inhibitors have become marketed drugs for lowering cholesterol levels (Figure 11.5).

These HMG-CoA reductase inhibitors, commonly referred to as statins, have accounted for the majority of prescriptions for cholesterol-lowering drugs worldwide. All the statins in clinical use are analogues of the substrate HMG-CoA with an HMG-like moiety, which may be present in an inactive lactone form in the prodrugs (Figure 11.5.). These statins are classified in two groups according to their molecular structures. Type I statins, including lovastatin and simvastatin, are lactone prodrugs originally isolated from fungi. They are enzymatically hydrolyzed *in vivo* to produce the active drug. Type II statins are all synthetic products with larger groups attached to the HMG-like moiety. All the statins are competitive with respect to HMG-CoA and noncompetitive with respect to NADPH, a cosubstrate of the reaction. Crystal structures of HMG-CoA reductase complexed with six different statins showed that the statins occupy the HMG-binding region, but do not extend into the NADPH site. The orientation and bonding interactions of the HMG-moiety of the statins resemble those of the substrate complex. However, from combination crystal structures, binding thermodynamics, and SAR studies it is clear that the 5′-hydroxyl group of the acidic side-chain acts as a mimetic of the tetrahedral intermediate of the reduction reaction. The multiple H-bonds between the C–5–OH of the statins and the HMG-CoA reductase active site contribute significantly to the tight binding of the statin inhibitors. Strictly speaking, the HMG-CoA reductase inhibitors are not products of rational design; rather they were identified

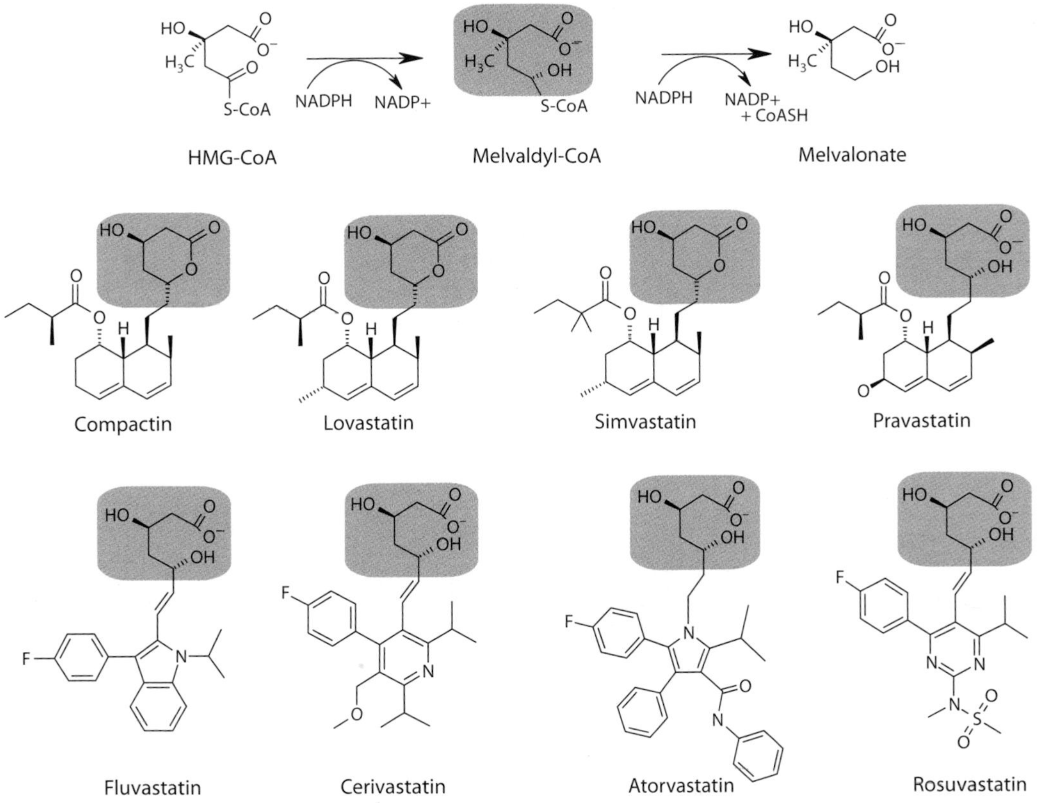

FIGURE 11.5 Structures of HMG-CoA reductase reaction substrate, tetrahedral intermediate, product, and the statin inhibitors. Compactin, Lovastatin, and Simvastatin are type I statins. All other statins are type II statins. The melvaldyl tetrahedral intermediate that is mimicked in all statins is shaded in gray.

through natural product screening and analoging of the natural product hits. Nevertheless, it is quite clear that all statins share a common strategy for inhibiting their target: tetrahedral intermediate state mimicry.

11.4.2.2 Inhibitors of Purine Nucleoside Phosphorylase

Purine nucleoside phosphorylase (PNP) catalyzes the phosphorolysis of 6-oxypurine nucleosides and deoxynucleosides. In humans, the PNP pathway is the only route for deoxyguanosine degradation and genetic deficiency in this enzyme leads to profound T-cell-mediated immunosuppression. Inhibition of PNP has applications in treating aberrant T lymphocyte activity, which is implicated in T-cell leukemia and autoimmune diseases. The challenge to inhibitor design for PNP arises from the abundance of the enzyme in human tissues. It has been shown that near complete inhibition of PNP (>95%) is required for significant reduction in T-cell function. Structural-based inhibitor design produced some inhibitors with K_d values in the nanomolar range. However, clinical evaluations showed that these inhibitors did not produce sufficient inhibition of PNP to be effective anti-T-cell therapies. Much more potent PNP inhibitors were later designed with the aid of transition state analysis. In theory, a perfect transition state inhibitor of PNP should bind with a K_d value of approximately 10^{-17} M (10 attomolar). The structure of the transition state for human PNP was determined by Schramm and coworkers by measuring kinetic isotope effects. Their studies revealed a transition state with significant ribooxycarbenium character (Figure 11.6). Based on the features of this transition state, compounds with picomolar affinity to PNP were synthesized. Among them,

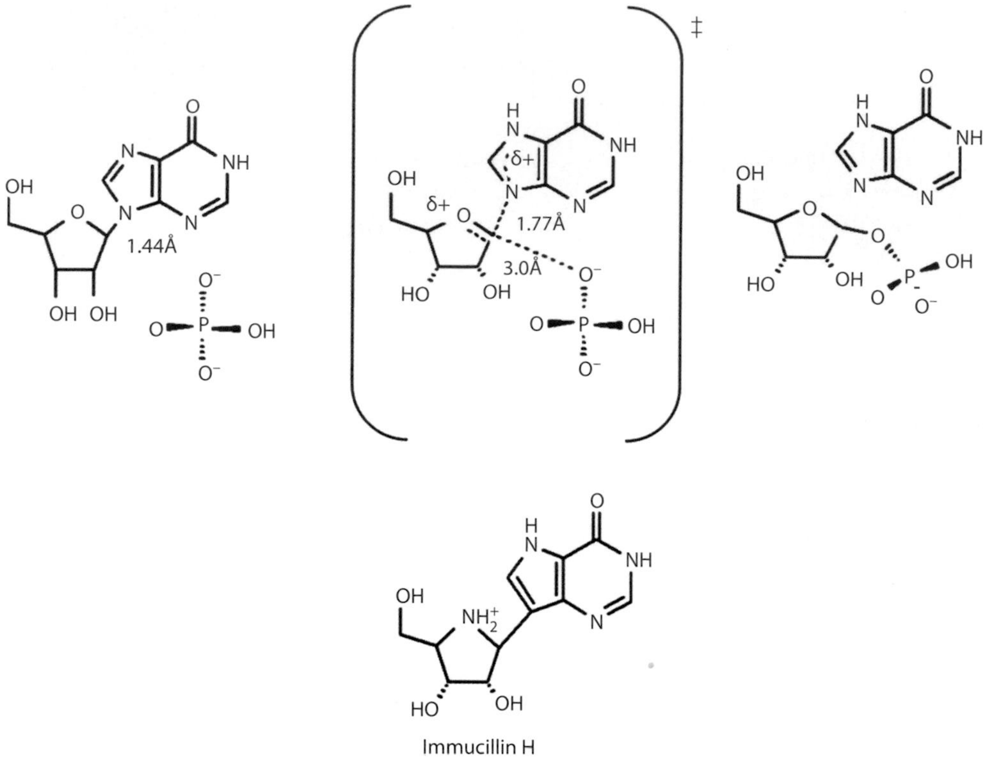

FIGURE 11.6   The structures of PNP reaction substrate inosine, transition state, and reaction product (top panel) and transition state-based inhibitor Immucillin H (bottom panel).

Immucillin H was a 56 pM inhibitor of human PNP with good potency against cultured human T-cell lines in the presence of deoxyguanosine. Currently, Immucillin H is in phase II clinical trials for the treatment of leukemia.

11.5   BIOSTRUCTURE-BASED DESIGN

In the preceding section of this chapter we established the fundamental importance to drug discovery of a deep, mechanistic understanding of the reaction mechanism of an enzyme target. While this can be accomplished by the application of mechanistic enzymology, it can be facilitated greatly by the knowledge of the three-dimensional structure of the protein, obtained via biostructure-based technologies such as computational biochemistry, NMR spectroscopy, and x-ray crystallography. Visualization of the detailed architecture of an enzyme's active site, in complex with a small-molecule inhibitor, can be an important driver in the optimization of a medicinal chemistry effort. The structural insights thus obtained allow for improvements in target potency, selectivity, and inhibitor physicochemical properties, all of which are paramount in establishing inhibitor SAR.

The term "rational drug design" is often used to describe the application of structure-guided drug discovery approaches. Over the past two decades, several drugs have been made available to patients as a result of advances in protein crystallography and other structural methods. For example, more than 40 compounds have entered clinical trials whose discovery was reliant upon a biostructure-based approach, and as of 2007, at least 10 of these have been approved by regulatory agencies. In this section, we exemplify how a detailed understanding of the topography of a ligand–enzyme complex can provide a basis for the design of better inhibitors and can complement enzymological studies to rationalize their biochemical mode of action.

11.5.1 STRUCTURE-BASED DESIGN OF PROTEIN KINASE INHIBITORS

Owing to their central roles in mediating cellular signaling pathways, protein kinases are increasingly important targets for treating a number of diseases. In particular, many of the over 500 kinases encoded by the human genome function to regulate tumor cell proliferation, migration, and survival, rendering them attractive targets for chemotherapeutic intervention in the treatment of cancer. Despite their diversity, all protein kinases catalyze the transfer of the γ-phosphate of ATP to the hydroxyl group of serine, threonine, or tyrosine residues on specific proteins. Their catalytic domains reflect this singular function in that they share a common feature called the protein kinase fold, which includes a highly conserved ATP-binding pocket, flanked by N-terminal and C-terminal lobes. The ATP-binding site has been the major focus of inhibitor design; owing to its high degree of conservation, however, selectivity has been a major challenge for inhibitors that target this binding site of protein kinases. The use of biostructure-based approaches has therefore been of great importance in the optimization of targeted anticancer therapies.

X-ray crystallographic studies have indicated that the catalytic activity in most kinases is controlled by an "activation loop," which adopts different conformations depending upon the phosphorylation state of serine, threonine, or tyrosine residues within the loop. In kinases that are fully active, the loop is thought to be stabilized in an open conformation as a result of phosphorylation, allowing a β-strand within the loop to serve as a platform for substrate binding. While the "active" conformation of the loop is very similar in all known structures of activated kinases, there is great variability in the loop conformation in the inactive state of kinases. In this inactive-like conformation, the loop places steric constraints, which preclude substrate binding.

One of the first protein kinase inhibitors developed as a targeted cancer therapy is imatinib (Gleevec®; Novartis Pharmaceuticals, Basel, Switzerland). Imatinib has been used with remarkable success to treat patients with chronic myeloid leukemia (CML), which is a malignancy resulting from the deregulated activity of Abl due to a chromosomal translocation that gives rise to the breakpoint cluster region-abelson tyrosine kinase oncogene (BCR-ABL). Imatinib inhibits the tyrosine kinase activity of Bcr-Abl and it is considered as a frontline treatment for CML by virtue of its high degree of efficacy and selectivity. Together with biochemical studies, crystallographic studies of the interaction of imatinib with the Abl kinase domain have revealed that imatinib binds to the Bcr-Abl ATP-binding site preferentially when the centrally located activation loop is not phosphorylated, thus stabilizing the protein in an inactive conformation (Figure 11.7). In addition, imatinib's interactions with the NH_2-terminal lobe of the kinase appear to involve an induced-fit mechanism, further adding to the unique structural requirements for optimal inhibition. One of the most interesting aspects of this interaction is that the specificity of inhibition is achieved despite the fact that residues that contact imatinib in Abl kinase are either identical or very highly conserved in other Src-family tyrosine kinases. Thus, despite targeting the relatively well conserved nucleotide-binding pocket of Abl, studies have shown that imatinib achieves its high specificity by recognizing the distinctive inactive conformation of the Abl activation loop. Biostructure-based methods have had a further impact on more recent efforts to design second-generation therapies targeting imatinib-resistant mutations in Bcr-Abl kinase that have been identified in CML patients. It is very likely that these new inhibitors will have substantial clinical utility in the treatment of imatinib-resistant CML; continued exploration of the structural details of the interactions between these compounds and the mutant kinase are still necessary, as resistance remains an inevitable consequence of such drug treatment regimens.

The three catalytically active receptor tyrosine kinases (RTKs) of the ErbB family represent another attractive target for the treatment of a variety of cancers: epidermal growth factor receptor (EGFR, also known as ErbB1), ErbB2 (also known as HER2/*neu*), and ErbB4. These RTKs are large, multidomain proteins that contain an extracellular ligand-binding domain, a single transmembrane domain, and a cytoplasmic domain responsible for the tyrosine kinase activity. Ligand binding to the extracellular domain induces the formation of receptor homo- and heterodimers, which leads to activation of the tyrosine kinase activity and subsequent phosphorylation of the cytoplasmic tail.

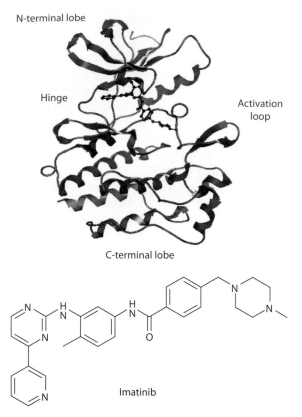

FIGURE 11.7   The structure of Imatinib (Gleevec) in complex with the catalytic domain of cAbl (PDB 1iep).

A number of ErbB-targeted molecules have already reached the market, with a number of others in various stages of clinical investigation. Two of these molecules, erlotinib (Tarceva®, Genetech, Inc. and OSI Pharmaceuticals, Inc. San Francisco, CA) and lapatinib (Tykerb®, GlaxoSmithKline plc. Brentford, U.K.), share a common 4-anilinoquinazoline core, yet their ErbB inhibition profiles and mechanisms of action are clearly differentiated on the basis of biochemical and crystallographic studies. For example, while erlotinib is a potent and selective inhibitor of EGFR only (K_i^{app} = 0.4 nM), lapatinib exhibits potent activity against both EGFR and ErbB2, with estimated K_i^{app} values of 3 and 13 nM, respectively. In addition to its dual kinase activity profile, lapatinib can be distinguished further from erlotinib in that it has a prolonged off-rate from its kinase targets compared to the very fast off-rate from EGFR of erlotinib. This translates to a half-life of dissociation of 300 min for the lapatinib-EGFR complex. Importantly, in cellular washout experiments this slow off-rate correlates with a prolonged inhibition of receptor tyrosine phosphorylation in tumor cells (see Section 11.6 for additional information).

An evaluation of the binding mode of lapatinib, based on the crystal structure of the compound in complex with EGFR, suggests a rationale for its long target residence time compared to other 4-anilinoquinazoline inhibitors. Not surprisingly, the quinazoline ring was observed to be hydrogen-bonded to the flexible hinge region between the NH_2 and COOH-terminal lobes of the kinase, but there are variations in the key H-bonding interactions compared to those revealed in the erlotinib-EGFR structure. These differences indicate that lapatinib binds to a relatively closed form of this binding site, whereas erlotinib binds to a more open form. In addition, the ATP-binding pocket of the lapatinib-EGFR complex has a larger back pocket than the apo-EGFR or erlotinib-EGFR structures owing to a shift in one end of the C-helix. This enlarged back pocket accommodates

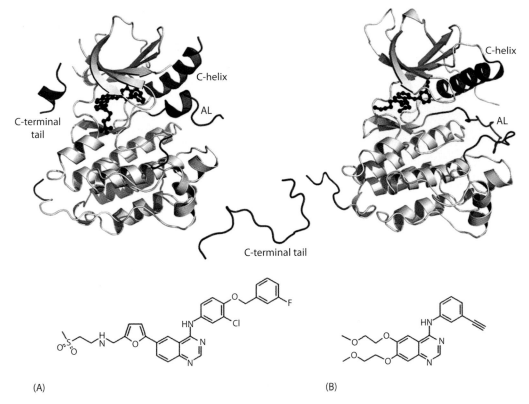

FIGURE 11.8 Protein–ligand binding modes of (A) lapatinib and (B) erlotinib in complex with EGFR (AL, activation loop). (Courtesy of Lisa M. Shewchuk.)

the 3-fluorobenzyloxy group of lapatinib (Figure 11.8). The structural change is significant because it results in the loss of a highly conserved Glu738–Lys721 salt bridge, which is an important regulatory mechanism of kinases, functioning to ligate the phosphate groups of ATP. The net result of these structural differences is that the activation loop in the lapatinib-EGFR structure adopts a conformation that is reminiscent of that found in inactive kinases, while the erlotinib-EGFR structure displays the activation loop in an active conformation. These effects provide a potential molecular rationale for the prolonged residence time of lapatinib on its target, which in turn may result in the observed duration of drug activity in cells. In total, these elegant structural and biochemical studies have important implications for the discovery of novel, targeted signal transduction inhibitors, and suggest that subtle differences in kinase inhibitor structure can have a profound impact on the binding mode, kinetics, and cellular activity.

11.5.2 STRUCTURE-BASED DESIGN OF HIV PROTEASE INHIBITORS

Perhaps the greatest impact of structure-based design on the identification of novel medicines has been in the treatment of AIDS, the etiologic agents of which are human immunodeficiency virus type 1 and type 2 (HIV-1 and HIV-2). These retroviruses encode relatively simple genomes consisting of three open reading frames (ORFs), *gag*, *pol*, and *env*. The *gag* gene encodes the structural capsid, nucleocapsid, and matrix proteins, while the *env* gene is processed by multiple alternative splicing events to yield regulatory proteins. The *pol* ORF encodes the essential viral enzymes necessary for viral replication: RT, integrase, and protease (PR). HIV-1 PR is an aspartyl protease that is required for proteolytic processing of the Gag and Gag-Pol polyprotein precursors to yield

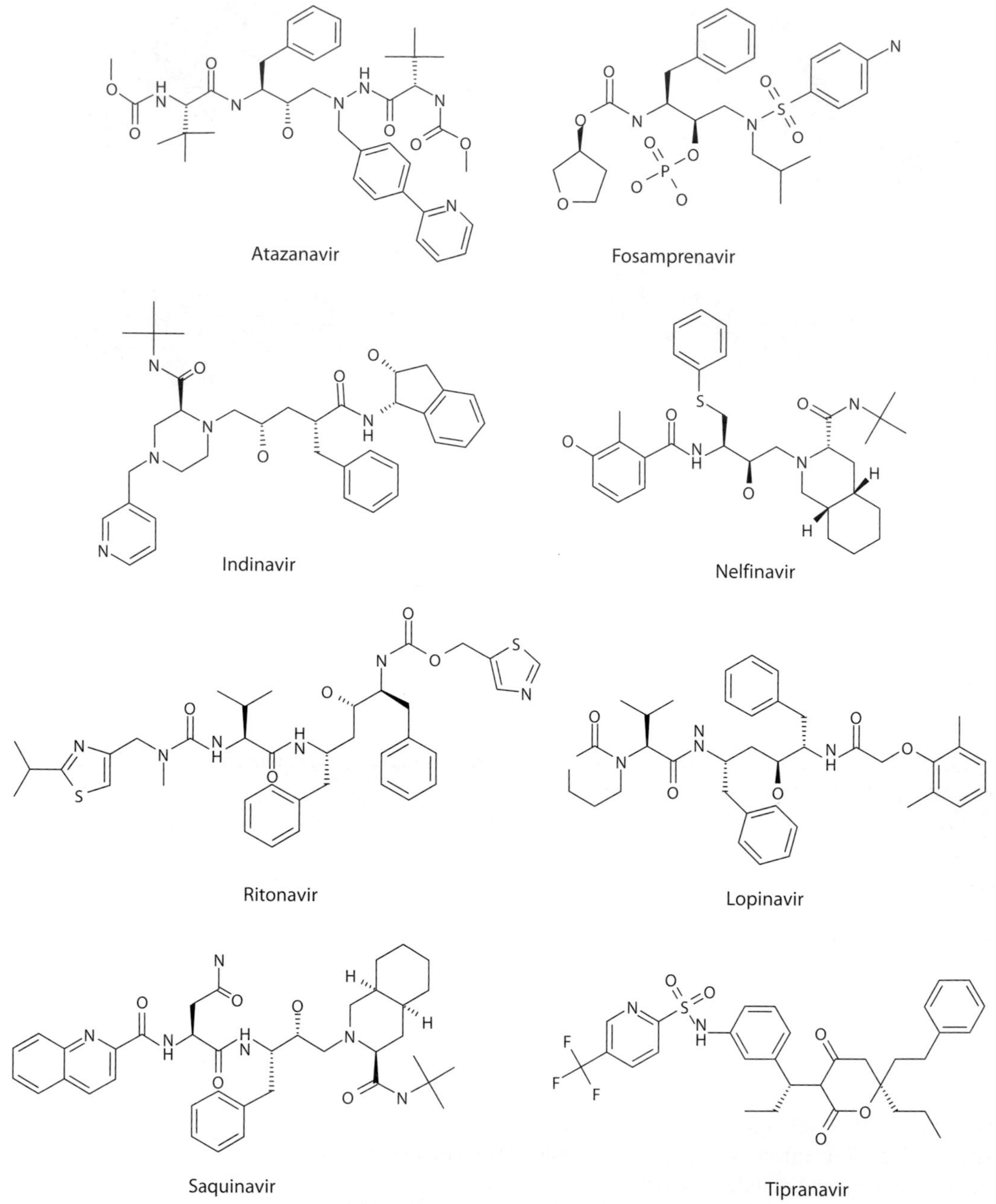

FIGURE 11.9 Clinically approved HIV-1 protease inhibitors for the treatment of AIDS.

the viral enzymes and structural proteins, and it is absolutely indispensable for proper virion assembly and maturation. For this reason it has been an important target for the discovery of anti-HIV therapeutics, and indeed there are at least eight drugs in current clinical use whose antiviral mode of action is by potent inhibition of the HIV protease (Figure 11.9).

One of the major driving forces behind the rapid progress in the identification of HIV protease inhibitors to combat AIDS has been the intense investigation of the structure of the enzyme, particularly in complex with a number of different inhibitors. HIV-1 PR is a dimer comprised of two polypeptide chains of 99 amino acids, each contributing a single catalytic aspartate residue within the active site that lies at the dimer interface. This active site is covered by two symmetric flaps whose dynamic motions allow entry and exit of polypeptide substrates. For each of the different substrates, three to four amino acids on each side of the scissile bond are thought to be involved in binding to the substrate cavity. Since there is little similarity in the primary sequence of the cleavage sites of each of the protease substrates, binding specificity is thought to be driven by the conservation in the secondary structure surrounding the cleavage sites. All of the inhibitors currently used to treat HIV infection are competitive in nature and bind to the protease active site.

Saquinavir was the first HIV protease inhibitor available for the treatment of AIDS, and its design was based on a strategy using a transition state mimetic. A distinguishing feature of HIV PR is its ability to cleave Tyr-Pro and Phe-Pro sequences found in the viral substrates, as mammalian endopeptidases are unable to cleave peptide bonds followed by a proline. A rational inhibitor design approach based on this property offered hopes of identifying inhibitors selective for the viral enzyme. Since reduced amides and hydroxyethylamine isosteres most readily accommodate the amino acid moiety of Tyr-Pro and Phe-Pro in the HIV substrates, they were chosen for further interrogation. Systematic substitutions were explored on a minimum peptidic pharmacophore, and one compound containing an (S,S,S)-decahydro-isoquinoline-3-carbonyl (DIQ) replacement for proline exhibited a K_i value of 0.12 nM at pH 5.5 for HIV-1 PR and <0.1 nM for HIV-2 PR. The interactions of this compound, later named saquinavir, with HIV-1 PR were studied crystallographically (Figure 11.10). The compound was shown to bind to the enzyme in an extended conformation with the carbonyl of the DIQ group binding to a water molecule that connects the inhibitor with the flap regions. These studies shed much light on the binding mode of the first HIV PR inhibitors and set the stage for further exploration of novel compounds with improved properties.

The availability of new HIV protease inhibitors represented a great triumph in the fight against AIDS, but it was only a matter of time before the selective pressure of antiretroviral therapy led to the emergence of HIV strains harboring drug-resistant mutations against protease inhibitors. One of the primary mutations first noted in protease inhibitor-resistant strains was in Val82 of HIV-1 PR. Crystallographic and modeling studies suggested that the binding of protease inhibitors like ritonavir might be compromised due to the loss of hydrophobic interactions between the isopropyl side chain of Val 82 of the enzyme and the isopropyl substituent projecting from the 2 position of the P3 thiazolyl group of ritonavir. This functionality was substituted to identify an inhibitor whose activity was less dependent on interaction with Val 82, and the optimization that was supported by modeling studies led to the identification of ABT-378, later named lopinavir. This novel inhibitor had extraordinary potency against wild-type and mutant HIV PR ($K_i = 1.3–3.6$ pM) *in vitro*, and maintained activity against ritonavir-resistant mutants of HIV-1. Lopinavir, as a combination drug

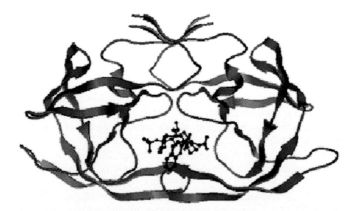

FIGURE 11.10 The crystal structure of HIV-1 protease in complex with the inhibitor saquinavir (PDB 1hxb).

with ritonavir, is known as Kaletra, and is an important salvage drug for patients who have failed primary therapy with other protease inhibitors.

11.6 CONCLUDING REMARKS

The structural details of drug-binding pocket interactions, gleaned from crystallographic and other biophysical methods, can provide a rich source of information for inhibitor optimization. Historically, SAR through structure-based methods has been limited by the time and protein demands of x-ray crystallography. These limitations, however, are rapidly diminishing due to significant advances in the technologies of protein expression and protein crystallography. In particular, robotic methods for crystallization trials have significantly reduced the time required for obtaining crystals of a target protein in complex with multiple inhibitory compounds. Likewise, the more routine use of high-energy beam sources has facilitated structure determinations from crystals that would otherwise not be sufficient for diffraction studies. These advances provide the basis for greater reliance on structural biology as a common tool during the iterative process of lead optimization.

Conformational dynamics within the drug-binding pockets of enzymes is a common feature, dictated by the chemistry of enzyme catalysis. Hence, binding pocket structures are not static; rather, they often change in response to encounters with inhibitory molecules. Thus, as described in this chapter, there can often be a temporal component to enzyme–inhibitor affinity. Advances in kinetic methodologies and instrumentation have made the determination of inhibition kinetics more facile, so that such measurements can be a routine part of the SAR of lead optimization. It is therefore no longer necessary to rely solely on equilibrium measures of inhibitor-binding affinity, such as IC_{50} and K_i values, for lead optimization. Instead, routine measurements of enzyme–inhibitor association and dissociation rates are becoming practical, with throughput that makes these measurements germane to drug discovery. Hence, increased attention is being paid to the importance of understanding these kinetic components of drug–target interactions, and their potential impact on clinical efficacy. For example, the duration of drug efficacy *in vivo* has been suggested to depend in part on the duration of the drug–target complex; this is experimentally measured as the residence time, which is the reciprocal of the dissociation rate constant for the drug–target complex. Drugs that demonstrate long residence times, especially when this exceeds the pharmacokinetic half-life of the drug, may significantly extend the pharmacodynamic efficacy of a drug *in vivo*, and may also ameliorate the potential for adverse events. Future drug discovery efforts may thus be focused not merely on optimization of inhibitor affinity, but also on the extension of residence time (see Further Readings).

FURTHER READINGS

Copeland, R. A. (2000) *Enzymes: A Practical Introduction to Structure, Mechanism and Data Analysis*, 2nd edn. Wiley, Hoboken, NJ.

Copeland, R. A. (2005) *Evaluation of Enzyme Inhibitors in Drug Discovery: A Guide for Medicinal Chemists and Pharmacologists*. Wiley, Hoboken, NJ.

Copeland, R. A., Pompliano, D. L., and Meek, T. D. (2006) Drug-target residence time and its implications for lead optimization. *Nat. Rev. Drug Discov.*, **5**: 730–739.

Nagar, B., Bornmann, W. G., Pellicena, P., et al. (2002) Crystal structures of the kinase domain of c-Abl in complex with the small molecule inhibitors PD173955 and imatinib (STI-571). *Cancer Res.* **62**: 4236–4243.

Roberts, N. A., Martin, J. A., Kinchington, D., et al. (1990) Rational design of peptide-based HIV proteinase inhibitors. *Science* **248**: 358–361.

Robertson, J. G. (2005) Mechanistic basis of enzyme-targeted drugs. *Biochemistry* **44**, 5561–5571.

Schramm, V. L. (2005) Enzymatic transition states: Thermodynamics, dynamics and analogue design. *Arch. Biochem. Biophys.* **433**, 13–26.

Wood, E. R., Truesdale, A. T., McDonald, O. B., et al. (2004) A unique structure for epidermal growth factor receptor bound to GW572016 (Lapatinib): Relationships among protein conformation, inhibitor off-rate, and receptor activity in tumor cells. *Cancer Res.*, **64**: 6652–6659.

12 Receptors: Structure, Function, and Pharmacology

Hans Bräuner-Osborne

CONTENTS

12.1   INTRODUCTION

Communication between cells is mediated by compounds such as neurotransmitters and hormones, which, upon release, will activate receptors in the target cells. This communication is of pivotal importance for many physiological functions and dysfunction in cell communication pathways often have severe consequences. Many diseases are caused by dysfunction in the pathways and in these cases, drugs designed to act at the receptors have beneficial effects. Thus, receptors are very important drug targets.

The first receptors were cloned in the mid-1980s and since then hundreds of receptor genes have been identified. Based on the sequence of the human genome it is currently estimated that more than 1000 human receptors exist. Almost all receptors are heterogeneous, meaning that several receptor subtypes are activated by the same signaling molecule. One such example is the excitatory neurotransmitter glutamate. As shown in Figure 12.1, the amino acid sequence of the glutamate receptors vary and the receptors form subgroups, which, as will be discussed in Chapter 15, share pharmacology.

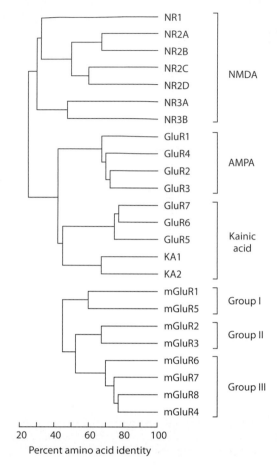

FIGURE 12.1 Phylogenetic tree showing the amino acid sequence identity between cloned mammalian glutamate receptors. The subgroups according to receptor pharmacology have been noted. The NMDA, AMPA, and kainic acid receptors belong to the superfamily of ligand-gated ion channels whereas the metabotropic glutamate receptors (mGluR1-8) belong to the superfamily of GPCRs.

The same signaling molecule can act on both G protein-coupled receptors (GPCRs) and ligand-gated ion channels (Figure 12.1). One of the reasons for the heterogeneity is that it allows cells to be regulated in subtle ways. For example, whereas the fast synaptic action potential is initiated by glutamate receptors of the ligand-gated ion channel family, these receptors are themselves regulated by slower and longer acting glutamate receptors from the GPCR family. The action on these two receptor families is shared by a number of other neurotransmitters such as gamma-aminobutyric acid (GABA) (Chapter 15), acetylcholine (Chapter 16), and serotonin (Chapter 18).

12.1.1 Synaptic Processes and Mechanisms

Receptors are located in a complex, integrated, and highly interactive environment, which can be further illustrated by the processes and mechanisms of synapses (Figure 12.2). The synapses are key elements in the interneuronal communication in the peripheral and in the central nervous system (CNS). In the CNS, each neuron has been estimated to have synaptic contact with several thousand other neurons, making the structure and function of the CNS extremely complex.

The receptor is activated upon release of the signaling molecule and it is, evidently, equally important to stop the signaling again. This is often achieved by transporters situated in the vicinity

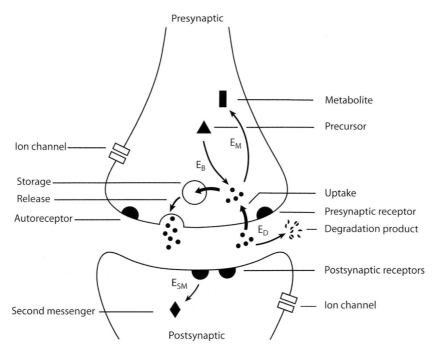

FIGURE 12.2 Generalized schematic illustration of processes and mechanisms associated with an axosomatic synapse in the CNS. E, enzymes; E_M, metabolic; E_B, biosynthetic; E_D, degradation; E_{SM}, second messenger; (•) neurotransmitter.

of the receptor, which remove the signaling molecule from the extracellular to the intracellular space, where it is either stored or metabolized. The blockade of a transporter or a metabolic enzyme will cause an elevation of the extracellular concentration of the signaling molecule and lead to increased receptor activation, and transporters and metabolic enzymes can thus be viewed as indirect receptor targets. Synaptic functions may also be facilitated by the stimulation of the neurotransmitter biosynthesis, for example, by administration of a biochemical precursor. Transport mechanisms in synaptic storage vesicles (Figure 12.2) are also potential sites for pharmacological intervention. Autoreceptors normally play a key role as a negative feedback mechanism regulating the release of certain neurotransmitters, making this class of presynaptic receptors therapeutically interesting.

Pharmacological stimulation or inhibition of the earlier mentioned synaptic mechanisms are, however, likely to affect the function of the entire neurotransmitter system. Activation of neurotransmitter receptors may, in principle, represent the most direct and selective approach to the stimulation of a particular neurotransmitter system. Furthermore, activation of distinct subtypes of receptors operated by the neurotransmitter concerned may open up the prospect of highly selective pharmacological intervention. Nevertheless, indirect mechanisms of targeting receptors via regulation of the level of the endogenous agonist at the site-of-action remains an important pharmacological principle, which has also been applied outside the synapse as exemplified by compounds increasing insulin release and preventing GLP-1 breakdown.

Direct activation of receptors by full agonists may result in rapid receptor desensitization (insensitive to activation). Partial agonists are less liable to induce receptor desensitization and may therefore be particularly interesting for neurotransmitter replacement therapies. Desensitization may be a more or less pronounced problem associated with the therapeutic use of receptor agonists, whereas receptor antagonists, which in other cases have proved useful therapeutic agents, may inherently cause receptor supersensitivity. The presence of allosteric binding sites at certain receptor complexes, which may function as physiological modulatory mechanisms, offer unique prospects of selective and flexible pharmacological manipulation of the receptor complex concerned. While some receptors are associated with ion channels, others are coupled to second messenger systems. The key steps in such enzyme-regulated multistep intracellular systems (Figure 12.2), also including regulation of gene transcription by second messengers, represent novel targets for therapeutic interventions.

12.2   RECEPTOR STRUCTURE AND FUNCTION

Receptors have been divided into four major superfamilies: GPCRs, ligand-gated ion channels, tyrosine kinase receptors, and nuclear receptors. The first three receptor superfamilies are located in the cell membrane and the latter family is located intracellularly.

Our understanding of ligand–receptor interactions and receptor structure has increased dramatically during the previous decade, not least due to the rapidly growing number of 3D crystallographic structures that have been determined of either full receptors or isolated ligand-binding domains. Thus today, structures of partial or full receptors of all four receptor superfamilies have been determined. Clearly, the information obtained from 3D structures of ligand-binding domains in the presence of ligands is very valuable for rational drug design (see Chapter 2). Likewise, knowledge about receptor mechanisms can be used to, e.g., design allosteric modulators interfering with receptor activation.

12.2.1   G Protein-Coupled Receptors

GPCRs are the largest of the four superfamilies with some estimated 1000 human receptor genes. Approximately 50% of these are taste- and odor-sensing receptors that are not of immediate interest for the pharmaceutical industry but are of interest, for example, fragrance manufactures. Nevertheless, it is estimated that 30% of all currently marketed drugs act on GPCRs and the superfamily thus remains a very important target for drug research. It is fascinating to note the very broad variety of signaling molecules or stimuli that are able to act via this receptor superfamily, including tastes, odors, light (photons), ions, monoamines, nucleotides, amino acids, peptides, proteins, and pheromones.

The GPCRs are also referred to as seven transmembrane (7TM) receptors due to the seven alpha-helical transmembrane segments found in all GPCRs (Figure 12.3) and the fact that the receptors can also signal via G protein independent pathways (see later). The GPCRs have been further subdivided into family A, B, and C based on their amino acid sequence homology. Thus receptors

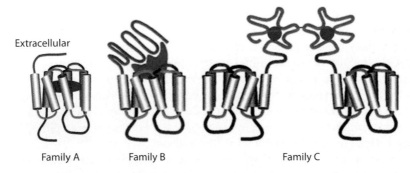

FIGURE 12.3   The superfamily of GPCRs. All GPCRs contain seven α-helical transmembrane segments and are thus also called seven transmembrane (7TM) receptors. Cartoon of the three families showing the typical orthosteric binding site (agonist in red); family A receptors bind the agonist in the 7TM region, family B receptors bind the agonist in both the 7TM region and the extracellular amino-terminal domain, and dimeric family C receptors bind the agonist exclusively in the extracellular amino-terminal domain. (Adapted from Ji, T. et al., *J. Biol. Chem.*, 273, 17299, 1998.)

within family A are closely related to each other than to receptors in family B and C and the like. This grouping also coincides with the way ligands binds to the receptors. Thus, as illustrated in Figure 12.3, the endogenous signaling molecules generally bind to the transmembrane region of family A receptors (e.g., acetylcholine, histamine, dopamine, serotonin, opioid, and cannabinoid GPCRs, Chapters 16 through 19), to both the extracellular loops and amino-terminal domain of family B receptors (e.g., glucagon and GLP-1 GPCRs) and exclusively to the extracellular amino-terminal domain of family C receptors (e.g., glutamate and GABA GPCRs, Chapter 15).

The intracellular loops of GPCRs interact with G proteins. As illustrated in Figure 12.4, the G proteins are trimeric consisting of G_α, G_β, and G_γ subunits. Receptor activation will cause an interaction of the receptor with the trimeric $G_{\alpha\beta\gamma}$-protein, catalyzing an exchange of GDP for GTP in the G_α subunit whereupon the G protein disassociate into activated G_α and $G_{\beta\gamma}$ subunits. Both of these will then activate effector molecules such as adenylate cyclase or ion channels (Figure 12.4). 16 G_α, 5 G_β, and 12 G_γ subunits have been identified in humans and like the receptors they form groups based on the amino acid homology and the effectors they interact with.

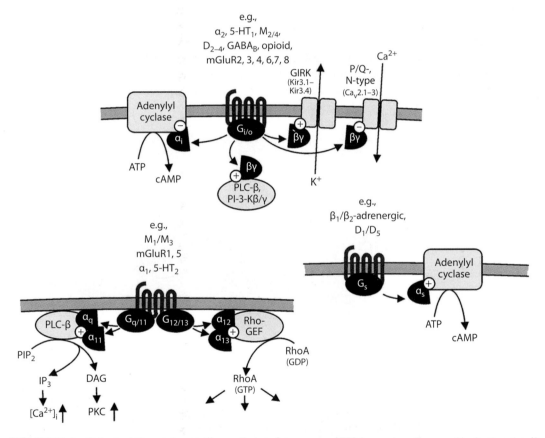

FIGURE 12.4 Principal G protein coupling pathways for a range of 7TM receptors discussed in further detail in Chapters 15 through 19. Receptor activation will catalyze an exchange of GDP for GTP in the α-subunit, which leads to activation and separation of the α- and $\beta\gamma$-subunits. Both of these will modulate downstream effectors. α_{1-2} and β_{1-2}, adrenergic receptor subtypes; D_{1-5}, dopamine receptor subtypes 1–5; GIRK, G protein-regulated inward rectifier potassium channel; 5-$HT_{1,2}$, serotonin receptor subtypes 1 and 2; M_{1-5}, muscarinic acetylcholine receptor subtypes 1 to 5; mGluR1–8, metabotropic glutamate receptor subtypes 1 to 8; PLC-β, phospholipase C-β; PI-3-K, phosphoinositide-3-kinase; PIP_2, phosphatidylinositol 4,5-bisphosphate; IP_3, inositol 1,4,5-trisphosphate; DAG, diacylglycerol; PKC, protein kinase C; Rho-GEF, Rho-guanine nucleotide exchange factor. (Adapted from Wettschreck, N. and Offermanns, S., *Physiol. Rev.*, 85, 1159, 2005. With permission.)

Most GPCRs desensitize quickly upon activation via phosphorylation of specific serine/threonine residues in the intracellular loops and/or C-terminal by kinases such as G protein-coupled receptor kinases (GRKs). Once phosphorylated, β-arrestin molecules will bind to the receptor and cause arrest of the G protein-mediated signaling and induce internalization. Recent evidence has shown that β-arrestins can activate the tyrosine kinase pathway directly leading to non-G protein mediated cellular effects (Figure 12.5). In some cases, it has even been possible to develop ligands that selectively activate the β-arrestin pathway without activating the G proteins. Such ligands will induce different cellular effects than ligands activating both signaling pathways.

Recent evidence has shown that some if not all GPCRs exist as dimeric or even oligomeric complexes. As shown in Figure 12.3, family C receptors dimerize via a covalent cystein-bridge, which leads to either homo- or heterodimers. The latter is, for example, the case for GABA$_B$ receptors, which are formed by heterodimerization of GABA$_{B1}$ and GABA$_{B2}$ receptor subunits whereas, e.g., metabotropic glutamate receptors homodimerize. Whether family A and B receptors also homo- or heterodimerize have been heatedly debated in the literature and only a few examples have been convincingly shown to be of physiological importance. One such case is in the field of opioid receptors (Chapter 19), where it has been shown that the κ, μ, and δ subtypes can form pharmacologically distinct receptor subtypes by heterodimerization.

Collectively, the fact that one GPCR can activate several signaling pathways and heterodimerize to create additional subtypes has greatly complicated our view of receptor function. From a medicinal chemistry point of view it is interesting to note that in some cases it has been possible to develop ligands selectively targeting a specific signaling pathway or heterodimer. This has opened up not only new possibilities but also new challenges in drug design.

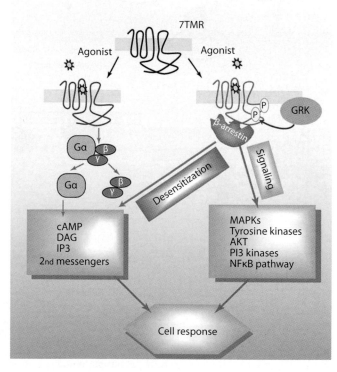

FIGURE 12.5 The G protein and β-arrestin signaling pathway of 7TM receptors (7TMR). Agonist activation of 7TMR's initiates the classical G protein cascade (see Figure 12.4 for further details) and rapid receptor phosphorylation by GRKs. The latter leads to the recruitment of β-arrestins, which causes desensitization and internalization of the receptor and activation of tyrosine kinase pathways. (Adapted from Lefkowitz, R.J. and Shenoy, S.K., *Science*, 308, 512, 2005.)

12.2.2 LIGAND-GATED ION CHANNEL RECEPTORS

Ligand-gated ion channel receptors can be divided into two major groups, namely, the Cys-loop and ionotropic glutamate receptor families. The latter family is exclusively excitatory, the former are either excitatory (serotonin and nicotinic acetylcholine receptors) or inhibitory (glycine and GABA receptors) by influx of Na^+/Ca^{2+} or Cl^- ions, which will hypo- or hyperpolarize the cell, respectively (see Chapter 13 for further details).

12.2.2.1 The Cys-Loop Receptor Family

The nicotinic acetylcholine receptor, at the nerve–muscle synapse, is the best understood Cys-loop receptor, which, upon acetylcholine binding, allow as many as 10,000 potassium and sodium ions per millisecond to pass through the channel. As shown in Figure 12.6, the receptor consists of two acetylcholine binding α_1 subunits and three other subunits (β_1, γ, and δ) that form a pentameric pore in the cell membrane. The pore itself is lined with five α-helices (termed M2), one from each of the five receptor subunits, which have a kink in the middle of the membrane spanning part. This bend is the gate of the receptor, which in the closed state points toward the channel. Agonist binding to the extracellular part of the α-subunits induces local conformational changes that are then relayed through the receptor subunits and ultimately leads to rotation of the pore-lining α-helices and channel opening.

Recently, several high-resolution 3D structures of acetylcholine-binding protein (AChBP), a water-soluble homolog of the ligand-binding domain of nicotinic acetylcholine receptors from the snail *Lymnaea stagnalis*, have been solved in the presence of various ligands (Figure 12.6). These structures have shown that agonists bind in the interface between the subunits and provided detailed insight into the ligand–receptor interactions. For example, all endogenous agonists of the Cys-loop family contain an amine, which, according to the AChBP structures, is interacting with a cluster of aromatic residues via π-cation bonding.

Most Cys-loop receptors form heteropentamers (e.g., the neuromuscular nicotinic acetylcholine receptor described earlier), but some can form homopentamers (e.g., the nicotinic α_7 receptor).

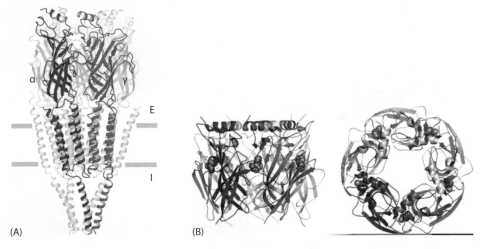

FIGURE 12.6 Structure of the family of Cys-loop ligand-gated ion channel receptors. (A) 3D structure of the neuromuscular nicotinic acetylcholine receptor that consists of five subunits (two α-, one β-, one γ-, and one δ-subunit) forming an ion-channel in the center. An α- and the γ-subunit is shown in red and blue, respectively. The receptor consists of an extracellular (E) ligand-binding domain, a transmembrane domain made of four α-helices, which is the gate of the receptor, and an intracellular (I) domain. (B) 3D structure of the acetylcholine-binding protein (AChBP) viewed from the side (left) and top (right). AChBP is a soluble protein from the snail *Lymnaea stagnalis*, which is homologous to the extracellular ligand-binding domain of the mammalian Cys-loop receptors. AChBP consists of five identical subunits (one shown in yellow and one in blue), which forms ligand-binding pockets in their interfaces shown here with nicotine bound (in pink). (Adapted from Unwin, N., *J. Mol. Biol.*, 346, 967, 2005; Celie, P.H.N. et al., *Neuron*, 41, 907, 2004. With permission.)

Numerous subunits for both nicotinic acetylcholine receptors and GABA$_A$ receptors have been cloned, which can theoretically heteromerize to a staggering high number of subunit combinations. However, in reality, only certain subunit combinations are present and even fewer combinations have therapeutic interest. The glycine and serotonin Cys-loop receptors have fewer subunits, which each can form homo- and heterodimers. Interestingly, some subunits are unable to form their part of the agonist binding pocket in either one or both sides of the two interfaces they participate in. Depending on their subunit composition, Cys-loop receptors can bind from two to five agonist molecules. For example, the neuromuscular nicotinic acetylcholine receptor binds two agonist molecules whereas the nicotinic α_7 receptor and AChBP can bind five agonist molecules (Figure 12.6). Whether all agonist binding sites need to be occupied in order to achieve receptor activation has yet to be demonstrated.

12.2.2.2 The Ionotropic Glutamate Receptor Family

The ionotropic glutamate receptor family comprise of the 15 NMDA, AMPA, and kainic acid receptors listed in Figure 12.1 and two orphan receptors (termed δ1-2) with unknown function. The name of the receptor family is a bit misleading as NR1 and NR3A-B actually has glycine as ligand (Chapter 15). Nevertheless, all 17 receptor subunits have the same overall structure: two large extracellular domains referred to as the N-terminal domain (NTD) and agonist-binding domain (ABD), a transmembrane domain (TMD) consisting of three transmembrane segments and a reentry loop and a C-terminal domain (CTD) (Figure 12.7). It is quite interesting to note the resemblance of the structures of the TMD with the amino-terminal domain of potassium channels (Chapter 13), respectively. Functional receptors are comprised of four subunits assembled around the ion channel. All NMDA receptors are heteromeric assemblies as NR1 together with either NR2 or NR3 subunits (forming glutamate or glycine receptors, respectively) whereas AMPA and kainic acid receptors can either be homo- or heteromeric assemblies.

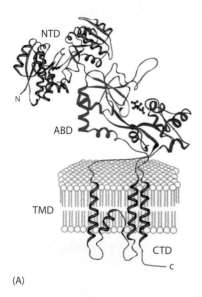

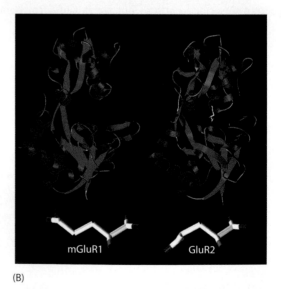

(A)   (B)

FIGURE 12.7 (A) Illustration of a single ionotropic glutamate receptor subunit with the location of the N-terminal domain (NTD), agonist-binding domain (ABD), transmembrane domain (TMD), and C-terminal domain (CTD) noted. The colored ABD with glutamate in the binding site is based on the crystal structure obtained of the soluble GluR2 binding core shown in (B). (Adapted from Bräuner-Osborne, H. et al., *J. Med. Chem.*, 43, 2610, 2000.) (B) Structure of the ABD of GluR2 in the open inactive form (left) and the closed active form with glutamate bound in the cleft (right). The difference in conformation of glutamate bound to mGluR1 and GluR2 are also shown. The structures were generated using the program "Swiss PDB viewer 3.5" with coordinates from Brookhaven Protein Data Base.

Recently, high-resolution 3D structures of the isolated ABD of the ionotropic glutamate receptor subunits NR1, NR2A, GluR2, GluR5, and GluR6 have been determined in the absence of ligands and with full and partial agonists, antagonists, and allosteric modulators. Overall these studies have shown that activation is initiated by closure of the ABD around the ligand, which is then relayed to the membrane spanning part of the receptor causing an opening of the channel pore (Figure 12.7). The plentitude of ABD structures has also provided a compelling insight into ligand–receptor interactions and has, for example, shown that the conformation of glutamate bound to various glutamate receptors is quite different as illustrated in Figure 12.7. Such information is very valuable in the design of glutamate receptor subtype selective compounds as will be discussed in further detail in Chapter 15.

12.2.3 TYROSINE KINASE RECEPTORS

As illustrated (refer to Figure 12.9) the tyrosine kinase receptors have a large extracellular agonist-binding domain, one transmembrane segment and an intracellular domain. The receptors can be divided into two groups: those that contain the tyrosine kinase as an integral part of the intracellular domain and those that are associated with a Janus kinase (JAK). Examples of the former group are the insulin receptor family and the epidermal growth factor (EGF) receptor family and examples of the latter are the cytokine receptor family such as the erythropoietin (EPO) receptor and the thrombopoietin (TPO) receptor. However, both groups share the same overall mechanism of activation: upon agonist binding two intracellular kinases are brought together, which will initiate autophosphorylation of tyrosine residues of the intracellular tyrosine kinase domain (Figure 12.8). This will attract other proteins (e.g., Shc/Grb2/SOS and STAT for the two receptor groups, respectively) that are also phosphorylated and this will initiate protein cascades and ultimately lead to regulation of transcriptional factors (e.g., Elk-1, Figure 12.8) and thus regulation of genes involved in, e.g., cell proliferation and differentiation. As described for the GPCRs, all the proteins in the intracellular activation cascades are heterogeneous leading to individual responses (i.e., regulation of different subset of genes) in individual cell types.

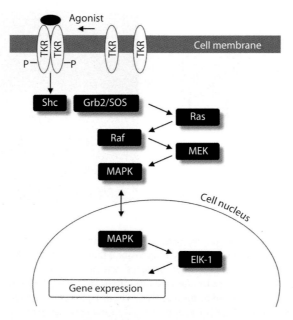

FIGURE 12.8 Cartoon of a protein cascade initiated by agonist binding to two tyrosine kinase receptors (TKR) causing autophosphorylation of the dimerized intracellular receptor domains. This causes activation of a cascade of intracellular proteins (abbreviated Shc, Grb2/SOS, Ras, Raf, MEK, and MAPK), which ultimately leads to activation of transcription factors (e.g., Elk-1) and thus regulation of gene expression.

Albeit the tyrosine kinase receptors share the overall activation mechanism, the family has turned out to be rather heterogeneous with respect to the structure and ligand–receptor interaction. Some of the receptors exist as monomers (e.g., the EGF receptor family) in the absence of agonist whereas others exist as covalently linked dimers (e.g., the insulin receptor family) or noncovalently linked dimers (e.g., the EPO receptor). In case of the monomers, agonist binding to either one or both subunits will bring the two receptor subunits together, and thereby initiate the autophosphorylation. In case of the preformed inactive dimers, agonist binding will cause a conformational change in the receptor, which brings the two intracellular kinases together and thus initiate the autophosphorylation. One of the best understood examples in this regard is the EPO receptor of which the 3D structure of the extracellular agonist-binding domain has been determined in the absence and in the presence of EPO (Figure 12.9). In the absence of EPO the domain is a dimer in which the ends are too far apart for the JAKs to reach each other. EPO binds to the same amino acids on the receptor that forms the dimer interface and thereby tilts the two receptor subunits. This brings the JAKs close together and initiate the autophosphorylation (Figure 12.9).

12.2.4 NUCLEAR RECEPTORS

Nuclear receptors are cellular proteins and are thus not embedded in the cell membrane like the previously described receptors. In contrast to the membrane bound receptors, they bind small lipophilic compounds and function as ligand-modulated transcription factors. The nuclear receptors have been classified according to the type of hormone they bind. Thereby, receptors have been divided into those which bind steroids (glucocorticoids, progestestins, mineralocorticoid androgens, and estrogens) and steroid derivatives (vitamin D_3), nonsteroids (e.g., thyroid hormone, retinoids, and prostaglandines), and orphan receptors for which the physiological agonist has yet to be discovered. The receptor family is relatively small (~50 subtypes) of which 50% still belongs to the group of orphan receptors.

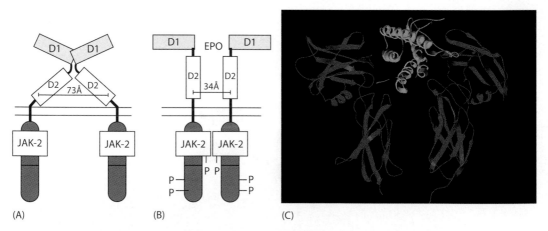

FIGURE 12.9 Cartoon of the activation mechanism of the erythropoietin (EPO) receptor, which belong to the JAK/STAT receptor class of the superfamily of tyrosine kinase receptors. (A) The receptor is dimerized in the inactive conformation by interaction of amino acids, which are similar to those involved in binding of EPO and the intracellular JAKs are kept too far apart to initiate autophosphorylation. (B) Binding of EPO to the dimer interface tilts the structure and brings the JAKs in close proximity, which initiates the autophosphorylation. (C) The actual structure of EPO (in cyan) bound to the extracellular receptor domains of the EPO receptor (in green). The structure was generated using the program "Swiss PDB viewer 3.5" with coordinates from Brookhaven Protein Data Base. (Adapted from Wilson, K.S. et al., *Curr. Opin. Struc. Biol.* 9, 696, 1999. With permission.)

The nuclear receptors consist of a ligand-binding domain, a DNA binding domain, and a trans-activation domain. Upon activation, two receptors dimerize, as homo- or heterodimers, and bind to specific recognition sites on the DNA. Coactivators will then associate with the dimeric receptor and initiate transcription of the target gene(s). Each receptor recognizes specific DNA sequences, also known as the hormone response elements, which are located upstream of the genes that are regulated. 3D high-resolution structures of both ligand- and DNA-binding domains have been determined. In drug research, the main focus has been on the structures of the ligand-binding domains, which, for several receptors, have been determined in the absence and in the presence of ligands.

12.3 RECEPTOR PHARMACOLOGY

12.3.1 RECOMBINANT VERSUS IN SITU ASSAYS

The previous decade has had a profound impact on how receptor pharmacology is performed. As mentioned in the introduction, receptor cloning was initiated in the mid-1980s and today the majority of receptors have been cloned. Thus, it is now possible to determine the effect of ligands on individual receptor subtypes expressed in recombinant systems rather than on a mixture of receptors in, e.g., an organ. This is very useful given that receptor selectivity is a major goal in terms of decreasing side effects of drugs and the development of useful pharmacological tools that can be used to elucidate the physiological function of individual receptor subtypes. Furthermore, recombinant assays allow one to assay cloned human receptors, which would otherwise not have been possible. Most receptors are more than 95% identical between humans and rodents, but due to the small differences in primary amino acid sequence there have been cases of drugs developed for rats rather than for humans, because the compounds were active on the rat receptor but not on the human receptor.

It should be noted that the use of organ and whole animal pharmacology is still required. As previously noted, the cellular effects of receptor activation depend on the intracellular contents of the proteins involved in, e.g., the signaling cascades. These effects can only be determined when the receptor is situated in its natural environment rather than in a recombinant system. In most situations, both recombinant and *in situ* assays are thus used to fully evaluate the pharmacological profile of new ligands. Furthermore, once a compound with the desired selectivity profile has been identified in the recombinant assays, it is important to confirm that this compound has the predicted physiological effects in, e.g., primary nonrecombinant cell lines, isolated organs and/or whole animals.

12.3.2 BINDING VERSUS FUNCTIONAL ASSAYS

Binding assays were used as the method of choice for primary pharmacological evaluation, mainly due to the ease of these assays compared to functional assays that generally required more steps than binding assays. However, several factors have changed this perception: (1) biotechnological functional assays have evolved profoundly and have decreased the number of assay steps and increased the throughput, (2) functional assay equipment has been automated, (3) ligand binding requires a high-affinity ligand, which for many targets identified in genome projects simply does not exist, (4) binding assays are unable to discriminate between agonists and antagonists, and (5) binding assays will only identify compounds binding to the same site as the radioactively labeled tracer.

The Fluorometric Imaging Plate Reader (FLIPR™) illustrates this development toward functional assays. Cells transfected with a receptor coupled to increase in intracellular calcium levels (e.g., a $G_{\alpha q}$ coupled GPCR or a Ca^{2+} permeable ligand-gated ion channel) are loaded with the dye Fluo-3, which in itself is not fluorescent. However, as shown in Figure 12.10, the dye becomes fluorescent when exposed to Ca^{2+} in the cell in a concentration-dependent manner. In this manner, ligand concentration–response curves can be generated on the FLIPR very fast as it automatically reads all wells of a 96-, 384-, or 1534-well tissue culture plate. Many other functional assays along these lines have been developed in recent years.

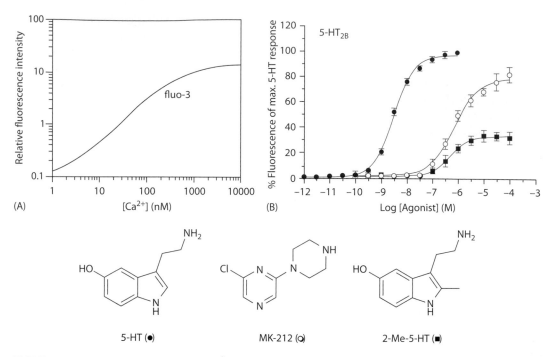

FIGURE 12.10 (A) Relation between Ca^{2+} concentration and relative fluorescence intensity of the fluorescent probe fluo-3. (B) The 5-HT_{2B} receptor subtype belong to the superfamily of GPCRs and are coupled to increase in inositol phosphates and intracellular Ca^{2+}. Cells expressing 5-HT_{2B} receptors were loaded with fluo-3 and the fluorescence was determined upon exposure to the endogenous agonist 5-HT (●) and the partial agonists MK-212 (○) and 2-Me-5-HT (■) on a FLIPR™. (Adapted from Jerman, J.C. et al., *Eur. J. Pharmacol.*, 414, 23, 2001.)

12.3.3 PARTIAL AND FULL AGONISTS

Agonists are characterized by two pharmacological parameters: potency and maximal response. The most common way of describing the potency is by measuring the agonist concentration, which elicit 50% of the compound's own maximal response (the EC_{50} value). The maximal response is commonly described as percent of the maximal response of the endogenous agonist. The maximal response is also often described as efficacy or intrinsic activity, which was defined by Stephenson and Ariëns, respectively. Compounds, such as 2-Me-5-HT and MK-212 in Figure 12.10, show a lower maximal response than the endogenous agonist and are termed partial agonists. The parameters potency and maximal response are independent of each other and on the same receptor it is thus possible to have, e.g., a highly potent partial agonist and a low potent full agonist. Both parameters are important for drug research, and it is thus desirable to have a pharmacological assay system that is able to determine both the potency and the maximal response of the tested ligands.

12.3.4 ANTAGONISTS

Antagonists do not activate the receptors, but block the activity elicited by agonists and accordingly they are only characterized by the parameter affinity. The most common way of characterization of antagonists is by competition with an agonist (functional assay) or a radioactively labeled ligand (binding assay). In both cases, the antagonist concentration is increased and displaces the agonist or radioligand, which are held at a constant concentration. It is then possible to determine the concentration of antagonist that inhibits the response/binding to 50% (the IC_{50} value). The IC_{50} value can then be transformed to affinity (K) by the Cheng–Prusoff equation:

Functional assay:

$$K = IC_{50}/(1 + [Agonist]/EC_{50}) \qquad (12.1)$$

where
[Agonist] is the agonist concentration
EC_{50} is for the agonist in the particular assay

Binding assay:

$$K = IC_{50}/(1 + [Radioligand]/K_D) \qquad (12.2)$$

where
[Radioligand] is the radioligand concentration
K_D is the affinity of the radioligand

It is important to observe that the Cheng–Prusoff equation is only valid for competitive antagonists. The Schild analysis is often used to determine whether an antagonist is competitive or noncompetitive. In the Schild analysis the antagonist concentration is kept constant while the agonist concentration is varied. For a competitive antagonist this will cause a rightward parallel shift of the concentration–response curves without a reduction of the maximal response (Figure 12.11A). The degree of right shifting is determined as the dose ratio (DR), which is the concentration of agonist

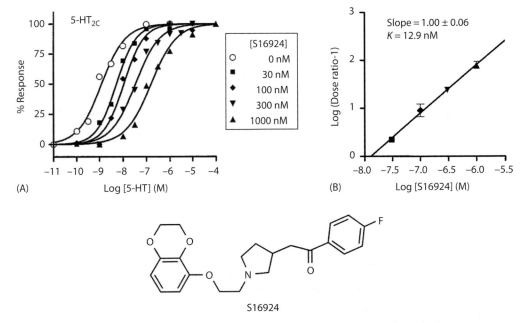

FIGURE 12.11 Schild analysis of the competitive antagonist S16924 on cells expressing the 5-HT$_{2C}$ receptor. (A) Concentration–response curves of the agonist 5-HT were generated in the presence of varying concentrations of S16924. Note the parallel right shift of the curves and the same level of maximum response. (B) DRs are calculated and plotted as a function of the constant antagonist concentration generating a straight line with a slope of 1.00 ± 0.012. These results and the observations from (A) are in agreement with a competitive interaction and the antagonist affinity can thus be determined by the intercept of the abscissa; $K = 12.9$ nM. (Adapted from Cussac, D. et al., *Naunyn Schiedbergs Arch. Phamacol.*, 361, 549, 2000. With permission.)

giving a particular response in the presence of antagonist divided by the concentration of agonist that gives the same response in the absence of antagonist. Typically one will chose the EC_{50} values to calculate the DR. In the Schild analysis the log (DR-1) is depicted as a function of the antagonist concentration (Figure 12.11B). When the slope of the curve equals 1 it is a sign of competitive antagonism and the affinity can then be determined by the intercept of the abscissa. When the slope is significantly different from 1 or the curve is not linear it is a sign of noncompetitive antagonism, which invalidates the Schild analysis.

As shown in the example in Figure 12.11, five concentration–response curves are generated to obtain one antagonist affinity determination, illustrating that the Schild analysis is rather work intensive compared to, e.g., the transformation by the Cheng–Prusoff equation where one inhibition curve generates one antagonist affinity determination. However, the latter cannot be used to determine whether an antagonist is competitive or noncompetitive, which is the advantage of the Schild analysis. When testing a series of structurally related antagonists one would thus often determine the nature of antagonism with the Schild analysis for a couple of representative compounds. If these are competitive antagonists, it is reasonable to assume that all compounds in the series are competitive and thus determine the affinity of these compounds by using the less work intensive Cheng–Prusoff equation.

12.3.5 CONSTITUTIVELY ACTIVE RECEPTORS AND INVERSE AGONISM

Most receptors display no basal activity or only minor activity but some receptors display increased basal activity in the absence of agonist that has been referred to as constitutive activity. Interestingly, it has been shown that inverse agonists can inhibit this elevated basal activity, which contrast antagonists that inhibit agonist-induced responses but not the constitutive activity (Figure 12.12A).

The examples of important constitutively active receptors include the human ghrelin receptor and several viral receptors that display constitutive activity when expressed in the host cell. This latter group includes the ORF-74 7TM receptor from human herpesvirus 8 (HHV-8), which show a marked increased basal response when expressed in recombinant cells (Figure 12.12B). ORF-74 is homologous to chemokine receptors and does indeed bind chemokine ligands. As shown in Figure 12.12B, chemokines display a wide range of activities on the receptor from full agonism (e.g., $GRO\alpha$) to full inverse agonism (e.g., IP10), which correlates with the angiogenic/angiostatic effects of the chemokines. In 1994, it was demonstrated that HHV8 infection is the cause of Kaposi's sarcoma, which is a multifocal angioproliferative cancer disease mainly affecting AIDS patients.

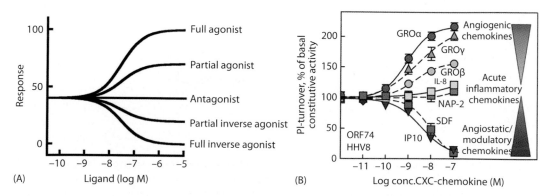

FIGURE 12.12 (A) The nomenclature of ligand efficacies and schematic illustration on their concentration-dependent effects on constitutive activity. (B) Ligand regulation of the constitutively active ORF-74 receptor from Human Herpesvirus 8 (HHV8). ORF-74 is a GPCR coupled to phosphatidylinositol (PI) turnover, which is regulated by a variety of human chemokines ranging from full agonism by $GRO\alpha$ to full inverse agonism by IP10. (Adapted from Rosenkilde, M., *Neuropharmacology*, 48, 1, 2005. With permission.)

Recently, it was further demonstrated by generation of transgenic mice expressing ORF-74 that it is indeed the constitutively activated receptor, which alone is the cause of Kaposi's sarcoma. Clearly, it would be highly desirable to develop selective inverse agonists of ORF-74, which would very likely inhibit or even prevent the development of Kaposi's sarcoma in HHV8 infected humans.

Constitutive activity can also be caused by somatic mutations. The known examples include constitutively activating mutations in the thyrotropin receptor and the luteinizing hormone receptor, which leads to adenomas, and the rhodopsin receptor, which leads to night blindness. Moreover, in this case it is conceptually possible to alleviate the diseases by the development of inverse agonists.

12.3.6 ALLOSTERIC MODULATORS

Allosteric modulators can both be stimulatory or inhibitory (noncompetitive antagonists) and typically these compounds bind outside the orthosteric binding site (binding site of the endogenous agonist). Allosteric modulators have a number of potential therapeutic benefits compared to agonists and competitive antagonists, which has led to significant increased pharmaceutical interest in recent years. This increased interest has also been fueled by the development of functional high-throughput screening assays, which has made it possible to screen for allosteric modulators (see Section 12.3.2).

The allosteric modulators mentioned in the following text act through allosteric mechanisms as evident from the fact that they do not displace radiolabeled orthosteric ligands. Furthermore, their activity is dependent on the presence of agonists as they do not activate the receptors by themselves. The fact that they bind outside of the orthosteric ligand-binding pocket often leads to increased receptor subtype selectivity. Evolutionary pressure has led to conservation of the orthosteric binding site at different subtypes, as radical mutations would severely impact the binding properties. Thus, it is often seen that the orthosteric binding site is much more conserved than the remaining part of the receptor and accordingly, ligands binding to an allosteric site have a higher chance of being selective. Likewise, the allosteric ligands will have a different pharmacophore than the endogenous ligand, which might improve, e.g. bioavailability. For example, ligands acting at the orthosteric site of the $GABA_A$ receptor need a negatively charged acid function and a positively charged basic function, which greatly impairs the transport through biomembranes, whereas allosteric ligands such as the benzodiazepine Diazepam (Chapter 15) does not have any charged groups and show excellent bioavailability. It is well known that many agonists, particularly full agonists, lead to desensitization and internalization of receptors. Unlike agonists, the positive modulators should prevent the development of tolerance (as seen for, e.g. morphine), because they avoid the prolonged receptor activation leading to desensitization and internalization. The fact that the receptors are stimulated in a more natural way by positive modulators rather than the prolonged receptor activation caused by agonists may also lead to a difference in physiological effects, which may or may not be an advantage.

12.3.6.1 Negative Allosteric Modulators (Noncompetitive Antagonists)

As noted in the previous section, the Schild analysis is very useful to discriminate between competitive and noncompetitive antagonists, and an example of the latter is shown in Figure 12.13A. CPCCOEt is a selective antagonist at the mGluR1 receptor, and the Schild analysis clearly demonstrates that the antagonism is noncompetitive due to the depression of the maximal response (compare with Figure 12.11). As noted previously, glutamate binds to the large extracellular amino-terminal domain whereas CPCCOEt has been shown to bind to the extracellular part of the 7TM domain. CPCCOEt does not hinder binding of glutamate to the extracellular domain, but hinder the conformational change leading to receptor activation.

12.3.6.2 Positive Allosteric Modulators

Positive allosteric modulation can be achieved through several mechanisms. For example, benzodiazepines positively modulate the $GABA_A$ receptor by increasing the frequency of channel opening (Chapter 15). Positive modulation can also be obtained by blocking receptor desensitization as exemplified by cyclothiazide (Chapter 15).

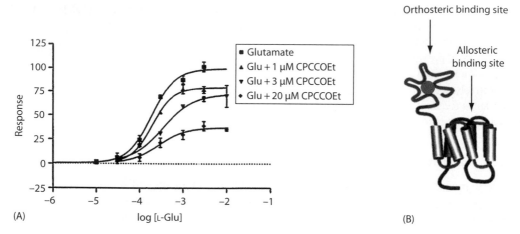

(A)   log [L-Glu]   (B)

FIGURE 12.13 Schild analysis of the noncompetitive antagonist CPCCOEt on cells expressing the metabotropic glutamate receptor subtype mGluR1. (A) Concentration–response curves of the agonist glutamate (L-Glu) were generated in the presence of varying concentrations of CPCCOEt. In contrast to the Schild analysis shown in Figure 12.10, a clear depression of the maximal response is seen with increasing antagonist concentrations. This shows that the antagonist is noncompetitive. (Adapted from Litschig, S. et al., *Mol. Pharmacol.*, 55, 453, 1999. With permission.) (B) Cartoon showing overall structure of a family C receptor with the orthosteric (endogenous agonist) and allosteric binding sites pointed out.

12.4 CONCLUDING REMARKS

The previous decade of receptor research has provided many breakthroughs in our understanding of receptor structure, function, and pharmacology. The many new 3D structures of either full receptors or important domains have provided detailed knowledge about ligand–receptor interactions and receptor activation mechanisms. It has been shown that most receptors can activate several different signaling pathways, which may also be selectively activated/inhibited by drugs. Finally, inverse agonism and allosteric modulation have pointed to novel ways that receptors can be regulated *in vivo*. Collectively, these new developments have created the foundation for structure-based drug design and new concepts of pharmacological intervention.

FURTHER READINGS

Bond, R.A. and IJzerman, A.P. 2006. Recent developments in constitutive receptor activity and inverse agonism, and their potential for GPCR drug discovery. *Trends Pharmacol. Sci.* 27:92–96.

Bräuner-Osborne, H., Egebjerg, J., Nielsen, E.Ø., Madsen, U., and Krogsgaard-Larsen, P. 2000. Ligands for glutamate receptors: Design and therapeutic prospects. *J. Med. Chem.* 43:2609–2645.

Bräuner-Osborne, H., Wellendorph, P., and Jensen, A.A. 2007. Structure, pharmacology and therapeutic prospects of family C G-protein-coupled receptors. *Curr. Drug Targets.* 8:169–184.

Gronemeyer, H., Gustafsson, J.Å., and Laudet, V. 2004. Principles for modulation of the nuclear receptor superfamily. *Nat. Rev. Drug Discov.* 3:950–964.

Ji, T.H., Grossmann, M., and Ji, I. 1998. G Protein-coupled receptors. I. Diversity of receptor-ligand interactions. *J. Biol. Chem.* 273:17299–17302.

Lefkowitz, R.J. 2007. Seven transmembrane receptors: Something old, something new. *Acta Physiol.* 190:9–19.

Lefkowitz, R.J. and Shenoy, S.K., 2005. *Science* 308:512.

Madsen, U., Bräuner-Osborne, H., Greenwood, J.R., Johansen, T.N., Krogsgaard-Larsen, P., Liljefors, T., Nielsen, M., and Frølund, B. 2005. GABA and glutamate receptor ligands and their therapeutic potential in CNS disorders. In *Drug Discovery Handbook*, ed. S.C. Gad, pp. 797–907. New York: Wiley.

McKay, M.M. and Morrison, D.K. 2007. Integrating signals from RTKs to ERK/MAPK. *Oncogene*. 26:3113–3121.

Ridge, K.D. and Palczewski, K. 2007. Visual rhodopsin sees the light: Structure and mechanism of G protein signaling. *J. Biol. Chem.* 282:9297–9301.

Ward, C.W., Lawrence, M.C., Streltsov, V.A., Adams, T.E., and McKern N.M. 2007. The insulin and EGF receptor structures: New insights into ligand-induced receptor activation. *Trends Biochem. Sci.* 32:129–137.

13 Ion Channels: Structure and Function

Søren-Peter Olesen and Daniel B. Timmermann

CONTENTS

13.1 INTRODUCTION

Ion channels form pores through the cell membrane, which are permeable to the small physiological ions Na^+, K^+, Ca^{2+}, and Cl^-. The channels can open and close and thereby turn the flux of the charged ions through the cell membrane on and off. By this mechanism, the ion channels govern the fast electrical activity of the cells. Additionally, ion channels control Ca^{2+} influx and regulate responses as diverse as muscle contraction, neuronal signaling, hormone secretion, cell division, and gene expression. The opening of the channels is subject to regulation by physiological stimuli such as changes in membrane potential and ligand binding. Ion channels also lend themselves to pharmacological modulation and constitute important targets for drug treatment of diverse diseases including cardiac arrhythmia, arterial hypertension, diabetes, seizures, and anxiety.

13.1.1 Ion Channels Are Pores through the Cell Membrane

The cell membrane is impermeable to the small ions since they are charged. The ions polarize the water molecules around them and carry a shell of hydration water rendering them insoluble in the hydrophobic phospholipid membrane. Ion transport in and out of cells has to occur through specialized molecules, allowing the cells to compose a specific intracellular ion-milieu, which in many ways is different from the extracellular ion-milieu, e.g., there is more than a 10-fold gradient in the Na^+- and K^+-concentrations and a 10,000-fold gradient for Ca^{2+} across the cell membrane (Table 13.1).

The membrane proteins establishing these gradients are transporters such as the Na–K ATPase pumping three Na^+ out and two K^+ into the cell while consuming one ATP molecule. Other transporters are the Ca–ATPases pumping Ca^{2+} out of the cell or into the endoplasmic reticulum, and secondary active transporters such as the Na–Ca exchanger not using energy themselves but exploiting the gradients created by the ATPases. The transporters typically move 0.1–10 ions/ms each, they show saturation kinetics like enzymes, and they slowly build up the ion gradients.

The ion channels are different in many ways. They form water-filled pores through the cell membrane once they open, and permeation through the channels is only limited by diffusion. The transport is very fast, in the range of 10^4–10^5 ions/ms, and the opening of ion channels may change the membrane potential by 100 mV within less than 1 ms. Ion channels are thus in an ideal position to govern the fast electrical activity of cells.

TABLE 13.1
Typical Intra- and Extracellular Ion Concentrations and Corresponding Equilibrium Potentials

Ion	Intracellular Concentration (mM)	Extracellular Concentration (mM)	Equilibrium Potential (mV)
Ca^{2+}	0.0001	1–2	+120
Na^+	10	145	+70
Cl^-	10	110	−62
K^+	140	4	−93

13.1.2 Ion Currents Change the Electrical Membrane Potential

In biological tissue electrical currents are conducted by the movement of ions. Bulk movements of ions in organs give rise to large currents resulting in voltage differences that can be measured on the body surface as was first done by Willem Einthoven in 1901 when he recorded human electrocardiograms. At the cellular level, the nature of excitable ion currents through the cell membrane was demonstrated by Hodgkin and Huxley in 1953 using a preparation of the squid axon. This giant nerve axon is about 1 mm in diameter, i.e., about 1000-fold thicker than human axons allowing electrodes to be positioned on either side of the membrane and the ion compositions on both sides to be controlled. Using this method the authors showed that selective movement of Na^+ ions into the cell followed by an efflux of K^+ ions is the basis for the electrical activity in nerve cells. The ions move passively across the cell membrane when the permeability increases, and the direction of the movement is determined by the combined chemical and electrical forces acting on them.

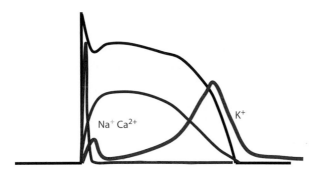

FIGURE 13.1   Cardiac action potential and time-course of selective Na$^+$, Ca^{2+}, and K$^+$ currents.

These two forces are generated by the concentration gradient and by the electrical field generated by the membrane potential, respectively. The ion movement will stop once the two forces equal each other, which happens at the so-called equilibrium potential. This potential is determined by the ion distribution across the membrane, and for the typical intra- and extracellular ion concentrations the equilibrium potentials are shown in Table 13.1.

The effects on the membrane potential of activation on selective ion channels are shown in Figure 13.1. The cardiac action potential (AP) is initialized by opening of voltage-gated Na$^+$ channels, and the influx of the positively charged Na^{2+} ions leads to a fast positive shift in the membrane potential (depolarization). Subsequently voltage-gated Ca^{2+} channels are opened and the influx of Ca^{2+} ions keeps the membrane potential depolarized. Fast K$^+$ channels are activated early in the response and attenuate the depolarization, but the key role of the K$^+$ currents is to terminate the AP after about 350 ms when numerous K$^+$ channels open and the outflow of the positively charged K$^+$ ions mediate the repolarization.

The consequence of the sequential opening of Na$^+$-, Ca^{2+}-, and K$^+$-selective channels, is thus that the cell membrane potential will be pulled in the direction of the equilibrium potential for these ion species, i.e., about +70 mV for Na$^+$, +120 mV for Ca^{2+}, and −93 mV for K$^+$ (Table 13.1). Often the cell does not fully reach the equilibrium potential as shown for the cardiac action potential, since several types of channels are usually open at the same time. Likewise the impact on the membrane potential of physiological or pharmacological ion channel block or activation depends on the presence of other simultaneous conductances and is not just linearly correlated to the number of ion channels being affected. Thus, it can be complicated to predict the functional effect of modulating ion channel function, and extensive target validation studies have to be conducted to establish the anticipated role of an ion channel subtype in an organ.

13.1.3   Gating of Ion Channels

While it was clear to Hodgkin and Huxley that a sequential increase in Na$^+$ and K$^+$ membrane conductances underlies the neuronal action potential, their method could not reveal the nature of the conductance pathway. This had to await another technological breakthrough. In 1976, Neher and Sakmann reported the opening and closing of single acetylcholine-gated ion channels in striated muscle using a method by which they electrically isolated a patch of membrane *in situ* with a glass pipette. The method was called patch-clamp with reference to the patch of tissue and the clamp of the transmembrane (TM) voltage used to generate the electrical driving force. Since then the method has been extensively used to describe the characteristics and function of ion channels in all cells. Initially endogenous currents in cells were measured, but following the cloning area the combination of this functional method and heterologous expression of cloned ion channels has been a strong combination in the target-driven drug discovery process.

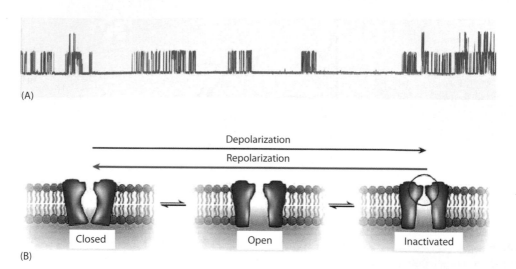

FIGURE 13.2 (A) Single channel recording of BK-type potassium channel. The baseline shown the closed state and the upward deflections are opening of single channels. The current through the channels is about 20 pA. (B) Opening, closing, and inactivation of ion channel. (Modified from Sanguinetti, M.C. and Tristani-Firouzi, M., *Nature*, 440, 463, 2006.)

Patch-clamp studies of single ion channels have shown that the duration of channel opening ranges from a few μs to several hundred ms when exposed to a ligand or a voltage change (Figure 13.2). In the absence of stimulus they either open less frequently or stay closed. The opening and closing of ion channels is called gating, and at the single channel level it is described by the distribution of open- and closed-times. Currents through all ion channels in the cell membrane can also be measured by ripping a hole in the cell, which makes it possible to voltage-clamp the whole cell. This whole-cell current depicts the sum of hundreds of ion channels, and the kinetics of the current reflect the average open- or closed-times of the channels.

The gating is a dynamic process reflecting structural changes in the channel protein. The opening of a channel is preceded by conformational changes, and to the extent that these changes result in movement of charged segments of the molecule (voltage sensors), it can be followed by measurement of small gating currents. Once the channel goes into the open state, the electrical current carried by ions through the channel can be recorded with a resolution of about 1 pA (10^{-12} A). The activation of single channels is a discrete event, and as seen from the recordings in Figure 13.2 the ion channels are either fully closed or fully open. Thus, it is possible to follow the movements between the two conformational states with an amazing time- and current-resolution.

The various types of ion channels gate differently: some channels open only transiently whereas others stay open as long as the stimulus exists. Stimuli for ion channel activation are either (1) a change in the membrane potential, (2) a change in the concentration of extracellular ligands (neurotransmitters), (3) a change in the concentration of intracellular ligands (Ca^{2+}, H^+, cyclic nucleotides, or G-protein subunits), or (4) mechanical stimulation (e.g., stretch). Once the channels are exposed to the electrical or chemical stimuli they open or activate, and when the stimulus is removed, the channels close in an opposite process called deactivation. A number of channel types do however also close in the presence of the stimulus. It is a general physiological phenomenon that continued stimulation of a signal process results in a decreasing output. This functional closure of ion channels in the presence of stimulus is called inactivation and can occur either by parts of the channel protein plugging the open pore after a short delay, by collapse of the pore, or by decreased coupling between ligand-binding- and pore-domains (Figure 13.2B).

13.1.4 MOLECULAR STRUCTURES OF ION CHANNELS

Ion channels are present in all cells, and these naturally occurring channels have been extensively characterized with respect to gating kinetics, voltage- and ligand-sensitivity, pharmacology, and other parameters. In addition many ion channel types exhibit high affinity (pM or nM) to a number of toxins derived from scorpions, snakes, snails, or other animals, so toxins have been widely used to differentiate between the channels subtypes. The overall parameter used when describing an ion channel is its selectivity, i.e., whether it is selective to permeation of K^+, Na^+, Ca^{2+}, or Cl^-. Some channels are nonselective among cations. Since the selectivity is tightly coupled to the physiological function of the channels, this division is pragmatic and will be used in this chapter.

Following the sequencing of the human genome, 406 proteins with clear ion channel structure appeared. The characteristics of most of the cloned channels correspond well to the endogenous currents found in nerve, muscle, and other cells. The molecular constituents underlying other endogenous currents is however still debated, and these channels appear to be composed by several subunits from the same molecular family plus an additional number of accessory proteins. For voltage-gated channels the pore-forming subunits are denoted α-subunits, whereas the accessory subunits are called β, γ, or δ subunits.

13.1.5 ION CHANNELS AND DISEASE

The functional significance of specific ion channels in the body can be difficult to deduce from their molecular function, but it can be studied in organs or whole animals using pharmacological tools or selective toxins. Transgenic animals also provide valuable knowledge, but the most precise information about their role in humans has come from patients with diseases caused by dysfunctional ion channels. The diseases are typically caused by a point mutation in a single ion channel gene, and the diseases are jointly called channelopathies. The most frequent and well-known disease is cystic fibrosis, arising from a point mutation in the Cl^- channel CFTR. In Northern Europe, 5% of the population is heterozygous for a mutation in the CFTR gene, and the prevalence of the disease is 0.5‰. Several types of cardiac arrhythmia (long and short QT syndromes, Brugada syndrome, and Andersen syndrome) are caused by mutations in cardiac K^+, Na^+, and Ca^{2+} channels. Mutations in neuronal and muscular ion channel subtypes cause epilepsy, ataxia, and myotonia. Luckily most of these are rare, but their study has given invaluable information about the role of the ion channels in health and disease.

13.1.6 PHYSIOLOGICAL AND PHARMACOLOGICAL MODULATION OF ION CHANNELS

In addition to the main mechanisms for ion channel activation (voltage, ligands), the channels may also in some cases be modulated by small organic molecules. The ligand-gated ion channels exhibit an endogenous ligand-binding site, so compounds with similar functionalities can make potent drugs. The voltage-gated channels are not expected to naturally exhibit high-affinity binding sites, but may possess such as in the case of the dihydropyridine-binding site on the Ca^{2+} channel. Most drugs act as positive or negative modulators of the channel gating, but some may also just plug the pore as the local anesthetics blocking the neuronal Na^+ channels or the neuromuscular blockers acting on the nicotinic channel in the neuromuscular junction.

13.1.7 DRUG SCREENING ON ION CHANNELS

The center-stage role of ion channels in many physiological responses has been stressed by functional studies in cells, organs, and animals, by the emerging channelopathies as well as by the successful use of ion channel modulating drugs. Current drugs only target a dozen of the known channel subtypes, while most of the other 400 types are currently all being investigated as potential drug targets in the pharmaceutical industry. Drug-discovery projects today depend strongly on large-scale blind-screening for finding new chemical lead molecules. The only high-throughput, high-quality technology to be used for screening on every ion channel subtype is the newly developed automated patch-clamp technique. With this method, parallel recordings are performed by a robot on 50–100 arrays of ion channel expressing cells positioned on silicon chips. Smaller throughputs can be obtained on arrays of 8–10 frog eggs expressing the desired ion channels, but the pharmacology of some channels may be different in this nonmammalian system. Channels giving rise to changes in the intracellular Ca^{2+} concentration can be screened using fluorescent Ca^{2+} dyes in a 384 well fluorescent reader (FLIPR) (see Chapter 12.3.2), which may also be useful for channels causing slow voltage changes. The use of other screens is typically limited to specific ion channels such as rubidium or thallium flux through K^+ channels, or ligand binding to neurotransmitter-gated channels.

13.1.8 STRUCTURE OF VOLTAGE-GATED ION CHANNELS

The superfamily of voltage-gated ion channels encompasses more than 140 members and is one of the largest families of signaling proteins, following the G-protein-coupled receptors and protein kinases. The pore-forming α-subunits of voltage-gated ion channels are built upon common structural elements and come in four variations. The simplest version is composed of two TM segments connected by a membrane-reentrant pore-loop and having N- and C-termini on the inside (Figure 13.3). Four of such subunits form the channel. This architecture is typical for the so-called inward-rectifying K^+ channels (K_{ir}). It is found in a number of bacterial channels, suggesting it is the ancestor of the family. The second type is made by a concatenation of two such subunits, and the channel is formed by two double constructs. The third type is the 6-TM subunit, in which four extra membrane-spanning N-terminal domains including a voltage-sensor have been added to the basic 2-TM pore unit. Four of these 6-TM units form a channel. The group of 6-TM channels is rather large and includes the voltage-gated K^+ channels (K_v), the calcium-activated K^+ channels (K_{Ca}), the cyclic nucleotide-gated (CNG) channels, the hyperpolarization-gated channels (HCN), and the transient receptor potential (TRP) channels. Finally, the fourth channel structure type is made by concatenating four of the 6-TM subunits, making up a 24-TM subunit that forms the channel alone. This type is represented by the voltage-gated Na^+ and Ca^+ channels (Na_v and Ca_v). Within each of the four domains the six TM segments are denoted S1–S6.

Three different parts of the channels are responsible for the functions: ion permeation, pore gating, and regulation. The narrow part of the pore is called the selectivity filter, and this has been studied by high-resolution x-ray in crystallized K^+ channels giving valuable insight into the selectivity mechanism (Figure 13.4). The residues in the pore loop line the selectivity filter and their carbonyl groups act as surrogate-water implying that the chemical energy of the dehydrated K^+ ions entering

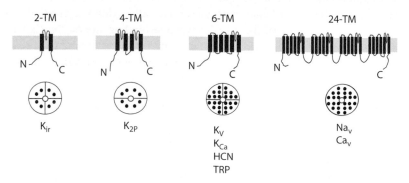

FIGURE 13.3 Topology of voltage-gated cation channels. (From Palle Christophersen.)

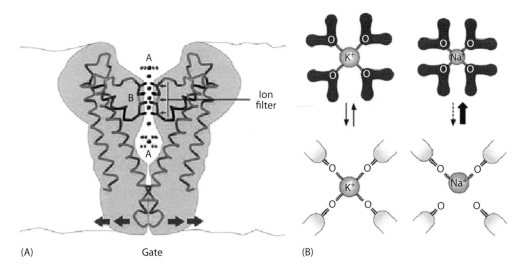

(A)   Gate   (B)

FIGURE 13.4   (A) Potassium channel structure with selectivity filter at the outer pore and gating mechanism at the inner pore. (B) Selectivity mechanism. The distance between the K^+ ions and the oxygen atoms is the same in water as in the selectivity filter enabling the K^+ ions to enter the pore at no energy cost. This is different for Na^+ ions, so they are excluded from the pore. (From Alberts, B. et al., *Molecular Biology of the Cell*, 2002 Garland Publishing Inc. and Doyle, D.A. et al. *Science* 280: 69–77, 1998.)

the pore is unchanged. By this means high selectivity and high permeability of the ions passing in single file is obtained. Although Na^+ ions are smaller than K^+ ions they will not enter the pore since it is energetically unfavorable.

13.2   PHYSIOLOGY AND PHARMACOLOGY OF VOLTAGE-GATED ION CHANNELS—POTASSIUM CHANNELS

The 2-TM K_{ir} channel family gives rise to six subtypes, which play diverse roles in the body. Many K_{ir} channels are open at resting membrane potential and clamp the potential at −70 and −90 mV in nerve and heart cells, respectively (e.g., $K_{ir}4$ and $K_{ir}2$). The $K_{ir}3$ channels are gated by binding of the βγ-subunit from the G_i-protein. This mechanism is important in the atria of the heart, where stimulation of the para-sympathetic vagus nerve leads to release of acetylcholine, activating the G_i protein and subsequently the $K_{ir}3$ channel to hyperpolarize the pacemaker cells.

The $K_{ir}6$ channels are expressed both in heart, vasculature, nerve, and in the pancreatic β-cells. This channel subtype can only be expressed in cells when it coassembles with its accessory subunit, the so-called sulfonyl urea receptors (SUR) of the ABC transporter family. The β-cell subtype is composed of 4 $K_{ir}6.2$ + 4 SUR1. Like other K_{ir} channels it is activated by binding of phosphatidylinositol-4, 5-bisphosphate (PIP_2). In contrast the complex is blocked by ATP binding to the internal surface of $K_{ir}6.2$ and activated by MgADP binding to the nucleotide-binding domains of SUR1. The channel complex is also denoted as the K_{ATP} channel and it is interesting for two reasons: it is a key regulatory protein in the β-cells coupling plasma glucose levels to insulin secretion, and the SUR has a well-exploited high-affinity drug-binding site.

Briefly, insulin secretion is regulated by the following mechanism: an increase in plasma glucose leads, through an increased ATP level in the β-cells, to block the K_{ATP} channel, depolarization, Ca^{2+}-influx, and insulin secretion (Figure 13.5). If this regulation is dysfunctional as in many type-2 diabetic patients, a similar functional effect can be obtained by directly blocking the K_{ATP} channel pharmacologically. The drug-binding site on SUR1 is on the inside of TM15 (plus partly on the inside of TM14), and the bulky substitution mutation S1237Y disrupts the site. Tolbutamide

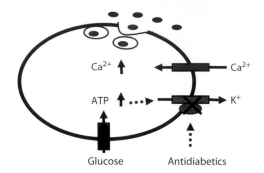

FIGURE 13.5 Ion channels in pancreatic β-cells and insulin secretion. The K channel subtype is called K_{ATP} and it is composed of the two molecular subunits $K_{ir}6.2$ and SUR1.

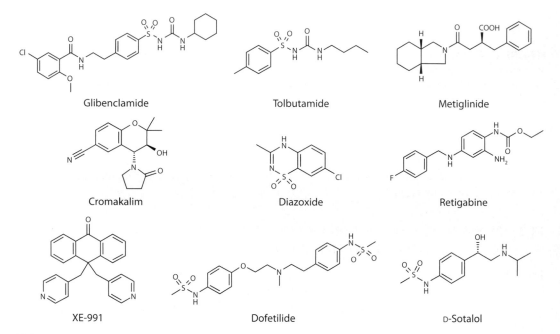

FIGURE 13.6 Structures of the K_{ATP} channel blockers glibenclamide, tolbutamide, and metiglinide; the K_{ATP} channel openers cromakalim and diazoxide; the K_V7 channel opener retigabine; the K_V7 channel blocker XE-991; the K_V11 channel blockers dofetilide and D-sotalol.

binds to this site only, whereas glibenclamide and metiglinide (Figure 13.6) binds to this as well as to a neighboring benzamido site. The latter low-affinity site is shared with the cardiac and vascular subunits SUR2A and SUR2B, respectively.

The cardiovascular side effects of the SUR-blockers are minimal whereas SUR-activators, such as cromakalim and diazoxide, which have been attempted primarily for the treatment of arterial hypertension had to be abandoned since they cause orthostatic hypotension and reflex tachycardia.

The K_v channels fall into 12 subfamilies, which are all gated by changes in the membrane potential, but they exhibit different kinetics. K_v channels can be composed of four different subunits from the same subfamily giving numerous possibilities for variations. Several K_v channel subfamilies are interesting drug targets. Retigabine is an activator of the $K_v7.2/3$ heteromultimeric channel being developed for the treatment of epilepsy, and XE-991 is a memory enhancing compound blocking the same channel (Figure 13.6).

Class III antiarrhythmics block K_v channels in the heart (K_v1, K_v4, and K_v11 subtypes) leading to a prolonged cardiac AP and termination of so-called reentry arrhythmia. Dofetilide, D-sotalol and other antiarrhythmics are selective for the K_v11 channels (hERG channels). These drugs show antiarrhythmic effects in some patients whereas they are proarrhythmic in others. The reason for the latter is that although the prolongation of the AP may terminate some arrhythmias, then blocking an important cardiac K^+ conductance being responsible for repolarizing the AP may destabilize the heart against triggered impulses (after depolarizations).

The K_v11 channel has a high-affinity binding site in the pore, which interacts with drugs of very different classes including antihistamines, antipsychotics, antidepressants, antibiotics, and many more. Proarrhythmia caused by drug binding to this site and channel block has been a major reason for withdrawal of drugs from the market and discontinued drug development projects, so the K_v11 channel has become a major cardiac safety pharmacology issue. The Ca-activated K^+ channels, K_{Ca}, are divided into three families depending on their single-channel conductance. They are gated by Ca^{2+} binding either directly to the channel or indirectly to a constitutively bound calmodulin. The channels are generally involved in attenuating the activity of a given cell by hyperpolarizing this, when the internal Ca^{2+} concentration rises.

TRP belong structurally to this group having six TM, a voltage sensor in S4, and a pore loop between S5 and S6. Despite the structural similarity to K_v channels the 28 different TRP subtypes can be either selective to Na^+/K^+, Mg^{2+}, or Ca^{2+}, and functionally they may associate with G-protein-coupled receptors, tyrosine kinases, or phospholipase C. A large number of TRP channel subunits have been cloned and based on their amino acid homology they can be divided into the TRPV-, TRPC-, and TRPM-families.

Due to their relatively recent discovery, only few TRP-subtype selective ligands have been identified. But there is no doubt that the one TRP channel that has attracted the most attention as a potential drug target is the TRPV1-channel. This ion channel is activated by heat but also by capsaicin, a constituent of chili pepper and TRPV1 is indeed responsible for the "hot" sensation induced by ingesting chili. TRPV1 has also been found to be upregulated in various animal models of chronic pain and selective antagonists of TRPV1 reduce pain sensation in these models. Selective antagonists of TRPV1 are currently undergoing clinical trials in patients suffering from different types of chronic pain.

13.3   VOLTAGE-GATED CALCIUM CHANNELS

13.3.1   STRUCTURE AND MOLECULAR BIOLOGY

The discovery of voltage-gated calcium channels (Ca_v) was originally made in the 1950s, through an investigation of crab leg muscle contraction. These experiments revealed that both membrane depolarization and muscle contraction depend on extracellular calcium ions, inferring that the muscle cells posses some membrane molecules enabling calcium to selectively permeate. By use of electrophysiological techniques, it was later found that a variety of functionally distinct Ca_vs exist and that these ion channels are also expressed in nerve cells.

Functionally, Ca_vs are closed at the resting membrane potential (i.e., −50 to −80 mV), but are activated by depolarization. Two distinct classes of Ca_v-mediated currents can be distinguished by this feature: high-voltage-activated calcium currents, requiring membrane potentials of ca. −20 to +10 mV to activate and low-voltage activated currents, which activate at much more negative membrane potentials, typically −50 to −40 mV. Following activation, Ca_vs inactivate in the presence of sustained membrane depolarization, although the speed of inactivation can vary from ~50 ms to several seconds. Therefore, different types of Ca_vs can be distinguished on the basis of biophysical, i.e., activation and inactivation characteristics, and on pharmacological properties. Voltage-activated calcium currents, measured in native tissues, have traditionally been classified as L-, N-, P/Q-, or R-type or T-type currents (see Table 13.2).

TABLE 13.2

Ca$_v$ Channel Terminology and Properties

Channel Subtype	Ca$_v$1	Ca$_v$2	Ca$_v$3
Former names	L-type	Ca$_v$2.1 = P/Q type Ca$_v$2.2 = N type Ca$_v$2.3 = R type	T-type
Activation threshold	High voltage	High voltage	Low voltage
Blocker	Dihydropyridines Phenylalkylamines Benzothiazepines	Ca$_v$2.1 blockers: ω-conotoxin MVIIC, ω-agatoxin IVA Ca$_v$2.2: blockers: ω-conotoxin GVIA, ω-conotoxin MVIIA Ca$_v$2.3 blocker: SNX-482	Mibefradil R-(−)-efonidipine Kurtoxin

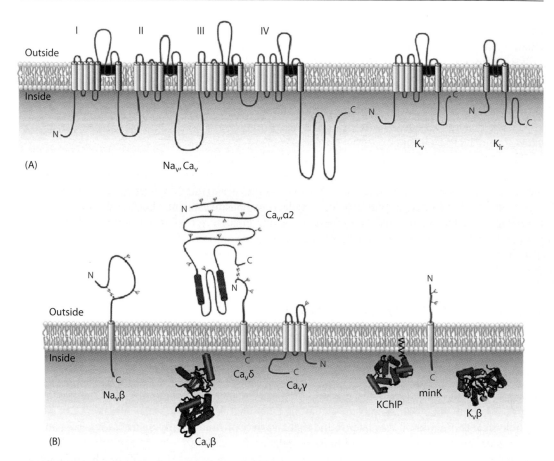

FIGURE 13.7 Overview of the membrane topology of voltage-gated ion channel α-subunits. (A) The voltage-sensing S4 TM segments (green) contain several positively charged amino acid residues and the segments that constitute the ion channel pore (shown in red) are the S5, S6, and pore loop segments. (B) Membrane topology of auxiliary subunits of Na$_v$, Ca$_v$, and K$_v$ ion channels. (From Catterall, W.A. et al., *Toxicon*, 49, 124, 2007. With permission from Elsevier.)

The major component of the Ca$_v$ is the large α$_1$-subunit, consisting of ~2000 amino acid residues. This subunit has 24 TM segments, arranged in four linked homologous domains (I–IV), each comprising six TM α-helices (S1–S6), including the positively charged voltage-sensing S4 segments, and the S5–S6 pore loops, with the pore loops and S6 segments believed to line the channel lumen; the structure of the α$_1$- and other Ca$_v$ subunits is schematically shown in Figure 13.7A.

Ca$_v$s are several 1000-fold selective for Ca^{2+} ions over Na$^+$ and K$^+$ and this amazing selectivity is created by a ring of four negatively charged glutamic acid residues projecting into the ion channel pore, one such residue being contributed by each of the four pore loops. Ten different α_1-subunit types have been cloned and based on their amino acid homology, these have been divided into three distinct families (Ca$_v$1, Ca$_v$2, and Ca$_v$3) that display 30%–50% amino acid identity with each other. Within each family there are three to four members (Ca$_v$1.1–Ca$_v$1.4, Ca$_v$2.1–Ca$_v$2.3, and Cav3.1–Cav3.3) that each show a much higher degree of sequence identity (~80%) with each other. The Ca$_v$1.1-subunit is only expressed in skeletal and cardiac muscle, and the Cav1.4-subunit is exclusively expressed in retina. The other α_1-subunits are widely expressed in many tissues, in particular the peripheral and central nervous system (CNS) as well as many types of endocrine cells.

When expressed alone, the α_1-subunit can form a functional ion channel. But native Ca$_v$s are multisubunit complexes in which the α_1-subunit interacts with a β-, an $\alpha_2\delta$- and sometimes a γ-subunit (Figure 13.7B). The role of these subunits is to promote incorporation of Ca$_v$ into the cell membrane and to modulate the functional properties of Ca$_v$s.

13.3.2 Physiological Roles of Voltage-Gated Calcium Channels

Ca^{2+} is an important second messenger molecule in eukaryotic cells where it initiates muscle contraction, neurotransmitter release, and activates many types of protein kinases. Many homeostatic mechanisms operate to keep intracellular [Ca^{2+}] < 100 nM under resting conditions. Outside the cell, [Ca^{2+}] is 1–2 mM, creating a 10,000-fold concentration gradient. The Ca^{2+}-equilibrium potential is > +100 mV so Ca^{2+} always flows into a cell, when Ca$_v$s are activated by depolarization. While the primary function of voltage-gated Na$^+$ and K$^+$ channels is to produce depolarization/repolarization of the cell membrane, voltage-gated Ca^{2+} channels should be thought of as "gatekeepers" of calcium entry into excitable cells.

In muscle tissue, the binding of Ca^{2+} to the protein troponin C allows myosin-mediated sliding of actin-filaments, leading to shortening of muscle fibers. In skeletal muscle, the calcium necessary for this process actually comes from the sarcoplasmic reticulum and is released from this into the cytoplasm via ryanodine receptors. In this particular context, the Ca$_v$ functions as a voltage-sensor for the process—a direct interaction between the Ca$_v$1.1 α_1-subunit and the ryanodine receptors then activates the Ca^{2+} release.

Ca$_v$s are also very important in cardiac and smooth muscles, where direct Ca^{2+}-influx through the Ca$_v$ itself provides the Ca^{2+} necessary for muscular contraction. In cardiac muscle, Ca$_v$1.2 or Ca$_v$1.3 is responsible for the plateau-phase of the cardiac action potential, which is important for cardiac muscle contraction and for regulation of the heart rate, so dihydropyridines are used for treatment of hypertension and cardiac arrhythmia. Ca$_v$3.1- and Ca$_v$3.2-subunits are found in the sino-atrial nodes where they play important roles for cardiac pacemaking.

The release of neurotransmitters from synaptic nerve terminals is triggered by influx of Ca^{2+} ions via Ca$_v$2.1- (P/Q-type) or Ca$_v$2.2- (N-type) subunits, which are expressed in all nerve terminals. When neuronal action potentials travel down the axon and reach the nerve terminal, they provide the depolarization necessary for activation of Ca$_v$s leading to Ca^{2+}-influx. The Ca$_v$2.1 and Ca$_v$2.2 subunits bind directly to proteins of the protein-machinery involved in membrane fusion of neurotransmitter-containing vesicles.

A similar role of Ca$_v$s is found in various endocrine cells such as the pancreatic β-cells in which ATP-mediated closing of K$_{ATP}$-channels leads to cellular depolarization, activation of Ca$_v$1.3 channels, and release of insulin-containing vesicles (Figure 13.5).

13.3.3 Pharmacology of Voltage-Gated Calcium Channels

There are two types of inhibition of Ca$_v$ function, namely, blockade of the ion channel pore and allosteric modulation of ion channel function. An example of pore blockade is cadmium (Cd^{2+}), which produces nonselective inhibition of all type of Ca$_v$s. The mechanism behind this effect is that Cd^{2+} binds to the ring of four glutamates in the selectivity filter of the pore with much higher affinity than Ca^{2+} itself and thus blocks the pore. Most of the peptide toxins, which block Ca$_v$-subtypes with high specificity, also act by producing pore block. Allosteric modulation, on the other hand, is exemplified by the dihydropyridines, which selectively affect members of the Ca$_v$1-family. The binding site for these compounds is located away from the pore and their mechanism of action relies on modification of the gating characteristics of the channel.

13.3.3.1 Ca$_v$1-Family (L-Type Currents)

The best characterized group of Ca$_v$ modulators is the so-called organic calcium blockers or calcium antagonists, comprising phenylalkylamines (e.g., verapamil), benzothiazepines (e.g., diltiazem), and the dihydropyridines (e.g., nifedipine; Figure 13.8). Several dihydropyridines are widely used clinically for the treatment of cardiovascular disorders such as hypertension, angina pectoris, and cardiac arrhythmia.

The organic calcium blockers bind with high affinity and selectivity to α_1-subunits of the Ca$_v$1-family, and act as allosteric modulators. This is highlighted by the fact that among the

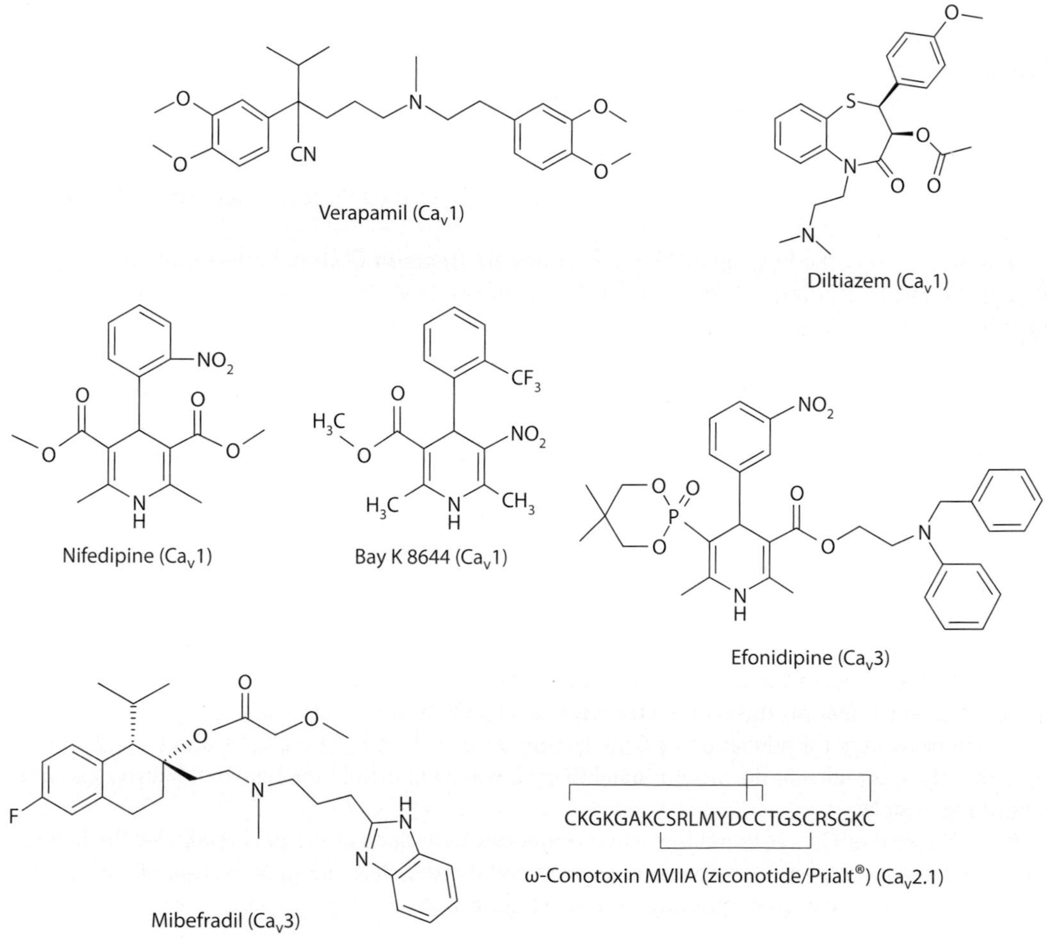

FIGURE 13.8 Chemical structure of drugs acting as blockers of Ca$_v$1 (L-type) and Ca$_v$3 (T-type) channels and the amino acid sequence of the highly specific peptide blocker of Ca$_v$2.2 (N-type) channels, ω-conotoxin MVIIA.

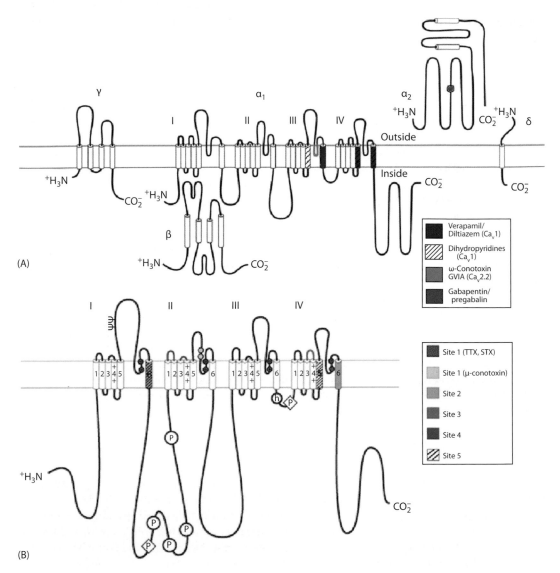

FIGURE 13.9 Overview of the binding sites of toxins and drugs acting at (A) Ca_v and (B) Na_v channels. ([A] From Catterall, W.A. et al., *Pharmacol. Rev.*, 57, 385, 2005. With permission from Elsevier; (B) From Catterall W.A. et al., *Toxicon*, 49, 124, 2007. With permission from Elsevier.)

dihydropyridine-type compounds, positive modulators of Ca_v1 have also been identified, e.g., the compound Bay K 8644 (Figure 13.8).

Amino acid residues important for the binding of these compounds have been identified through mutagenesis studies and are located in the S5 and S6 segments of domains III and IV of the α_1-subunit (Figure 13.9).

Organic calcium blockers bind with a much higher affinity to the inactivated conformations of the Ca_vs, relative to the closed conformation, thereby trapping the receptors in the inactivated state. Therefore, inhibition of Ca_vs by these compounds has been termed "use-dependent:" the rate and extent of Ca_v inhibition will increase with channel activation frequency. Use-dependence is generally considered to be an attractive quality of ion channel inhibitors, since only the highly active channels—presumably the ones responsible for a given disorder—will be inhibited, while less frequently activated channels are spared, thereby reducing the risk of side effects.

13.3.3.2   Ca$_v$2-Family (N-, P/Q-, and R-Type Current)

Within this family, the Ca$_v$2.2-subunit (N-type current) has attracted the most attention as potential drug target. The most efficient inhibitors of N-type currents are peptide toxins isolated from the venom of fish-eating marine snails that use these toxins to paralyze their prey. The category includes the 25–30 amino acid residue peptides ω-conotoxin GVIA and ω-conotoxin MVIIA (Figure 13.8), which bind to Ca$_v$2.2 with very high affinity and selectivity. Binding of ω-conotoxin GVIA mainly occurs to residues located in the pore loop region of domain III, suggesting that this toxin acts as a pore blocker of the Ca$_v$2.2-subunit.

The reason for the pharmacological interest in Ca$_v$2.2 is that these channels are responsible for neurotransmitter release in neural pathways relaying pain signals to the brain. Although ω-conotoxins are poorly suited for use as drugs because of their lack of biomembrane permeability, ω-conotoxin MVIIA (Prialt®) was recently approved for use in humans. Since the drug has to be given through an intrathecal catheter to circumvent the blood–brain barrier, the clinical use of ω-conotoxin MVIIA is limited to severe pain in patients suffering from terminal cancer or AIDS. A selective, nonpeptide Ca$_v$2.2 blockers that can be administered orally has so far not been identified, despite significant efforts, and finding the "dihydropyridines of Cav2.2 calcium channels" therefore still remains an open challenge!

Ca$_v$2.1 channels (P/Q-type current) are generally involved in neurotransmitter release in most synapses throughout the brain. Ca$_v$2.1 can be selectively blocked by peptide toxins from either *Conus* snails (ω-conotoxin MVIIC) or from spider venom (ω-agatoxin IVA) (Table 13.2). From a drug discovery point of view, however, these Ca$_v$s are not of great interest, since their widespread role in neurotransmitter release predicts severe toxicity as a consequence of channel inhibition.

The function(s) and pharmacology of Ca$_v$2.3 channels (R-type current) are not well understood. A peptide toxin, SNX-482, isolated from tarantula venom, has been found to act as a selective blocker of Ca$_v$2.3-channels.

13.3.3.3   Ca$_v$3-Family (T-Type Current)

Certain small-molecule compounds appear to act as moderately selective blockers of Ca$_v$3. The vasodilating compound mibefradil (Figure 13.8), which has been used widely for treatment of hypertension and angina pectoris, inhibits Ca$_v$3.1–Ca$_v$3.3 channels in a use-dependent way with ~10-fold selectivity over Ca$_v$1.2 channels. Moreover, certain novel dihydropyridine compounds (e.g., R-(−)-efonidipine, Figure 13.8) inhibit Ca$_v$3 channels up to ~100-fold more potently compared to Ca$_v$1 channels. It is not yet known exactly how these compounds interact with Ca$_v$3, but this family of ion channels seems to have a great potential as drug targets for treatment of cardiovascular disease. Certain classical antiepileptic compounds, such as ethosuximide, phenytoin, and zonisamide exert their antiepileptic action at least partly via inhibition of Ca$_v$3 channels. Substances such as nickel ions (Ni^{2+}), *n*-octanol, and the diuretic amiloride display moderate selectivity for Ca$_v$3 channels over the other Ca$_v$ channel types. Kurtoxin is a scorpion venom toxin, which produces potent and selective blockade of Ca$_v$s containing Cav3.1- and Ca$_v$3.2- but not Ca$_v$3.3-subunits.

13.3.3.4   Auxiliary Subunits

The drugs gabapentin and the more recently developed pregabalin are used clinically for the treatment of epilepsy and neuropathic pain. Their mechanism of action was not understood before the discovery that gabapentin binds with extremely high affinity to the $\alpha_2\delta$-subunit of Ca$_v$s. Functionally, gabapentin and pregabalin decrease the amplitude of calcium currents partially without producing the complete blockade seen with Ca$_v$ inhibitors targeting the α_1-subunit. Both Ca$_v$2.1 and Ca$_v$2.2 are involved in mediating the effects of gabapentin/pregabalin. Both drugs are nontoxic, which may be related to their partial blocking effect.

13.4 VOLTAGE-GATED SODIUM CHANNELS

13.4.1 STRUCTURE AND MOLECULAR BIOLOGY OF VOLTAGE-GATED SODIUM CHANNELS

Functionally, Na_vs are closed at the resting membrane potential and open when the membrane becomes depolarized, activation requiring membrane potentials of −70 to −30 mV, with some variation between different Na_v types. Most Na_vs inactivate within ~1–10 ms in the presence of sustained depolarization. In certain types of neurons, a more persistent Na_v current with slow inactivation has also been identified.

At the molecular level, Na_vs are composed of a large (~2000 amino acid residues) α-subunit, which is structurally similar to the $α_1$-subunit of Ca_vs, and forms the ion conducting pore (Figure 13.7A). The high selectivity for Na^+ over K^+ is due to the composition of the ion selectivity filter, which consists of two rings of amino acids with each of the four homologous domains contributing one amino acid to each ring: outer ring with Glu-Glu-Asp-Asp and inner ring with Asp-Glu-Lys-Ala. The rapid inactivation of most Na_vs is explained by the cytoplasmic domain III–IV linker (Figure 13.9B h motiv), which functions as a "hinged lid," that simply swings in to occlude the intracellular mouth of the pore.

Nine different Na_v α-subunits (Na_v1.1–Na_v1.9) have been cloned and all these display >50% amino acid identity with each other, so they compose one subfamily. The Na_v1 family has most likely arisen from a singe ancestral gene and that their present diversity reflects gene duplication events and chromosomal rearrangements occurring late in evolution.

By analogy to the Ca_vs, functional Na_vs can be formed from expression of α-subunits alone although native Na_vs are protein complexes composed by α-subunits and auxiliary subunits. Only a single class of auxiliary Na_v subunits (β-subunits) has been identified. β-Subunits are composed of a large extracellular part, through which it interacts with the α-subunit (Figure 13.7B) and a small C-terminal portion consisting of a single TM segment. The function of the β-subunits can be divided into: (1) modulation of the functional properties of Na_vs, (2) enhancement of membrane expression, and (3) mediating interactions between Na_vs and extracellular matrix proteins as well as various signal transduction molecules.

13.4.2 PHYSIOLOGICAL ROLES OF VOLTAGE-GATED SODIUM CHANNELS

The biological importance of Na_vs relies on their ability to cause depolarization of cell membranes. Most of the Na_v α-subunits are capable of detecting even very small increases in membrane potential and this makes the Na_vs activate, and subsequently inactivate, on a ms time scale. This combination of high sensitivity toward depolarization and very rapid gating kinetics makes Na_vs perfect for initiating and conducting action potentials.

Na_v1.1, Na_v1.2, Na_v1.3, or Na_v1.6 subunits are expressed in virtually all neurons within the CNS, in particular, at the base and along the entire length of the axon. When an excitatory synaptic signal (e.g. glutamate, released by a neighboring neuron, acting on AMPA receptors, see Chapter 15) is received, this generates a small depolarization of the neuronal membrane in the dendrites and cell body. This rather modest depolarization is sufficient for activating Na_vs at the initial segment of the axon, leading to the generation of an AP. Once the AP reaches the nerve terminal, this will activate Ca_vs, leading to release of neurotransmitter. The importance of these Na_vs for AP initiation and conduction is also highlighted by the fact that point mutations in the genes encoding Na_v1.1, Na_v1.2, and Na_v1.3, which alter their functional properties, have been linked to certain forms of epilepsy.

Dorsal root ganglion (DRG) neurons are important for transmitting sensory signals, including pain, from the periphery to the CNS. Sensory stimulation leads to generation and conduction of action potentials in DRG neurons and these APs are mediated by Na_vs. Na_vs of DRG neurons contain the Na_v1.7, Na_v1.8, and Na_v1.9 subunits, which are almost exclusively expressed in these neurons. It has also been shown that expression of these α-subunits is altered in a complex fashion in animal models of inflammatory and neuropathic pain. From a therapeutic point of view, these α-subunits are therefore of particular interest, since compounds capable of selectively blocking $Na_v1.7$–$Na_v1.9$ channels could have great potential as analgesics.

13.4.3 PHARMACOLOGY OF VOLTAGE-GATED SODIUM CHANNELS

A large number of natural products (peptides and alkaloids) have been found to bind Na_vs with high affinity. Radioligand-binding, photoaffinity-labeling, and mutagenesis techniques have been used to identify the regions of the α-subunit to which these substances bind. Six binding sites for these toxins are therefore used to provide the conceptual framework for understanding the pharmacology of Na_vs (Table 13.3; Figure 13.9B). Given the high degree of homology between the Na_v1-subunits, very few examples of subunit-selective toxins are known. The substances mentioned in Table 13.3 thus bind to nearly all Na_v1 subunits. Most toxins act as gating modifiers, and only tetrodotoxin and saxitoxin binding to site 1 are pore blockers.

In addition to these different toxins, a number of clinically used drug molecules are known to exert their pharmacological action through inhibition of Na_v function. Consistent with the physiological roles of Na_vs, these drugs include antiepileptic compounds (carbamazepine, lamotrigine, and phenytoin), local anesthetic and analgetic compounds (lidocaine), and drugs used to treat cardiac arrhythmia (class I antiarrhythmics including quinidine, lidocaine, mexiletine, and flecainide).

TABLE 13.3
Toxin-Binding Sites on Na_v Channels

Site No.	Site Location	Toxins Binding to Site	Mechanism of Action
1	Selectivity filter of pore	Tetrodotoxin, saxitoxin	Pore block
2	Interface between the S6 segments of domains I and IV	Plant alkaloid toxins: grayanotoxin, batrachotoxin, and veratridine	Inhibition of inactivation and channel opening at resting potential
3	Outer pore loop regions of domains I and IV	Sea anemone peptide toxins and α-scorpion toxins	Slow inactivation
4	Extracellular S3–S4 loop close to the voltage sensor	Large β-scorpion peptide	Enhance opening at negative membrane potential
5	Interface between the IS6 and IVS5 segments	Plant alkaloids ciguatoxins and brevetoxins	Enhance activation and inhibit inactivation
6	Unknown	δ-Conotoxins	Slow inactivation

13.5 CHLORIDE CHANNELS

Most cells have anion channels and the primary ion permeating these is Cl^-. The ClC channel family members are involved in transepithelial transport, acidification of synaptic vesicles, and endocytotic trafficking. The ClC proteins are unique and are widely expressed in intracellular organelles and not in the plasma membrane. Surprisingly, some of the ClC proteins show characteristics of ion transporters. The CFTR Cl^- channel is an ABC protein involved in transepithelial transport, and dysfunction of this Cl^- channel causes cystic fibrosis. The CFTR protein has been extensively used in attempts to establish a gene therapy for cystic fibrosis without success. Ca^{2+}-activated and volume-activated Cl^- currents have been characterized in native cells, but these have not been identified at a molecular level yet. The pharmacology of Cl^- channels is currently quite poor.

13.6 LIGAND-GATED ION CHANNELS

Na_v and K_v channels are essential for generation and conduction of neuronal action potentials and Ca_v channels are essential for converting action potentials into neurotransmitter release—but it is the ligand-gated ion channels which receive the chemical signals of synaptic transmission and convert them into the electrical signals that initiate action potentials. See Chapter 12 for further details about ligand-gated ion channels.

FURTHER READINGS

Alberts, B., Johnson, A., Lewis, J., Raff, M., Roberts, K., and Walter, P. 2002. *Molecular Biology of the Cell*, 4th edn. Garland Publishing Inc., New York.

Ashcroft, F. 2000. *Ion Channels and Disease*. Academic Press, London, U.K.

Catterall, W.A. 2000. From ionic currents to molecular mechanisms: The structure and function of voltage-gated sodium currents. *Neuron* 26: 13–25.

Catterall, W.A. et al. 2005. International Union of Pharmacology. Compendium of voltage-gated ion channels. *Pharmacol. Rev.* 57: 385–540.

Catterall, W.A. et al. 2007. Voltage-gated ion channels and gating modifier toxins. *Toxicon* 49: 124–141.

Dalby-Brown, W., Hansen, H., Korsgaard, M.G., Mirza, N., and Olesen, S.-P. 2006. Kv7 channels: Function, pharmacology and channel modulators. *Curr. Top. Med. Chem.* 6: 999–1023.

Doyle, D.A., Cabral, J.M., Pfuenzner, R.A., Kuo, A., Gulbis, J.M., Cohen, S.L., Chait, B.T., and MacKinnon, R. 1998. The structure of the potassium channel: Molecular basis of K^+ conduction and selectivity. *Science* 280: 69–77.

Hille, B. 2001. *Ionic Channels of Excitable Membranes*. 3rd edn. Sinauer Associates, Sunderland, MA.

Nardi, A. and Olesen, S.-P. 2008. BK channel modulators: A comprehensive overview. *Curr. Med. Chem.* 15: 1126–1146.

Nilius, B., Owsianik, G., Voets, T., and Peters, J.A. 2007. Transient receptor potential cation channels in disease. *Physiol. Rev.* 87: 165–217.

Sanguinetti, M.C. and Tristani-Firouzi, M. 2006. Herg potassium channels and cardiac arrhythmia. *Nature* 440: 463–469.

14 Neurotransmitter Transporters: Structure and Function

Claus J. Loland and Ulrik Gether

CONTENTS

14.1 INTRODUCTION

The transport of solutes across membranes is of fundamental importance for all living organisms and is mediated via specific integral membrane proteins. The transport processes are often energetically coupled, either directly through the hydrolysis of ATP by the transport protein itself or indirectly by the use of transmembrane ion gradients that enable the transport of the substrate against its concentration gradient. A vast amount of different transport proteins are found in both prokaryotic and eukaryotic organisms, transporting everything from nutrients and metabolites to ions, drugs, proteins, toxins, and transmitter molecules. If ion channels (see Chapter 13) are excluded, major classes of transport proteins in humans encompass ATP-driven ion pumps (e.g., the ubiquitously expressed Na–K ATPase), ATP-binding cassette (ABC) transporters (e.g., the cystic fibrosis transmembrane conductance regulator and the multidrug resistance transporter p-glycoprotein), cytochrome B-like proteins, aquaporins (water transporters), and the solute carrier superfamily (SLC) (http://www.bioparadigms.org).

The immense functional heterogeneity among transporters is illustrated by the fact that the SLC gene family alone consists of nothing less than 46 different subfamilies (http://www.bioparadigms.org/slc/menu.asp). These include several types of plasma membrane transporters, such as, high-affinity glutamate transporters (SLC1), sodium-glucose cotransporters (SLC5), sodium-coupled neurotransmitter transporters (SLC6), and a variety of ion exchangers (SLC8+9+24). The SLC gene family also includes intracellular vesicular transporters, such as the vesicular glutamate transporters (SLC17), the vesicular monoamine transporters (SLC18), and the vesicular inhibitory amino acid transporters (SLC32).

Notwithstanding the huge number of transport proteins present in the human body, relatively few of them are targets for the action of drugs. It might even be argued that transport proteins are relatively overlooked as drug targets in spite of their critical physiological functions and some real "success stories," such as, inhibitors of the gastric ATP-driven proton pump, used against peptic ulcers, and inhibitors of monoamine transporters, used against depression/anxiety disorders (see Chapter 18). In this chapter, we focus on the monoamine transporters and then on the neurotransmitter transporters belonging to the SLC6 family (also named neurotransmitter:sodium symporters or Na-/Cl-dependent transporters) (Table 14.1). Indeed, the SLC6 transporters represent important targets for several drugs including not only medicines used against depression/anxiety but also against

TABLE 14.1
SLC6 Gene Family Neurotransmitter Transporters

	Endogenous Substrates	Synthetic Substrates	Potent Inhibitors	Therapeutic Use/Potential
Dopamine transporter (DAT)	Dopamine Norepinephrine Epinephrine	Amphetamine MPP+	CFT, GBR12,909 benztropine, mazindol, RTI-55, Cocaine, Zn^{2+}	ADHD (amphetamines), Parkinsonism? (inhibitors)
Serotonin transporter (SERT or 5-HTT)	5-HT	p-Cl-Amphetamine MDMA ("ecstasy")	Citalopram, escitalopram, fluoxetin, paroxetin, sertraline, imipramine, cocaine, RTI-55	Depression, anxiety, OCD (inhibitors)
Norepinephrine transporter (NET or NAT)	Dopamine Norepinephrine Epinephrine	Amphetamine MPP+	Nisoxetine, nortriptyline, desipramine, duloxetine, venlafaxine, mazindol	Depression (inhibitors)
Glycine transporter 1 (GlyT1)	Glycine	—	(R)NFPS (ALX5407), NPTS, Org24598	Schizophrenia? Psychosis? Dementia? (inhibitors)
Glycine transporter 2 (GlyT2)	Glycine	—	ALX1393, ALX1405 Org25543	Anticonvulsant? Analgesic? (inhibitors)
GABA transporter 1 (GAT-1)	GABA	—	Tiagabine, SKF89976A, THPO, exo-THPO	Epilepsy (tiagabine)
GABA transporter 2 (GAT-2 equivalent to mouse GAT-3)	GABA beta-Ala	—	—	?
GABA transporter 3 (GAT-3 equivalent to mouse GAT-4)	GABA beta-Ala	—	SNAP5114	?
GABA transporter 4 (GAT-4, equivalent to BGT1 or mouse GAT-2)	GABA Betaine	—	EF-1502, NNC052090, THPO Zn^{2+}	Epilepsy? (inhibitors) EF-1502

Note: The known neurotransmitter transporters belonging to the SLC6 gene family with their respective substrates, inhibitors, and potential therapeutic use.

Abbreviations: CFT, 2β-carbomethoxy-3β-(4-fluorophenyl)tropane; MDMA, 3, 4-methylenedioxymethamphetamine; MPP+, 1-methy l-4-phenylpyridinium; (R)NFPS, N-[3-(40-fluorophenyl)-3-(40-phenylphenoxy) propyl]sarcosine; NPTS, N-[3-phenyl-3-(40-(4-toluoyl) phenoxy)propyl]sarcosine; THPO, (4,5,6,7-tetrahydroisoxazolo[4,5-c] pyridine-3-ol); ADHD, attention deficit hyperactivity disorder; OCD, obsessive-compulsive disorder. Exo-THPO: 4-amino-4,5,6,7-tetrahydrobenzo[d]isoxazol-3-ol.

obesity and epilepsy as well as drugs of abuse such as cocaine, amphetamine, and "ecstasy." Of interest, high-resolution crystal structures have recently become available of bacterial homologues of SLC6 transporters opening up entirely new possibilities for understanding how these transporters operate at a molecular level and how their function can be altered by different types of drugs.

14.2 NEUROTRANSMITTER TRANSPORTERS BELONGING TO THE SLC6 FAMILY

The availability in the synaptic cleft of the neurotransmitters dopamine, serotonin, norepinephrine, glycine, and γ-amino butyric acid (GABA) is tightly regulated by specific transmembrane transport proteins belonging to the SLC6 family (Figure 14.1 and Table 14.1). The transport proteins are situated either in the presynaptic membrane or on the surface of adjacent glial cells, where they mediate the rapid removal of the released neurotransmitters and thereby terminate their effect at the pre- and postsynaptic neurons. Inside the presynaptic nerve endings, specific vesicular transporters sequester the neurotransmitters into vesicles, making them ready for subsequent release into the synaptic cleft upon the arrival of the next stimulus. The plasma membrane neurotransmitter transporters serve three main purposes: first, the transport proteins increase the rate by which the released neurotransmitters are cleared from the synaptic cleft. This rapid removal of the released neurotransmitters allows for 100-fold faster termination of neurotransmission than is possible with simple diffusion. Second, reuptake may prevent diffusion of the neurotransmitters away from the

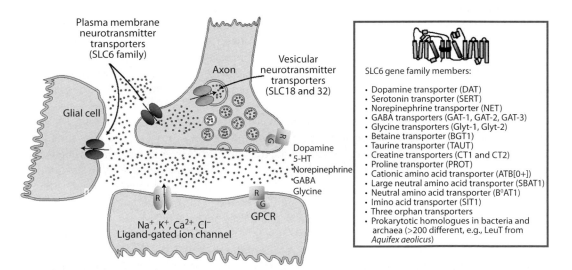

FIGURE 14.1 The role of neurotransmitter transporters in synaptic signaling. Neurotransmitters are sequestered into synaptic vesicles through vesicular monoamine transporters (VMAT1–2) belonging to the SLC18 gene family or through vesicular inhibitory amino acid transporters (VIAAT) belonging to the SLC32 gene family. Upon arrival of an axon potential, the synaptic vesicle releases its content of the neurotransmitter into the synaptic cleft by fusion of the vesicle with the plasma membrane. The neurotransmitter exerts its effects by activating ionotropic receptors, (ligand-gated ion channels), such as GABA_A receptors, glycine receptors, and 5-HT_3 receptors or via G-protein-coupled receptors (GPCRs) such as dopamine receptors, adrenoceptors, 5-HT receptors, and metabotropic GABA_B receptors. The fast removal of the neurotransmitter from the synaptic cleft is governed by the neurotransmitter transporter belonging to the SLC6 family located on the presynaptic neuron (DAT, SERT, NET, GlyT2, GAT-1, and GAT-2) or on glia cells (GlyT-1, GAT-1, GAT-2, and GAT-3). The neurotransmitter taken up by the presynaptic neuron allows recycling with a presumed savings in synthetic cost.

synapse of their release, thereby minimizing chemical crosstalk between adjacent synapses. Third, transporters allow recycling by reuptake of transmitters into the nerve terminal with presumed savings in synthetic cost. The crucial physiological role of the neurotransmitter transporters has been cemented by gene knockout experiments. In case of, e.g., the dopamine transporter (DAT), the disruption of the transporter gene in mice revealed the unequivocal importance of this carrier in the control of locomotion, growth, lactation, and spatial cognitive function.

The SLC6 transporters include not only the transporters of neurotransmitters (Table 14.1, Figure 14.1) but also transporters of amino acids, metabolites (creatine), and osmolytes (betaine and taurine) (Figure 14.1). Moreover, a large number of homologues have been identified in archaea and bacteria. The function of the majority of these transporters is still unknown; however, a few of them have been identified as amino acid transporters, such as, e.g., the leucine transporter LeuT$_{Aa}$ from the *Aquifex aeolicus* bacterium and the tyrosine transporter Tyt1 from *Fusobacterium nucleatum*.

At the molecular level, the SLC6 family transporters operate as Na$^+$ dependent cotransporters that utilize the transmembrane Na$^+$ gradient to couple the "downhill" transport of Na$^+$ with the "uphill" transport (against a concentration gradient) of their substrate from the extracellular to the intracellular environment. The transport process is so efficient that, e.g., the serotonin transporter (SERT) can accumulate internal serotonin (5-HT) to concentrations 100-fold higher than the external medium when appropriate ion gradients are imposed. Most SLC6 transporters are also cotransporters of Cl$^-$ and, accordingly, SLC6 transporters have been referred to as the family of Na$^+$/Cl$^-$-dependent transporters.

14.2.1 Structures and Mechanisms of SLC6 Transporters

It is generally believed that SLC6 transporters function according to an alternating access model, which suggests a transport mechanism in which, at any given time, only the substrate-binding site is accessible to either the intracellular or the extracellular side of the membrane. Thus, at all times an impermeable barrier exists between the binding site and one side of the membrane, but the barrier can change from one side of the binding site to the other, giving the site alternate access to the two aqueous compartments that the membrane separates. A prerequisite for this model is the existence of both external and internal "gates," i.e., protein domains that are capable of occluding access to the binding site of the substrate from the external and internal domains, respectively (Figure 14.2).

In the absence of high-resolution structural information, however, it was, for a long time, only possible to speculate about the molecular basis of the transport process. A major breakthrough came when the bacterial homologue, LeuT$_{Aa}$, which displays 20%–25% sequence identity to its mammalian counterparts, was successfully crystallized and the structure solved at high resolution (1.65 Å). The transporter was crystallized with substrate and Na$^+$ bound to the transporter. The x-ray diffraction pattern revealed a protein containing 12 transmembrane segments (TMs) in a unique fold and with a binding site for L-leucine buried inside the center of the protein (Figure 14.3). The diffraction pattern also revealed an unexpected structural repeat in the first 10 TMs that relates TM1–5 with TM6–10 around a pseudo-twofold axis of symmetry located in the plane of the membrane. The binding pockets for leucine and Na$^+$ are formed by TM1, TM3, TM6, and TM8. TM3 and TM8 are long helices that are related by the twofold symmetry axis and are strongly tilted (~50°) (Figure 14.3). TM1 and TM6 are characterized by unwound breaks in the helical structure in the middle of the lipid bilayer. These breaks expose main carbonyl oxygen and nitrogen atoms for direct interaction with the substrate.

The LeuT$_{Aa}$ structure was crystallized in a conformation in which access to the substrate-binding site is closed from both the intracellular and extracellular environments, i.e., the predicted external and internal "gates" appear closed in the structure, and, hence, the structure likely represents an intermediate state between the outward facing conformation (where the substrate-binding site is exposed to the extracellular environment) and the inward facing conformation (where it is exposed to the intracellular milieu).

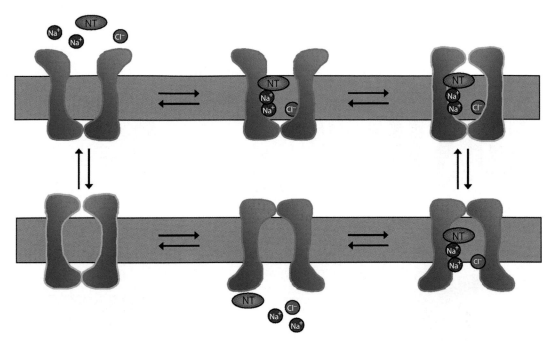

FIGURE 14.2 The alternating access model. The mechanism by which the neurotransmitter transporters translocate the substrate from the extracellular environment to the cytosol can be explained by the alternating access model. Without a neurotransmitter (NT) or ions (Na^+, Cl^-), the transporter resides in an outward facing conformation where the neurotransmitter-binding site is only accessible to the external environment. Upon binding of the solutes, the transporter undergoes a conformational change, first closing the outer gate excluding access to the binding site and subsequently opening the inner gate, allowing access to the binding site from the cytosol. The low Na^+ concentration in the cytosol allows the release of the Na^+ ions, which also causes a release of its substrate. The release of solutes again closes the inner gate and opens the outer gate, making the neurotransmitter transporter ready for another translocation cycle to occur. NT, neurotransmitter.

To accommodate shifts between outward- and inward-facing conformations it was suggested that the transport process involves major movements of TM1 and TM6 relative to TM3 and TM8. On the extracellular side, these movements were proposed to be controlled by an external gate that involves residues in TM1, TM3, TM6, and TM10, with a charged pair between Arg30 (TM1) and Asp404 (TM10) being of particular importance (Figure 14.3). Opening and closing of the external gate are also likely to involve conformational rearrangements of the extracellular loops (ECLs). ECL4 is especially interesting because it forms a lid that extends into the center of the transporter and interacts with residues in, for example, TM1. In this way, ECL4 covers the substrate-binding site from the outside without being in direct contact with the substrate (Figure 14.3). Thus, it is conceivable that opening the transporter to the outside involves a major conformational rearrangement of ECL4. This is in agreement with previous studies involving the engineering of Zn^{2+}-binding sites and application of the substituted-cysteine accessibility method in the corresponding part of the DAT, SERT, and GABA transporter (GAT)-1.

The intracellular gate of $LeuT_{Aa}$ is predicted to comprise ~20 Å of ordered protein structure, involving in particular the intracellular ends of TM1, TM6, and TM8. A key residue in the predicted gate is Tyr268 at the cytoplasmic end of TM6, which is conserved in all transporters of this class (Figure 14.3). The tyrosine is positioned below the substrate-binding site at the cytoplasmic surface of the protein and forms a cation–π interaction with an arginine in the N-terminus just below TM1 that forms a salt bridge with an aspartate at the cytoplasmic end of TM8 (Asp369). A likely possibility would be that opening of the gate to the inside will require disruption of this set of interactions.

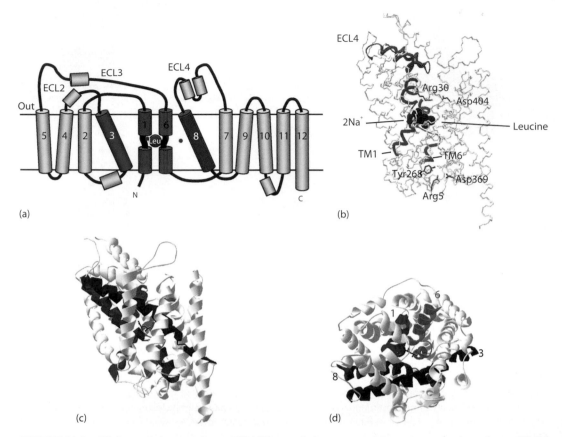

FIGURE 14.3 High-resolution structure of Na$^+$/Cl$^-$-coupled neurotransmitter transporter homologue (SLC6 family) from *A. aeolicus* (LeuT$_{Aa}$). (a) Schematic representation of transmembrane topology. The binding pocket for leucine and Na$^+$ is formed by TM1, TM3, TM6, and TM8 (TMs highlighted in distinct colors; Na$^+$ denoted as blue dots). (b) View of LeuT$_{Aa}$ parallel to the membrane, highlighting important structural features. The central parts of TM1 (green) and TM6 (light blue) are characterized by unwound breaks in the helical structure. These breaks expose main carbonyl oxygen and nitrogen atoms for direct interaction with the substrate (red). The two Na$^+$ ions (dark blue) also interact with the unwound part of TM1 and TM6 and have a key role in stabilizing this structure and the leucine-binding site. Access to the substrate-binding site from the external medium is predicted to be controlled by an external gate that in particular involves a charged pair between Arg30 (TM1) and Asp404 (TM10) as well as conformational rearrangements of the ECL 4 (yellow), which could form a lid that excludes the substrate-binding site from the extracellular space. A key residue in the predicted intracellular gate is Tyr268 at the bottom of TM6. This residue is conserved in all transporters of this class. Tyr268 interacts with Arg5 in the bottom of TM1, which again interacts with Asp369 at the bottom of TM8, with both residues also highly conserved throughout the SLC6 family. (c) View parallel to the membrane. (d) View from the extracellular side. The TMs forming the substrate-binding site are highlighted in distinct colors. (From Gether, U. et al., *Trends Pharmacol. Sci.*, 27, 375, 2006. With permission.)

In agreement with this, recent experimental observations in the DAT in conjunction with computational simulations strongly support such a role of the interaction network and that the mechanism is highly conserved among all SLC6 transporters (Figure 14.3).

A major reorganization on the intracellular side during translocation is further supported by a marked increase in accessibility of cysteines introduced in the cytoplasmic half of TM5 of SERT to a membrane permeant cysteine-reactive reagent in the presence of serotonin. The observations indicated that when the transporter becomes inward-facing, the cytoplasmic half of TM5, which is occluded in the LeuT structure, lines an aqueous pathway leading from the binding pocket to the cytoplasm.

Taken together, based on the LeuT$_{Aa}$ structure and available functional data, it seems reasonable to conclude that SLC6 transporters follow an alternating access model. However, the mechanism of transport by LeuT$_{Aa}$ is probably distinct from those suggested for other ion-coupled transporters; hence, the mechanistic predictions clearly differ from those involving movements of two symmetrical hairpins reaching from the extracellular and intracellular environments, respectively, that were offered for sodium-coupled glutamate transporters based on a crystal structure of a bacterial member of this transporter family. Similarly, the suggested mechanism differs from the "rocker-switch" type mechanism proposed for Lac permease and the glycerol-3-phosphate transporter, two other recently crystallized transport proteins that mediate proton-coupled secondary active transport.

14.2.2   THE BINDING SITES FOR NA⁺ AND CL⁻: IMPORTANCE IN SUBSTRATE BINDING AND TRANSLOCATION

The binding and cotransport of sodium ions is a feature that probably serves several purposes: first, the sodium ion(s) serve as a driving force for the translocation of the substrate against its electrochemical gradient; second, the ions coordinate the binding of the substrate to the transporter; and third, the ions might function as conformational guides, ensuring that the transporter undergoes the proper conformational changes during the translocation cycle.

SLC6 transporters bind and translocate one to three sodium ions during the translocation of one substrate molecule. In addition, several of the transporters bind and cotransport one chloride ion during one cycle, although this is not a ubiquitous feature of all transporters in the family. The SERT appears to be special because it also mediates the countertransport of one potassium ion during one transport cycle. The high-resolution structure of LeuT$_{Aa}$ provided for the first time insight into the possible localization of the sodium-binding sites in SLC6 transporters by showing two distinct sodium-binding sites adjacent to the substrate-binding site. The two sodium ions appeared to have a key role in stabilizing the LeuT$_{Aa}$ core, the unwound structures of TM1 and TM6, and the bound leucine molecule. One sodium ion (designated Na1) was found to possess an octahedral coordination, with one coordinate to the carboxyl group of leucine, thereby providing a possible structural link for the coupling of Na⁺ and solute fluxes. The other sodium ion (Na2) is positioned between the TM1 unwound region and TM8, about 7.0 Å from Na1 and not directly involved in coordinating the bound leucine (Figure 14.3). Notably, with the conservative substitution of a serine for a threonine, all residues coordinating both Na1 and Na2 are conserved from LeuT$_{Aa}$ to the mammalian transporters.

The LeuT$_{Aa}$ does not possess any apparent Cl⁻-binding site, and accordingly the transport of leucine is not dependent on the presence of Cl⁻. However, the use of homology modeling and energy minimization of the GAT-1 based on the LeuT$_{Aa}$ structure, a potential chloride-binding site in this transporter, was elegantly identified. A cavity in the GAT-1 was found where the chloride ion may interact with the hydrogen atoms of the amide group of Gln291 and of the hydroxyl groups of Ser331, Ser295, and Tyr86 (GAT-1 numbering). The model was experimentally verified in part by the introduction of a negatively charged amino acid in position 331 (S331D/E), rendering both the net flux and the exchange of GABA largely chloride independent. Equivalent mutations introduced in the mouse GABA transporter-4 and the DAT also result in a chloride-independent transport, whereas the reciprocal mutations in LeuT$_{Aa}$ and in Tyt1 convey these transporters from displaying chloride-independent substrate binding to chloride-dependent binding. Furthermore, the transport rate of GABA increased by lowering the intracellular pH, and thereby likely increasing the protonation during the return step of the glutamate inserted in position 331 of the GAT-1. This result suggests that in the wild-type transporter, the chloride ion is a substrate for the GAT and is released to the cytosol in contrast to simply binding to the protein throughout the entire translocation cycle. The requirement of the negative charge during the translocation of GABA, but not during the return step, suggests that the role of chloride is mainly to compensate for the multiple positive charges that enable accumulation of the substrate against huge concentration gradients.

14.2.3 Substrate Specificity and Binding Sites in SLC6 Neurotransmitter Transporters

The biogenic amine transporters include DAT, the norepinephrine transporter (NET), and the SERT. Among these, DAT and NET display marked overlapping selectivity for dopamine and norepinephrine; hence, DAT transports dopamine and norepinephrine with similar efficacy, and the apparent affinity for dopamine is only a few folds higher than that for norepinephrine (Figure 14.4). Moreover, NET transports dopamine with 50% of the efficacy seen for norepinephrine, and the apparent affinity for norepinephrine is even a few folds higher than that for dopamine. Accordingly, their classification as DAT and NET seems primarily determined by their localization to dopaminergic and noradrenergic neurons, respectively, rather than by their distinct substrate specificity. In contrast, the SERT displays high specificity toward 5-HT although it has been shown that the SERT can transport dopamine if present in very high concentrations. The GABA transporters (GAT-1 to GAT-3) and the glycine transporters (GlyT-1 and -2) all display high specificity for their respective endogenous substrates; however, GAT-2 and GAT-3 can also transport beta-alanine, the only beta-amino acid that occurs naturally, and GlyT1 can transport the naturally occurring N-methyl-derivative of glycine, sarcosine.

The availability of the LeuT$_{Aa}$ structure has provided the first reliable hypothesis for the location of the primary substrate-binding site in this class of transporters. The most striking feature of

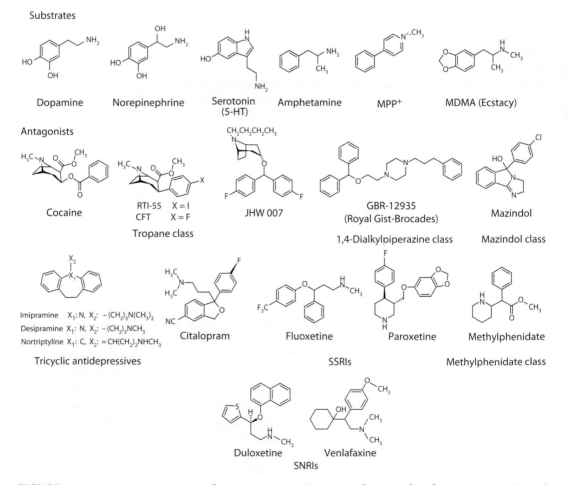

FIGURE 14.4 Chemical structures of most common substrates and antagonists for the DAT, NET, and SERT. SSRIs, selective serotonin reuptake inhibitors; SNRIs, serotonin–norepinephrine reuptake inhibitors.

the leucine-binding site is that it exposes main-chain atoms from the unwound regions of TM1 and TM6 make most of the contacts with the α-amino and α-carboxy groups of the bound leucine. The unwound regions of TM1 and TM6 allow direct hydrogen-bonding partners as well as orientating the α-amino and α-carboxy groups so they can bind close to the ends of the helical segments and establish α-helix dipole interactions. In addition and as mentioned earlier, one of the Na^+ ions makes direct contact with the carboxyl group of the leucine. Notably, this interaction cannot occur in transporters of the biogenic amines, dopamine, 5-HT, and norepinephrine, which do not possess a carboxyl group; however, the glycine in position 24 of $LeuT_{Aa}$ (which is conserved among the amino acid transporters) is replaced with aspartate in these transporters (Asp79 in the DAT). According to the $LeuT_{Aa}$ structure, this aspartate is predicted to be in the immediate proximity of Na^+ and, thus, can probably substitute for the missing carboxyl group of the substrate. Studies on DAT and SERT also support the hypothesis that the aspartate in TM1 plays an important role in the coordination of the protonated primary amine in dopamine, and 5-HT. In GAT-1 the residue in this position is a glycine as it is in the $LeuT_{Aa}$. Hence, GABA is probably coordinated in the same way as leucine in the $LeuT_{Aa}$. Indeed, the residue seems to be important for the binding of GABA to the GAT-1.

Right above the leucine molecule in the $LeuT_{Aa}$ structure, a tyrosine in TM3 (Tyr108) forms via its hydroxyl a hydrogen bond with the main-chain amide nitrogen of Leu25 in TM1. This interaction could function as a latch to stabilize the irregular structure near the unwound region in TM1 and may even be the first determinant of the closure of the extracellular gate. This hypothesis is even more interesting in light of the fact that the tyrosine is strictly conserved among all SLC6 family members and has been implicated in the substrate binding and transport of GAT-1 and SERT. Of further interest, recent homology modeling studies in DAT and SERT suggest that in these transporters, the hydroxyl group of the tyrosine does not form a hydrogen bond with the main-chain amide nitrogen of Leu25 but with the central aspartate in TM1, believed also to interact directly with the monoamine substrates (Asp79 in DAT and Asp98 in SERT and corresponding to position 24 in $LeuT_{Aa}$). In support of this hypothesis and thereby of the role of the hydrogen bond in stabilizing the substrate-binding site, mutation of the tyrosine in DAT (Tyr156) to phenylalanine decreases apparent dopamine around 10-fold and decreases the maximum uptake capacity by ~50%.

It is also important to note that the identification of residues shaping the leucine-binding site in $LeuT_{Aa}$ illuminates the determinants of substrate specificity in the eukaryotic homologues. In the SERT, for example, residues at equivalent positions to those surrounding the isopropyl moiety of leucine in $LeuT_{Aa}$ are replaced with smaller amino acids to accommodate the larger serotonin molecule. Correspondingly, homology modeling of the glycine transporters GluT1 and GlyT2 suggests together with mutational analysis that the substrate specificity is determined by a few key residues and that the ability of GlyT1 but not GlyT2 to transport sarcosine in addition to glycine is determined by a single residue difference between GlyT1 and GlyT2.

14.3 DRUGS TARGETING BIOGENIC AMINE TRANSPORTERS: SPECIFICITY, USE, AND MOLECULAR MECHANISMS OF ACTION

The biogenic amine transporters, DAT, NET, and SERT, are targets for a wide variety of drugs. Overall, these drugs can be classified as either pure inhibitors that block substrate binding and transport, or as substrates that in addition to competing with the endogenous substrate are also transported themselves.

14.3.1 COCAINE, BENZTROPINE, AND OTHER TROPANE CLASS INHIBITORS

The most thoroughly studied class of inhibitors of biogenic amine transporters is the "tropane" class, with cocaine as the most well-known member (Figure 14.4). Cocaine is a moderately potent antagonist inhibiting the function of all three transporters nonselectively. However, earlier correlative studies as well as studies on genetically modified mice suggest that presynaptic DAT is the primary target for cocaine's stimulatory action. DAT knockout mice are insensitive to the administration of cocaine and, moreover, knockin mice expressing a DAT mutant incapable of binding cocaine shows insensitivity to cocaine administration. It is, therefore, the current view that the rapid increase in extracellular dopamine concentration elicited by cocaine inhibition of DAT produces the psycho-motor stimulant and reinforcing effect that underlie cocaine abuse.

Some closely related cocaine analogues possess higher potency toward the biogenic amine transporters and, thus, have been more suitable than cocaine itself in experimental setups (e.g., radi-oligand binding assays) directed toward understanding the pharmacological properties of the trans-porters. Important examples include CFT (2β-carbomethoxy-3β-(4-fluorophenyl)tropane or WIN 35,428) and RTI-55 ((−)-2β-carbomethoxy-3β-(4-iodophenyl)tropane or β-CIT) (Figure 14.4). Both compounds display nanomolar affinity for the biogenic amine transporters; however, while CFT shows selectivity for DAT over NET and SERT, RTI-55 shows selectivity for SERT and DAT over NET.

For compounds of the tropane class, the tropane ring and the 2β-carbomethoxy group are crucial for their affinity. An exception for this rule is the benztropine class. This group of tropanes lack the 2β-carbomethoxy group but still bind DAT with high affinity. Instead of the 2β-carbomethoxy group, the benztropines contain a diphenylmethoxy moiety. Recently, there has been increasing focus on benztropine analogues. Several of these compounds posses similar or even higher affin-ity and greater selectivity for the DAT than cocaine. The compounds tested so far readily cross the blood–brain barrier and produce increases in extracellular levels of dopamine for even longer durations than cocaine. Nonetheless, several of these DAT inhibitors are less effective than cocaine as behavioral stimulants. Furthermore, one benztropine analogue, JHW 007, has been found to potently antagonize the behavioral effects of cocaine (Figure 14.4). Assuming a correlation between behavioral effects of cocaine in laboratory animals and abuse potential in humans, these findings suggest JHW 007 as a potential lead for development of cocaine abuse pharmacotherapeutics. The reason for this discrepancy in the stimulating effect between cocaine and the benztropines has been suggested at least in part to be related to different pharmacodynamic properties of the compounds. Interestingly, recent studies suggest that while cocaine and cocaine analogues bind and stabilize an outward facing conformation of the transporter, benztropine analogues bind and stabilize a more closed conformation of the transporter. It is possible that binding to the open and likely more prev-alent outward facing conformation of DAT conformation results in a faster on-rate, which may facilitate faster inhibition of DAT function and thereby a more rapid rise in extracellular dopamine concentration. In contrast, binding to a more closed and predicted less prevalent conformation of the transporter may result in a slower on-rate of the compound and thereby a slower rise in dopamine levels and a less stimulatory effect.

The molecular mode of interaction of cocaine and analogues with DAT has long been the subject of speculation. In particular, it has been debated whether or not the cocaine-binding site in DAT overlaps with that of dopamine. If inhibition of dopamine uptake by cocaine is the result of an allosteric mechanism, it would be possible, at least in theory, to generate a cocaine antagonist for treatment of cocaine addiction that might block cocaine binding without affecting dopamine trans-port. An experimentally validated molecular model of the cocaine-binding site in the DAT has been reported (Figure 14.5). The DAT model was generated on the basis of the LeuT$_{Aa}$ structure followed by molecular docking of dopamine, cocaine, and other inhibitors into the model. The docking procedure revealed a binding site for cocaine and cocaine analogues that was deeply buried between TM1, 3, 6, and 8, and overlapped with the binding site for dopamine. There were, however, also significant differences between the binding modes: cocaine and cocaine analogues displayed a unique interaction with Asn157 and, moreover, the binding mode of cocaine/cocaine analogues dis-torted the conformation of the binding (Tyr156) that (as described earlier) plays a key role in closing the dopamine-binding pocket to the extracellular environment through formation of a stabilizing hydrogen bond with Asp79 in TM1 (Figure 14.5). In the cocaine-binding model, Tyr156 is pushed away by the 2β-methylester substituent of cocaine resulting in disruption of the hydrogen bond and a conformation of the binding pocket that is more open to the outside. This is in agreement with

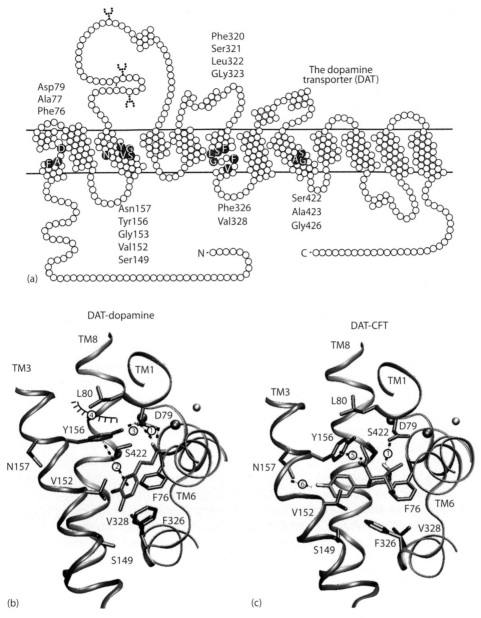

FIGURE 14.5 Models of DAT/ligand complexes. (a) Two-dimensional schematic representation of the human dopamine transporter (hDAT). The colored circles denote residues that interact with either dopamine or the cocaine analogue CFT in the molecular models. Red circles, side chain interaction; orange, only backbone interaction. (b) Docked dopamine and (c) CFT in DAT. TMs 1, 3, 6, and 8 are shown in various shades of blue; the other TMs and intra- and extracellular loops have been removed for clarity. The ligands are shown in green. Sodium and chloride ions are shown as purple and salmon spheres, respectively. The protonated amine of dopamine forms a salt-bridge with the Asp79 side chain (motif 1 in b). A polar interaction is also predicted between the amine of CFT and Asp79 (motif 1 in c). Dopamine further engages in hydrogen bonds with exposed backbone carbonyls of the unwound regions of TM1 and TM6. The binding of dopamine results in an additional aromatic–aromatic interaction between the catechol ring and Tyr156 (motif 2). Tyr156 also forms a hydrogen bond with Asp79 (motif 3) and a hydrophobic interaction with Leu80 (motif 4). In the CFT model, Tyr156 interacts in an edge-to-face manner with the methylester subtituent on the tropane ring (motif 5) leading to reorientation of the Tyr156 side chain and disruption of the hydrogen bond between Tyr156 and Asp79 seen in the dopamine model. Finally, there is CFT-specific hydrogen-bond interaction between the nitrogen of Asn157 and the fluoride atom of CFT (motif 6). (Modified from Beuming, J. et al., *Nat. Neurosci.*, 11, 780, 2008. With permission.)

other results suggesting that cocaine locks the DAT in an outward facing conformation. The binding models have been substantiated by extensive mutagenesis of the proposed interacting residues. Specifically, the buried nature of the cocaine/cocaine analogue-binding site has been validated by trapping the radiolabeled cocaine analogue [^{3}H]CFT in the transporter either through cross-linking of engineered cysteines or with an engineered Zn^{2+}-binding site situated extracellular to the predicted common binding pocket.

In support of the proposed binding site for cocaine and benztropines, photoaffinity labeling studies in DAT with, e.g., the cocaine analogue [^{125}I]RTI-82 and the benztropine analogue [^{125}I]MFZ 2–24 have identified the possible major binding domains to be located in TM1 and 6, respectively. Although this technique does not permit detailed insight into the precise nature of the binding site, the results suggest that the binding sites are positioned in the same location as predicted by the binding models described earlier. In the SERT the most convincing data regarding the binding site for cocaine and cocaine analogues show that a combined mutation of Tyr95 in TM1 and Ile172 in TM3 markedly decrease the affinity for cocaine and RTI-55, suggesting that these residues possess a direct interaction very similar in the two transporters.

14.3.2 AMPHETAMINE AND OTHER NONENDOGENOUS SUBSTRATES

Several nonendogenous compounds are substrates of the biogenic amine transporters and are used either as medication, drugs of abuse, or biochemical tools. Amphetamine and derivatives thereof, e.g., metamphetamine, p-chloroamphetamine, and 3,4-methylenedioxymetamphetamine (MDMA or ecstacy) are a class of psychostimulants that are transported by DAT, NET, and SERT (Figure 14.4). Methamphetamine preferentially acts on the DAT and NET while p-chloroamphetamine and MDMA have higher specificity for the SERT. This is supported by analyses of mice deficient in either DAT or SERT, i.e., DAT knockout mice are hyperactive and do not respond to amphetamine, while SERT deficient mice display locomotor insensitivity to MDMA. Interestingly, amphetamines do not only increase the synaptic concentration of dopamine DAT by competing with dopamine for uptake via DAT but also by promoting reversal of transport resulting in efflux of dopamine via the transporter. This efflux dramatically increases the levels of extracellular dopamine and is believed to be of major importance for the psychostimulatory properties of amphetamines. Increasing evidence supports that this efflux is not just the result of "facilitated exchange," but also might involve a channel mode of the transporter. Furthermore, recent studies suggest that the efflux is dependent on binding to the DAT C-terminus of Ca^{2+}/calmodulin-dependent protein kinase α(CaMKIIα) that in turn facilitates phosphorylation of one or more serines situated in the distal N-terminus of the transporter.

As for the inhibitors, the binding sites for amphetamine and MDMA have also been investigated by molecular docking models suggesting an overlap of binding sites with dopamine.

14.3.3 ANTIDEPRESSANTS

The biogenic amine transporters are also targets for medicines used against depression and anxiety. The selective serotonin-reuptake inhibitors (SSRIs), such as citalopram, fluoxetin, paroxetin, and sertraline, are, as implicated by their name, potent and selective inhibitors of the SERT (Figure 14.4). Another class of antidepressants includes the so-called serotonin–norepinephrine reuptake inhibitors (SNRIs) or "dual action" antidepressants such as venlafaxine and duloxetine that are active at both SERT and NET (Figure 14.4). Finally, the classical though still often used tricyclic antidepressants (TCAs) are potent inhibitors of NET and/or SERT with imipramine and amitriptyline being approximately 10 times more potent on SERT, while desipramine is relatively a selective inhibitor of the NET (see also Chapter 18). Interestingly, the antiobesity drug sibutramine exerts its action via combined inhibition also of NET and SERT. Conceivably, this effect is achieved through a combination of an anorectic effect due increased extracellular serotonin levels and increased thermogenesis due to increased norepinephrine levels.

The binding sites for antidepressants at their main targets: the NET and SERT are poorly described. Remarkably, it was recently reported that TCAs, such as clomipramine, imipramine, and desipramine, have activity at the bacterial homologue $LeuT_{Aa}$; hence, it was observed that the compounds were capable of noncompetitively inhibiting substrate binding to the transporter. The effect was only seen with high concentrations of the compounds and, thus, their affinity for $LeuT_{Aa}$ is substantially lower than that observed for the NET/SERT. Nonetheless, it has been possibly to crystallize $LeuT_{Aa}$ in complex with these compounds. The structures showed that clomipramine, desipramine, and imipramine bind in an extracellular-facing vestibule about 11 Å above the occluded substrate-binding site, apparently stabilizing the extracellular gate in a closed conformation. The TCAs are cradled by the carboxy-terminal half of transmembrane helix 1 (TM1), the aminoterminal regions of TM6 and TM10, the approximate midpoint of TM3, and the sharp turn of ECL4.

The structures of TCA-bound $LeuT_{Aa}$ obviously raise the key question whether they describe a binding mode for inhibitors that can be generalized to their mammalian counterparts. Mutations suggested that desipramine might interact in a similar fashion with the SERT; however, previous mutagenesis has supported that SSRIs, such as citalopram, fluoxetine, and sertraline as well as the TCA clomipramine, bind deeper in the transporter structure in a site more close to the substrate-binding site, i.e., mutation of Tyr95 in TM1 and/or Ile172 in TM3 of SERT substantially decreased the affinity for these compounds. Most significantly, the combined mutation of Tyr95 (Y95F) and Ile172 (I172M) decreased transporter affinity ~10,000-fold for escitalopram. The recent evidence for a buried binding site for cocaine and related inhibitors in DAT (see earlier) also strongly argues against that the TCA-binding mode seen in $LeuT_{Aa}$ can be generalized to other transporters.

14.3.4 OTHER BIOGENIC AMINE TRANSPORTER INHIBITORS

The examples of additional biogenic amine inhibitors include the GBR (from Royal Gist-Brocades) analogues that are highly selective for DAT and mazindol (Figure 14.4) that inhibits NET with one and two orders of magnitude higher potency than DAT and SERT, respectively. Finally, the amphetamine derivative methylphenidate (Figure 14.4) is a potent blocker of primarily DAT and NET, and often used for treatment of narcolepsy and attention deficit hyperactivity disorder (ADHD). Not much is known about the molecular basis for the interaction of these compounds with the transporters.

14.4 INHIBITORS OF GLYCINE AND GABA TRANSPORTERS: SPECIFICITY, USE, AND MOLECULAR MECHANISM OF ACTION

Other SLC6 family transporters than the biogenic amine transporters are targets for drugs or for drug discovery. GAT-1 is, for example, the target for the antiepileptic drug tiagabine; however, the molecular basis for its interaction with GAT-1 is not known. Recently, the N-dithienyl-butenyl derivative of N-methyl-exo-THPO (4-methylamino-4,5,6,7-tetrahydrobenzo[d]isoxazol-3-ol) (EF-1502) has been shown to inhibit not only GAT-1 but also the betaine carrier (BGT-1) and to act as a very efficient anticonvulsant whose action is synergistic with that of tiagabine (see also Section 15.4). Thus, BGT-1 is likely to be an important antiepileptic drug target. The explanation for the observations might be related to a differential distribution of BGT-1 and GAT-1. While GAT-1 is localized to synaptic sites, BGT-1 is localized to astrocytes and possibly extrasynaptic loci in the neurons; hence the efficacy of EF-1502 owing to its interaction with BGT-1 could be explained by modulation of extracellular GABA concentrations at extrasynaptic sites (for further details about GABA receptors and transporters see Chapter 15).

The high-affinity glycine transporters (GlyT1 and GlyT2) might also represent interesting drug targets. Physiologically, GlyT1 appears to play a role in astroglial control of glycine availability at NMDA receptors whereas GlyT2 is likely to play a fundamental role in glycinergic inhibition as reflected in a lethal neuromotor deficiency in GlyT2 knockout mice. The putative role of GlyT1 in regulating glycine availability at NMDA receptors has warranted attempts to develop high-affinity inhibitors of GlyT1 as a novel class of antipsychotic drugs, i.e., blockade of the GlyT1 is envisioned to increase synaptic levels of glycine ensuring saturation of the glycine-B (GlyB) site at the NMDA receptor at which glycine acts as an obligatory coagonist. Importantly, a derivative of sarcosine [3-(4-fluorophenyl)-3-(4′-phenylpheroxy)]propylsarcosine (NFPS) has been shown to potentiate NMDA receptor-sensitive activity and to produce an antipsychotic-like behavioral profile in rats. Several GlyT1 and GlyT2 inhibitors have now been described; however, little is known about their mode of interaction with the transporters.

14.5 CONCLUSION

The SLC6 neurotransmitter transporters represent a prototypical class of ion-coupled membrane transporters capable of utilizing the transmembrane Na^+ gradient to couple "downhill" transport of Na^+ with "uphill" transport (against a concentration gradient) of their substrate from the extracellular to the intracellular environment. The transporters play key roles in regulating synaptic transmission in the brain by rapidly sequestering transmitters such as dopamine, norepinephrine, serotonin, GABA, and glycine away from the extracellular space. Moreover, they are targets for a wide variety of drugs including antidepressants, antiepileptics, and psychostimulants as well as they are subject to current drug discovery efforts. Only recently, high-resolution structural information became available for this class of transporters through crystallization of the bacterial homologue, $LeuT_{Aa}$. For the first time, this permitted insight into the tertiary structure of this family of transporters. The structure serves as an important framework for future studies aimed at deciphering the precise molecular details and dynamics of the transport process. The structure also serves as an important template for delineating the molecular determinants for drug binding to SLC6 neurotransmitter transporters. The first experimentally validated computational models of drug binding have now been published and provided the first insight into the exact molecular basis for drug action at these important proteins.

FURTHER READINGS

Beuming, T., Kniazeff, J., Bergman, M. L., Shi, L., Gracia, L., Raniszewska, K., Newman, A. H., Javitch, J. A., Weinstein, H., Gether, U., and Loland, C. J. 2008. The binding sites for cocaine and dopamine in the dopamine transporter are overlapping. *Nat. Neurosci.* 11: 780–789.

Gether, U., Andersen, P. H., Larsson, O. M., and Schousboe, A. 2006. Neurotransmitter transporters: Molecular function of important drug targets. *Trends Pharmacol. Sci.* 27: 375–383.

Kniazeff, J., Shi, L., Loland, C. J., Javitch, J. A., Weinstein, H., and Gether, U. 2008. An intracellular interaction network regulates conformational transitions in the dopamine transporter. *J. Biol. Chem.* 283: 17691–17701.

Reith, M. E. A. 2002. *Neurotransmitter Transporters: Structure, Function, and Regulation*, 2nd edn. Humana Press, Totawa, NJ.

Singh, S. K., Yamashita, A., and Gouaux, E. 2007. Antidepressant binding site in a bacterial homologue of neurotransmitter transporters. *Nature* 448: 952–956.

Yamashita, A., Singh, S. K., Kawate, T., Jin, Y., and Gouaux, E. 2005. Crystal structure of a bacterial homologue of Na^+/Cl^--dependent neurotransmitter transporters. *Nature* 437: 215–223.

Yernool, D., Boudker, O., Jin, Y., and Gouaux, E. 2004. Structure of a glutamate transporter homologue from *Pyrococcus horikoshii*. *Nature* 431: 811–818.

Zomot, E., Bendahan, A., Quick, M., Zhao, Y., Javitch, J. A., and Kanner, B. I. 2007. Mechanism of chloride interaction with neurotransmitter:sodium symporters. *Nature* 449: 726–730.

15 GABA and Glutamic Acid Receptor Ligands

Bente Frølund and Ulf Madsen

CONTENTS

15.1 INTRODUCTION

γ-Aminobutyric acid (GABA [**15.1**]) and (*S*)-glutamic acid (Glu [**15.2**]) are the major inhibitory and excitatory neurotransmitters, respectively, in the central nervous system (CNS) and form the basis for neurotransmission in the mammalian CNS. Given the fact that the majority of central neurons

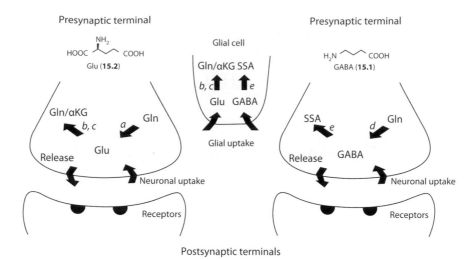

FIGURE 15.1   Schematic illustration of the biochemical pathways, transport mechanisms, and receptors at Glu and GABA operated neurons. Enzymes are indicated by the following: *a*, glutaminase; *b*, glutamine synthase; *c*, aspartate synthase; *d*, L-glutamic acid decarboxylase (GAD); *e*, GABA aminotransferase (GABA-AT).

are under excitatory and inhibitory controls by Glu and GABA, the balance between the activities of the two is of utmost importance for CNS functions. Both neurotransmitter systems are involved in the regulation of a variety of physiological mechanisms and dysfunctions of either of the two can be related to various neurological disorders in the CNS.

The transmission processes mediated by Glu and GABA are very complex and highly regulated. A general and simple model for the Glu and GABA neurotransmissions is shown in Figure 15.1. Glu and GABA are formed in their respective presynaptic nerve terminals and upon depolarization released into the synaptic cleft in a high concentration to activate postsynaptic ionotropic receptors that directly modify the membrane potential of the receptive neuron, generating an excitatory or inhibitory postsynaptic potential. This basic system is further modulated through G-protein-coupled receptors for a variety of neuroactive substances including Glu and GABA themselves. Subsequently Glu and GABA are removed from the synaptic cleft into surrounding neurons and glia cells via specialized transporters to restore the neurotransmitter balance. The reuptaken Glu and GABA are enzymatically metabolized to form glutamine (Gln) or α-ketoglutarate (αKG) and succinic acid semialdehyde (SSA), respectively.

15.2   THERAPEUTIC PROSPECTS FOR GABA AND GLUTAMIC ACID NEUROTRANSMITTER SYSTEMS

The therapeutic potentials of manipulating these neurotransmitter systems seem to be unlimited. Therefore, virtually all of the known molecular components of the GABA and Glu neurotransmitter systems have been considered as potential therapeutic targets. The therapeutic indications are numerous and include neurodegenerative disorders, e.g., Alzheimer's disease, Parkinson's disease, Huntington's chorea, epilepsy and stroke, and other neurologic disorders, e.g., schizophrenia, depression, anxiety, and pain. Furthermore narcolepsy, spasticity, muscle relaxation, and insomnia are among the vast number of therapeutic possibilities and finally cognitive enhancers can be mentioned as a much pursued therapeutic application.

GABA-based therapeutics have been in clinical use for some time, where the most successful therapeutic application to date involve the upregulation of GABA activity by the modulation of the ionotropic GABA$_A$ receptor, notably by benzodiazepines (BZD) and barbiturates. Vigabatrin (**15.3**)

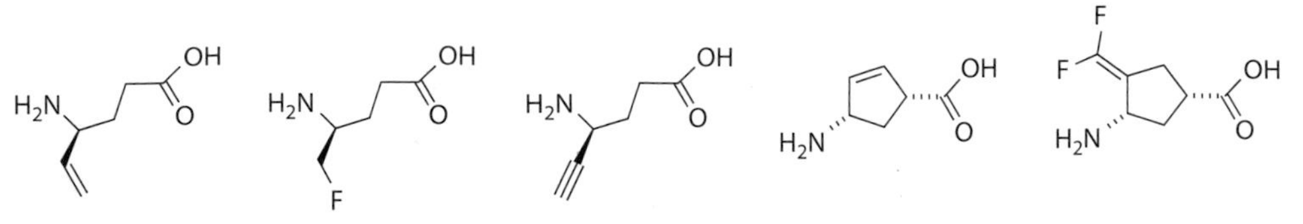

FIGURE 15.2   Structures of some GABA-AT inhibitors, compound **15.6** being inactive.

(Figure 15.2), a suicide inhibitor for the enzyme GABA-aminotransferase (GABA-AT) responsible for GABA degradation, is used clinically as an anticonvulsant. Elevation in extracellular GABA levels by the inhibition of the reuptake of GABA is effected by Tiagabine (**15.20**) (refer to Figure 15.4) marketed for the treatment of epilepsy and in preclinical studies for treatment of anxiety and insomnia. The G-protein-coupled GABA$_B$ receptor is the target for the antispastic drug baclofen (**15.41**) (refer to Figure 15.9).

Drugs targeting the Glu neurotransmitter system have been slower to emerge. Memantine (see Section 15.7.3) is used with some success for treatment of Alzheimer's disease and a few compounds with mixed mechanisms of action, including reduction of Glu release (through blockade of Na channels) are used for the treatment of migraine, epilepsy, and amyotrophic lateral sclerosis.

All of the Glu and GABA receptors and transporters are heterogeneous and may individually be involved in specific CNS disorders and disease conditions. These receptor/transporter subtypes are unevenly distributed in the CNS, which opens up the prospect of developing ligands selective for receptor/transporter subtypes with predominant location in different brain regions of therapeutic relevance.

15.3   GABA BIOSYNTHESIS AND METABOLISM

The GABA concentration is regulated by two pyridoxal 5′-phosphate (PLP) dependent enzymes, L-glutamic acid decarboxylase (GAD), which catalyzes the decarboxylation of Glu to GABA prior to release into the synaptic cleft, and GABA-AT, which degrades reuptaken GABA to SSA (Figure 15.1). This transamination step takes place within presynaptic GABA terminals as well as in surrounding glia cells.

15.3.1   INHIBITORS OF GABA METABOLISM

A number of mechanism-based inactivators of GABA-AT has been developed and has been shown to elevate the extracellular levels of GABA. These compounds are typically analogs of GABA, containing appropriate functional groups at C4 of the GABA backbone (e.g., **15.3**, **15.4**, and **15.5**). The functional group is converted by GABA-AT into electrophiles, which react with nucleophilic groups at or near the active site of the enzyme and thereby inactivate the enzyme irreversibly (Figure 15.3). The most effective of these, γ-vinyl-GABA (Vigabatrin, **15.3**), is clinically used as an anticonvulsant for the treatment of epilepsy.

The mechanism for inactivation of GABA-AT by Vigabatrin (**15.3**) is outlined in Figure 15.3. As shown, PLP is forming a Schiff base (**15.8**) with the terminal amino group of a lysine residue in the active site of GABA-AT. Transamination with **15.3** generates a new imine **15.9**, which undergoes a rate-determining enzyme-catalyzed deprotonation to give the imine **15.11** after reprotonation. In analogy with transamination reaction on GABA, **15.11** could be hydrolyzed to give the SAA analog **15.13** and pyridoxamine-5-phosphate **15.12**. However, **15.11** is a Michael acceptor electrophile, which undergoes conjugate addition by an active-site nucleophile X and the inactivated enzyme **15.14** is produced.

To optimize the effect of Vigabatrin a number of conformationally restricted GABA analogs has been developed. In contrast to the Vigabatrin analog **15.6**, which does not inactivate GABA-AT, the

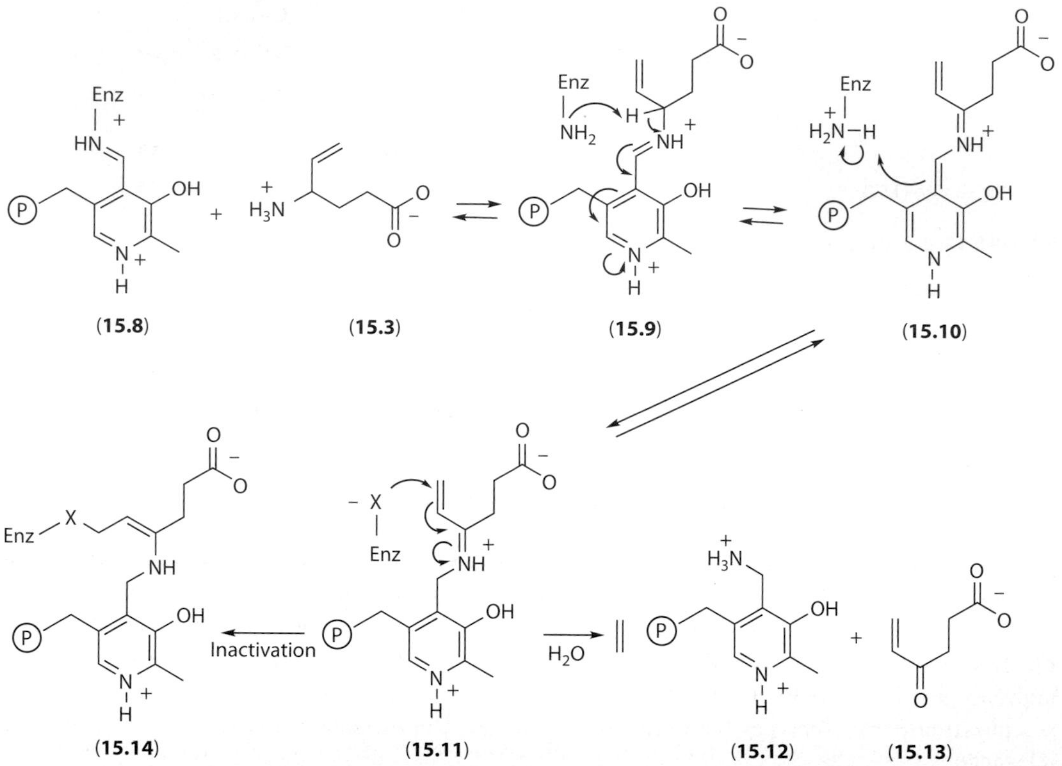

FIGURE 15.3 Proposed inactivation mechanism of GABA-AT by Vigabatrin (**15.3**). The cofactor PLP and an amino group from a lysine residue in GABA-AT (Enz) form a Schiff base (**15.8**), which reacts with Vigabatrin and eventually leads to inactivation of GABA-AT.

difluoromethylene analog **15.7** (Figure 15.2), is reported as a markedly more potent inactivator of GABA-AT than Vigabatrin.

15.4 GABA TRANSPORT

The GABA transporters belong to the family of Na$^+$/Cl$^-$ dependent transporters (SLC-6 gene family) that also include transporters for the neurotransmitters dopamine, serotonin, norepinephrine, and glycine (see Chapter 14). Four subtypes of GABA transporters have been identified in the mamalian CNS. For rat and human GABA transporters, the nomenclature is GAT-1, betaine/GABA-transporter-1 (BGT-1), GAT-2, and GAT-3.

15.4.1 INHIBITORS OF GABA TRANSPORT

The pharmacological inhibition of GABA transporters constitutes an attractive approach to increase the overall GABA neurotransmission. A selective blockade of glial uptake is believed to be optimal, as this will ensure an elevation of the GABA level in the presynaptic nerve terminals.

Nipecotic acid (**15.15**) and guvacine (**15.16**), competitive inhibitors and substrates for the GABA uptake, have been important lead structures for the development of a large number of lipophilic GABA uptake inhibitors. Introduction of a lipophilic moiety, such as 4,4-diphenyl-3-butenyl (DPB), on the nitrogen atom led to N-DPB-nipecotic acid (**15.19**) and related analogs, which are markedly more potent than the parent amino acids. These lipophilic compounds are able to cross the blood–brain

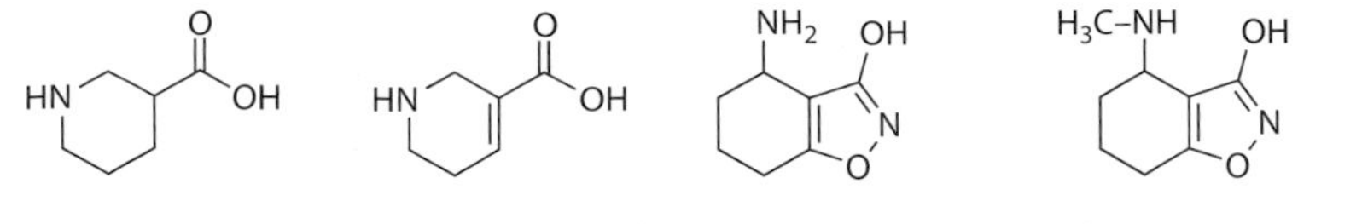

Nipecotic acid (**15.15**) Guvacine (**15.16**) *exo*-THPO (**15.17**) *N*-Me-*exo*-THPO (**15.18**)

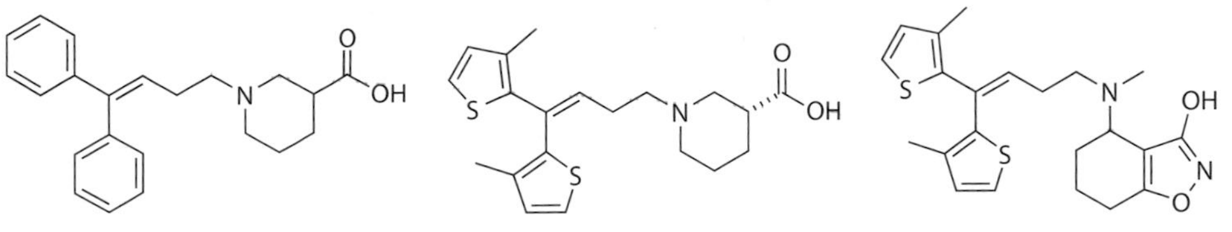

N-DPB-nipecotic acid (**15.19**) Tiagabine (**15.20**) EF-1502 (**15.21**)

FIGURE 15.4 Structures of some GABA transport inhibitors.

barrier and are potent anticonvulsants in animal models. Tiagabine (**15.20**), a structurally related compound, is now marketed as an add-on therapeutic agent for the treatment of epilepsy.

A highly glia-selective compound was discovered based on the structure of *exo*-THPO (**15.17**), where the monomethylated compound *N*-methyl-*exo*-THPO (**15.18**) proved to be the most selective inhibitor for glial vs. neuronal GABA uptake reported yet.

EF-1502 (**15.21**), developed as a hybrid of *exo*-THPO and Tiagabine, has similar potency at GAT-1 and BGT-1. An *in vivo* study of the anticonvulsant properties of the compound revealed a synergistic effect between EF-1502 and GAT-1-selective inhibitors, indicating a possible role for BGT-1 as a therapeutic target (Figure 15.4).

15.5 GABA RECEPTORS AND THEIR LIGANDS

GABA exerts its effects on the CNS via two different types of receptors: the ionotropic $GABA_A$ and $GABA_C$ receptors, mediating the fast synaptic transmission and the G-protein-coupled $GABA_B$ receptors, mediating the slower responses to GABA via coupling to second messenger cascades.

15.5.1 IONOTROPIC GABA RECEPTORS

The ionotropic GABA receptors belong to a superfamily of ligand-gated ion channels (Cys-loop receptors) that also includes the nicotinic acetylcholine, the glycine, and the serotonin ($5\text{-}HT_3$) receptors (see Chapter 12). Whether the $GABA_C$ receptor is a subgroup of the $GABA_A$ receptors or a distinct group of GABA receptors is still a matter of debate. The $GABA_A$ receptors are widely distributed in the CNS and involved in a wide variety of CNS functions, whereas the $GABA_C$ receptors predominantly are expressed in the retina and primarily implicated in visual processing. However, $GABA_C$ receptors have also been identified in some CNS regions, where they have been proposed to be involved in processes connected with sleep and cognition processes.

The ionotropic GABA receptors are transmembrane protein complexes composed of five subunits. So far, 19 human GABA receptor subunits have been identified, and they have been classified into α_{1-6}, β_{1-3}, γ_{1-3}, δ, ϵ, π, θ and ρ_{1-3} subunit classes (Figure 15.5). Each of the subunits consists of an amino-terminal domain and a transmembrane region formed by four transmembrane α-helices connected by intra- and extracellular loops. In the pentameric GABA receptor complex, the orthosteric site (i.e., the binding site for the endogenous ligand GABA) is formed at the interface between the terminal domains of two subunits, whereas the transmembrane regions of the subunits form the ion channel pore through which chloride ions can enter the cell upon activation. The $GABA_A$ receptors

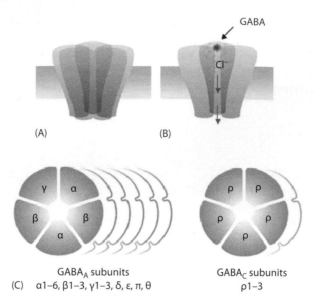

FIGURE 15.5 Schematic illustrations of (A) the pentameric structure of the ionotropic GABA receptors, (B) with indication of the GABA-binding site and the chloride ion channel, (C) and the multiplicity of ionotropic $GABA_A$ and $GABA_C$ receptors.

are heteromeric complexes, and although a wide range of different receptor combinations exists *in vivo*, $\alpha_1\beta_2\gamma_2$ combination is the predominant physiological $GABA_A$ receptor subtype. In contrast, the $GABA_C$ receptors are homomeric assemblies of five identical ρ subunits or pseudoheteromeric complexes comprising different ρ subunits.

The binding site for GABA and ligands for the orthosteric binding has been shown to be located at the interface of the α and β subunits in the $GABA_A$ receptor complex, whereas the binding site for BZD is located at the interface of the α and γ subunits.

There is still no 3D-structure available for the ionotropic GABA receptors. Thus, the understanding of the molecular architecture of the orthosteric-binding site in the ionotropic GABA receptors has to a large extent been based on the publication of crystals structures of acetylcholine-binding proteins (AChBP) from snails. These proteins display low but still significant amino acid sequence homologies with the amino-terminal domains of all ligand-gated ion channels within the Cys-loop family, including the ionotropic GABA receptors, and this homology has been exploited for the construction of homology models of this region in both $GABA_A$ and $GABA_C$ receptors. Such homology models offer an insight into the identities of the residues lining the binding pockets in the respective receptors. However, given the low level of sequence identity (~18%) between the AChBP and the ionotropic GABA receptors, it is not straightforward to use these models for the prediction of ligand affinity or structure–activity studies.

15.5.2 IONOTROPIC GABA RECEPTOR LIGANDS

The GABA-binding site has very distinct and specific structural requirements for recognition and activation. Thus, only a few different classes of structures have been reported. Within the series of compounds showing agonist activity at the $GABA_A$ receptor site are the selective agonists muscimol (**15.22**) and THIP (**15.23**), which have been used for the pharmacological characterization of the $GABA_A$ receptors. BMC (**15.24**) and SR95531 (**15.25**) are the classical $GABA_A$ receptor antagonists.

In the absence of a 3D-receptor structure, the relationship between the ligand structure and the binding/activity at the $GABA_A$ receptor has been extensively studied. On the basis of a

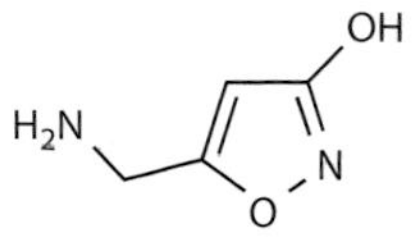

Muscimol (**15.22**)

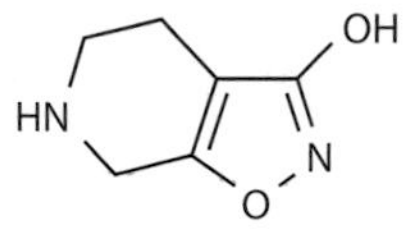

THIP (**15.23**)

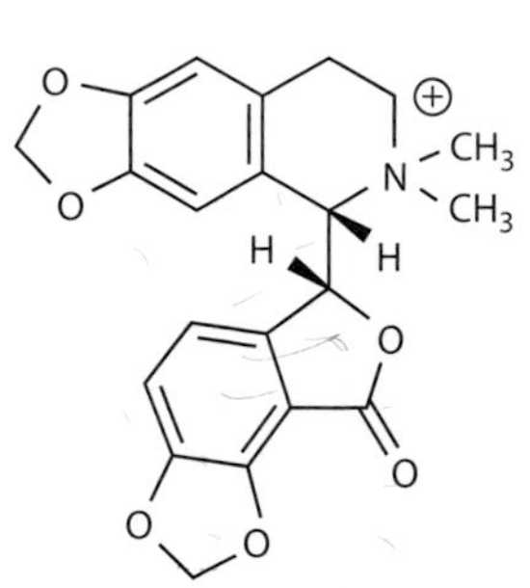

BMC (**15.24**)

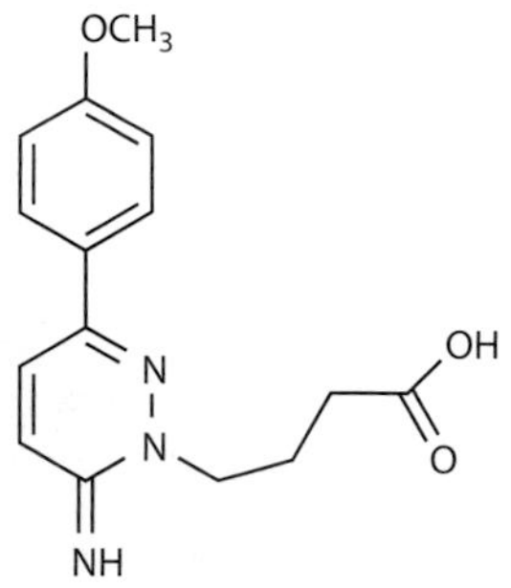

SR 95531 (**15.25**)

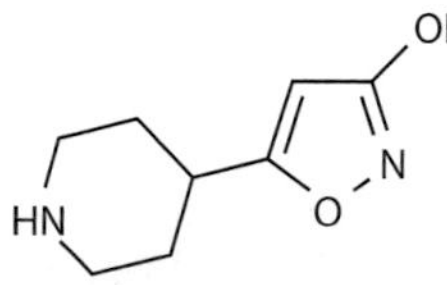

4-PIOL (**15.26**)

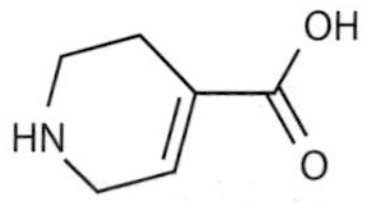

Isoguvacine (**15.27**)

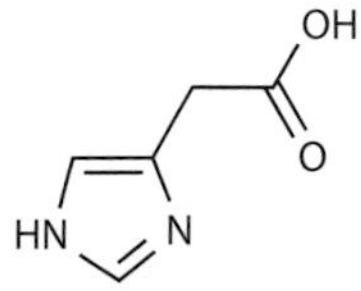

IAA (**15.28**)

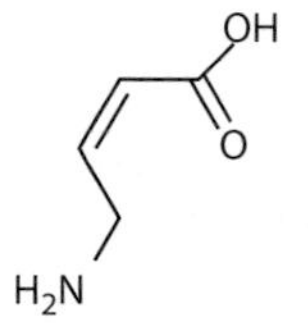

CACA (**15.29**)

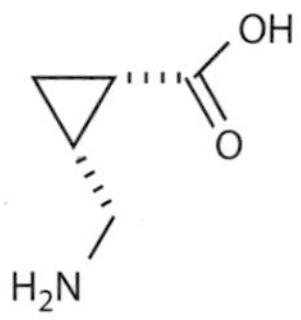

CAMP (**15.30**)

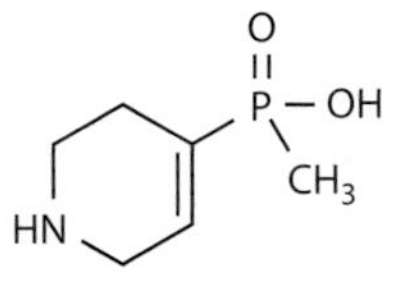

TPMPA (**15.31**)

FIGURE 15.6 Structures of $GABA_A$ and $GABA_C$ ligands.

hypothesis originating from the bioactive conformation of muscimol, the partial $GABA_A$ agonist 4-PIOL (**15.26**), and on pharmacological data for an additional series of $GABA_A$ ligands, a simple 3D-pharmacophore model for the orthosteric $GABA_A$ receptor ligands has been developed. The main features of this model are that the 3-hydroxyisoxazolol rings of muscimol and 4-PIOL do not overlap in their proposed binding modes and that the two compounds interact with different conformations of an arginine residue located at the $GABA_A$ recognition site. The space surrounding the ligands has been defined and the existence of a cavity of considerable dimensions in the vicinity of the 4-position of the 3-hydroxyisoxazolol moiety in the structure of 4-PIOL has been identified, whereas the corresponding position in muscimol is identified as "receptor essential volumes" (Figure 15.7). Based on this model, a series of selective and highly potent competitive antagonists have been developed including the compounds **15.32a–d**.

In contrast, structure–activity studies of ligands targeting the $GABA_C$ receptors have been very limited. *cis*-4-Aminocrotonic acid (CACA [**15.29**]) (Figure 15.6) has been the key ligand for the identification of the $GABA_C$ receptors. The compound is a moderately potent partial $GABA_C$ agonist and inactive at $GABA_A$ receptors, but it has been shown to effect GABA transport as well. In the search for selective $GABA_C$ receptor ligands, the folded conformation of CACA has been used as a scaffold for new compounds such as *cis*-2-aminomethyl cyclopropane carboxylic acid (CAMP [**15.30**]). (+)-CAMP has been reported to be a selective $GABA_C$ receptor agonist with potency in the mid-micromolar range, displaying only weak activity on the $GABA_A$ receptors. Finally the first antagonist capable of differentiating the $GABA_C$ receptors from both $GABA_A$ and $GABA_B$ receptors was TPMPA (**15.31**).

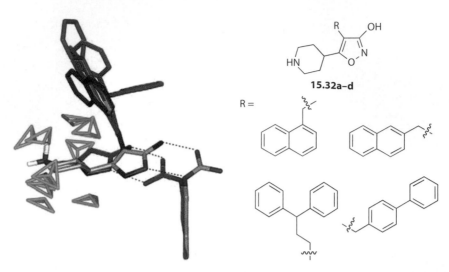

FIGURE 15.7 A superimposition of the proposed bioactive conformations of muscimol (**15.22**, green carbon atoms) and 4-PIOL (**15.26**, gray carbon atoms) binding to two different conformations of an arginine residue at the orthosteric-binding site. A series of 4-substituted 4-PIOL compounds (**15.32a–d**) are included illustrating the large space spanned by the 4-substituents. The tetrahedrons indicate receptor excluded volumes.

The overall molecular architecture of the orthosteric sites at the $GABA_A$ and $GABA_C$ receptors appear to be quite similar as most $GABA_A$ agonists display some agonist/antagonist activities at $GABA_C$ receptors as well. THIP (**15.23**), the standard $GABA_A$ agonist, has been shown to be a partial agonist at $GABA_A$ receptors and a competitive antagonist at $GABA_C$ receptors. Likewise, the $GABA_A$ agonists muscimol (**15.22**), isoguvacine (**15.27**), and imidazol-4-acetic acid (IAA [**15.28**]) act as partial $GABA_C$ agonists. However the fact that $GABA_A$ and $GABA_C$ receptors exhibit distinct antagonist profiles clearly indicates that orthosteric sites of these receptors are not identical.

Upregulation of GABA activity would, in general, be beneficial in various conditions including epilepsy, pain, anxiety, and insomnia. Direct activation of the ionotropic GABA receptors using $GABA_A$ agonists has for long not been anticipated as a useful therapeutic approach due to desensitization of the receptors. However, the $GABA_A$ agonist THIP has proven to be a potential drug in the treatment of insomnia (see Chapter 20).

15.5.3 MODULATORY AGENTS FOR THE GABA_A RECEPTOR COMPLEX

The $GABA_A$ receptor complex is the target for a large number of structurally diverse compounds, some of which are pharmacologically active and used clinically. These compounds include BZD, ethanol, general anesthetics, barbiturates, and neuroactive steroids, all of which act via a wide range of distinct allosteric-binding sites within the pentameric receptor complex.

The allosteric modulators exert their effects by binding to the $GABA_A$ receptor complex and affect GABA-gated chloride conductance. This modulation only takes place when GABA is present in the synaptic cleft, which could be preferable rather than a general receptor activation by exposure to a GABA agonist. Compounds within this group of modulators are marketed for the treatment of anxiety, epilepsy, insomnia, muscle relaxation, and anesthesia. Preclinical studies are going on with the focus on cognitive enhancement and schizophrenia as well.

The fact that receptor regions targeted by allosteric ligands typically are less conserved than the orthosteric sites, in general, opens up for development of subtype-selective modulators with more specific pathophysiological effects and reduced side effects.

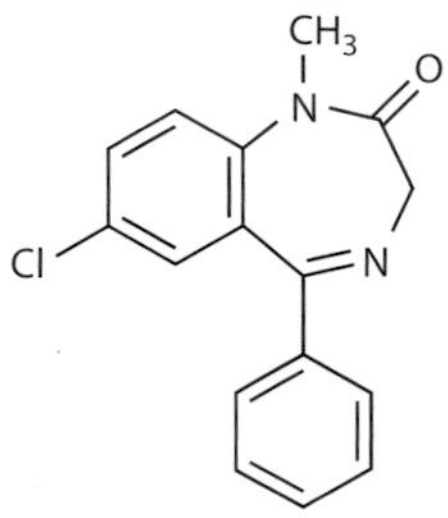

Diazepam (**15.33**) 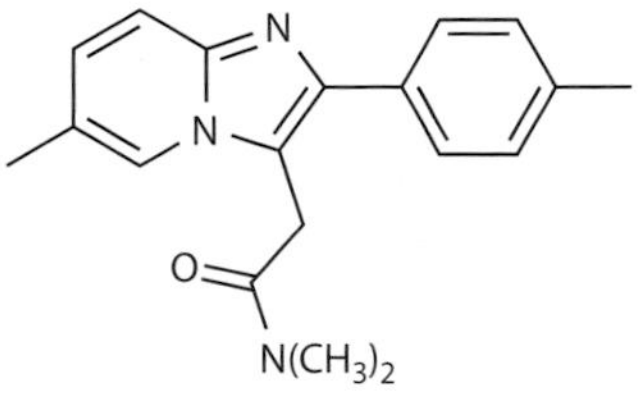 Zolpidem (**15.34**)

FIGURE 15.8 Structures of some ligands for the benzodiazepine site.

Among the modulatory sites at the $GABA_A$ receptor complex, the BZD site is the far most studied to date. The pharmacological profiles of ligands binding to the BZD site span the entire continuum from full and partial agonists, through antagonists, to partial and full inverse agonists. Antagonists do not influence the GABA-induced chloride flux, but antagonizes the action of BZD agonists as well as of inverse agonists.

With the improved knowledge of the subtypes of $GABA_A$ receptors and their influence on BZD pharmacology (see Chapter 20), it has become clear that subtype-selective ligands for the modulatory sites could provide more specific pharmacological profiles compared to that of the traditional BZD. Research based on this knowledge is focused on development of hypnotics (α_1-selective), nonsedating anxiolytics (α_2- and α_3-selective), antipsycotics (α_3-selective), and cognition-enhancement (α_5-selective inverse agonist). In spite of intensive efforts in this area unselective BZD ligands like diazepam (**15.33**) and a few α_1 preferring ligands, including zolpidem (**15.34**), are still the most important BZD ligands in the market (Figure 15.8).

15.5.4 GABA_B RECEPTOR LIGANDS

The $GABA_B$ receptors belong to the subfamily C, which also comprises the G-protein-coupled Glu receptors (see later sections and Chapter 12). The $GABA_B$ receptors exist as heterodimers consisting of two subunits, $GABA_{B1}$ and $GABA_{B2}$. The former contains the GABA-binding domain, whereas $GABA_{B2}$ provides the G-protein-coupling mechanism. The diversity in this class of receptors arises from the two $GABA_{B1}$ splice variants, $GABA_{B1a}$ and $GABA_{B1b}$, which together with $GABA_{B2}$ form the two physiological receptors. Activation of the G-protein-coupled receptor causes a decrease in calcium levels, an increase in potassium membrane conductance and inhibition of cAMP formation. The resulting response is thus inhibitory and leads to hyperpolarization and decreased neurotransmitter release.

The $GABA_B$ receptors are selectively activated by baclofen (**15.35**), of which the (*R*)-form is the active enantiomer. Baclofen was developed as a liphophilic derivative of GABA, in an attempt to enhance the blood–brain barrier penetrability of the endogenous ligand. Among the limited number of $GABA_B$ receptor agonists, the phosphinic acid GABA bioisostere, CGP27492 (**15.36**), is the most potent reported to date, being approximately 10-fold more potent than GABA.

Phaclofen (**15.37**) and saclofen (**15.38**), the phosphonic acid and sulfonic acid analogs of baclofen, respectively, were the first $GABA_B$ antagonists reported. In an attempt to improve the pharmacology and pharmacokinetics of the $GABA_B$ agonist phosphinic acid analogs mentioned above, a series of selective and highly potent $GABA_B$ antagonists, including compound **15.39**, capable of penetrating the blood–brain barrier after systemic administration was discovered (Figure 15.9).

Predominant effects of $GABA_B$ agonists are muscle relaxation, but also various neurological and psychiatric disorders, including neuropathic pain, anxiety, depression, absence epilepsy, and drug

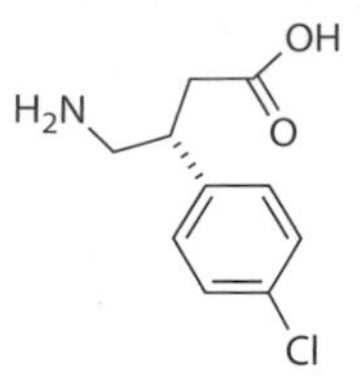

(R)-Baclofen (**15.35**) 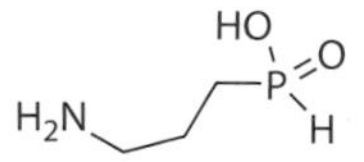   CGP27492 (**15.36**) 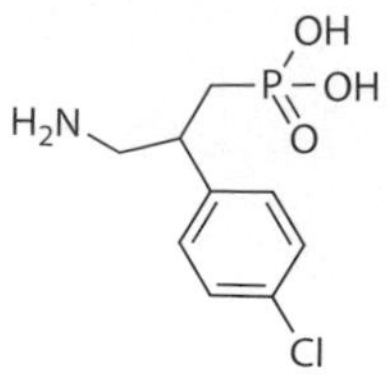   Phaclofen (**15.37**)

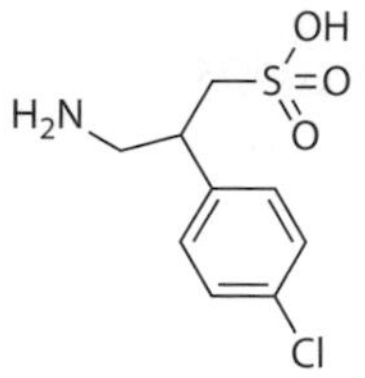

Saclofen (**15.38**) 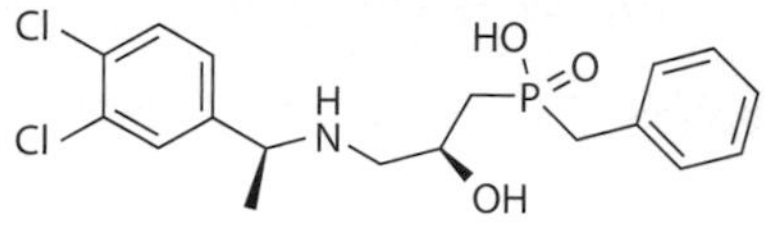   CGP-55845 (**15.39**)

FIGURE 15.9   Structures of ligands for the GABA$_B$ receptor.

addiction are potential targets for GABA$_B$ agonist therapy. However, the use of GABA$_B$ agonists has been limited due to serious side effects such as sedation, tolerance, and muscle weakness following systemic administration.

15.5.5 Ligands Differentiating the GABA$_A$ and GABA$_C$ Receptors

IAA (**15.31**) is a naturally occurring metabolite of histamine. The compound has various neurologi-cal effects, believed to be mediated by the central GABA$_A$ receptors. It penetrates the blood–brain barriers on systemic administration and is therefore advantageous from a bioavailability perspective compared to the other known standard ligands for the ionotropic GABA receptors.

Like other GABA analogs, IAA displays activities on the GABA$_A$ as well as on the GABA$_C$ receptors, being a partial agonist of both groups of receptors. In an attempt to deduce the structural determinants for the activity of the respective receptor groups, a series of IAA analogs have been synthesized.

The introduction of even small substituents in the 2-position of IAA was found to have detri-mental effects on the activities of both receptor classes, suggesting that there is little space in the orthosteric sites around this position in the IAA molecule (Figure 15.10). In contrast to the lack of activity in the 2-substituted IAA analogs (**15.42**), several of the 5-substituted IAA analogs, **15.40** and **15.41**, retained the agonist properties at ρ_1 GABA$_C$ receptors while exhibiting no activity at the $\alpha_1\beta_2\gamma_{2S}$ GABA$_A$ receptors (Figure 15.10).

The 5-Me-IAA analog (**15.40**) was docked into receptor-models of the GABA$_A$ $\alpha_1\beta_2$ interface and into the orthosteric site on ρ_1 GABA$_C$ receptors based on the bioactive conformations of the ligand deduced from the previously mentioned pharmacophore model (see Section 15.5.2). The resulting ligand orientation and receptor interactions are show in Figure 15.11. According to the models, the main difference in the vicinity of the ligands in the orthosteric sites of the $\alpha_1\beta_2\gamma_2$ and the ρ_1 receptors is a threonine residue (Thr129) in the α_1 subunit and a serine (Ser168) residue in the equivalent position in ρ_1. The smaller size of the Ser168 residue in the ρ_1 receptor makes it possible for the orthosteric site to accommodate substituents in the 5-position of IAA. A mutagenesis study based on the above men-tioned ligand–receptor docking experiments verified the Thr129 residue in the α_1 subunit of the $\alpha_1\beta_2\gamma_2$ GABA$_A$ receptor and the corresponding Ser168 residue in ρ_1 receptor as major molecular determinants for the observed differences in agonist potencies between the two receptors.

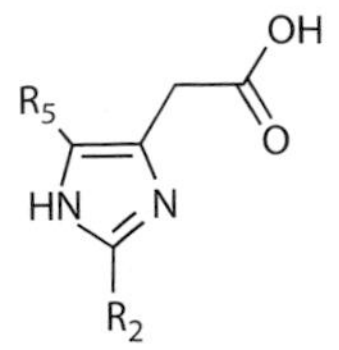

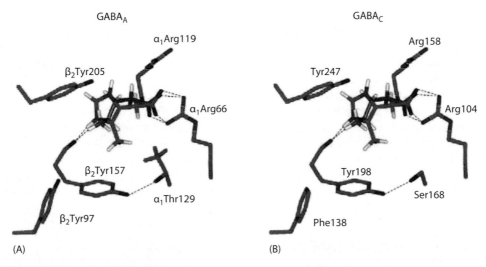

	$\alpha_1\beta_2\gamma_{2S}$ expressed in *Xenopus* oocytes	ρ_1-HEK293 cell line
	EC$_{50}$ (μM)	EC$_{50}$ (μM)
(+)-CAMP (**15.30**)	>1000	39.7
IAA (**15.28**)	310	13
5-Me-IAA (**15.40**)	>1000	22
5-Ph-IAA (**15.41**)	>1000	420
2-Me-IAA (**15.42**)	>1000	>1000

$R_2, R_5 = H$ (IAA (**15.28**))

$R_2 = H, R_5 = Me$ (5-Me-IAA (**15.40**))

$R_2 = H, R_5 = Ph$ (5-Ph-IAA (**15.41**))

$R_2 = Me, R_5 = H$ (2-Me-IAA (**15.42**))

FIGURE 15.10 Functional data of IAA (**15.28**) and analogs (**15.40**, **15.41**, and **15.42**) from *Xenopus* oocytes expressing $\alpha_1\beta_2\gamma_2$ receptors using two-electrode voltage-clamp recordings or a hρ_1-HEK293 cell line in the FLIPR membrane potential assay.

FIGURE 15.11 GABA and 5-Me-IAA (**15.40**) docked into the orthosteric sites of models of (A) the GABA$_A$ receptor and (B) the GABA$_C$ receptor. The ligands GABA and 5-Me-IAA are shown using orange and green carbon atoms, respectively. Hydrogen atoms other than those on the ligands are omitted for clarity. Proposed hydrogen bond interactions are shown as dashed lines. (Adapted from Madsen et al., *J. Med. Chem.*, 50, 4147, 2007.)

15.6 GLUTAMATE—NEUROTRANSMITTER AND EXCITOTOXIN

Glu is ubiquitously distributed in high concentrations in the CNS and Glu serves other important functions apart from being the major excitatory neurotransmitter, e.g., as building block in proteins and precursor for the neurotransmitter GABA. A very important aspect of Glu functions is the fact that high concentrations of Glu is neurotoxic, which led to the term excitotoxicity even before Glu was recognized as a neurotransmitter. Excitotoxicity describes the ability of all Glu receptor agonists to excite neurons and at the same time being neurotoxic, if the neurons are exposed to the agonist for too long a period and/or exposed to a high concentration of the agonist.

15.6.1 RECEPTOR CLASSIFICATION AND UPTAKE MECHANISMS

The targets at the Glu receptor system are receptors, allosteric sites, uptake mechanisms, and Glu metabolism (see Figure 15.1). However the primary focus has been on the ligands and their receptor sites due to the fact that most research has been aimed at lowering Glu activity in relation to the mechanisms of neurodegenerative diseases mentioned previously.

The Glu receptors are divided into two main classes, the ionotropic and the metabotropic Glu receptors (iGluRs and mGluRs), both of these covering three different receptor classes. The three iGlu classes are named by the selective agonists N-methyl-D-aspartic acid (NMDA), 2-amino-3-(3-hydroxy-5-methyl-4-isoxazolyl)propionic acid (AMPA), and kainic acid (KA) receptors. These are further subdivided into subtypes, NR1, NR2A-D, NR3A,B for NMDA receptors; GluR1–4 for AMPA receptors; and GluR5–7, KA1,2 for KA receptors (see phylogenetic tree in Chapter 12.1). iGluRs are tetrameric in structure forming an ion channel fluxing Na, K, and Ca ions upon opening leading to depolarization of the cell membrane and excitation of the neurons. The mGluRs modulates the activity of neurons and are G-protein-coupled receptors, also named 7TM receptors as described in Chapter 12. The mGluRs consists of mGluR1–8 and are divided into Groups I, II, and III based on pharmacology, signal transduction mechanisms, and amino acid sequences. Group I consists of mGluR1,5 and stimulates phosholipase C, whereas Group II (mGluR2,3) and Group III (mGluR4,6–8) inhibits the formation of cyclic AMP.

A number of modulatory sites have been recognized on iGluRs as well as on mGluRs and have been given substantial attention after the limited success with especially iGluR antagonists for neuroprotection.

The Glu transporters, excitatory amino acid transporters (EAAT), have been studied extensively in recent years and 2 glial and 3 neuronal subtypes have been characterized; EAAT1,2 and EAAT3–5, respectively. These have different distributions, EAAT2 is the major transporter for Glu in the forebrain, EAAT3 is the major neuronal transporter in the brain and spinal cord, EAAT4 in the cerebellum, and EAAT5 in the retina. EAAT3–5 is located in postsynaptic terminals, but a splice variant of EAAT2 may be the transport system for presynaptic Glu uptake. Furthermore three subtypes of Glu transporters exist on synaptic vesicles (VGLUT1–3) with the function of packing Glu into vesicles for subsequent release from the presynaptic terminal. The enzymatic systems for metabolism of Glu has only been studied to a limited extent in the context of therapeutic potential and will not be discussed here.

15.7 IONOTROPIC GLUTAMATE RECEPTOR LIGANDS

The development of selective ligands for the different receptor classes within iGluRs has been going on for almost three decades and especially within the last decade the search for subtype-selective ligand has taken speed. Within some areas a vast number of potent and selective agents have been developed, whereas for other areas there is still an urgent need for good ligands, with the purpose of characterizing the functions of the respective receptors/subtypes to give an understanding of the physiological and pathophysiological roles of these. Only a limited number of ligands will be discussed and one specific example of a development project will be briefly described.

Extensive knowledge about the structure of the ligand-binding domain of iGluRs have been obtained during the last decade by a number of x-ray structure determinations. These structures include the apostructure (ligand-binding domain without a ligand) and structures with Glu, other agonists, antagonists, partial agonists, and modulatory ligands in a ligand-binding construct of different iGluR subtypes (see also Figures 2.11 and 12.7). The detailed structural information gained by these structures is used extensively in the design of new ligands for the various GluRs as exemplified in Section 15.9.

15.7.1 NMDA RECEPTOR LIGANDS

NMDA receptors include a number of different binding sites, and thus several potential targets for therapeutic attack. NMDA receptors are unique among ligand-gated ion channels, in that they require two different agonists for activation, Glu (**15.2**) and glycine (**15.58**), and at the same time membrane depolarization in order to relieve a blockade by Mg^{2+}. In the normal brain, NMDA receptors are fundamental to development and function, because of their involvement in synaptic plasticity and neuronal signaling processes, including mechanisms of learning and memory. Furthermore, NMDA receptor-induced neurotoxicity is intimately involved in a number of neuronal disorders as previously mentioned. Functional NMDA receptors are heteromers, typically consisting of NR1- and NR2-subunits in a tetrameric structure. The Glu-binding site is located on NR2 subunits and the glycine-binding site on NR1 subunits.

NMDA receptors are unusual among Glu recognition sites inasmuch as the majority of the more potent ligands, both agonists and antagonists, posses an *R*-configuration about the α-amino acid center. NMDA (**15.43**) itself represents the only known agonist in which *N*-methylation does not lead to reduced affinity. Other potent NMDA agonists have been developed, particularly by replacement of the distal acidic group and/or by conformational restriction of the three essential functional groups, namely, the α-amino group, the α-carboxyl group, and the ω-acidic moiety. Among the potent NMDA agonists with different distal acidic groups are tetrazolylglycine (**15.44**) and (*R*)-AMAA (**15.45**), exemplifying two widely used carboxyl bioisosteric groups, the tetrazole and the 3-isoxazolol, respectively. (*R*)-AMAA (**15.45**) and other Glu ligands have been developed using the naturally occurring neurotoxin ibotenic acid (**15.46**) as a lead. Ibotenic acid (**15.46**) is, apart from being a potent NMDA agonist, a potent agonist of some mGluR subtypes and a somewhat weaker agonist at other Glu receptor types (Figure 15.12).

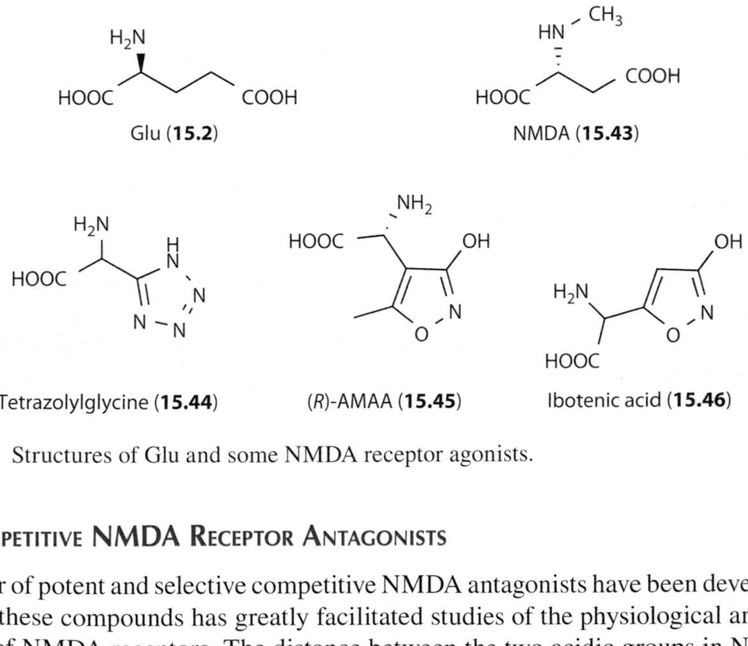

FIGURE 15.12   Structures of Glu and some NMDA receptor agonists.

15.7.2 Competitive NMDA Receptor Antagonists

A large number of potent and selective competitive NMDA antagonists have been developed, and the availability of these compounds has greatly facilitated studies of the physiological and pathophysiological roles of NMDA receptors. The distance between the two acidic groups in NMDA antagonists is typically one or three C–C bonds longer than in Glu. Many potent ligands have successfully been developed using ω-phosphonic acid analogs, such as (*R*)-APV (**15.47**), as lead structures. Combination of an ω-phosphonate group, a long carbon backbone, and conformational restriction has led to different series of potent antagonists. Conformational restriction has been achieved by the use of double bonds (CGP 39653 (**15.48**)), ring systems (CGS19755 (**15.49**)), and bicyclic structures (LY235959 (**15.50**)). These antagonists have shown very effective neuroprotective properties in various *in vitro* models. However, many of these compounds suffer from poor BBB penetration. LY233053 (**15.51**) represents another class of antagonists with a tetrazole ring as the terminal acidic group. Substitution of the tetrazole by a phosphono group has limited effect on the *in vitro* activity and shows improved bioavailability (Figure 15.13).

(*R*)-APV (**15.47**)

CGP 39653 (**15.48**)

CGS19755 (**15.49**)

LY235959 (**15.50**)

LY233053 (**15.51**)

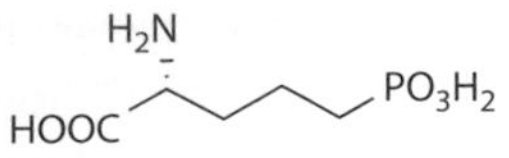

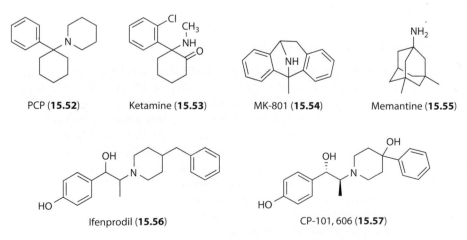

FIGURE 15.13 Structures of some competitive NMDA receptor antagonists.

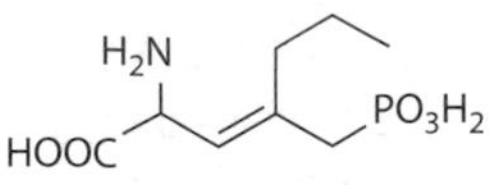

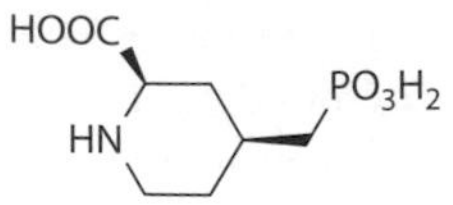

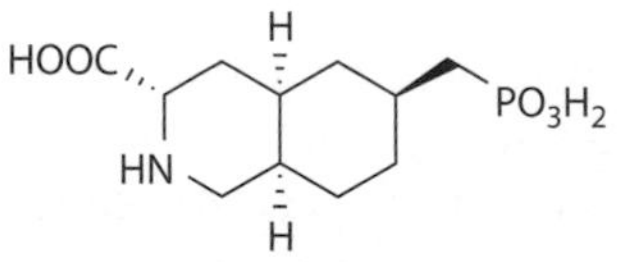

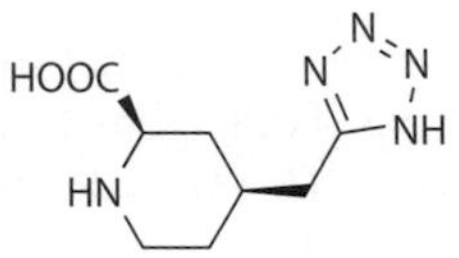

15.7.3 Uncompetitive and Noncompetitive NMDA Receptor Antagonists

The dissociative anesthetics PCP (**15.52**) and ketamine (**15.53**) block the NMDA receptor channel in a use-dependent manner. Thus, initial agonist activation of the channel is a prerequisite in order for such uncompetitive antagonists to gain access to the binding site, which is situated within the ion channel. The antagonists eventually become trapped within the ion channel and this may result in very slow kinetics. MK-801 (**15.54**) has been developed as a very effective uncompetitive NMDA antagonist and has been extensively investigated to probe the therapeutic utility of such compounds, notably for the treatment of ischemic insults such as, stroke. MK-801 (**15.54**) and related high-affinity ligands have, however, shown severe side effects, including psychotomimetic effects, neuronal vacuolization, and impairment of learning and memory. Ligands with lower affinity, such as memantine (**15.55**), have shown improved therapeutic indexes. Memantine (**15.55**) is being used in the treatment of AD and Parkinson's disease, and may also have potential in the treatment of AIDS dementia. The fast kinetics and low affinity of memantine (**15.55**) compared to MK-801 (**15.54**) may explain the absence of severe side effects (Figure 15.14).

Several compounds with noncompetitive activity at NMDA receptors have been described, and these compounds most likely do not bind to the same site as the uncompetitive ligands. Ifenprodil (**15.56**)

PCP (**15.52**)

Ketamine (**15.53**)

MK-801 (**15.54**)

Memantine (**15.55**)

Ifenprodil (**15.56**)

CP-101, 606 (**15.57**)

FIGURE 15.14 Structures of some uncompetitive (**15.52–15.55**) and noncompetitive (**15.56** and **15.57**) NMDA receptor antagonists.

and CP-101,606 (**15.57**) represent an important series of noncompetitive NMDA receptor antagonists. These compounds are active in ischemia models and as anticonvulsants and antinociceptive agents. Early clinical trials with these analogs have been disappointing due to unwanted effects. The side effect profiles do however seem to be significantly improved as compared to, e.g., competitive NMDA antagonists.

15.7.4 THE GLYCINE COAGONIST SITE

The excitatory coagonist site for glycine (**15.58**) at the NMDA receptor is named the glycine$_B$ receptor. This receptor site is different from the inhibitory glycine receptors found primarily in the spinal cord of the mammalian CNS, where glycine activates strychnine-sensitive ionotropic receptors named as glycine$_A$ receptors. Glycine$_B$ receptors seem to modulate the level of activity at NMDA receptors. A certain concentration level of glycine is always present in the synapse. Thus, Glu activates the NMDA receptors, whereas the level of glycine can modulate this activity and possibly control receptor desensitization. (*R*)-Serine (**15.59**) is a potential endogenous agonist at glycine$_B$ receptors.

Limited success of competitive NMDA receptor antagonists as therapeutic agents has focused attention on the glycine$_B$ site. (*R*)-Cycloserine (**15.60**) and (*R*)-HA-966 (**15.61**) are partial glycine agonists, capable of penetrating the BBB after systemic administration. (*R*)-Cycloserine (**15.60**) has shown promising effects in the treatment of schizophrenia and AD, and partial agonists may have therapeutic advantages as compared to full antagonists in terms of fewer side effects. A number of glycine$_B$ antagonists have also been developed. L-689,560 (**15.63**) displays high potency, and is derived from the endogenous compound kynurenic acid (**15.62**), the first glycine$_B$ antagonist reported (Figure 15.15).

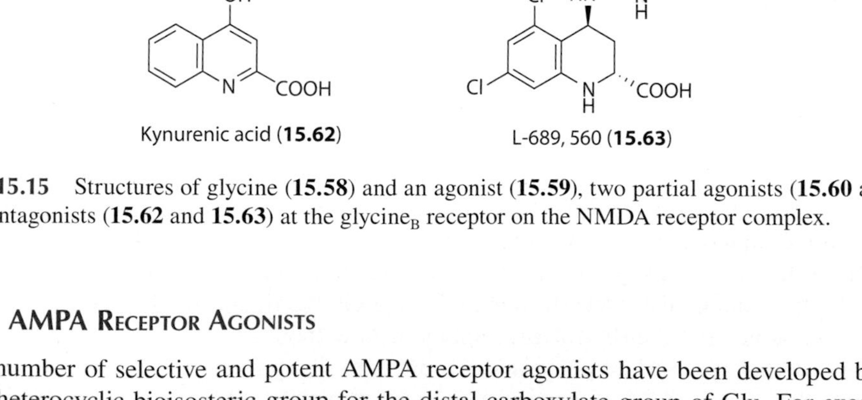

Glycine (**15.58**)   (*R*)-Serine (**15.59**)   (*R*)-Cycloserine (**15.60**)   (*R*)-HA-966 (**15.61**)

Kynurenic acid (**15.62**)   L-689, 560 (**15.63**)

FIGURE 15.15 Structures of glycine (**15.58**) and an agonist (**15.59**), two partial agonists (**15.60** and **15.61**) and two antagonists (**15.62** and **15.63**) at the glycine$_B$ receptor on the NMDA receptor complex.

15.7.5 AMPA RECEPTOR AGONISTS

A large number of selective and potent AMPA receptor agonists have been developed by substituting a heterocyclic bioisosteric group for the distal carboxylate group of Glu. For example, the heterocycles 1,2,4-oxadiazole-3,5-dione, 3-isoxazolol, and uracil, as represented by quisqualic acid (**15.64**), AMPA (**15.65**), and (*S*)-willardiine (**15.66**), respectively, have been incorporated into numerous AMPA receptor agonists. The natural product quisqualic acid (**15.64**) was the first agonist in use for pharmacological characterization of AMPA receptors, but due to nonselective action it was later replaced by AMPA (**15.65**).

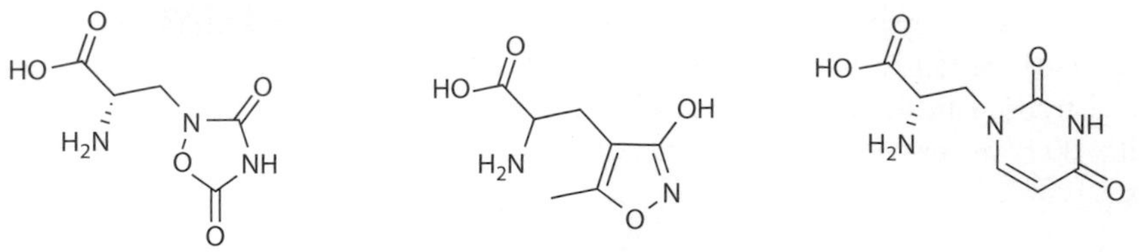

Quisqualic acid (**15.64**)   AMPA (**15.65**)   (*S*)-Willardiine (**15.66**)

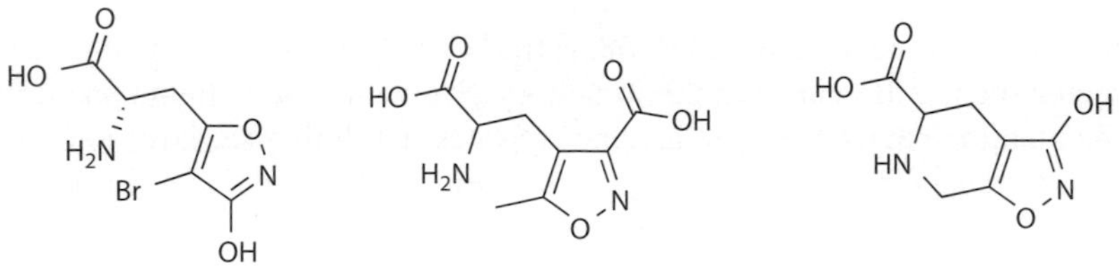

(*S*)-Br-HIBO (**15.67**)   ACPA (**15.68**)   5-HPCA (**15.69**)

FIGURE 15.16   Structures of AMPA (**15.65**) and some AMPA receptor agonists.

The isoxazole-based Glu homolog (*S*)-Br-HIBO (**15.67**) also shows AMPA receptor subtype selectivity, preferring GluR1 over GluR3 in receptor-binding and functional assays. Replacing the 3-isoxazolol group of AMPA by a 3-carboxyisoxazole unit gives the Glu homolog ACPA (**15.68**), which is a selective AMPA receptor agonist that is more potent than AMPA. The potent excitatory AMPA receptor activity of ACPA (**15.68**) has been shown to reside with the *S*-enantiomer.

Conformational restriction of the skeleton of Glu has played an important role in the design of selective GluR ligands. However, only few structurally rigid AMPA receptor-selective Glu analogs have been reported. One such example is the cyclized analog of AMPA, 5-HPCA (**15.69**), which recently has been resolved. Interestingly, the pharmacological effects of 5-HPCA (**15.69**) reside exclusively with the *R*-enantiomer, in striking contrast to the usual stereoselectivity trend among AMPA receptor agonists (Figure 15.16).

15.7.6   COMPETITIVE AND NONCOMPETITIVE AMPA RECEPTOR ANTAGONISTS

Early pharmacological studies on AMPA and KA receptors were hampered by the lack of selective and potent antagonists. The discovery of the quinoxaline-2,3-diones CNQX (**15.70**) and DNQX (**15.71**) was a breakthrough since these compounds are quite potent antagonists, although nonselective. Subsequently, the more potent analog NBQX (**15.72**) was shown to be neuroprotective in cerebral ischemia and to have improved AMPA receptor selectivity compared to CNQX (**15.70**). However, NBQX (**15.72**) failed in clinical trials because of nephrotoxicity due to a limited aqueous solubility, but nonetheless has become a valuable tool for research. DNQX (**15.71**) has played a key role in elucidating the binding mode of competitive antagonists, as it was the first antagonist cocrystallized with the GluR2 ligand-binding domain. Attempts to improve the aqueous solubility of such antagonists without losing activity at AMPA receptors, by introducing appropriate polar substituents onto the quinoxaline-2,3-dione ring system have been highly successful, and have resulted in very potent AMPA receptor antagonists, as exemplified by ZK200775 (**15.73**).

Another series of potent and selective competitive AMPA receptor antagonists based on the isantin oxime skeleton includes NS 1209 (**15.74**), which shows long-lasting neuroprotection in animal models of ischemia and an increased aqueous solubility compared to NBQX (**15.72**). At least two classes of amino acid-containing compounds, based on decahydroisoquinoline-3-carboxylic

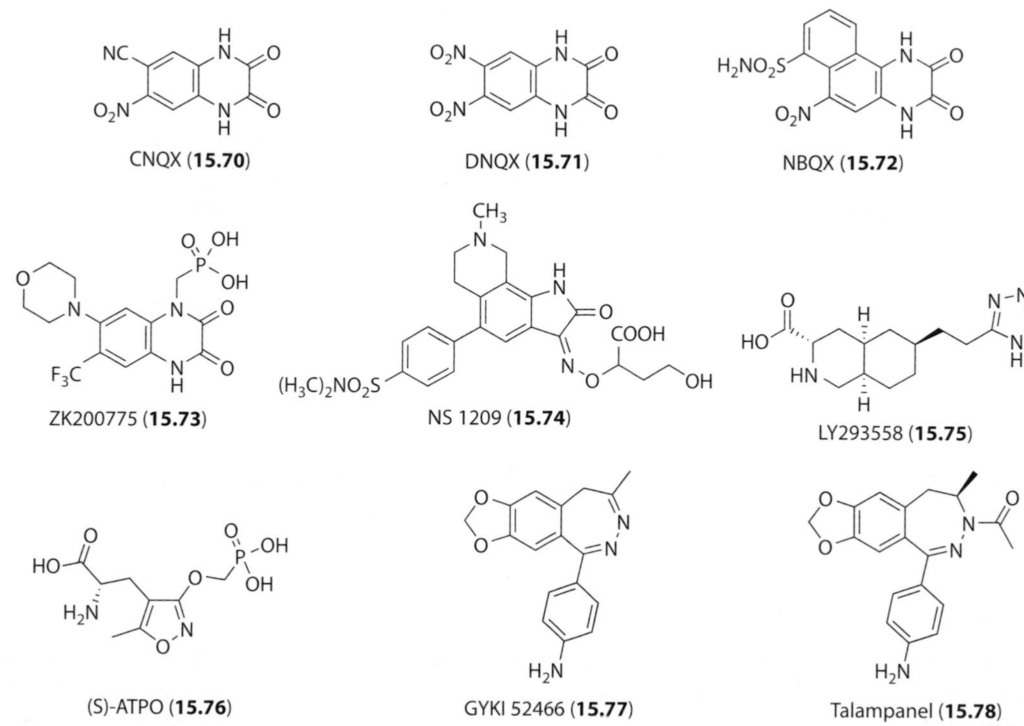

FIGURE 15.17 Structures of some competitive (**15.70–15.76**) and noncompetitive (**15.77** and **15.78**) AMPA receptor antagonists.

acid and AMPA, have been found to be competitive AMPA receptor antagonists. LY293558 (**15.75**), a member of the former class, is systemically active although it shows significant antagonist effects at KA receptors in addition to its potent AMPA receptor blocking effects. The AMPA receptor antagonist (*S*)-ATPO (**15.76**), which was designed using AMPA as a lead structure, has like LY293558 a carbon backbone longer than that which normally confers AMPA receptor agonism.

The 2,3-benzodiazepines, such as, GYKI 52466 (**15.77**) and Talampanel (**15.78**), represent a class of noncompetitive AMPA receptor antagonists that have enabled the effective pharmacological separation of AMPA and KA receptor-mediated events. These compounds appear to bind to sites distinct from the agonist recognition site, and are thus negative allosteric modulators. Talampanel (**15.78**), currently under clinical development as a treatment for multiple sclerosis, epilepsy, and Parkinson's disease, may inhibit AMPA receptor function even in the presence of high levels of Glu (Figure 15.17).

15.7.7 MODULATORY AGENTS AT AMPA RECEPTORS

The agonist induced desensitization of AMPA receptors can be markedly inhibited by a number of structurally dissimilar AMPA receptor potentiators known as AMPA-kines, including aniracetam (**15.79**), cyclothiazide (CTZ) (**15.80**), and in particular CX-516 (**15.81**), which has been shown to improve memory function in aged rats. These AMPA-kines positively modulate ion flux via stabilization of receptor subunit interface contacts and subsequent reduction in the degree of desensitization. A series of more potent arylpropylsulfonamide-based AMPA-kines have been identified, including LY395153 (**15.82**) (Figure 15.18).

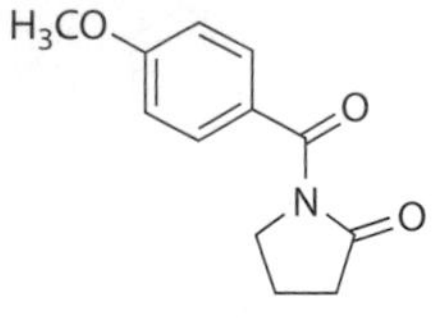

Aniracetam (**15.79**) 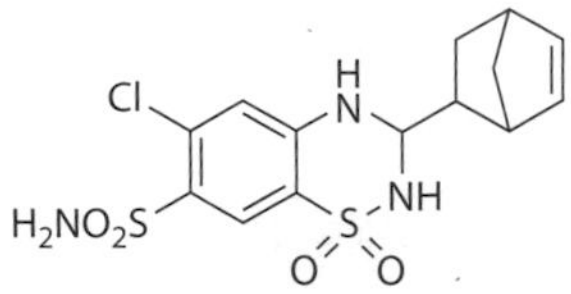   CTZ (**15.80**) 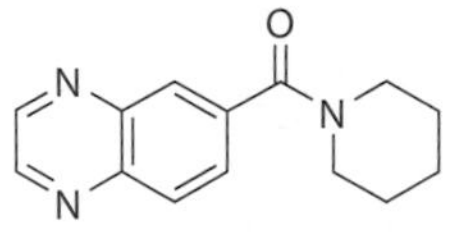   CX-516 (**15.81**)

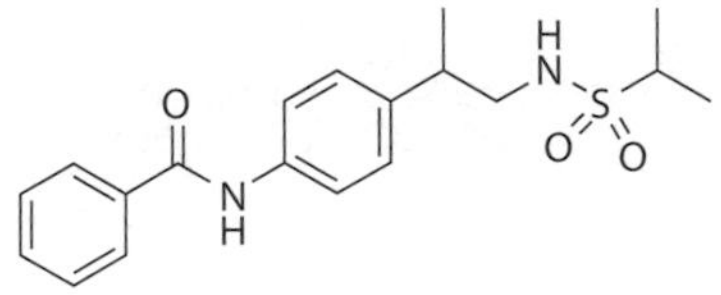

LY395153 (**15.82**)

FIGURE 15.18   Structures of some positive allosteric modulators of AMPA receptors.

15.7.8   KA Receptor Agonists and Antagonists

The pharmacology and pathophysiology of KA receptors are far less well understood than for AMPA receptors. However, identification of selective agonists and competitive antagonists has developed the field of KA receptor research during recent years, and has provided insight into the roles of these receptors in the CNS. For a number of years, KA (**15.83**) and domoic acid (**15.84**) have been used as standard KA receptor agonists despite their activities at AMPA receptors, characterized by nondesensitizing responses at these receptors. (*S*)-ATPA (**15.85**) and (*S*)-5-I-willardiine (**15.86**) are more selective KA receptor agonists, and these compounds exhibit some selectivity for the low-affinity KA receptor subtype GluR5 compared to GluR6. (*S*)-ATPA (**15.85**) and (*S*)-5-I-willardiine (**15.86**) are structurally related to potent AMPA agonists discussed in earlier sections, illustrating that the structural characteristics required for activation of GluR1–4 and GluR5 receptors are quite similar. However, the presence of the relatively bulky and lipophilic *tert*-butyl- or iodo-substituents of these compounds is apparently the major determinant of the observed receptor selectivity.

Among the four possible stereoisomers of the 4-methyl substituted analog of Glu, only the 2*S*,4*R*-isomer (**15.87**) shows selectivity for KA receptors. Replacement of the 4-methyl group of (2*S*,4*R*)-Me-Glu (**15.87**) by a range of bulky, unsaturated substituents containing alkyl, aryl, or heteroaryl groups has yielded a number of interesting GluR5 receptor-selective compounds including LY339434 (**15.88**). LY339434 shows approximately a 100-fold selectivity for GluR5 over GluR6 and no affinity for GluR1, 2, or 4 receptors.

Whereas a large number of selective competitive AMPA receptor antagonists have been identified, only a few selective KA receptor antagonists have been reported. One of the first reported KA receptor-preferring antagonists was the isantin oxime, NS 102 (**15.89**), which shows some selectivity toward low affinity [³H]KA sites as well as antagonist effect at homomeric GluR6. However, low aqueous solubility has limited the use of NS 102 (**15.89**) as a pharmacological tool. A number of decahydroisoquinoline-based acidic amino acids, including LY382884 (**15.90**), have been characterized as competitive GluR5-selective antagonists that exhibit antinociceptive effects.

More recently, a series of arylureidobenzoic acids have been reported as the first compounds with noncompetitive antagonist activity at GluR5. The most potent ligands, exemplified by compound **15.91**, exhibit more than 50-fold selectivity for GluR5 over GluR6 or the AMPA receptor subtypes (Figure 15.19).

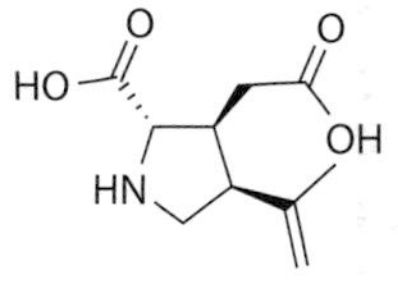

KA (**15.83**)

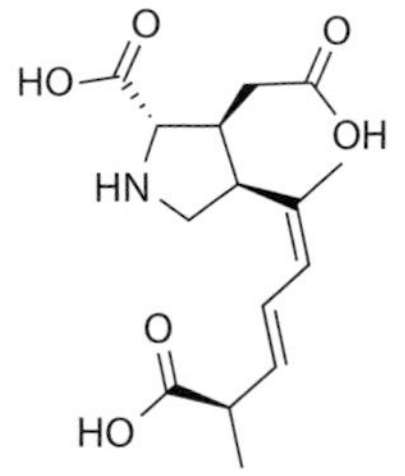

Domoic acid (**15.84**)

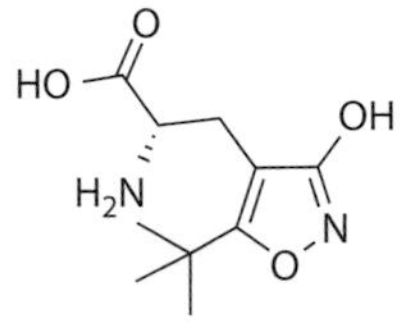

(*S*)-ATPA (**15.85**)

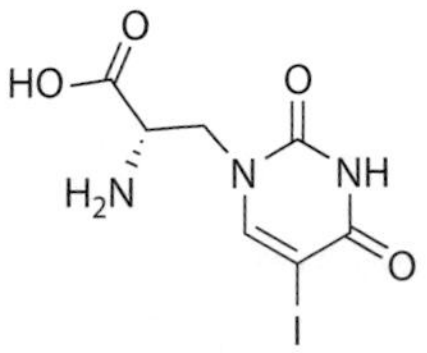

(*S*)-5-I-Willardiine (**15.86**)

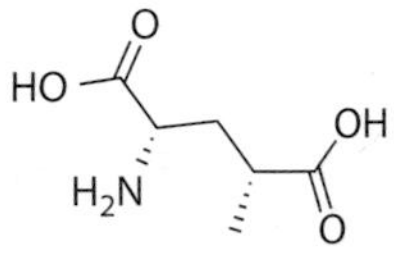

(2*S*,4*R*)-Me-Glu (**15.87**)

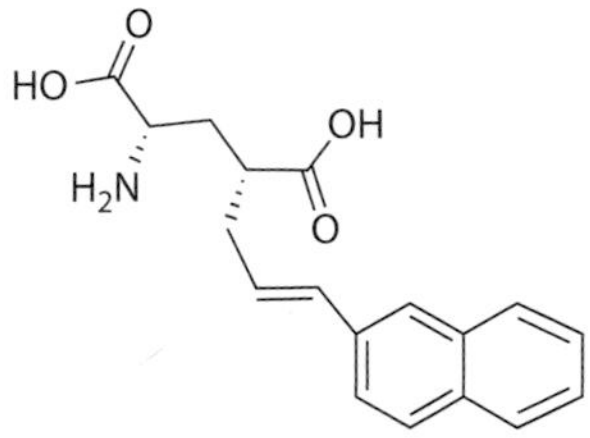

LY339434 (**15.88**)

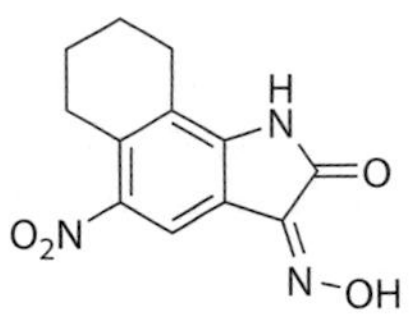

NS 102 (**15.89**)

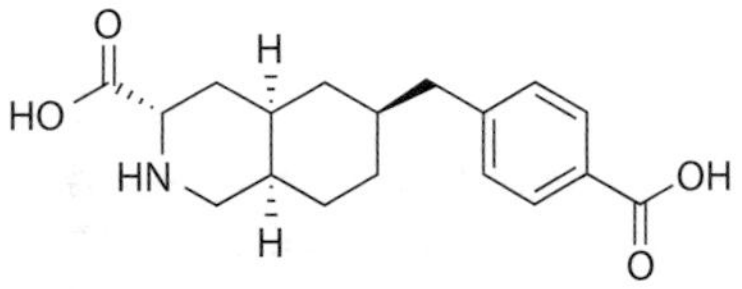

LY382884 (**15.90**)

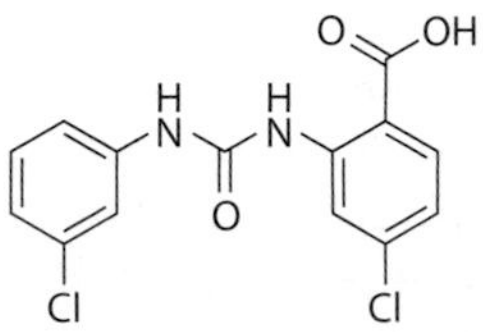

(**15.91**)

FIGURE 15.19 Structures of KA (**15.83**) and some KA receptor agonists (**15.84–15.88**), two competitive (**15.89** and **15.90**) and one noncompetitive antagonist (**15.91**).

15.8 METABOTROPIC GLUTAMATE RECEPTOR LIGANDS

The cloning of the mGluRs and the evidence, which has subsequently emerged on their potential utility as drug targets in a variety of neurological disorders, have encouraged medicinal chemists to design ligands targeted at the mGluRs. In analogy to the iGluRs, several x-ray structures of a mGluR ligand-binding construct (see Figure 12.7) including different ligands have been obtained and afforded important structural knowledge of value, e.g., in the design of ligands.

15.8.1 Metabotropic Glutamate Receptor Agonists

The first agonist to show selectivity for mGluRs over iGluRs was (1*S*,3*R*)-ACPD (**15.92**) which has been used extensively as a template for the design of new mGluR ligands. Introduction of a nitrogen atom in the C4 position of **15.92** gave (2*R*,4*R*)-APDC (**15.93**) which displays an increased potency for group II receptors compared to the parent compound while losing affinity for group I and III receptors.

LY354740 (**15.94**) displays low nanomolar agonist potency at mGluR2 and mGluR3, low micromolar agonist potency at mGluR6 and mGluR8, while showing no activity at the remaining mGluRs.

ABHxD-I (**15.95**) displays potent agonist activity, comparable to Glu, at all three mGlu groups. This observation has been of key importance in developing early models of the mGluR-binding site. Compound **15.95** is quite a rigid molecule, which adopts a conformation corresponding to an extended conformation of Glu. The observation that the compound is a potent agonist for all three mGluR groups led the suggestion that Glu adopts the same extended conformation at all three receptor groups, and that group selectivity is thus not a consequence of different conformations but rather a consequence of other factors such as, steric hindrance.

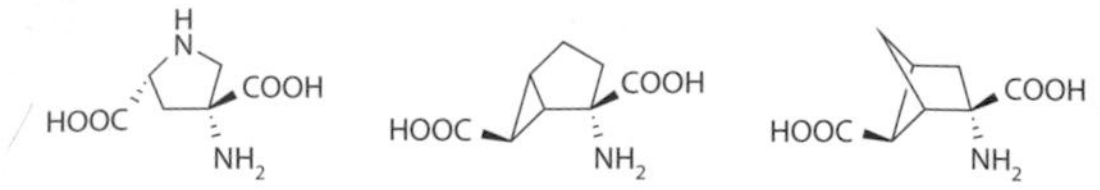

(2R,4R)-APDC (**15.93**) LY354740 (**15.94**) ABHxD-I (**15.95**)

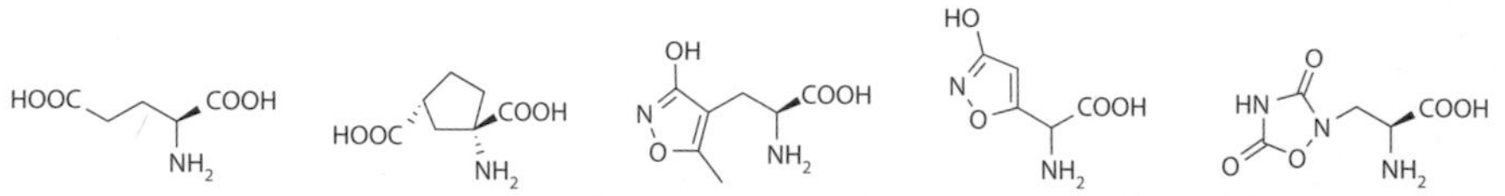

Glu (**15.2**) (1S,3R)-ACPD (**15.92**) AMPA (**15.65**) Ibotenic acid (**15.46**) Quisqualic acid (**15.64**)

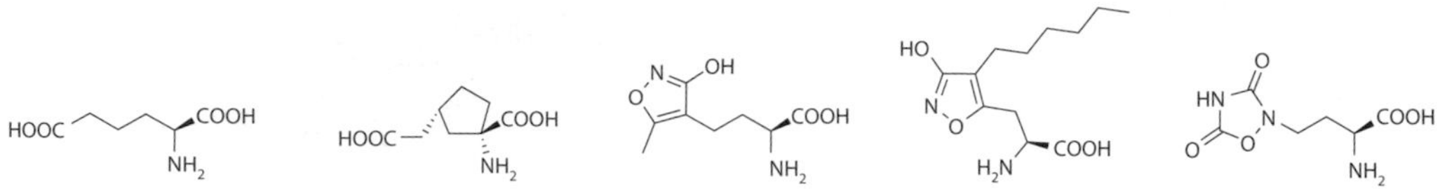

(S)-2-Aminoadipic acid (**15.96**) (1S,3R)-Homo-ACPD (**15.97**) (S)-Homo-AMPA (**15.98**) (S)-Hexyl-HIBO (**15.99**) (S)-Homo-quis (**15.100**)

FIGURE 15.20 Structures of some mGluR agonists (upper row) and some Glu analogs acting at iGluRs and/or mGluRs (middle row) and the corresponding homologs acting selectively at mGluRs (lower row).

Apart from Glu itself (1*S*,3*R*)-ACPC (**15.92**), ibotenic acid (**15.46**) and quisqualic acid (**15.64**) were among the first potent metabotropic agonists, though fairly nonselective. Synthesis of homologs of these and other Glu analogs afforded compounds with more selective activity at mGluRs. Thus, (*S*)-aminoadipic acid (**15.96**) was shown to be a mGluR2 and mGluR6 agonist, (1*S*,3*R*)-homo-ACPD (**15.97**) a Group I agonist, whereas (*S*)-homo-AMPA (**15.98**) showed specific activity at mGluR6, and no activity at neither iGluRs nor at other mGluRs. A number of HIBO analogs including (*S*)-hexyl-HIBO (**15.99**) show group I antagonistic activity and (*S*)-homo-quis (**15.100**) is a mixed group I antagonist/group II agonist. The effect of backbone extension of different Glu analogs is often unpredictable, but chain length is nevertheless a factor of importance (Figure 15.20).

15.8.2 Competitive Metabotropic Glutamate Receptor Antagonists

One of the first potent mGluR antagonists to be reported was (*S*)-4CPG (**15.101**), and it has been used extensively as a template for designing further potent and selective antagonists at mGluR1. The α-methylated analog, (*S*)-M4CPG (**15.102**), is an antagonist at both mGluR1 and mGluR2. It has been shown that the antagonist potency is increased by methylation at the 2-position of the phenyl ring. Thus (+)-4C2MPG (**15.103**) is approximately fivefold more potent than the nonmethylated parent compounds. It is notable that most 4-carboxyphenylglycines show selectivity for the mGluR1 subtype with no or weak activities at the closely related mGluR5 subtype. One exception to this rule is (*S*)-hexyl-HIBO (**15.99**), which is equipotent as an antagonist at mGluR1 and mGluR5.

α-Methylation has been widely used to derive antagonists from agonists. Maintaining the selectivity profiles as of their parent compounds, MAP4 (**15.104**) and MCCGI (**15.105**) antagonize mGluR2 and mGluR4, respectively, albeit with significantly reduced antagonist potency compared to the parent agonist.

Substituting agonists with bulky, lipophilic side chains has been a much more successful approach to the design of potent antagonists. Two of the early compounds in this class are 4-substituted analogs of Glu such as **15.107** and **15.108**, which are potent and specific antagonists for mGluR2 and mGluR3. Interestingly, compounds with small substituents in the same position, such as (2*S*,4*S*)-Me-Glu (**15.106**), are more potent agonists at mGluR2 than Glu, with some activity at mGluR1 but without appreciable activity at mGluR4. Thus by increasing the bulk and lipophilicity at the 4-position to give such "flyswatter" substituents, the selectivity for group II is retained, and even increased, but the compounds are converted from agonists to antagonists. One of the most potent

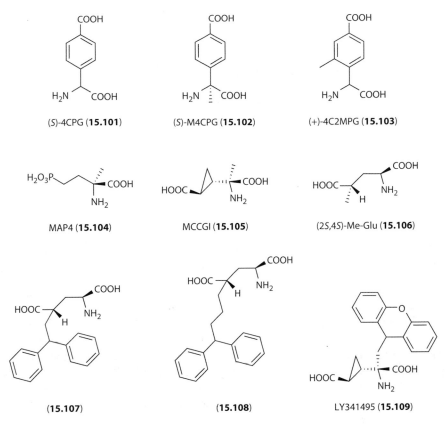

FIGURE 15.21   Structures of some mGluR ligands.

compounds of this type is LY341495 (**15.109**) with a xanthylmethyl substituent. However LY341495 (**15.109**) also shows affinity for other subtypes, especially mGluR8 (Figure 15.21).

It can be concluded that in their antagonized state receptors from all three mGluR groups can accommodate quite large and lipophilic side chains in a variety of positions. Furthermore, compared with small α substituents such as methyl groups which most often confer antagonists with reduced potency, the large "flyswatter" substituents in most cases confer antagonists with increased potency.

15.8.3   ALLOSTERIC MODULATORS OF METABOTROPIC GLUTAMATE RECEPTORS

CPCCOEt (**15.110**) is a nonamino acid compound with no structural similarity with Glu and acts as a noncompetitive group I-selective antagonist at the 7TM region rather than the agonist-binding site. A number of other nonamino acid mGluR antagonists have been discovered, e.g., BAY36–7620 (**15.111**) and EM-TBPC (**15.112**) which are potent mGluR1 specific antagonists acting at the 7TM domain.

The two compounds SIB-1893 (**15.113**) and MPEP (**15.114**) have been reported to be potent and selective, noncompetitive antagonists at mGluR5. Like CPCCOEt (**15.110**), MPEP (**15.114**) has been shown to act at the 7TM region rather than the agonist-binding site. MPEP (**15.114**) also antagonizes NMDA receptors with low micromolar potency, which has led to the design of the analog MTEP (**15.115**), which is slightly more potent than **15.114** as an antagonist at mGluR5 and with no NMDA antagonist activity. SIB-1893 (**15.113**) and MPEP (**15.114**) also act as positive allosteric modulators at mGluR4. The allosteric effect is dependent upon Glu activation, and the compounds are thus unable to activate the mGluR4 receptor directly. Instead, the compounds enhance the response mediated by Glu, causing a leftward shift of concentration–response curves and an increase in the maximum response (Figure 15.22).

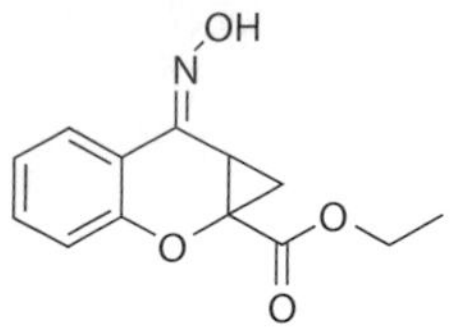

CPCCOEt (**15.110**) 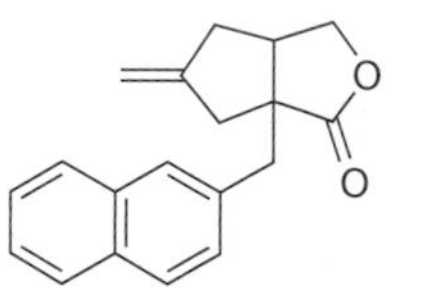   BAY36-7620 (**15.111**) 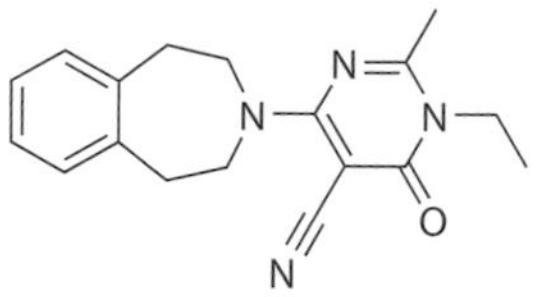   EM-TBPC (**15.112**)

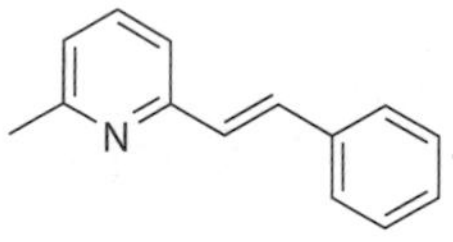

SIB-1893 (**15.113**) 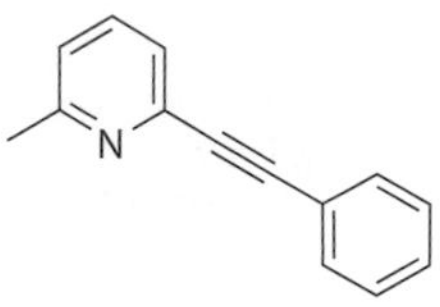   MPEP (**15.114**) 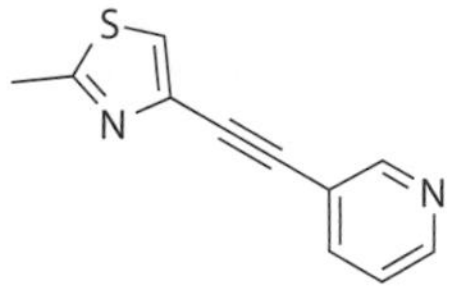   MTEP (**15.115**)

FIGURE 15.22 Structures of some noncompetitive mGluR antagonists and positive allosteric modulators.

15.9 DESIGN OF DIMERIC POSITIVE AMPA RECEPTOR MODULATORS

Many receptors, including the Glu receptors, exist as dimers or higher oligomers and this creates the possibility of having ligand-dimers, which can bind to two binding sites simultaneously. Dimeric ligands have been developed in many different receptor areas and have led to compounds with not only improved potency, but also improved selectivity, solubility, and pharmacokinetic properties can be observed. In spite of numerous examples of dimeric ligands with improved pharmacology compared to their monomeric analogs, no structural evidence have previously been presented for the simultaneous binding of such ligands to two identical binding sites. However, such evidence have been obtained for a dimeric positive allosteric modulator at AMPA receptors.

CTZ (**15.80**), see Section 15.7.7, is a positive allosteric modulator at AMPA receptors and an x-ray structure of CTZ in complex with the GluR2-binding construct showed a symmetrical binding of two CTZ molecules in two identical binding sites close to each other. Another study showed a number of biarylpropylsulfonamide analogs (**15.116**) with good activity as positive modulators at the CTZ site, and these structures were used as templates for the design of a symmetrical dimeric ligand. By use of computer modeling, different symmetric dimeric ligands with two propylsulfonamide moieties, a biphenyl linker, and different alkyl substituents were constructed and tested for binding by computer docking. This led to the proposal of dimer **15.118** as a potential ligand to bind, in a symmetrical mode, to two adjacent CTZ-binding sites, with an expected affinity three orders of magnitude better than the monomeric ligand **15.117**. Upon synthesis of the two enantiomers of monomer **15.117** and the three stereoisomers of the dimer **15.118** (*R,R*-, *S,S*-, and mesoform), these were tested for activity at cloned AMPA receptors expressed in oocytes by electrophysiological experiments. (*R,R*)-**15.118** proved to be the most potent compound with $EC_{50} = 0.79\,\mu M$ at GluR2 compared to $EC_{50} = 1980\,\mu M$ for the monomer (*R*)-**15.117**. The maximal potentiation of Glu responses for the monomer as well as for the dimer was in the order of 800%–1000%, showing that they are both effective potentiators by blockade of AMPA receptor desensitization. Obviously the dimeric compound was dramatically more potent than the monomer, more than three orders of magnitude, and a similar pattern was observed for the other less active enantiomer (Figure 15.23).

In addition to this was (*R,R*)-**15.118** cocrystallized with the GluR2-binding construct, and from the obtained x-ray structure it was shown that (*R,R*)-**15.118** simultaneously binds to two identical modulatory binding sites at the AMPA receptors. This is shown in Figure 15.24 illustrating the binding of (*R,R*)-**15.118** in comparison with two molecules of CTZ proving the simultaneous binding of a dimeric ligand to two identical binding sites.

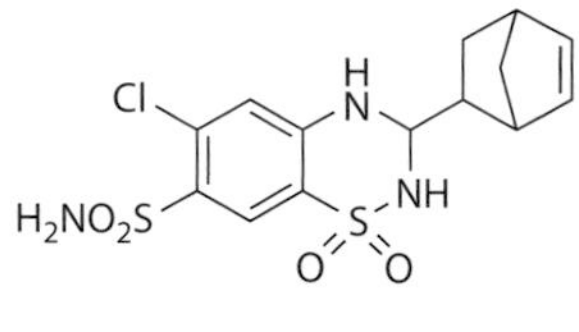

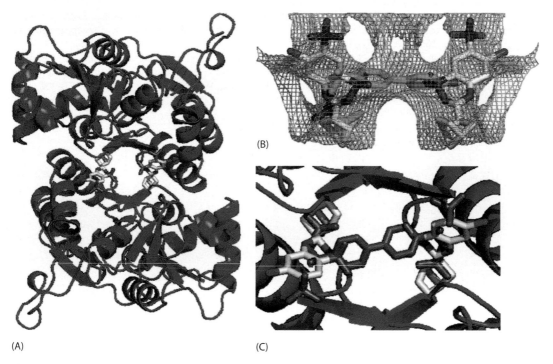

CTZ (**15.80**)

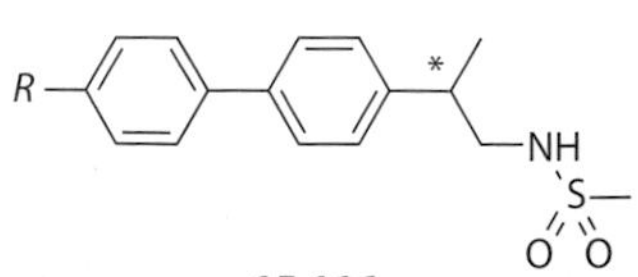

15.116

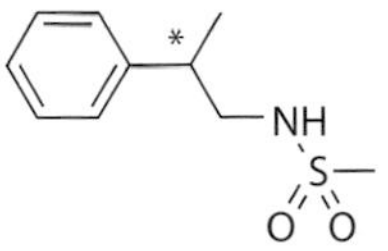

15.117

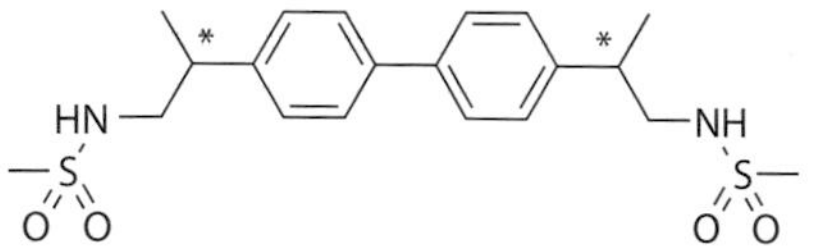

15.118

FIGURE 15.23 Structure of CTZ (**15.80**), and other positive allosteric modulators at AMPA receptors including the symmetric dimeric analog **15.118**.

(A)

(B)

(C)

FIGURE 15.24 (A) X-ray structure of the upper part of the extracellular amino terminal domain of the GluR2 receptor in complex with CTZ (**15.80**) (white carbon atoms). (B) Illustration of the binding of dimer **15.118** (cyan carbon atoms) compared to the binding of CTZ (white carbon atoms) in the calculated cavity formed by the binding pocket surface (side-view). (C) Close-up-view of the x-ray structure of the GluR2 receptor in complex with dimer **15.118** (brown carbon atoms) compared to CTZ (white carbon atoms).

15.10 CONCLUDING REMARKS

The cloning of the GABA and Glu receptor subtypes and their pharmacological characterization has been of great importance to the development of the field. For the Glu area a large number of crystal structure determinations has afforded valuable information about these receptors, their mechanism of action and the 3D information about binding sites is used extensively for the design and development of new ligands. Similar detailed information about GABA receptors are still awaited. Many selective ligands have been developed as experimental tools and have been important for the understanding of many brain functions and pathophysiological aspects. However, the success as new therapeutic agents, especially within the Glu area, is still small. The knowledge about what subtypes of receptors is involved in the neurological disorders is still limited, and further elucidation of this and development of new subtype selective ligands may eventually lead to new and better therapeutic agents.

FURTHER READINGS

A.C. Foster and J.A. Kemp, Glutamate- and GABA-based CNS therapeutics, *Curr. Opin. Pharmacol.* 2006, 6: 7–17.

B.H. Kaae, K. Harpsøe, J.S. Kastrup, A.C. Sanz, D.S. Pickering, B. Metzler, R.P. Clausen, M. Gajhede, P. Sauerberg, T. Liljefors, and U. Madsen, Structural proof of a dimeric positive modulator bridging two identical AMPA receptor-binding sites, *Chem. Biol.* 2007, 14: 1294–1303.

J.N.C. Kew and J.A. Kemp, Ionotropic and metabotropic glutamate receptor structure and pharmacology, *Psychopharmacology* 2005, 179: 4–29.

C. Madsen, A.J. Jensen, T. Liljefors, U. Kristiansen, B. Nielsen, C.P. Hansen, M. Larsen, B. Ebert, B. Bang-Andersen, P. Krogsgaard-Larsen, and B. Frølund, 5-Substituted imidazole-4-acetic acid analogues: Synthesis, modeling and pharmacological characterization of a series of novel γ-aminobutyric acid receptor agonists. *J. Med. Chem.* 2007, 50: 4147–4161.

U. Madsen, H. Bräuner-Osborne, J.R. Greenwood, T.N. Johansen, P. Krogsgaard-Larsen, T. Liljefors, M. Nielsen, and B. Frølund, GABA and Glutamate receptor ligands and their therapeutic potential in CNS disorders, in S.C. Gad (ed.) *Drug Discovery Handbook*, Wiley, Hoboken, 2005, pp. 797–907.

16 Acetylcholine

Anders A. Jensen and Povl Krogsgaard-Larsen

CONTENTS

16.1 ALZHEIMER'S DISEASE

Alzheimer's disease (AD) is a degenerative disorder of the human central nervous system (CNS), which in most cases manifests itself in mid-to-late adult life with progressive cognitive memory and intellectual impairments, leading invariably to death usually within 7–10 years after the diagnosis. AD afflicts 2%–3% of individuals at the age of 65, with an approximate doubling of incidence for every 5 years of age afterward. While age is the dominant risk factor in AD, genetic and epidemiological factors are also important determinants of the development of the disorder.

Clinical diagnosis of AD is based on the progressive impairment of memory and at least one other cognitive dysfunction, be it "aphasia" (difficulty with language), "apraxia" (difficulty with complex movements), "agnosia" (difficulty with identifying objects) or impaired executive functioning (making everyday decisions), and on the ability to exclude other diseases that also cause dementia. The ability to diagnose AD is very dependent on the progression of the disease but in mid or late stages of the disease the clinical accuracy in the diagnosis is very high (~90%). However, a diagnosis of AD can only be made definitely based on a direct pathological examination of brain tissue derived from biopsy or autopsy. The typical macroscopic picture observed in brain tissue from AD patients is massive atrophy of cortical and hippocampal brain regions, and at the microscopic level a widespread cellular degeneration and loss of neocortical neurons are observed together with the pathological hallmarks of the disease: the presence of amyloid "plaques" and neurofibrillary "tangles" (Figure 16.1).

The major component of the amyloid plaques is the amyloid β ($A\beta_{40/42}$) peptide, which is a self-aggregating, 40/42 amino acids long peptide derived from the proteolytic cleavage of the amyloid

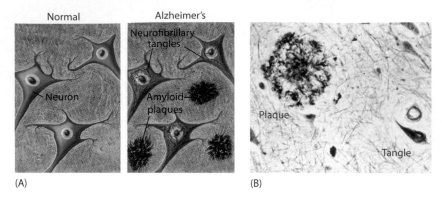

(A)   (B)

FIGURE 16.1 Amyloid plaques and neurofibrillary tangles. (A, Reprinted courtesy of Alzheimer's disease Research, a program of the American Health Assistance Foundation. www.ahaf.org/alzheimers/.; B, Reprinted courtesy of Science Foto Library.)

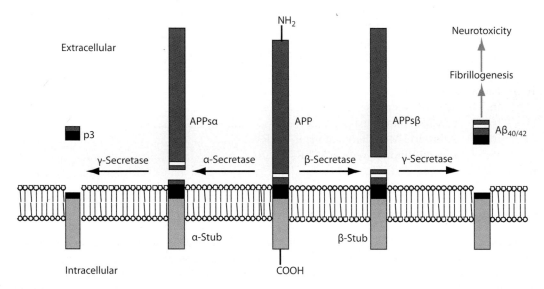

FIGURE 16.2 The amyloidogenic pathway.

precursor protein (APP), a transmembrane protein of unknown function. The APP is cleaved in two different regions of its extracellular amino-terminal domain by the enzymes α-secretase and β-secretase, and subsequently the remaining transmembrane protein sections, the α- and β-stubs, are cleaved by γ-secretase, giving rise to the peptides p3 and $A\beta_{40/42}$, respectively (Figure 16.2). In particular, the $A\beta_{42}$ peptide undergoes oligomerization and deposition, leading to microglial and astrocytic activations, oxidative stress, and progressive synaptic injury.

The neurofibrillary tangles are bundles of paired helical filaments constituted mainly by the tau protein, a widely expressed protein from the microtubule-associated family. Under normal conditions, tau maintains microtubule stability inside the cell but in AD the protein exists in a phosphorylated form, which aggregates into tangled clumps. The formation of the tangles reduces the number of tau proteins that are able to bind and stabilize the microtubules, which thus disintegrate, ultimately leading to cytoskeletal degeneration and neuronal death.

The complex and multifactorial pathogenesis of AD is not fully understood, and throughout the years several theories have been formed to explain the molecular mechanisms underlying the disease. In the 1970–1980s, the "cholinergic hypothesis" for AD was formulated based on the observed dramatic loss of cholinergic markers, such as choline acetyltransferase and acetylcholinesterase, in the AD brain, as this reflects a massive degeneration of cholinergic neurons. However, since the loss of cholinergic markers is a relatively a delayed event in the development of AD and since subsequent studies failed to demonstrate a causal relationship between cholinergic dysfunction and AD progression, the "cholinergic hypothesis" was abandoned as an explanation of AD's pathogenesis. Instead, it seems that the aberrant production and deposition of plaques and tangles in AD is not only a disease marker, as the pathways leading to the formation of these aggregates seem to play a causal role in the pathogenesis of AD. At present, the "amyloid cascade theory" is the dominant etiological paradigm in the AD field, and the theory is supported by a substantial amount of histopathological, biochemical, genetic, and animal model data. However, it is still debated whether the tau tangles or the amyloid plaques are the primary cause of the neurodegeneration in AD, and the links between the β-amyloid and tau and between the formation of plaques and tangles in the disease are still elusive.

The complexity of the etiology of AD is reflected in the numerous strategies applied over the years in the attempts to develop clinical efficacious drugs against the disorder. The formulation of the "cholinergic hypothesis" spawned the clinical testing of drugs targeting the acetylcholine (ACh) neurotransmitter system in the late 1980–1990s, the overall rationale being to augment cholinergic signaling to compensate for the degeneration of cholinergic neurons. Four out of five drugs currently approved for clinical treatment of AD are acetylcholinesterase inhibitors (AChEIs), and thus this drug class still represents the predominant clinical treatment of AD (see the following). In Figure 16.3 selected structures of noncholinergic ligands studied in the context of AD treatment is shown. Several potent inhibitors of the two enzymes mediating the formation of the $A\beta_{40/42}$ peptide, β-secretase, and γ-secretase, have been developed, for example, the γ-secretase inhibitors BMS 299897 (**16.1**) and LY-374973 (**16.2**). Other lines of research have focused on the development of molecules capable of inhibiting the aggregation processes leading to the formation of plaques and tangles. In view of the excitotoxicity in AD, considerable efforts have been put into the studies of calcium channel blockers, protease inhibitors, and glutamate receptor antagonists (see Chapter 15), and the uncompetitive NMDA receptor antagonist memantine (**16.3**) has recently become the first noncholinergic drug for the treatment of AD to be introduced on the market (Namenda®). On the other hand augmentation of AMPA receptor signaling (see Chapter 15) by a class of compounds

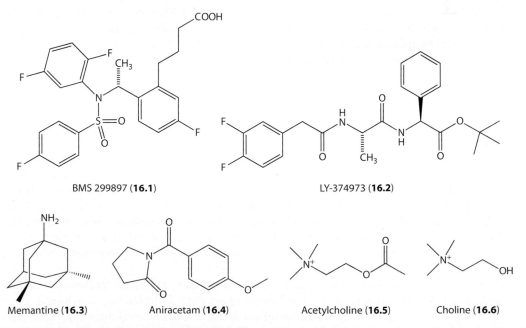

FIGURE 16.3 Chemical structures of acetylcholine, choline, and some noncholinergic ligands developed for the treatment of AD.

called ampokines such as aniracetam (**16.4**) has also been pursued as a way to treat AD. Finally, the effects of steroids and nonsteroidal anti-inflammatory drugs (NSAIDs) on the inflammation and of antioxidants on the oxidative damages observed in AD are also being investigated.

16.2 CHOLINERGIC SYNAPTIC MECHANISMS AS THERAPEUTIC TARGETS

The neurotransmitter acetylcholine (ACh, **16.5**) is found throughout the body, where it regulates a wide range of important functions. In the periphery, cholinergic signaling is, for example, of key importance for cardiac function, gastric acid secretion, gastrointestinal motility, and smooth muscle contractions. In the CNS, cholinergic neurotransmission is involved in numerous processes underlying cognitive functions, learning and memory, arousal, reward, motor control, and analgesia.

The cholinergic synapse and the complex events underlying cholinergic neurotransmission are depicted in Figure 16.4. ACh exerts its physiological effects via signaling through two distinct receptor classes: muscarinic ACh receptors (mAChRs) and nicotinic ACh receptors (nAChRs), which mediate the metabolic (slow) and the fast response to ACh, respectively. Once ACh is released into the synaptic cleft, two cholinesterases, acetylcholinesterase (AChE, EC 3.1.1.7) and butyrylcholinesterase (BuChE, EC 3.1.1.8), are responsible for its conversion into choline (**16.6**), which subsequently is taken up into the presynaptic terminal.

16.3 CHOLINESTERASES

The events underlying cholinergic neurotransmission depicted in Figure 16.4 are not that different from those in other neurotransmitter systems, as these also involve the biosynthesis and storage of the neurotransmitter in synaptic vesicles, synaptic release, and activation of different receptor classes and reuptake by transporter systems. The only extraordinary feature of the cholinergic synapse is the presence of two synaptic enzymes converting the neurotransmitter into its precursor in order for it to be taken back up by the presynaptic terminal. The AChE and BuChE belong to the "α/β hydrolase fold protein" superfamily comprising serine hydrolases such as cholinesterases, carboxylesterases, and lipases. Both cholinesterases are present in cholinergic synapses in the CNS, in the parasympathic synapses in the periphery, and in the neuromuscular junction. Whereas AChE is selective for ACh hydrolysis, BuChE accommodates and degrades several other substrates, including numerous neuroactive peptides. Of the two enzymes most attention has been paid to AChE, since it is responsible for ~80% of the total cholinesterase activity in the brain and has a remarkable high turnover (in the 10^4 s^{-1} range) compared to BuChE.

16.3.1 CHOLINESTERASE INHIBITORS

The physiological significance of AChE activity is reflected by the observation that it is targeted by numerous "natural" and synthetic toxins, ranging from snake and insect venoms to pesticides and nerve gasses used in chemical warfare. The efforts in the design of AChEIs in medicinal chemistry have been greatly facilitated by the availability of crystal structures of AChE complexed with ligands. Based on the nature of their activity, AChEIs can be divided into two main classes: (1) irreversible organophosphorus inhibitors and (2) reversible inhibitors. Compounds such as dyflos (**16.7**) and sarin (**16.8**) belong to the former class, which due to the irreversible nature of their action are characterized by having a long duration of action in the body, since AChE activity only is restored after resynthesis of the enzyme.

The reversible AChEIs were the first drugs developed for the symptomatic treatment of AD, and the drug class is still dominating the field. Inhibition of synaptic cholinesterase activity has proven to be efficacious in the treatment of AD, as the effect of this amplification of the natural spatial and temporal tone of ACh-mediated signaling seems to be preferable to the constant stimulus resulting from direct activation of mAChRs or nAChRs by agonists.

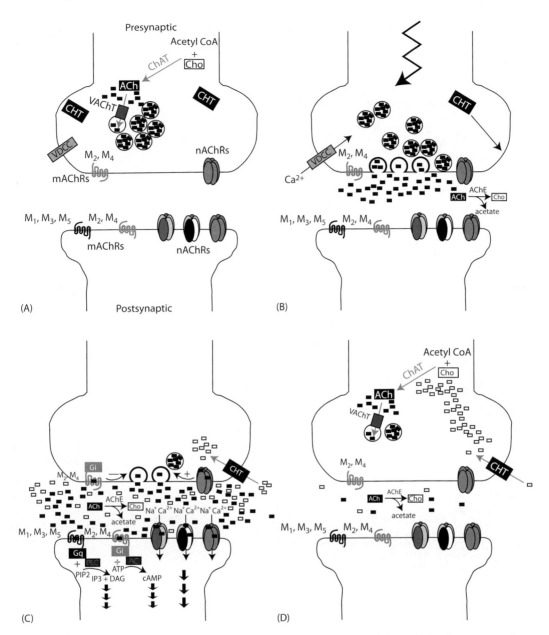

FIGURE 16.4   The cholinergic synapse and the events during synaptic firing. (A) In the presynaptic terminal, acetylcholine (ACh) is synthesized by "choline acetyltransferase" (ChAT) through acetylation of the precursor choline (Cho), the source of the acetyl groups being acetyl Coenzyme A (acetyl CoA). The synthesized ACh is packaged into synaptic vesicles by uptake by the vesicular ACh transporter (VAChT). (B) Upon stimulation of the presynaptic neuron, cytoplasmic Ca^{2+} concentrations are elevated due to the influx of the dication trough voltage-dependent calcium channels (VDCCs), and this causes the ACh-containing vesicles to fuse with the plasma membrane and release ACh into the synaptic cleft. (C) Here, ACh exerts its effects via activation of mAChRs and nAChRs. Activation of the postsynaptic nAChRs and mAChRs elicits the fast excitatory and the slow metabolic signaling of ACh, respectively, whereas activation of presynaptic receptors inhibits or augments the synaptic release of ACh, thus constituting negative and positive feedback mechanisms. Concurrently with its binding to the receptors, ACh signaling is being terminated by "acetylcholinesterase" (AChE) and "butyrylesterase," the two enzymes in the synaptic cleft, which converts ACh back to Cho. Furthermore, the choline transporter (CHT) is transported from the inside of the presynaptic terminal to the plasma membrane. (D) Having reached the membrane, CHT transports Cho back into the presynaptic terminal, where it once again is used in the synthesis of ACh. With the degradation of ACh and the subsequent removal of Cho from the synaptic cleft, the cholinergic neuron returns to its resting state.

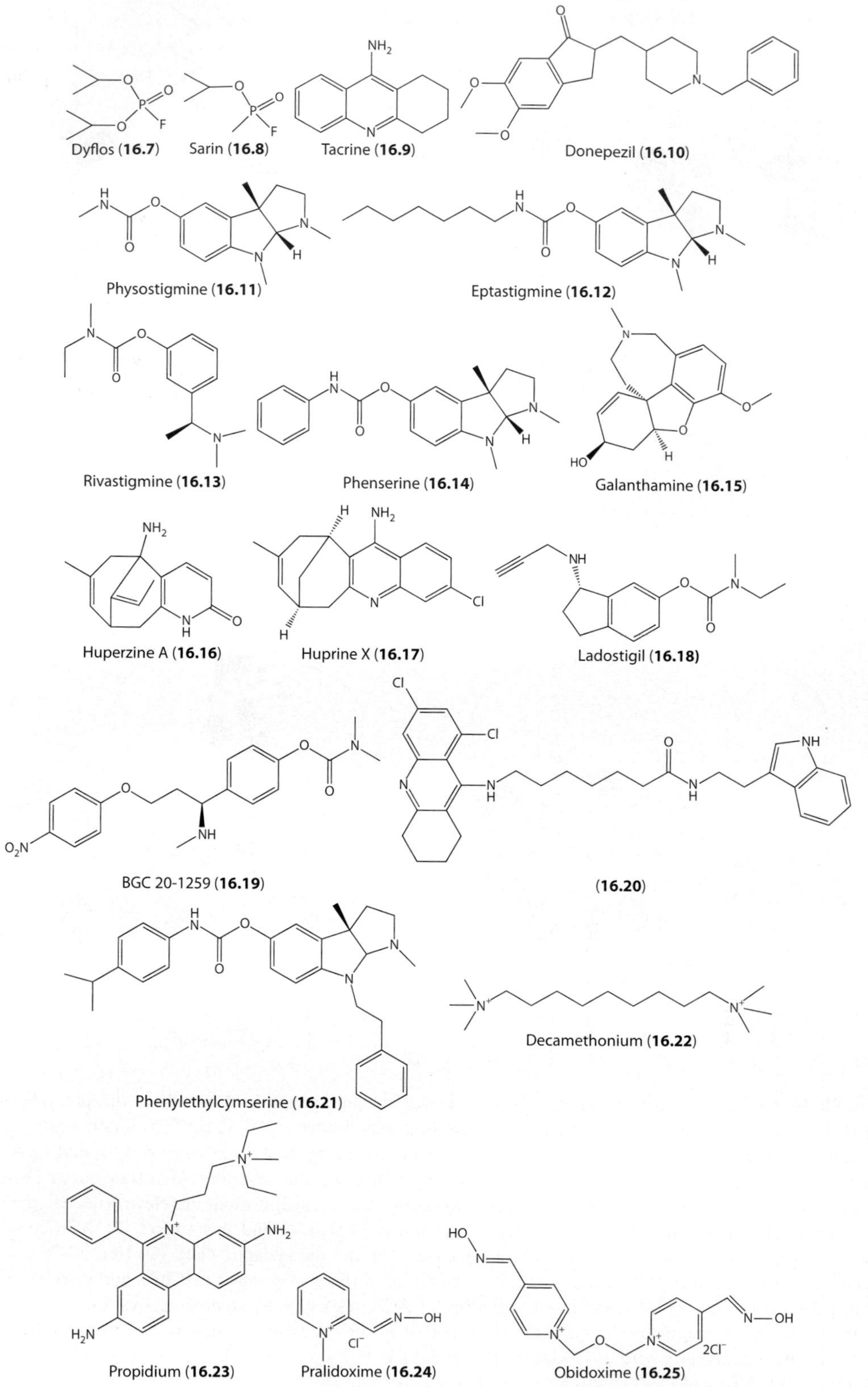

FIGURE 16.5 Chemical structures of selected cholinesterase inhibitors (**16.7–16.23**) and the AChE reactivators **16.24** and **16.25**.

In 1993, 9-amino-1,2,3,4-tetrahydroacrine (tacrine, **16.9**) was marketed as the first AChEI against AD. It is a reversible, nonselective AChE/BuChE inhibitor, which also displays activities at monoamine oxidases, potassium channels, and mAChR and nAChR subtypes. In fact, the "dirty" profile of tacrine has been proposed to contribute to its therapeutic effects. In 1997, the piperidine-based ligand donepezil (**16.10**) was rationally designed and marketed for treatment of AD. Donepezil inhibits AChE in a reversible and noncompetitively manner and displays a significant selectivity for AChE over BuChE. Medicinal chemistry explorations into the "carbamate–stigmine" structure of physostigmine (**16.11**) from the Calabar bean (*Physostigma venenosum*) has given rise to several important analogs, including eptastigmine (**16.12**), rivastigmine (**16.13**, marketed in 2000), and phenserine (**16.14**, currently in phase III trials), which all exhibit inhibitory activities at both AChE and BuChE. Galanthamine (**16.15**), a phenantrene alkaloid originally isolated from *Galanthus nivalis*, was marketed for treatment of AD in 2001. The natural product huperzine A (**16.16**) has been isolated from the Chinese folk medicine *Huperzia serrata*, and it is a potent AChEI with no activity at the BuChE.

Substantial efforts in medicinal chemistry have gone into the optimization of existing AChEIs. Developed hybrid compounds combining structural components from two AChEIs, such as the tacrine/hyperzine A hybrid, huprine X (**16.17**), have displayed higher inhibitory potencies, and in some cases different kinetic properties and/or AChE/BuChE selectivities, than their parent compounds. In other hybrids, substructures of AChEIs have been combined with molecular components of other drugs hereby giving rise to novel compounds, where the AChE activity has been supplemented with activities at other neurotransmitter systems. For example, rivastigmine (**16.13**) constitutes the template of ladostigil (**16.18**), where the structure has been combined with the propylargyl group of the MAO-B inhibitor rasagiline, and of BCG 20-1259 (**16.19**), an inhibitor of both AChE and the serotonin transporter. Another popular strategy has been the development of bivalent ligands, such as the bivalent tacrine-indole ligand **16.20**, which displays a low picomolar IC_{50} value at AChE. Finally, the observation that regulation of the synaptic ACh concentrations appears to become more and more dependent on BuChE as AD progresses has inspired the development of completely selective BuChEIs, including several cymserine analogs (**16.21**) (Figure 16.5).

16.3.2 Substrate Catalysis of the AChE and Ligand Binding to It

At the molecular level, AChE is a 537 amino acids long protein composed of a 12-stranded mixed β-sheet surrounded by 14 α-helices (Figure 16.6A). The hydrolysis of ACh in AChE takes place at the bottom of a long and narrow gorge lined with numerous aromatic amino acid residues that penetrates half into the enzyme. The active site is located ~20 Å from the surface of the enzyme and is composed of two subsites. In the "catalytic anionic site" the choline moiety of ACh is stabilized by a cation–π interaction between the quaternary amino group of ACh and the aromatic ring system of the Trp[84] residues with minor contributions from the Glu[199] and Phe[330] residues, whereas the "esteratic subsite" contains a typical serine-hydrolase catalytic triad consisting of the residues Ser[200], His[440], and Glu[327] (Figure 16.6A). In addition, binding is stabilized by interactions between the carbonyl oxygen and the acetyl group of ACh with neighboring residues in AChE. Another binding site for ACh, the "peripheral anionic site" (PAS), is located on the surface of the enzyme at the entrance to the gorge, ~20 Å above the active site and contains the Trp[70], Asp[72], Tyr[121], Trp[279], and Phe[331] residues (Figure 16.6A). In addition to PAS being involved in processes important for other aspects of AChE function, binding of ACh to PAS represents the first step in the ACh catalysis, as the trapping of the substrate on its way to the active site enhances the catalytic efficiency of the process. From the PAS, ACh is transferred to the active site, where the catalysis occurs (Figure 16.7A).

The complex processes underlying ACh catalysis is reflected in the diverse binding modes displayed by the reversible AChEIs both in terms of their binding sites and the mechanisms underlying their inhibition. The "carbamoylating" inhibitors, compounds **16.11–16.14**, all contain a carbamate group, which analogously to the ester group in ACh can be hydrolyzed by AChE (Figure 16.7B). Thus, these AChEIs are split into their carbamate moiety and their stigmine (**16.11**, **16.12**, and **16.14**) or dimethylamino-α-methylbenzyl (**16.13**) moieties interacting with the Ser[200] residue and

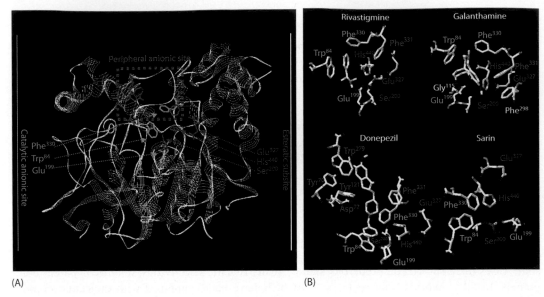

FIGURE 16.6 AChE structure and AChEI-binding modes. (A) The 3D structure of AChE. The localization of the PAS and the residues constituting the catalytic anionic site and the esteratic site in the enzyme are indicated. (B) The active site of AChE in crystal structures of the enzymes complexes with AChEIs. The binding modes of reversible inhibitors rivastigmine, galanthamine, and donepezil to the AChE and the phosphorus AChE conjugate formed after interaction with the irreversible inhibitor sarin.

the anionic site, respectively (exemplified by rivastigmine in Figure 16.6B). However, in contrast to the very fast (microseconds) hydrolysis of acetate from the Ser^{200} residue following the hydrolysis of ACh, the dissociation of the carbamate group is very slow (minutes) (Figure 16.7B). Thus, the inhibition exerted by this AChEI class arises from the active site in AChE being occupied and unable to bind ACh for a considerable time after the initial binding of the inhibitor. In particular, for rivastigmine (**16.13**) the duration of this reactivation phase is considerable, and the compound is often termed a "pseudo-irreversible" inhibitor.

Other AChEIs, such as tacrine (**16.9**) and galanthamine (**16.15**), act like "true" inhibitors, as they are not hydrolyzed by AChE but instead compete with ACh for the active site in the enzyme, interacting with residues in both the anionic and esteratic subsites (Figure 16.6B). Donepezil (**16.10**), on the other hand, targets the gorge connecting the active site with the surface of the enzyme, the dimethoxy-indanone and benzyl piperidine moieties of the inhibitor interacting with Trp^{279} in the PAS and with Trp^{84} and Phe^{330} in the anionic subsite of the active site, respectively (Figure 16.6B). Bisquaternary inhibitors, such as decamethonium (**16.22**), bind in a similar fashion, and the remarkably enhanced potencies displayed by bivalent ligands such as **16.20** is also attributed to their ability to target two binding sites in AChE. Finally, propidium (**16.23**) and fasciculin-2, a 61-amino acid peptide isolated from the venom of *Dendroaspis* (mamba) species, inhibit AChE noncompetitively by binding exclusively to the PAS.

Irreversible organophosphorus AChEIs such as **16.7** and **16.8** all act at the active site of AChE, as they form covalently attached phosphorus conjugates with the Ser^{200} residue in the esteratic subsite of the enzyme, thereby disrupting its catalytic mechanism (Figures 16.6B and 16.7C). Treatment of the phosphorus AChE conjugate with so-called reactivators, oxime-based medical antidotes such as pralidoxime (**16.24**) and obidoxime (**16.25**), can restore AChE function, but being unable to pass the blood–brain barrier (BBB) they cannot reverse the central effects of organophosphate poisoning.

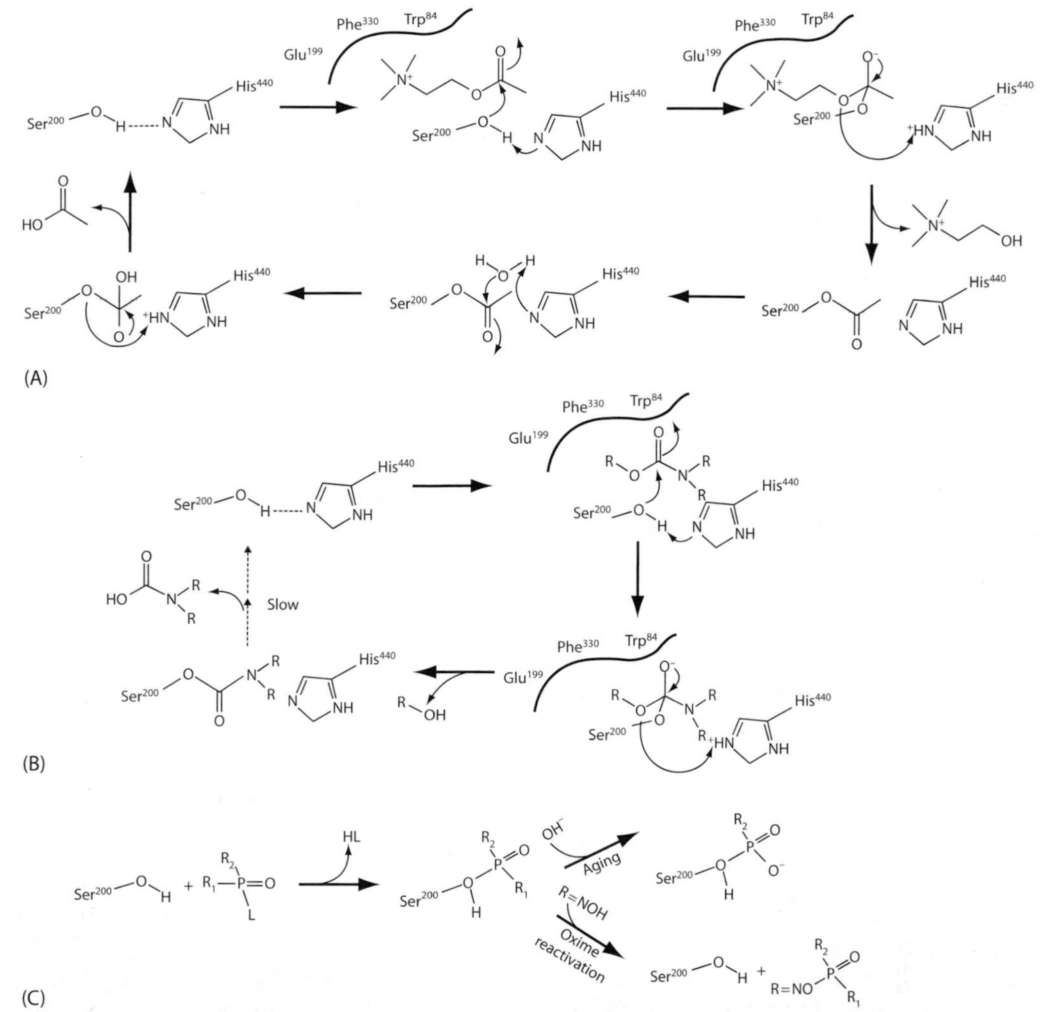

(A)

(B)

(C)

FIGURE 16.7 Substrate catalysis of the AChE and inhibition of it. (A) Catalysis of ACh in the AChE. (B) Inhibition of AChE by a reversible carbamoylating AChEI. (C) Formation of the phosphorus AChE conjugate by organophosphorus AChEIs, and the following "aging" and oxime reactivation processes. R_1, O-alkyl or amid; R_2, alkyl, O-alkyl or amid; L, leaving group.

16.4 MUSCARINIC ACh RECEPTORS

The mAChRs belong to family A of the superfamily of G-protein coupled receptors (GPCRs) (see Chapter 12). Hence, the muscarinic component of cholinergic signaling is mediated by intracellular second messenger cascades initiated by the coupling of the activated mAChRs to G-proteins and other intracellular proteins such as β-arrestins. Five mAChR subtypes have been identified, termed M_1–M_5. The M_1, M_3, and M_5 mAChRs are coupled to $G_{\alpha q}$-proteins and the resulting stimulation of phospholipase C and intracellular release of Ca^{2+}, whereas the M_2 and M_4 subtypes are coupled to $G_{\alpha o}$ and $G_{\alpha i}$ proteins associated with inhibition of adenylate cyclase and a reduction of the intracellular levels of cAMP.

The mAChRs are expressed abundantly in both the CNS and in the peripheral tissues. In recent years, studies of "mAChR knock out mice," where the expression of one of the five mAChR subtypes have been eliminated have shed light on the functions of the respective mAChRs and the therapeutic perspectives in selective targeting of these individual mAChRs.

The M_2 mAChR is widely expressed in the CNS and in the peripheral nervous system (PNS). Besides being the far most abundant mAChR in the heart, where it mediates the cholinergic regulation of the heart rate, the M_2 and M_3 receptors are the key subtypes when it comes to the cholinergic component of the smooth muscle contraction. In the CNS, M_2, and the other G_i-coupled mAChR, M_4, function as autoreceptors and heteroreceptors. In contrast to the widespread distribution of M_2, M_4 is predominantly centrally expressed, and besides its potential as a target for the treatment of pain, the subtype has been shown to regulate striatal dopamine release.

Similar to the M_2 mAChR, the M_3 subtype is expressed throughout the CNS and the PNS and in organs innervated by parasympathetic nerves. M_3 knockout mice display a substantial reduction in body weight due to a reduced food intake compared to wild-type mice, and this phenotype has been ascribed to their lack of M_3 receptors in hypothalamus, a key brain region for regulation of appetite. M_1 is the most abundantly expressed mAChR subtype in the forebrain, where it is predominantly expressed at the postsynaptic termini. Although the M_1 knockout mice do not exhibit significantly impaired cognitive impairment, substantial pharmacological evidence indicate that the receptor is involved in cognitive processes underlying learning and memory. Furthermore, M_1 knockout mice exhibit epileptic symptoms, and inhibition of the receptor could hold prospects in the treatment of Parkinson's disease. The fifth mAChR subtype, M_5, is expressed in considerably lower levels than the other mAChRs in both the CNS and PNS.

16.4.1 mAChR Agonists

The compounds **16.26–16.30** in Figure 16.8 are classical mAChR agonists displaying no significant selectivity for any of the five subtypes. Muscarine (**16.26**), which like ibotenic acid and muscimol (see Chapter 15) is a constituent of *Amanita muscaria*, has defined the mAChR family because of its selectivity for these receptors over the nAChRs. Stabilization of the ester moiety of ACh as a carbamate group yields carbachol (**16.27**), which not only is nonselective when it comes to the mAChRs but also is equipotent as a nAChR agonist. The naturally occurring heterocyclic agonist pilocarpine (**16.28**) is widely used as topical miotic for the control of elevated intraocular pressure associated with glaucoma. However, the bioavailability of pilocarpine is low, and the compound does not appear to penetrate the BBB. The potent agonist oxotremorine (**16.29**) has been used extensively as a lead compound for structure–activity studies giving rise to a plethora of mAChR ligands spanning the entire efficacy range from full agonists over partial agonist to antagonists.

Because of the high expression levels of the postsynaptic M_1 mAChRs in the brain regions affected in AD, substantial medicinal chemistry efforts have been put into the development of agonists selectively targeting this subtype. Arecoline (**16.31**), a constituent in areca nuts (the seeds of *Areca catechu*), is a cyclic "reverse ester" bioisostere of ACh, containing a tertiary amino group. **16.31** is only partially protonated at physiological pH, which is an advantage in terms of BBB penetration and CNS availability. A considerable number of analogs of **16.31** have been developed, including xanomeline (**16.32**), where the labile ester moiety of **16.31** has been replaced with the more metabolically stable thiadiazole. **16.32** has become the prototypic "M_1 selective agonist" but in functional assays the compound only exhibits a preference for M_1 over the other subtypes. In support of this, the *in vitro* profile of **16.32** has not been translated into an acceptable safety margin in patients, and due to cholinergic side effects clinical development of the compound for treatment of AD has been discontinued. Interestingly, in these clinical trials **16.32** have been found to improve the behavioral disturbances and hallucinations often observed in AD patients, an effect that subsequently has been ascribed to an effect on dopaminergic signaling via the M_4 subtype. Several other "M_1-selective/preferring" agonists, including **16.33–16.35**, have been in development for the treatment of AD but most of them have been faced with serious cholinergic side effects in their clinical development, suggesting that they might not be sufficiently subtype-selective. AC-42 (**16.36**) is a partial M_1 agonist identified in a high-throughput screening, and it bears no significant structural resemblance with ACh or other orthosteric mAChR ligands (Figure 16.8).

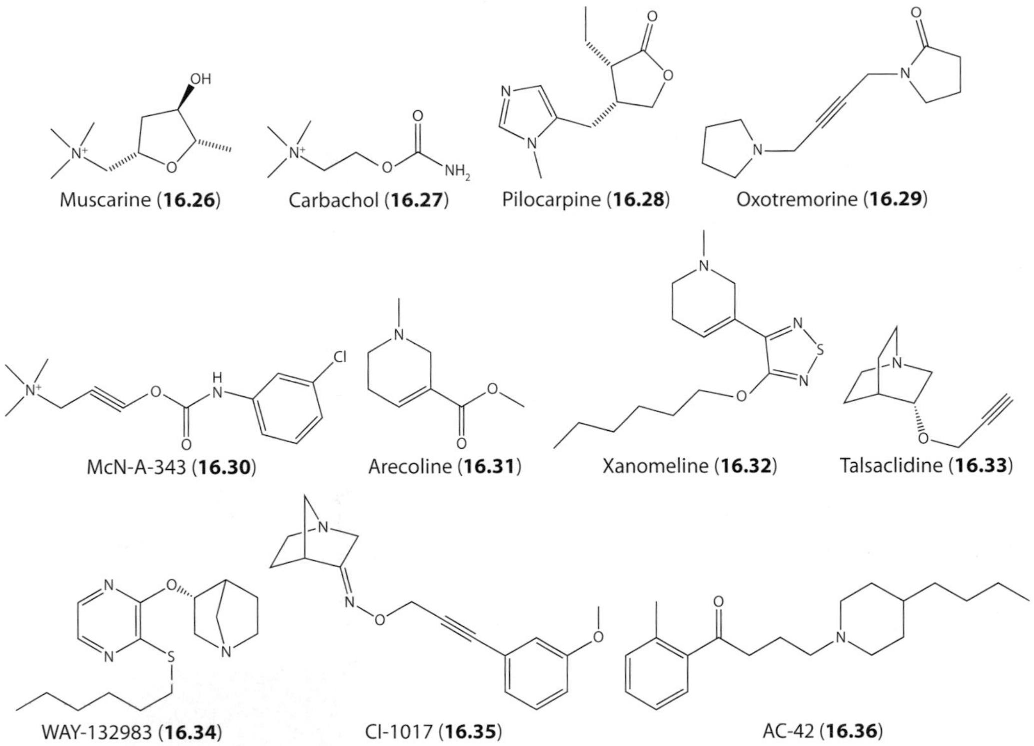

FIGURE 16.8   Chemical structures of mAChR agonists.

In contrast to its quite potent M_1 activity, AC-42 exhibits no agonist or antagonist activities at any of the other four mAChR subtypes, and thus, it is the only completely M_1-selective agonist reported to date. An AC-42 analog is currently in clinical trials for the treatment of glaucoma.

16.4.2   mAChR Antagonists

Analogous to the few true subtype-selective mAChR agonists available, it has been quite difficult to develop subtype-selective antagonists. The alkaloids atropine (16.37) and scopolamine (16.38) found in "the deadly nightshade" *Atropa belladonna* are potent mAChR antagonists, and like *N*-methylscopolamine (NMS, 16.39) and quinuclidinyl benzilate (QNB, 16.40) they have been highly important tools in the explorations of mAChR function over the years (Figure 16.9). In contrast to these nonselective antagonists, pirenzepine (16.42) has displayed higher binding affinities for mAChRs in the brain over mAChRs in the heart, suggesting a preference for M_1 over M_2 subtypes, although this "selectivity" has been less impressive when the compound has been tested at the cloned mAChRs. Similarly, so-called M_2 selective antagonists such as AF-DX-116 (16.43), with its remarkable structural similarity to 16.42, and SCH 57790 (16.44) only display 20–40-fold M_2/M_1 selectivity in binding assays. In contrast, the three peptides m1-toxin, m2-toxin (16.45), and m4-toxin isolated from the venom of the green mamba, *Dendroaspis augusticeps*, are potent and highly selective antagonists of M_1, M_2, and M_4, respectively. In recent years, however, the modest M_2 preference of 16.44 has formed the basis for extensive SAR studies, and this work has resulted in numerous small molecular weight compounds, including 16.46, displaying subnanomolar binding affinities to M_2 and >1000-fold selectivity over the other mAChRs. Several of these have entered clinical trials for AD, where the major question will be whether the desired effects of inhibition of presynaptic M_2 autoreceptors in the CNS can be separated from the unwanted side effects caused by antagonism of peripheral M_2 receptors.

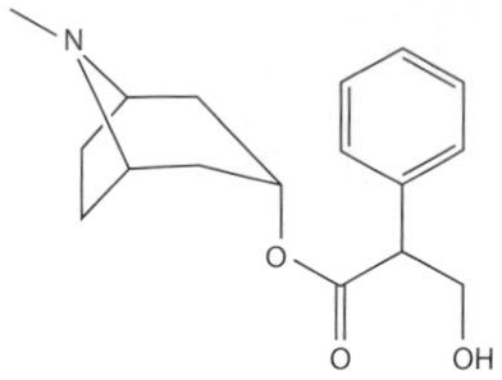

Atropine (**16.37**) 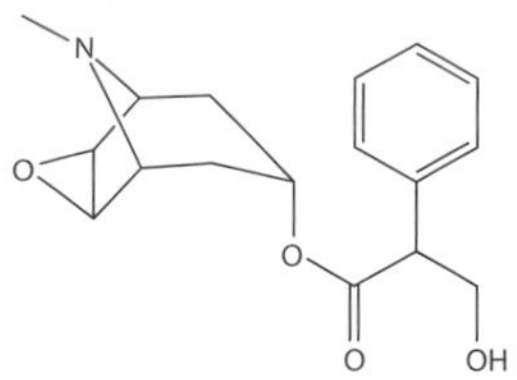   Scopolamine (**16.38**) 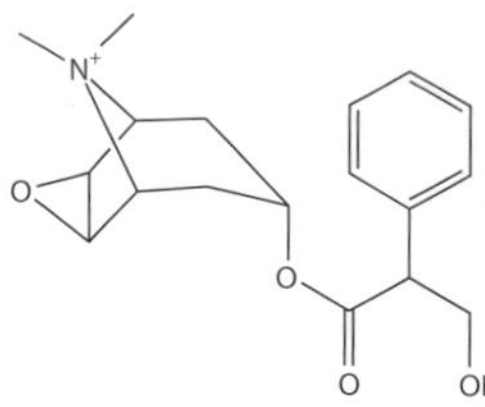   NMS (**16.39**) 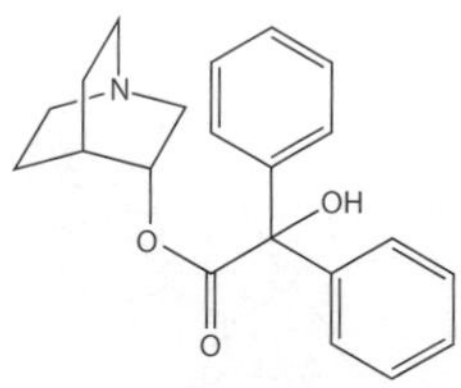   QNB (**16.40**)

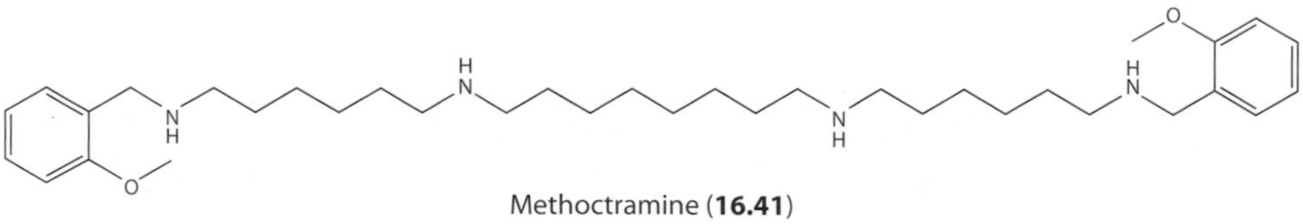

Methoctramine (**16.41**)

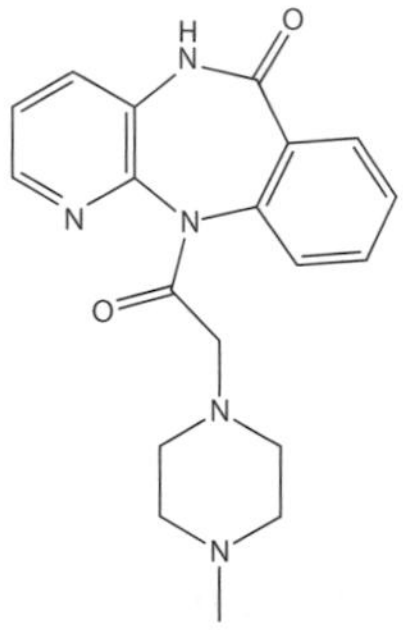

Pirenzepine (**16.42**) 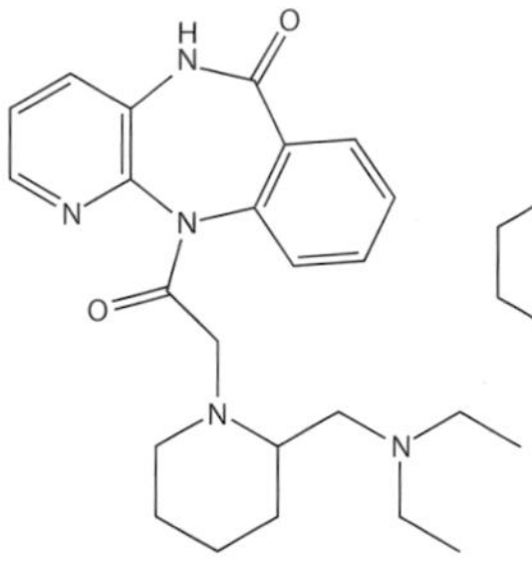   AF-DX-116 (**16.43**) 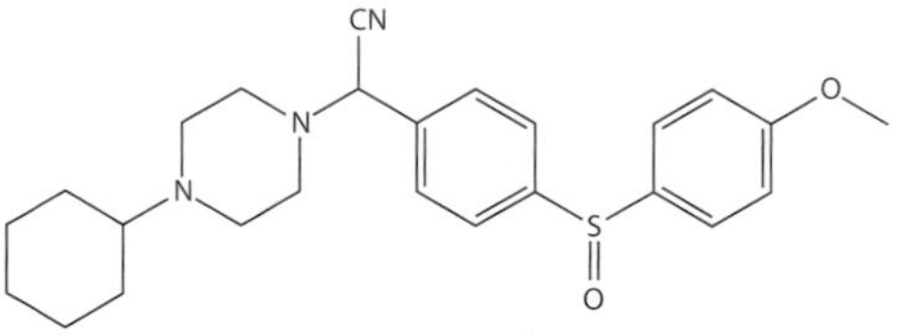   SCH 57790 (**16.44**)

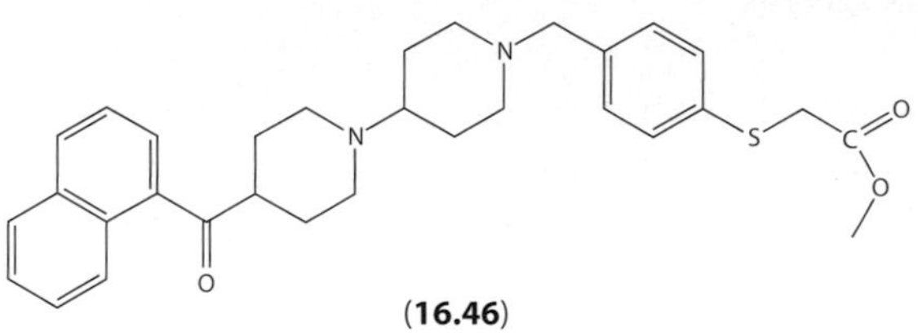

(**16.46**)

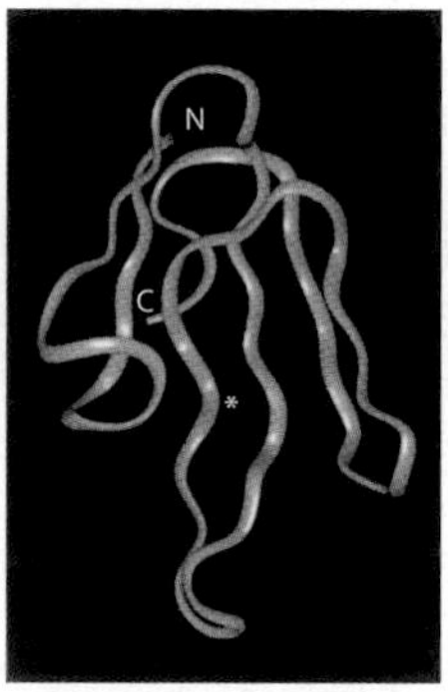

m2-toxin (**16.45**)

FIGURE 16.9 Chemical structures of competitive mAChR antagonists. (Part of the figure is reprinted from Krajewski, J.L. et al., *Mol. Pharmacol.*, 60, 725, 2001. With permission.)

16.4.3 ALLOSTERIC MODULATORS OF mAChRs

The difficulties connected with the development of agonists or competitive antagonists capable of discriminating between mAChR subtypes have prompted the search for and identification of "allosteric modulators" of the receptors (Chapter 12). As can be seen from the examples given in Figure 16.10, these compounds (**16.47–16.51**) are structurally completely different from ACh and other orthosteric mAChR ligands in a complete manner. Interestingly, several of these allosteric modulators also target other receptors or enzymes.

16.4.4 LIGAND BINDING TO THE mAChR

A considerable insight into the ligand-binding modes of mAChR ligands to their receptors has, over the years, been gained from mutagenesis studies. The orthosteric site in the mAChR (and other family A GPCRs) is localized in a cavity formed by transmembrane regions (TMs) 3, 5, 6, and 7 (Figure 16.11A). All orthosteric mAChR ligands contain a quaternifed ammonium group or an amino group

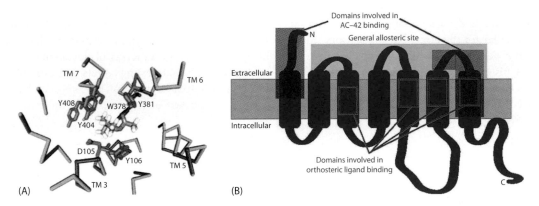

Gallamine (**16.47**)   Alcuronium (**16.48**)   WIN 26,577 (**16.49**)

Brucine (**16.50**)   W84 (**16.51**)

FIGURE 16.10   Chemical structures of allosteric modulators of mAChRs.

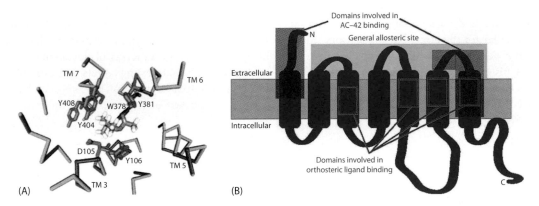

(A)   (B)

FIGURE 16.11   (A) Binding mode of ACh to the orthosteric site of the M_1 mAChR (observed from the extra-cellular side). (B) Localization of orthosteric and allosteric binding sites in the mAChR.

that is positively charged at physiological pH, and orthosteric ligand binding to the mAChR is centered in the ionic and cation–π interactions formed by this group to an aspartate residue in TM3 and aromatic residues in TM6 and TM7. This key interaction is supplemented by hydrogen bonds and van der Waals interactions between residues in TM3, TM5, TM6, and TM7 of the mAChR and other parts of the orthosteric ligand.

The wide range of structurally diverse allosteric mAChR modulators all seem to bind to a general allosteric site localized just above the orthosteric site in the receptor, constituted by the upper helix turns of the seven TMs and their three interconnecting extracellular loops (Figure 16.11B). Finally, in agreement with its novel structure and selectivity profile, the M_1-selective agonist AC-42 has been demonstrated to bind to an allosteric site in this receptor composed of residues in amino-terminal domain, TM1, and TM7 of the receptor (Figure 16.11B).

16.5 NICOTINIC ACh RECEPTORS

The nAChRs belong to the superfamily of ligand-gated ion channels termed "Cys-loop receptors" (Chapter 12). The receptors are complexes composed of five subunits forming an ion pore through which Na^+ and Ca^{2+} ions can enter the cell when the receptor is activated, resulting in depolarization of the neuron and increased intracellular Ca^{2+} concentrations.

To date 17 different nAChR subunits have been identified (Figure 16.12A). The "muscle-type nAChR" is composed of α_1, β_1, δ, and γ/ε subunits and is localized postsynaptically at the neuromuscular junction (Figure 16.12B). The receptor is a key mediator of the electrical transmission across the anatomical gap between the motor nerve and the skeletal muscle, thus creating the skeletal muscle tone. Hence, antagonists of this receptor are used clinically as muscle relaxants during anesthesia. The "neuronal nAChRs" are heteromeric or homomeric complexes composed of the α_2–α_{10} and β_2–β_4 subunits (Figure 16.12B), and they are located at presynaptic and postsynaptic densities in autonomic ganglia and in cholinergic neurons throughout the CNS. Equally important to the overall contribution of nAChRs to cholinergic neurotransmission are the roles of nAChRs as autoreceptors and heteroreceptors regulating the synaptic release of ACh and other important neurotransmitters such as dopamine, noradrenalin, serotonin, glutamate, and GABA.

The 12 neuronal nAChR subunits display considerable different expression patterns in the CNS, and this combined with the ability of the subunits to assemble into a vast number of different combinations characterized by significantly different pharmacological profiles give rise to a large degree of heterogeneity in the native receptor populations. The two predominant physiological neuronal nAChRs are the heteromeric $\alpha_4\beta_2$ subtype and the homomeric α_7 receptor. The $\alpha_4\beta_2$ subtype constitutes >90% of the high-affinity binding sites for nicotine in the brain and is the most obvious nAChR candidate in the treatment of AD and nicotine addiction. The homomeric α_7 nAChR is characterized by its low binding affinities for the classical nAChR ligands, by fast desensitization kinetics, and by a remarkable high Ca^{2+} conductance. In recent years, the α_7 nAChR has attracted considerable attention as a drug target for the treatment of states of inflammation, the fibromyalgia syndrome and various forms of pain, and modulation of α_7 signaling appears to be beneficial in particular for the cognitive and sensory impairments observed in schizophrenia. Although the two major neuronal nAChR subtypes have attracted most of the attention in terms of development of nAChR-based therapeutics, several of the "minor" subtypes are also interesting targets. For example, the α_6 subunit is localized exclusively in the midbrain, and α_6-containing nAChR subtypes have been shown to regulate synaptic dopamine release in striatum making them interesting in relation to Parkinson's disease (Chapter 17).

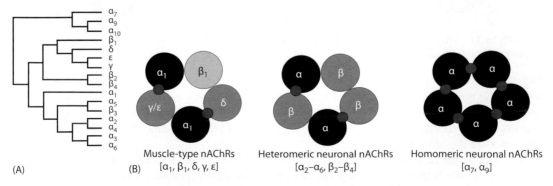

FIGURE 16.12 The nAChR family. (A) A phylogenetic tree over the nAChR family. (B) The multiple nAChR complexes formed by the 17 subunits. The localization of the orthosteric sites in the respective receptors is indicated.

16.5.1 nAChR Agonists

As in the case of the AChE and mAChR fields, medicinal chemistry development of nAChR ligands has been greatly influenced by the plethora of ligands isolated from natural sources. From a therapeutic perspective, there has been most interest in ligands augmenting nAChR signaling (i.e., agonists or allosteric potentiators). The classical nAChR agonists **16.52–16.57** have all been isolated from natural sources, and while they are selective for the nAChRs over the mAChRs, they are rather nonselective within the nAChR family (Figure 16.13). Because of their potent agonism and high BBB penetration, (*S*)-nicotine (**16.52**) and (±)-epibatidine (**16.53**) have been the lead compounds for the vast majority of nAChR agonists developed over the years. Modifications of the linker between the pyridine and pyrrolidine rings and introduction of new ring systems in **16.52** have resulted in several potent nAChR agonists, such as compounds **16.58–16.60**. Ring-opening of the pyrrolidine ring in (*S*)-nicotine has given rise to ispronicline (**16.61**) and several other potent analogs. Interestingly, these agonists display significant functional selectivity for the $\alpha_4\beta_2$ over all other nAChR subtypes and have entered clinical development as cognitive enhancers and analgesics. The 5-ethynyl nicotine analog, altinicline (**16.62**), is a partial agonist characterized by a modest functional preference for the $\alpha_4\beta_2$ nAChR over β_4-containing subtypes. **16.62** is a highly efficacious stimulant of dopamine release in nucleus accumbens and striatum and has been in clinical trial for Parkinson's disease.

(±)-Epibatidine (**16.53**) was originally isolated from the skin of the Ecuadorian frog *Epipedobates tricolor* and is by far the most potent of the classical nAChR agonists. The therapeutic interest in the compound was founded in studies showing that it is a nonaddictive analgesic, blocking pain 200 times more effectively than morphine. Because of the severe hypertension and neuromuscular paralysis observed upon epibatidine administration, however, several epibatidine analogs have been developed in the hope to isolate the analgesic effects from the side effects. Methylation of the basic amino group and substitution of the 2-chloropyridine ring in **16.53** with a 2-(pyridazin-4-yl) ring

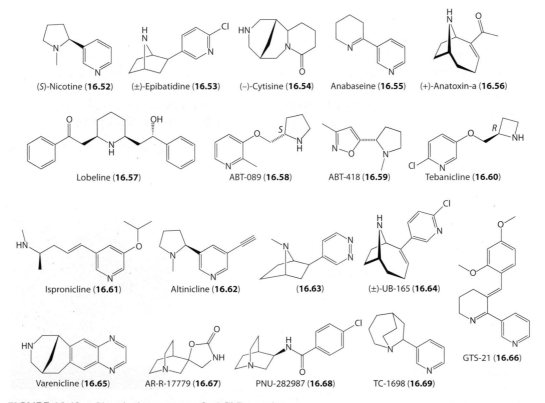

FIGURE 16.13   Chemical structures of nAChR agonists.

has resulted in **16.63**, which displays a significant functional preference as a potent agonist of $\alpha_4\beta_2$ and muscle-type nAChRs over other nAChRs. In contrast, (±)-UB-165 (**16.64**), a hybrid compound between epibatidine and anatoxin-a (**16.56**, isolated from the alga *Anabaena flos aquae*), is one of the few nAChR agonists available that preferentially activate the "minor" heteromeric nAChRs without concomitant activation of the major CNS subtypes.

(−)-Cytisine (**16.54**) from *Laburnum anagyroides* is a potent partial agonist of the heteromeric nAChRs. (−)-Cytisine and its analog varenicline (**16.65**) have recently been launched as smoking cessation aids. Further, the complex natural source compound lobeline (**16.57**) from *Lobelia inflata* is also under clinical development for treatment of smoking dependence.

Although the homomeric α_7 nAChR represents a low-affinity binding site for ACh and the natural source compounds **16.52–16.56**, it has been possible to develop potent and α_7 selective agonists from some of these leads. The toxin anabaseine (**16.55**) is isolated from marine worms and certain ant species, and it is a rather nonselective agonist displaying a somewhat higher efficacy at α_7 nAChR than at the heteromeric nAChRs. Introduction of conjugated aryl substituents in the 3-position of the pyridine ring of anabaseine has increased this selectivity, as exemplified by the prototypic α_7 agonist GTS-21 (**16.66**). The quinuclidine (1-azabicyclo[2.2.2]octane) ring system forms the scaffold in several α_7-selective agonists, including AR-R-17779 (**16.67**) and PNU-282987 (**16.68**), of which the latter has entered clinical trials for schizophrenia. Finally, replacement of the pyrrolidine ring in **16.52** with a azabicyclo[3.2.2.]nonane ring has provided the potent and selective α_7 agonist TC-1698 (**16.69**).

16.5.2 nAChR Antagonists

As described for the nAChR agonists, several competitive nAChR antagonists have been obtained from natural sources. The peptide toxin, α-bungarotoxin (**16.70**) from the Taiwan banded krait (*Bungarus multicinctus*) is a potent competitive antagonist of α_7 and muscle-type nAChRs, and methyllycaconitine (**16.71**), isolated from *Delphinum* and *Consolida* species, is a highly selective α_7 antagonist. In contrast to the selectivity of these two compounds, other classical competitive nAChR antagonists, including dihydro-β-erythroidine (DHβE, **16.72**), are far less discriminative between different nAChR subtypes (Figure 16.14).

In addition to the agonists derived from (*S*)-nicotine and (±)-epibatidine, several antagonists have emerged. Introduction of *n*-alkyl groups ranging from methyl to dodecyl ($C_{12}H_{25}$) at the pyridine nitrogen of (*S*)-nicotine have produced several potent albeit nonselective antagonists (exemplified by **16.73**). Furthermore, the pyridyl ether A-186253 (**16.74**), a **16.58** analog, displays high selectivity

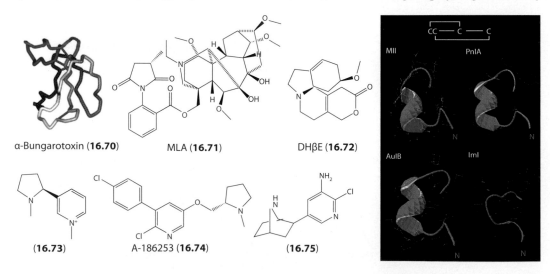

FIGURE 16.14 Chemical structures of competitive nAChR antagonists and 3D structures of α-bungarotoxin and four α-conotoxins.

for native $\alpha_4\beta_2^*$ nAChRs versus $\alpha_3\beta_4^*$ and α_7 receptors in binding assays. The epibatidine analog **16.75** is an antagonist with a K_i value of 1 pM to native $\alpha_4\beta_2^*$ nAChRs, making it the nAChR ligand with the highest binding affinity published to date.

Within the last decade, several small peptides of 12–20 amino acid residues, the so-called α-conotoxins, have been isolated from a family of predatory cone snails, the *Conus* snails. The peptides have turned out to be very interesting pharmacological tools, as they are highly subtype-specific in their antagonism of nAChRs. The 3D structures of the α-conotoxins are established by intramolecular disulfide bonds formed by the four highly conserved cysteine residues in the peptides, which are organized in different arrangements: $\alpha_{3/5}$, $\alpha_{4/3}$, $\alpha_{4/7}$, or $\alpha_{4/6}$, the nomenclature referring to the number of residues between the conserved cysteines (Figure 16.14). The subtype-selective activities of the respective α-conotoxins arise from the differences in the nonconserved residues. So far, α-conotoxins selective for nAChR subtypes α_7 (for example ImI), $\alpha_3\beta_2$ (for example MII and PnIA), and $\alpha_3\beta_4$ (AuIB) have been identified (Figure 16.14). Considering that the entire mollusk family is estimated to contain ~50,000 neuropharmacologically active toxins, additional subtype-selective nAChR antagonists are likely to be identified in the future.

16.5.3 ALLOSTERIC MODULATORS OF THE nAChRs

As it is the case with other ligand-gated ion channels, such as $GABA_A$ and NMDA receptors (see Chapter 15), the nAChRs are highly susceptible to allosteric modulation (see Chapter 12). Several endogenous ligands, such as steroids (for example **16.76**), 5-hydroxyindole (**16.77**), and Ca^{2+} and Zn^{2+} ions modulate signaling through the receptors (Figure 16.15). It is highly interesting to note that the $A\beta_{42}$ peptide has been found to be a potent noncompetitive inhibitor of $\alpha7$ nAChR signaling but the significance of this inhibition for AD remains to be elucidated.

Allosteric potentiators hold several advantages to regular agonists when it comes to the augmentation of nAChR signaling. First, analogous to the AChEIs they only exert their effect when ACh is

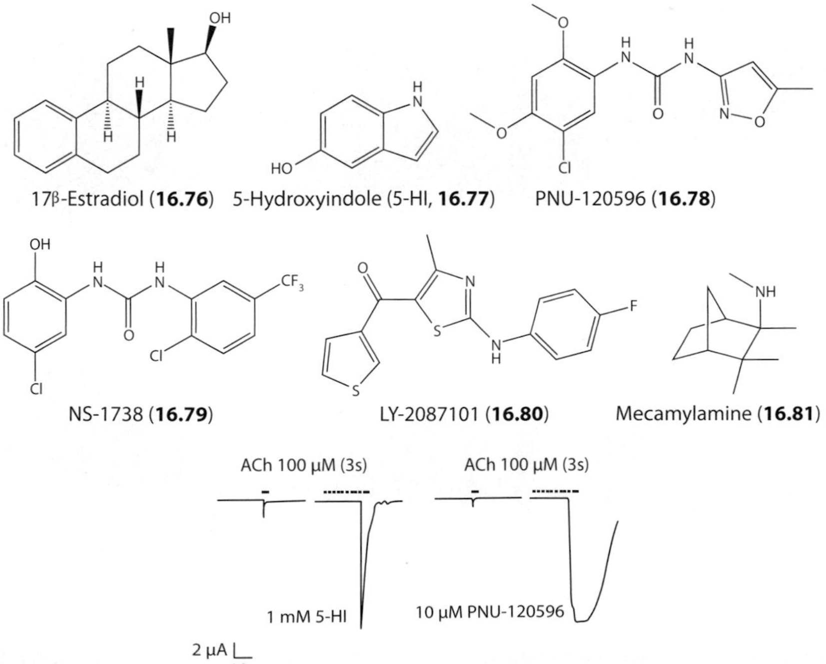

17β-Estradiol (**16.76**)   5-Hydroxyindole (5-HI, **16.77**)   PNU-120596 (**16.78**)

NS-1738 (**16.79**)   LY-2087101 (**16.80**)   Mecamylamine (**16.81**)

ACh 100 µM (3s)   ACh 100 µM (3s)

1 mM 5-HI   10 µM PNU-120596

2 µA ⌊
      10 s

FIGURE 16.15 Chemical structures of allosteric modulators of nAChRs and the potentiation of the ACh-induced α_7 nAChR signaling exerted by 5-hydroxyindole (5-HI) and PNU-120596. (Part of the figure is reprinted from Bertrand, D. and Gopalakrishnan, M., *Biochem. Pharmacol.*, 74, 1155, 2007. With permission.)

present in the synapse, and thus the stimulation of the receptors introduced by the allosteric modulator occurs in a physiological tone. Second, since the allosteric ligands target regions less conserved than the orthosteric site in different nAChR subtypes, they are more likely to be subtype-selective than orthosteric ligands. Finally, in contrast to agonists, which for the most parts mimic the activation kinetics of the endogenous agonist, allosteric potentiators can enhance nAChR signaling in many different ways, for example, by increasing the ion conductance of the receptor, by increasing the frequency of ACh-induced ion channel openings, or by reducing the desensitization rate.

The urea analogs PNU-120596 (**16.78**) and NS-1738 (**16.79**) are selective allosteric potentiators of the α_7 nAChR. Both compounds enhance the potency of as well as the maximal response elicited by ACh through the receptor, having no effect on receptor signaling in the absence of ACh. In addition to these effects, **16.78** also suppress the desensitization of α_7 (Figure 16.15) and can restore the activity in an already desensitized receptor. In contrast to the subtype-selective activities of these potentiators, LY-2087101 (**16.80**) potentiates the signaling of $\alpha_2\beta_4$, $\alpha_4\beta_4$, $\alpha_4\beta_2$, and α_7 nAChRs but not that of the muscle-type, $\alpha_3\beta_2$, and $\alpha_3\beta_4$ subtypes. Interestingly, the AChEIs physostigmine (**16.11**) and galanthamine (**16.15**) have also been shown to potentiate the ACh-evoked responses through several nAChR subtypes.

16.5.4 LIGAND BINDING TO THE nAChRs

The orthosteric sites of the nAChR are situated in the extracellular amino-terminal domain of the pentameric receptor complex, more specifically at the interfaces between α- and β-subunits in the heteromeric nAChR and between two α-subunits in the homomeric nAChR (Figure 16.16). Thus, the heteromeric and homomeric nAChRs contain two and five orthosteric sites, respectively (Figure 16.12). Similarly to the mAChR ligands, agonists, and almost all competitive antagonists

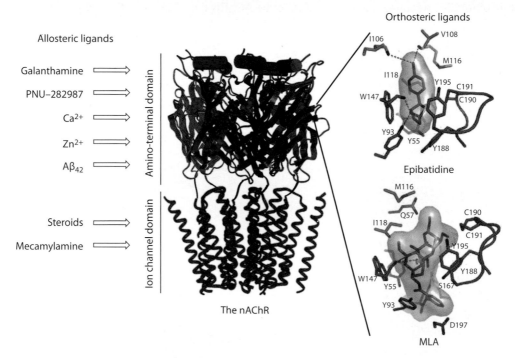

FIGURE 16.16 Ligand binding to the nAChR. Right: The binding modes of the agonist epibatidine and the competitive antagonist MLA to the orthosteric site in the amino-terminal domain of the receptor. (Part of the figure is reprinted from Hansen, S.B. et al., *EMBO J.*, 24, 3635, 2005. With permission.) Left: Allosteric modulators targeting the amino-terminal and ion channel domains of the nAChR complex. (Part of the figure is reprinted from Jensen, A.A. et al., *J. Med. Chem.*, 48, 4705, 2005. With permission.)

of the nAChRs possess a positively charged amino group, either in the form of a quaternary amino group or a protonated tertiary or secondary amino group (Figures 16.13 and 16.14). This amino group docks into an "aromatic box" formed by five aromatic residues facing the interface, predominantly from the α-subunit side, where the group forms a strong cation–π interaction with a tryptophan (W) residue, while the other four aromatic residues ensures optimal spatial orientation of the ligand for binding (exemplified by the agonist epibatidine and the competitive antagonist MLA in Figure 16.16). The so-called complementary binding component, i.e., the interactions between receptor and the rest of the ligand molecule, predominantly takes place to β-subunit side of the orthosteric site (Figure 16.16). Since the five aromatic residues constituting the "primary binding component" are highly conserved throughout the nAChR subunits, subtype-selectivity of orthosteric ligands most often arise from this "complementary binding component."

Galanthamine and physostigmine are believed to bind to an allosteric site in the amino-terminal domain of the α-subunit in the nAChR complex, thereby increasing the receptors affinity for the orthosteric agonist and/or the probability of ion channel opening. The potentiation and inhibition of nAChR signaling exerted by Ca^{2+} and the $A\beta_{42}$ peptide, respectively, also arise from binding to this domain (Figure 16.16). Conversely, the noncompetitive antagonist mecamylamine (**16.81**) binds to a site situated deep into the ion channel of the nAChR, where it blocks the influx of cations upon activation of the receptor. Finally, the modulation of nAChR function by steroids appears to originate from a binding site involving the small extracellular carboxy-termini of the nAChR subunits.

BIBLIOGRAPHY

Bertrand, D.; Gopalakrishnan, M. Allosteric modulation of nicotinic acetylcholine receptors. *Biochem. Pharmacol.* **2007**, 74(8), 1155–1163.

Clader, J. W.; Wang, Y. Muscarinic receptor agonists and antagonists in the treatment of Alzheimer's disease. *Curr. Pharm. Des.* **2005**, 11, 3353–3361.

Colletier, J. P.; Fournier, D.; Greenblatt, H. M.; Stojan, J.; Sussman, J. L.; Zaccai, G.; Silman, I.; Weik, M. Structural insights into substrate traffic and inhibition in acetylcholinesterase. *EMBO J.* **2006**, 25, 2746–2756.

Eglen, R. M.; Choppin, A.; Watson, N. Therapeutic opportunities from muscarinic receptor research. *Trends Pharmacol. Sci.* **2001**, 22, 409–414.

Hansen, S. B.; Sulzenbacher, G.; Huxford, T.; Marchot, P.; Taylor, P.; Bourne, Y. Structures of Aplysia AChBP complexes with nicotinic agonists and antagonists reveal distinctive binding interfaces and conformations. *EMBO J.* **2005**, 24, 3635–3646.

Jensen, A. A.; Frølund, B.; Liljefors, T.; Krogsgaard-Larsen, P. Neuronal nicotinic acetylcholine receptors: Structural revelations, target identifications and therapeutic inspirations. *J. Med. Chem.* **2005**, 48, 4705–4745.

Krajewski, J. L.; Dickerson, I. M.; Potter, L. T. Site-directed mutagenesis of m1-toxin1: Two amino acids responsible for stable toxin binding to M1 muscarinic receptors. *Mol. Pharmacol.* **2001**, 60, 725–731.

Lu, Z. L.; Saldanha, J. W.; Hulme, E. C. Seven-transmembrane receptors: Crystals clarify. *Trends Pharmacol. Sci.* **2002**, 23, 140–145.

Masters, C. L.; Cappai, R.; Barnham, K. J.; Villemagne, V. L. Molecular mechanisms for Alzheimer's disease: Implications for neuroimaging and therapeutics. *J. Neurochem.* **2006**, 97, 1700–1725.

Paterson, D.; Nordberg, A. Neuronal nicotinic receptors in the human brain. *Prog. Neurobiol.* **2000**, 61, 75–111.

Romanelli, M. N.; Gratteri, P.; Guandalini, L.; Martini, E.; Bonaccini, C.; Gualtieri, F. Central nicotinic receptors: Structure, function, ligands, and therapeutic potential. *Chem. Med. Chem.* **2007**, 2, 746–767.

Turner, R. S. Alzheimer's disease. *Semin. Neurol.* **2006**, 26, 499–506.

Wess, J.; Eglen, R. M.; Gautam, D. Muscarinic acetylcholine receptors: Mutant mice provide new insights for drug development. *Nat. Rev. Drug Disc.* **2007**, 6, 721–733.

17 Histamine Receptors

Iwan de Esch, Henk Timmerman, and Rob Leurs

CONTENTS

17.1 INTRODUCTION

Histamine is both an aminergic neurotransmitter and a local hormone and plays major roles in the regulation of several (patho) physiological processes. In biological systems histamine is synthesized from L-histidine by histidine-decarboxylase (HDC, Scheme 17.1). In the brain (where histamine acts as a neurotransmitter) the synthesis takes place in restricted populations of neurons that are located in the tuberomammillary nucleus of the posterior hypothalamus. These neurons project to most cerebral areas and have been implicated in several brain functions (e.g., sleep/wakefulness, hormonal secretion, cardiovascular control, thermoregulation, food intake, and memory formation). In peripheral tissues (where histamine acts as a local hormone), the compound is stored in mast cells, eosinophils, basophils, enterochromaffin cells, and probably also in some specific neurons. Once released, histamine is rapidly metabolized via N-methylation of the imidazole ring by the enzyme histamine *N*-methyltransferase (HMT) and by the oxidation of the amine function by diamine oxidase (DAO, Scheme 17.1).

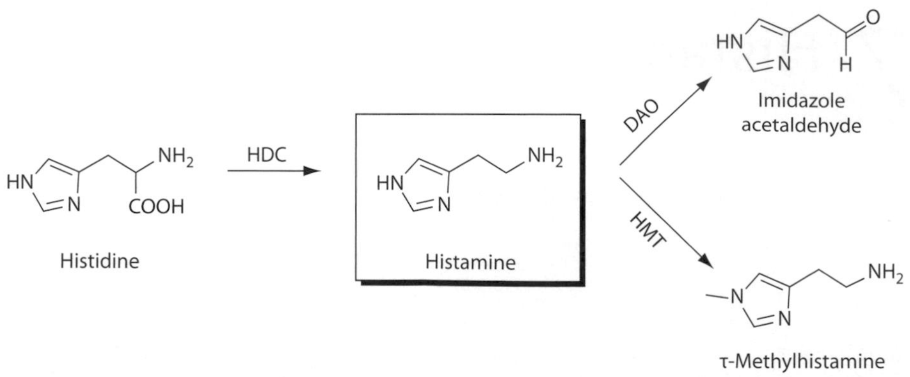

SCHEME 17.1 Synthesis and metabolism of histamine. HDC, histidine decarboxylase; DAO, diamine oxidase; HMT, histamine N-methyltransferase.

Some of the symptoms of allergic conditions in the skin and the airway system are known to result from histamine release after mast cell degranulation. In 1937, Bovet and Staub discovered the first compounds that antagonize these effects of histamine. From henceforth, there has been intense research devoted toward finding novel ligands with (anti)histaminergic activity.

As more (patho) physiological processes that are mediated by histamine (e.g., stomach acid secretion and neurotransmitter release) were studied, it became apparent that the action of histamine is mediated by several subtype receptors. This resulted in the identification of the histamine H_2 receptor in 1966, the histamine H_3 receptor in 1983, and the histamine H_4 receptor in 2000 (note the interval of 17 years). The first three histamine receptor subtypes were discovered by classical pharmacological means, i.e., using subtype selective ligands that were developed by medicinal chemists. The histamine H_4 receptor was discovered using the human genome, nicely illustrating the impact of molecular biology and genomics in drug discovery.

In this chapter we will describe in detail the state-of-the-art knowledge on the molecular features of the histamine receptor proteins, the medicinal chemistry of the four histamine receptors, and the (potential) therapeutic applications of selective receptor ligands.

17.2 THE HISTAMINE H_1 RECEPTOR: MOLECULAR ASPECTS AND SELECTIVE LIGANDS

17.2.1 MOLECULAR ASPECTS OF THE HISTAMINE H_1 RECEPTOR PROTEIN

The H_1 receptor belongs to the large family of rhodopsin-like, G-protein coupled receptors (GPCRs). In 1991, the cDNA encoding a bovine H_1 receptor protein was cloned after an expression cloning strategy in *Xenopus* oocytes by Fukui and coworkers. The human H_1 receptor gene resides on chromosome three and the deduced amino acid sequence revealed a 491 amino acid protein of 56 kDa. Using the cDNA sequence encoding the bovine H_1 receptor, the cDNA sequences and intronless genes encoding the rat, guinea pig, human, and mouse H_1 receptor proteins were cloned soon thereafter. These receptor proteins are slightly different in length, highly homologous, and do not show major pharmacological differences. The stimulation of the H_1 receptor leads to the phospholipase C-catalyzed formation of the second messengers inositol 1,4,5-triphosphate (IP_3) and 1,2-diacylglycerol (DAG), which in turn lead to the mobilization of intracellular calcium and the activation of protein kinase C, respectively.

17.2.2 H_1 RECEPTOR AGONISTS

The modification of the imidazole moiety of histamine has been the most successful approach for obtaining selective H_1 agonists (Figure 17.1). The presence of the tautomeric N^π–N^τ system of the

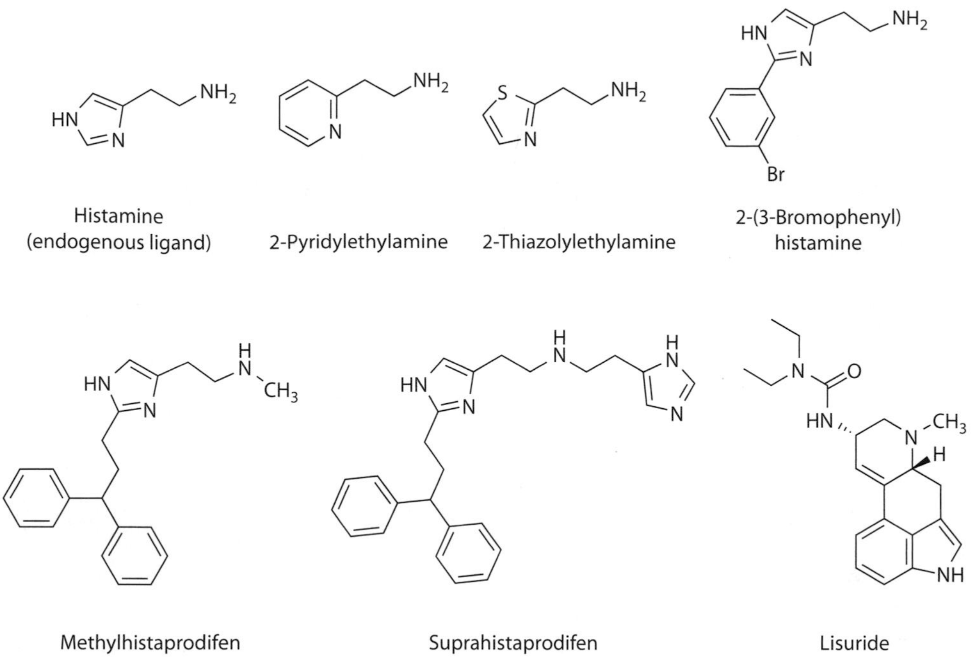

FIGURE 17.1 Histamine H_1 receptor agonists.

imidazole ring is not obligatory, as reflected by the selective H_1 agonists 2-pyridylethylamine and 2-thiazolylethylamine. Substitution of the imidazole ring at the 2-position will lead to relatively selective H_1 agonists. For example, 2-(3-bromophenyl)histamine is a relatively potent H_1 receptor agonist. Schunack and colleagues subsequently developed a series of so-called histaprodifens on the basis of the hypothesis that the introduction of a diphenylalkyl substituent on the 2-position of the imidazole ring yields high affinity agonists. This hypothesis was based on the realization that a diphenylmethyl group is a common feature of high-affinity H_1 antagonists (see Section 17.2.3). The introduction of the diphenylpropyl substituent at the 2-position of the imidazole ring and N-methylation of the ethylamine side chain results in the high potency agonist N-methyl-histaprodifen. Further modifications of the diphenylmethyl moiety were unsuccessful and indicated a considerable difference in structure-activity relationship (SAR) (and most likely receptor-binding site) of the diphenyl moieties of the histaprodifens and the structurally related H_1 antagonists. A further increase in H_1 receptor agonist potency was obtained by a bivalent ligand approach. Suprahistaprodifen, a dimer of histaprodifen and histamine is currently one of the most potent H_1 receptor agonists available. Surprisingly, recent high-throughput screening (HTS) of CNS-active drugs at the histamine H_1 receptor has identified the nonimidazole ergot derivative lisuride as another high affinity H_1 receptor agonist.

17.2.3 H_1 Receptor Antagonists

The first antihistamines were identified and optimized by exclusively studying *in vivo* activities. This might be the explanation why several compounds originally reported as antihistamines were later on developed for other applications; e.g., the first so-called tricyclic antidepressants (e.g., doxepin) are often also very potent H_1 antagonists. More modern approaches, using genetically modified cells expressing the human H_1 receptor, currently provide more in-depth information on the molecular mechanism of actions. All therapeutically used H_1 antagonists, in fact, act as inverse agonists (see Chapter 12) and favor an inactive conformation of the GPCR protein. In view of the detectable level

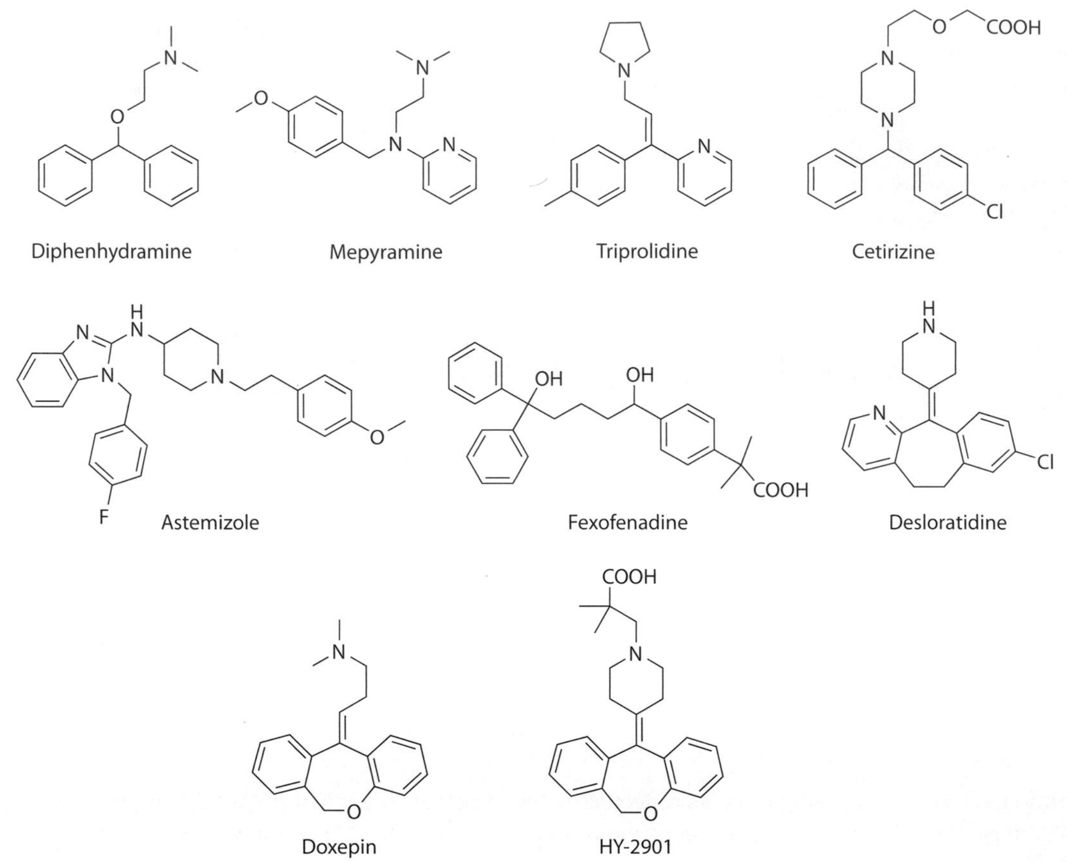

Diphenhydramine   **Mepyramine**   **Triprolidine**   **Cetirizine**

Astemizole   **Fexofenadine**   **Desloratidine**

Doxepin   **HY-2901**

FIGURE 17.2 Histamine H_1 receptor antagonists (inverse agonists).

of spontaneous activity of the H_1 receptor (i.e., receptor signaling without agonist, also known as constitutive GPCR activity), all the H_1 antagonists tested so far inhibit the constitutive activation of e.g., nuclear factor-κB (NF-κB).

Of the many first generation histamine blockers, diphenhydramine (Figure 17.2) is considered as the archetype. The compound is known in medicine as Benadryl®, the first antihistamine successfully used in man. Other compounds of this class are, e.g., mepyramine and triprolidine. These compounds are highly potent H_1 antagonists and very useful both for pharmacological investigations and medicinal use. The so-called classical "antihistamines" easily penetrate the brain and are therefore also useful for *in vivo* CNS studies.

The early antihistamines had two major drawbacks: they were ligands for several targets; especially the antimuscarinic effects caused unpleasant side effects (e.g., dry mouth). By careful structural modifications, it is possible to obtain selective antihistamines. Another drawback of the first generation H_1 antagonists was that the compounds show strong sedating effects, to such a level that some of them are still used as sleeping aids.

The notion that sedation is caused by a blockade of H_1 receptors in the brain, sparked the search for nonbrain penetrating compounds. Minor structural modifications resulted in a number of new, nonsedating H_1 antagonists (e.g., cetirizine, astemizole, fexofenadine, and desloratidine), also referred to as the second-generation H_1 blockers. Interestingly, the first of such compounds were more or less found by chance and it took quite some time to understand why the compounds did not manifest CNS effects. It is now understood that these compounds act as substrates of the P-glycoprotein (PgP) transport system in the blood–brain barrier. Consequently, these compounds are actively transported out of the brain and thereby are not able to occupy significant amounts of brain H_1 receptors. The new class of compounds, including terfenadine (later on due to HERG-blockade replaced by its active metabolite fexofenadine), cetirizine (now replaced by the L-enantiomer), and loratidine (now replaced by its desoxy-active metabolite, desloratidine) reached as antiallergics the blockbuster status.

Although effective in treating allergic reactions, the second-generation H_1 antagonists do not display significant antiinflammatory effects. Currently, research focuses on compounds also targeting inflammation; compounds having besides H_1 blocking properties antagonizing also, e.g., LTB_4 or blocking the synthesis of leukotrienes, have the interest of pharmaceutical companies. Recently, the combined blockade of H_1 and H_4 receptors (see below) has also been indicated as an attractive new approach. Interestingly, since the turn of the century the interest in the sleep promoting effects of histamine H_1 receptor antagonists has increased. Especially, the "old" derivative doxepin, a compound that blocks the H_1 receptor and also the H_2 receptor is used as a sleep inducer. An analog of doxepin, HY-2901, has been shown to have interesting properties for use as a sleep inducer and is currently under clinical evaluation as sleep aid.

17.2.4 Therapeutic Use of H_1 Receptor Ligands

The histamine H_1 receptor is a well-established drug target and has been thoroughly studied for decades. The first- but especially the second-generation antihistamines are clinically very successful and are widely available drugs. The main indications are hay fever, allergic rhinitis, and conjunctivitis as well as comparable allergic affections; the application for asthmatic conditions does not seem to be of much use. The first generation antihistamines are still used as in over the counter (OTC) sleep aids or antiflu combination pills. As indicated before, currently interest in sleep aids is increasing and new molecules are being developed (e.g., HY-2901).

17.3 THE HISTAMINE H_2 RECEPTOR: MOLECULAR ASPECTS AND SELECTIVE LIGANDS

17.3.1 Molecular Aspects of the Histamine H_2 Receptor Protein

The fact that the "antihistamines" did not antagonize histamine-induced effects at the stomach and the heart, led in 1966 to the proposal by Ash and Schild of two distinct histamine receptors: the H_1 and H_2 receptors. This hypothesis became generally accepted in 1972 when Black and his coworkers at Smith, Kline & Beecham presented burimamide and related compounds. These ligands antagonize the effects of histamine on the stomach and the heart. Nowadays, the H_2 receptor is (as all histamine receptor subtypes) known to belong to the rhodopsin-like family of GPCRs. Using a polymerase chain reaction (PCR)-based method, based on the known sequence similarity of various GPCRs and gastric parietal mRNA, the H_2 receptor nucleotide sequence was elucidated. This DNA sequence encodes for a 359 amino acid GPCR receptor protein. Soon thereafter, the intronless genes encoding the rat, human, guinea pig, and mouse H_2 receptor were cloned by means of homology. The H_2 receptor proteins are slightly different in length, highly homologous, and do not show major pharmacological differences. Interestingly, several polymorphisms have been found in the human H_2 receptor gene and one of the mutations might be linked to schizophrenia.

The histamine H_2 receptor is positively coupled to the adenylate cyclase system via G_s proteins in a variety of tissues (e.g., brain, stomach, heart, gastric mucosa, and lungs). Moreover, cell lines recombinantly expressing the H_2 receptor show increases in cAMP following H_2 receptor activation.

17.3.2 H_2 Receptor Agonists

A simple modification of the histamine molecule has not been very successful to obtain selective and potent H_2 receptor agonists. A first step toward an H_2 receptor agonist was made with the

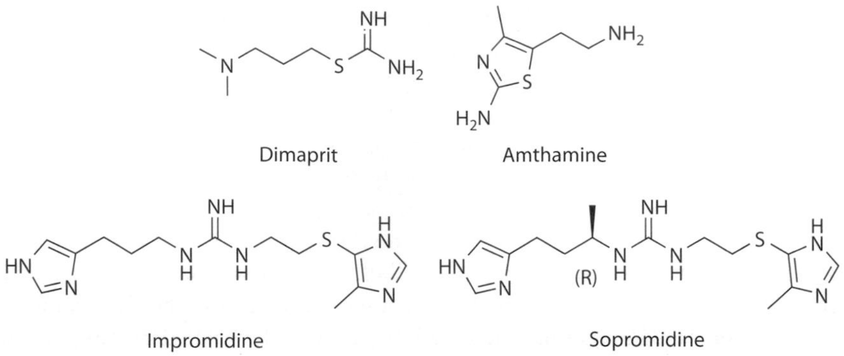

FIGURE 17.3 Reference histamine H_2 receptor agonists.

discovery of dimaprit (Figure 17.3), which was found during a search for H_2 receptor antagonists in a series of isothiourea derivatives. Dimaprit is an H_2 receptor agonist that is almost as
active as histamine at the H_2 receptor, but hardly displays any H_1 receptor agonism. Later it
was found that dimaprit is also a moderate H_3 receptor antagonist and a moderate H_4 receptor
agonist. Using dimaprit as a template, amthamine (2-amino-5-(2-aminoethyl)-4-methylthiazole)
was designed as a rigid dimaprit analog. Following the original suggestion of Green et al. that
the sulfur atom of dimaprit might act as a proton acceptor in a hydrogen bonding network with
the H_2 receptor (in analogy to the idea of the interaction of the imidazole ring with the receptor
protein), quantum chemical calculations by and synthesis of 2-aminothiazole analogs confirmed
this idea. Amthamine combines a high H_2 receptor selectivity with a potency, which is slightly
higher compared to histamine, both *in vitro* and *in vivo*. An H_2 receptor agonist that is more
potent than histamine is the guanidine derivative impromidine. This ligand actually combines
a rather high H_2 receptor affinity with a reduced efficacy. Impromidine also shows moderate H_1- and
potent H_3-receptor antagonistic and H_4-receptor agonistic activity. Interestingly, replacement
of the propyl-imidazole moiety of impromidine with an α-methyl-ethylimidazole group results
in the chiral analog (Figure 17.3). The $R(-)$-isomer, sopromidine, acts as a potent H_2 agonist,
whereas the $S(+)$-isomer is a weak H_2 antagonist. Both compounds posses only weak H_3 and H_4
antagonistic activities, making $R(-)$-sopromodine one of the most potent and selective H_2 agonist
to date.

17.3.3 H_2 Receptor Antagonists

The identification of N^α-guanylhistamine as a partial H_2 agonist in a gastric acid secretion model
led to the development of the relatively weak H_2 antagonist burimamide (Scheme 17.2) following
the replacement of the strong basic guanidine group by the noncharged, polar thiourea, and side
chain elongation. Years later, it was shown that burimamide is also an H_3 and H_4 receptor partial
agonist. As H_2 receptor antagonist, burimamide lacked oral activity in man most likely due to its
moderate potency. Nevertheless, burimamide was the lead for the development of selective and
clinically useful H_2 receptor antagonists, such as cimetidine. Over time, many H_2 antagonists have
been described; almost all of them possess two planar π-electron systems connected by a flexible chain. The 4-methylimidazole moiety of cimetidine can easily be replaced by other heterocyclic groups. Replacement by a substituted furan (e.g., ranitidine) or thiazole ring (e.g., famotidine)
leads to compounds that are usually more potent at the H_2 receptor than cimetidine. Moreover, the
replacement of the imidazole moiety also eliminates the undesired inhibition of cytochrome P-450.
Most H_2 antagonists are rather polar compounds, which do not readily cross the blood–brain barrier.

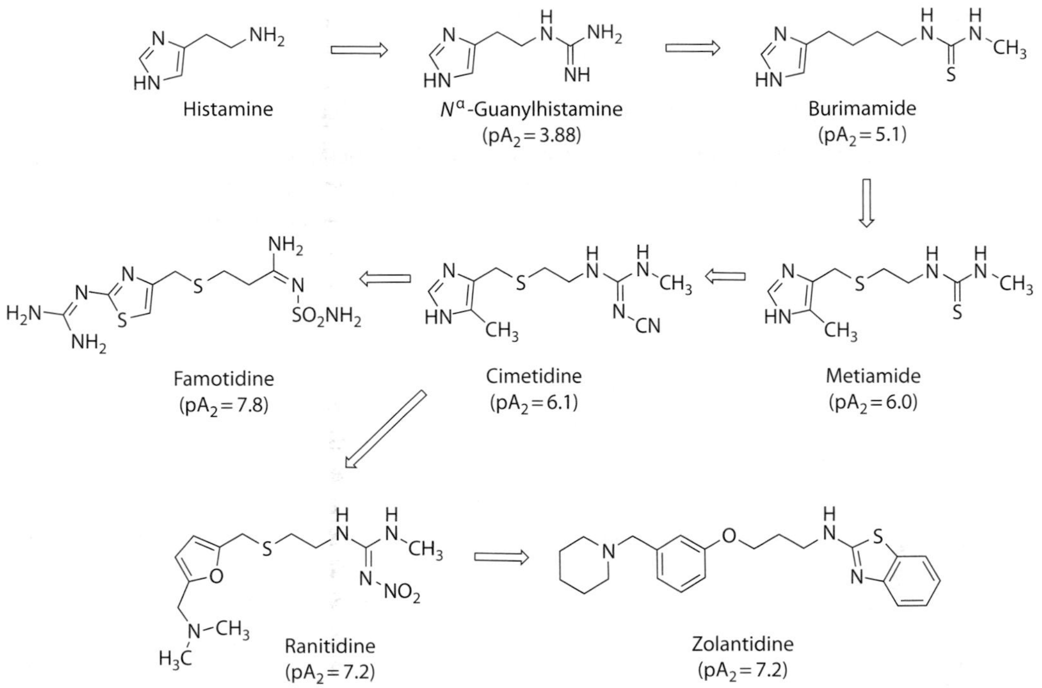

SCHEME 17.2 Stepwise structural modifications leading to the development of histamine H_2 receptor drugs to treat gastric ulcers.

The brain-penetrating H_2 antagonist zolantidine represents a rather nonclassical structure with the oxygen atom of the furan ring in ranitidine placed outside the aromatic ring and replacement of the polar group with a benzothiazole group (Scheme 17.2) and has become an important tool for *in vivo* CNS studies.

Like the H_1 receptor, the H_2 receptor was reported to be spontaneously active in transfected CHO cells. Based on this concept, many H_2 antagonists were reclassified: cimetidine, ranitidine, and famotidine are in fact inverse agonists, whereas burimamide acts in this model system as a neutral antagonist. This difference in pharmacological profile was also seen in a differential effect on H_2 receptor regulation; whereas long-term treatment with inverse agonists, like cimetidine, resulted in H_2 receptor upregulation, exposure to the neutral antagonist burimamide did not affect the receptor expression. Such receptor upregulation was also observed in rabbit parietal cells, resulting in acid hypersecretion after H_2 antagonist withdrawal. These data present a mechanistic explanation for the known tolerance induction by H_2 antagonist that sometimes occurs in man.

17.3.4 Therapeutic Use of H_2 Receptor Ligands

At the moment there is no clinical application of H_2 agonists, although sometimes histamine is used as a diagnostic aid in patients with stomach problems. In contrast, H_2 antagonists have proven to be very effective drugs to alleviate the symptoms of duodenal ulcers, stomach ulcers, and reflux oesophagitits. Nowadays, the blockbuster status of the H_2 antagonists has been strongly reduced with the introduction of the proton pump inhibitors, like omeprazole, to directly inhibit the gastric acid secretion and the eradication of *Helicobacter pylori* with antibiotics as actual cure, instead of symptomatic treatment.

17.4 THE HISTAMINE H₃ RECEPTOR: MOLECULAR ASPECTS AND SELECTIVE LIGANDS

17.4.1 MOLECULAR ASPECTS OF THE HISTAMINE H₃ RECEPTOR PROTEIN

The physiological role of histamine as a neurotransmitter became apparent in 1983, when Arrang and coworkers discovered the inhibitory effect of histamine on its own release and synthesis in the brain. This effect was not mediated by the known H₁ and H₂ receptor subtypes as no correlation with either the H₁ or the H₂ receptor activity of known histaminergic ligands was observed. Soon thereafter, the H₃ receptor agonist R-α-methylhistamine and the antagonist thioperamide (see Figures 17.4 and 17.5, respectively) were developed, thereby confirming that a new receptor

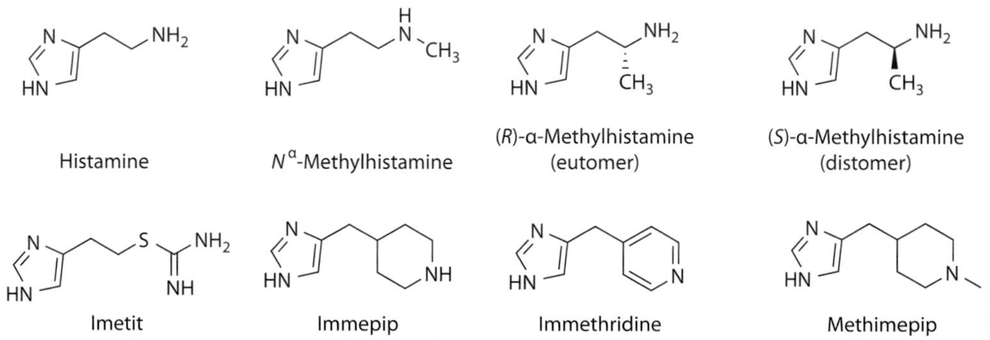

FIGURE 17.4 Reference histamine H₃ receptor agonists.

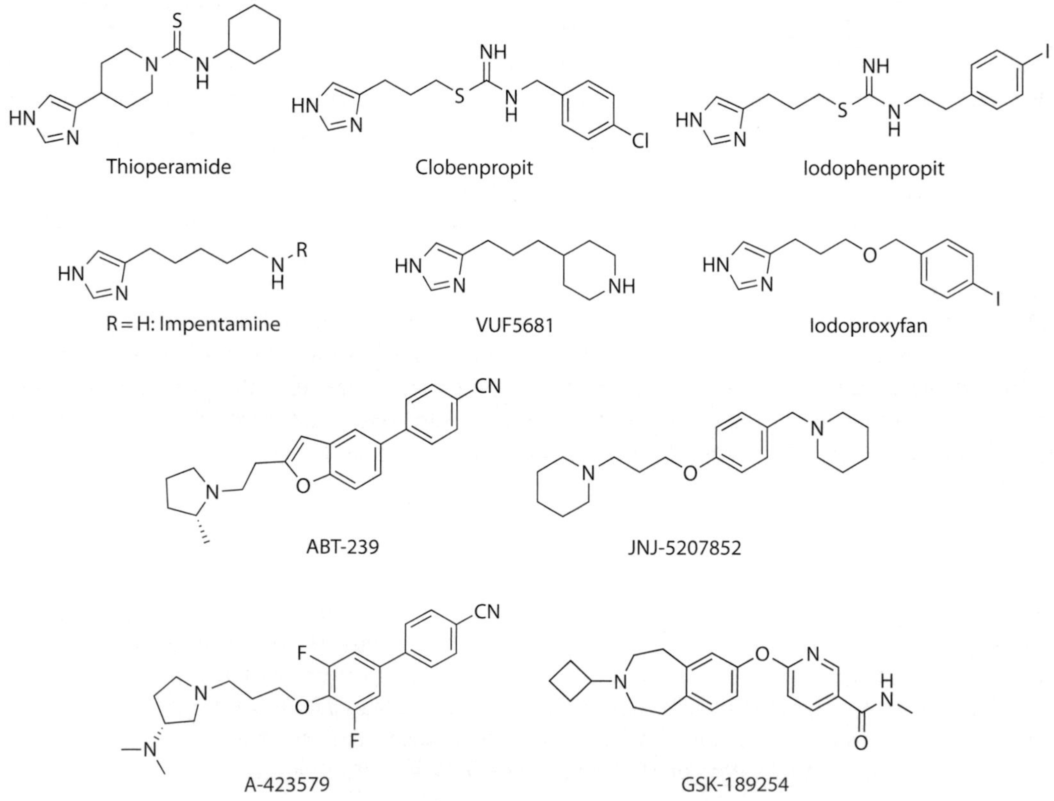

FIGURE 17.5 Histamine H₃ receptor antagonists and inverse agonists.

subtype regulates the release and synthesis of histamine. In addition, the H_3 receptor regulates the release of other important neurotransmitters, such as acetylcholine, serotonin, noradrenalin, and dopamine. Next to its high expression in certain regions of the human brain (for example, the basal ganglia, hippocampus, and cortical areas, i.e., the parts of the brain that are associated with cognition) the H_3 receptor is also present to some extent in the peripheral nervous system, e.g., in the gastrointestinal tract, the airways, and the cardiovascular system.

Initial efforts to identify the H_3 receptor gene, using the anticipated homology with the previously identified H_1 and H_2 receptor genes ended in vain. Eventually, the human H_3 receptor cDNA was identified by Lovenberg and his coworkers at Johnson & Johnson in 1999. In search for novel GPCR proteins using a homology search of commercial genome databases, a receptor with high similarity to the M_2 muscarinic acetylcholine receptor and high brain expression was identified. Expression of the gene and full pharmacological characterization established this protein as the histamine H_3 receptor. Cloning of the H_3 receptor genes of other species, including rat, guinea pig, and mouse, soon followed, and important H_3 receptor species differences have been identified. The H_3 receptor mRNA undergoes extensive alternative splicing, resulting in many H_3 receptor isoforms that have different signaling properties and expression profiles. Moreover, the H_3 receptor displays particularly high constitutive activity, which can also be observed *in vivo*, again leading to a reclassification of existing ligands into agonists, neutral antagonists, and inverse agonists.

The H_3 receptor signals via $G_{i/o}$ proteins as shown by the pertussis toxin sensitive stimulation of $[^{35}S]$-GTPγS binding in rat cortical membranes. The inhibition of adenylyl cyclase after stimulation of the H_3 receptor results in lowering of cellular cAMP levels and modulation of cAMP responsive element-binding protein (CREB) dependent gene transcription.

17.4.2 Histamine H_3 Receptor Agonists

At the H_3 receptor, histamine itself is an highly active agonist. Methylation of the amino function results in N^{α}-methylhistamine (Figure 17.4), a compound that is H_3 selective and even more active than histamine. Methylation of the α-carbon atom of the ethylamine side chain also increases the potency at the H_3 receptor. This increased activity resides completely in the *R*-isomer; the corresponding *S*-isomer is approximately 100-fold less potent. Since the methylation leads to highly reduced activity at both the H_1 and H_2 receptor, but still substantial activity at the H_4 receptor, *R*-(α)-methylhistamine is a moderately selective agonist at the H_3 receptor. In combination with its less active *S*-isomer, this compound has proven to be highly useful for the pharmacological characterization of H_3 receptor-mediated effects. For potent H_3 agonism, the amine function of histamine can be replaced by an isothiourea group, as in imetit. This compound is also very active *in vitro* and *in vivo*, as is *R*-(α)-methylhistamine. The basic group in the imidazole side chain can also be incorporated in ring structures. For example, immepip is a potent H_3 agonist that is effective *in vitro* and *in vivo*. Although the described first generation H_3 agonists were intensively used as reference ligands to study the H_3 receptor, all of them proved to have considerable activity for the recently discovered H_4 receptor. Therefore, a new generation of potent and selective H_3 agonists has been developed, most notably immethridine ($pEC_{50} = 9.8$; 300-fold selectivity over the H_4 receptor) and methimepip ($pEC_{50} = 9.5$; >10,000-fold selectivity over the H_4 receptor). These latter compounds are devoid of high H_4 receptor activity.

17.4.3 Histamine H_3 Receptor Antagonists

As with the first generation H_3 agonists, the first generation H_3 antagonists (all of them possessing an imidazole heterocycle) have considerable affinity for the more recently discovered histamine H_4 receptor. The first potent H_3 receptor antagonist (later reclassified as an inverse agonist) that was devoid of H_1 receptor and H_2 receptor activity was thioperamide (Figure 17.5). This compound has been used in many H_3 receptor studies as reference ligand and is active *in vitro* and *in vivo*

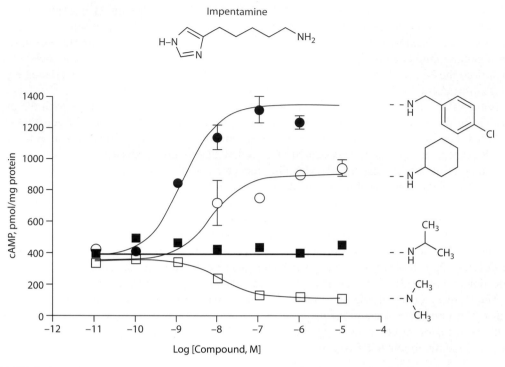

FIGURE 17.6 Alkylation of the primary amine function of impentamine leads to ligands that cover the complete spectrum of functional activity, i.e., agonism, neutral antagonism, and inverse agonism.

(the compound is able to penetrate the CNS). However, thioperamide displays some 5-HT$_3$ receptor antagonism and also is an inverse agonist at the H$_4$ receptor. Moreover, a remarkable H$_3$ receptor species differences can be demonstrated with thioperamide, as the compound has a 10-fold higher affinity for the rat H$_3$ receptor than for the human H$_3$ receptor. Based on the H$_3$ receptor agonist imetit, the highly potent H$_3$ inverse agonist clobenpropit was developed (pA$_2$ = 9.9). This compound also has some 5-HT$_3$ receptor activity and displays partial agonist activity at H$_4$ receptors. Impentamine is a potent histamine H$_3$ receptor partial agonist in SK-N-MC cells expressing human H$_3$ receptors. It has also been shown that small structural modifications of impentamine, i.e., alkylation of the primary amine moiety of impentamine with, e.g., methyl-, isopropyl-, and p-chlorobenzyl-groups results in ligands that cover the complete spectrum of functional activity, i.e., agonism, neutral antagonism, and inverse agonism (Figure 17.6). The compound VUF5681 (Figure 17.5) was reported as a neutral H$_3$ antagonist, not affecting the basal signaling of the histamine H$_3$ receptor. As such, it has proven to be a useful molecular tool in H$_3$ receptor studies, for example, when studying H$_3$ constitutive activity in the rat brain.

The imidazole-containing compounds have been very important in characterizing the H$_3$ receptor. Furthermore, similarity studies resulted in pharmacophore models (Figure 17.7) that explain the SAR of the different classes of imidazole-containing ligands and indirectly describe the ligand-binding site of the receptor.

Imidazole-containing ligands are associated with inhibition of cytochrome P-450 enzymes. Via this mechanism, the clearance of coadministered drugs can be compromised, leading to severe drug–drug interactions and extrapyramidal symptoms. Classic medicinal chemistry work, elegantly conducted by the team of Ganellin (already involved in the development of the H$_2$ antagonist cimetidine, *vide supra*) at University College London led to a first breakthrough in the search of nonimidazole H$_3$ antagonists, as illustrated in Scheme 17.3. The endogenous agonist histamine was once again taken as a lead structure. Attachment of a lipophilic group to the amine moiety led to

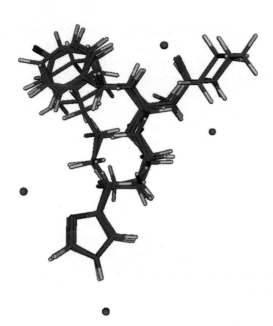

FIGURE 17.7 Pharmacophore model for imidazole-containing H_3 antagonists. Superposed are 10 different compounds. Carbon atoms in green, nitrogen atoms in blue, sulfur atoms in yellow, and hydrogen atoms in white. All imidazole rings are perfectly superposed. The aromatic heterocycle and the basic groups in the imidazole side chain can interact with a total of four predicted H-bonding groups of the receptor site (yellow sphere indicating a H-bonding donor atom of the site and purple indicating H-bonding acceptor atoms of the site. The lipophilic groups at the terminus of the side chain of the ligands are located in two distinct positions, suggesting that the H_3 receptor has two lipophilic pockets for ligand binding. These findings were later validated by several groups using receptor homology modeling.

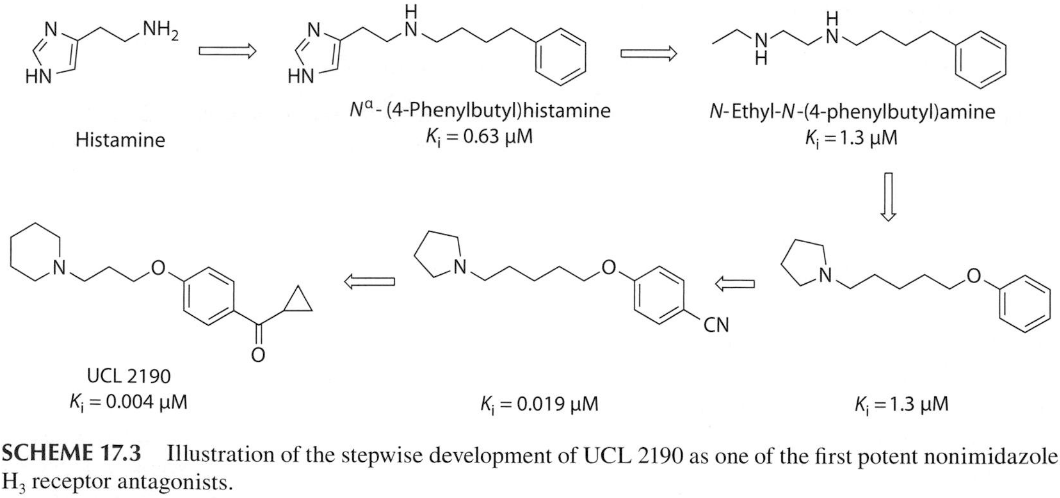

Histamine

N^α-(4-Phenylbutyl)histamine
$K_i = 0.63\ \mu M$

N-Ethyl-N-(4-phenylbutyl)amine
$K_i = 1.3\ \mu M$

UCL 2190
$K_i = 0.004\ \mu M$

$K_i = 0.019\ \mu M$

$K_i = 1.3\ \mu M$

SCHEME 17.3 Illustration of the stepwise development of UCL 2190 as one of the first potent nonimidazole H_3 receptor antagonists.

N^α-(4-phenylbutyl)histamine, a compound with H_3 antagonist activity. Replacement of the imidazole heterocycle, initially deemed essential for H_3 affinity led to N-ethyl-N-(4-phenylbutyl) amine with merely a twofold drop in affinity. Subsequent stepwise optimization of the structure for H_3 affinity, by systemically modifying the basic group, the linker and the aromatic moiety ultimately led to UCL 2190, a potent nonimidazole H_3 antagonist. Structural features of this compound, e.g., the amino-proxyphenyl substructure, reoccur in most H_3 medicinal chemistry programs that have been reported since.

Especially since the cloning of the H_3 receptor gene in 1999, the pharmaceutical industry has been actively exploring the potential of H_3 receptor ligands and many new antagonists/inverse agonists have been described. Typical examples are the inverse agonist ABT-239 and the neutral antagonist JNJ-5207852 (Figure 17.5). Interestingly, this latter compound is active in several models for cognition, but does not act as an appetite suppressant and has no effect on food intake. Other compounds, such as Abbott's A-423579, have good efficacy in obesity models, but lack clear procognitive effects. At present the differences in efficacy for distinct clinical applications of the different classes of H_3 ligands is not understood (e.g., involvement of different H_3 receptor isoforms) and subject of intense research.

17.4.4 THERAPEUTIC USE OF HISTAMINE H_3 RECEPTOR LIGANDS

Multiple lines of evidence indicate that the H_3 receptor is involved in numerous physiological processes and that this receptor bears potential as a promising drug target. A handful of applications for H_3 agonists has emerged from preclinical studies in the areas of migraine (modulating release of neurogenic peptides) and ischemic arrhythmias (modulating noradrenaline release). In migraine, the H_3 agonistic properties of N^α-methylhistamine have been reported to be beneficial in a Phase II trial. Intriguingly, both H_3 agonists and H_3 inverse agonist are claimed to have a beneficial activity when studied in preclinical obesity models. The full spectrum of diseases, where H_3 receptor mediated treatment might be applicable, is striking. H_3 antagonists and inverse agonists have been successfully used in animal models for narcolepsy, cognitive disorders, neuropathic pain, and others. In this respect, GSK-189254 is a remarkable H_3 ligand as it is in trials for three different diseases: neuropathic pain, narcolepsy, and dementia. It has proven a challenging task to predict in what specific preclinical model(s) a given structural series of H_3 antagonist or inverse agonist will be useful. Moreover, a full clinical validation of the promising role of H_3 receptor antagonists is still awaited.

17.5 THE HISTAMINE H_4 RECEPTOR: MOLECULAR ASPECTS AND SELECTIVE LIGANDS

17.5.1 MOLECULAR ASPECTS OF THE HISTAMINE H_4 RECEPTOR PROTEIN

Immediately following the cloning of the H_3 receptor gene, several groups identified the homologous H_4 receptor sequence in the human genome databases. Indeed, the H_4 receptor has high sequence identity with the H_3 receptor (31% at the protein level, 54% in the transmembrane domains). The H_3 and H_4 receptors are also similar in gene structure. The human H_4 receptor gene is present on chromosome 18q11.2 and the gene contains three exons that are interrupted by two large introns (like the H_3 receptor gene). To date, two H_4 receptor isoforms have been identified, but no functional role has been reported so far. Cloning of the genes that encode the mouse, rat, guinea pig, and pig H_4 receptors reveal only limited sequence homology with the human H_4 receptor. The H_4 receptor is mainly expressed in bone marrow and peripheral leukocytes and mRNA of the human H_4 receptor is detected in, e.g., mast cells, dentritic cells, spleen, and eosinophils. The H_4 receptor has a pronounced effect on the chemotaxis of several cell types that are associated with immune and inflammatory responses.

The H_4 receptor couples to $G_{i/o}$ proteins, thereby leading to a decrease in cAMP production and the regulation of CREB gene transcription. Furthermore, H_4 receptor stimulation affects the $G_{i/o}$ protein mediated activation of mitogen-activitated protein (MAP) kinase. Studying the increased $[^{35}S]GTP\gamma S$ levels in H_4 transfected cells, it has been shown that also the H_4 receptor is constitutively active.

17.5.2 HISTAMINE H_4 RECEPTOR AGONISTS

Most of the first generation imidazole-containing H_3 ligands have reasonable affinity for the H_4 receptor as well. The first imidazole-containing ligand that was reported to have some selectivity for the

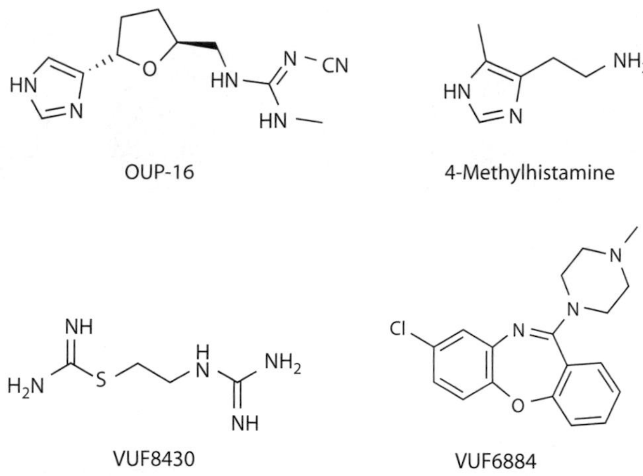

FIGURE 17.8 Histamine H_4 receptor agonists.

H_4 (40-fold) over the H_3 receptor is OUP-16 (Figure 17.8). This compound acts as a full H_4 agonist. More recently, the potent H_4 agonist 4-methylhistamine was discovered after the screening of a large number of histaminergic compounds. This compound was originally developed for an H_2 research program, but appears to be more than 100-fold more potent on the H_4 receptor than on any other histamine receptor subtype, including the H_2 receptor. VUF8430 was also reported as a potent H_4 agonist (pEC$_{50}$ = 7.3) with a complimentary selectivity profile, being 33-fold selective over the H_3 receptor. Again, VUF8430 was originally developed as a dimaprit analog in an H_2 research program. VUF6884 was developed as a clozapine analog with optimized histamine H_4 receptor affinity. This rigid compound is particularly useful for pharmacophore modeling studies (see below). Interestingly, VUF6884 is a full agonist on histamine H_4 receptors and an even more potent histamine H_1 receptor inverse agonist. Clozapine derivatives are well known as promiscuous GPCR ligands.

17.5.3 HISTAMINE H_4 RECEPTOR ANTAGONISTS

Potent and selective H_4 receptor antagonists are also emerging. For this histamine receptor subtype, the first nonimidazole ligands were found by successful HTS campaigns. The first reported neutral antagonist was derived from an indole-containing hit structure that was efficiently converted into JNJ7777120 (Scheme 17.4). This compound can currently be considered as an H_4 receptor reference ligand. Unfortunately, the compound has a poor stability in human and rat liver microsomes and

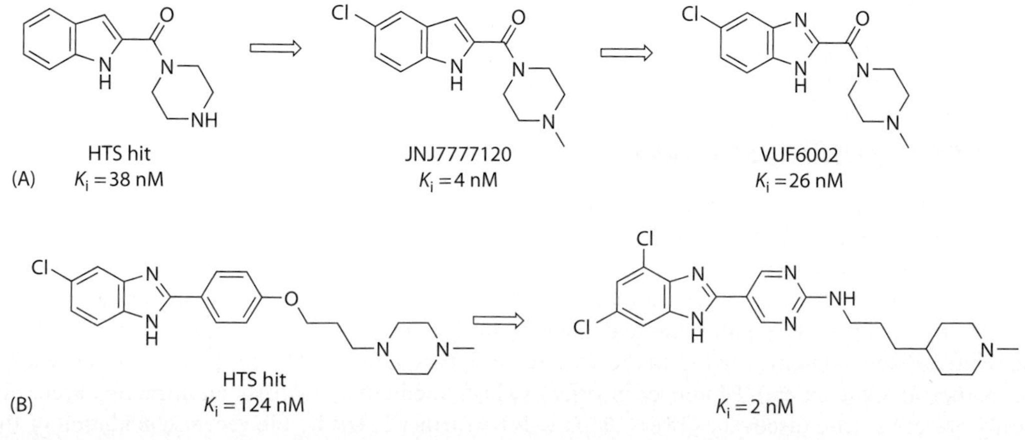

SCHEME 17.4 Illustration of two histamine H_4 receptor HTS hits (A and B) and subsequent hit optimization.

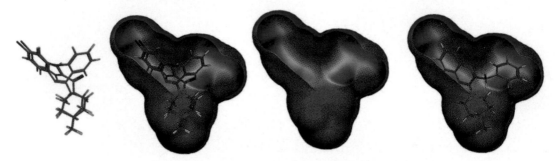

FIGURE 17.9 Pharmacophore modeling leading to the design of new ligands. Two reference histamine H_4 ligands (VUF6884 and JNJ7777120) were used to construct a pharmacophore model. This model indirectly describes the histamine H_4 receptor-binding site. Based on this model, novel and potent ligands that fit the binding pocket could be designed, e.g., VUF10148). Carbon atoms in green, nitrogen atoms in blue, oxygen atoms in red, and hydrogen atoms in white. Color coding surface: hydrogen-bonding region in purple, hydrophobic regions in yellow, and mild polar regions in blue.

a half-life of only 2 h in rats. The subsequently developed benzimidazole derivative JNJ10191584 is also a neutral H_4 antagonist. This compound is orally active *in vivo* and has improved liver microsomes stability but still a limited half-life. Also derived from a HTS hit, a series of 2-arylbenzimidazoles have been described as ligands with low nanomolar affinity for the H_4 receptor. In addition, rational approaches like pharmacophore modeling and subsequent ligand design (Figure 17.9) are being used to develop novel H_4 receptor ligands. Considering the number of H_4 receptor related patent applications that have recently been disclosed, it can be anticipated that many new H_4 ligands will be described in scientific literature in the near future.

17.5.4 THERAPEUTIC USE OF HISTAMINE H_4 RECEPTOR LIGANDS

The presence of the H_4 receptor on immunocompetent cells and cells of hematopoietic lineage suggests that this new histamine receptor subtype plays an important role in the immune system. This hypothesis is supported by the fact that IL-10 and IL-13 modulate H_4 receptor expression and that binding sites for cytokine-regulated transcription factors, like interferon-stimulated response element (ISRE), interferon regulatory factor-1 (IRF-1), NF-κB, and nuclear factor-IL6 (NF-L6), are present upstream of the H_4 gene. Considering the physiological role of the H_4 receptor, several applications are currently under preclinical investigation, including allergy and asthma, chronic inflammations such as inflammatory bowel disease (IBD) and rheumatoid arthritis. The H_4 receptor is also being associated with pruritus (itch) and is involved in the progression of colon cancer. At the moment, therapeutic applications are clearly anticipated for H_4 receptor antagonists (inverse agonists). In view of the strong interests of pharmaceutical industries more evidence for a therapeutic role of H_4 receptor ligands is soon to be expected.

17.6 CONCLUDING REMARKS

The medicinal chemistry of histamine receptors has so far been a very rewarding arena. Major blockbuster drugs have been developed on the basis of H_1 and H_2 receptor targeting. Expectations for ligands targeting the two latest additions to the histamine receptor family are currently also very high. Interestingly, for each of these receptor subtypes highly selective agonists and antagonists have been developed. The wide chemical diversity of the various selective receptor ligands reflects the relatively low homology between the various receptors (only the H_3 and H_4 receptors resemble each other to some extent). Moreover, it offers today's medicinal chemists an attractive arena for highly effective drug discovery efforts. This will be further aided by the recent elucidation of the x-ray structure of the beta$_2$ receptor, hopefully allowing future structure-based drug design.

FURTHER READINGS

Hancock, A.A. 2006. The challenge of drug discovery of a GPCR target: Analysis of preclinical pharmacology of histamine H$_3$ antagonists/inverse agonists. *Biochem. Pharmacol.* 71:1103–1113.

Hill, S.J., Ganellin, C.R., Timmerman, H., Schwartz, J.C., Shankley, N.P., Young, J.M., Schunack, W., Levi, R., and Haas, H.L. 1997. International Union of Pharmacology. XIII. Classification of histamine receptors. *Pharmacol. Rev.* 49:253–278.

Leurs, R., Bakker, R.A., Timmerman, H., and de Esch, I.J. 2005. The histamine H$_3$ receptor: From gene cloning to H$_3$ receptor drugs. *Nat. Rev. Drug Discovery* 4:107–120.

Thurmond, R.L., Gelfand, E.W., and Dunford, P.J. 2008. The role of histamine H$_1$ and H$_4$ receptors in allergic inflammation: The search for new antihistamines. *Nat. Rev. Drug Discovery* 7:41–53.

Zhang, M.Q., Leurs, R., and Timmerman, H. 1997. Histamine H$_1$-receptor antagonists. In M.E. Wolff (ed.), *Burger's Medicinal Chemistry and Drug Discovery*, 5th edn. New York: John Wiley & Sons, Inc., p. 495.

18 Dopamine and Serotonin

Benny Bang-Andersen and Klaus P. Bøgesø

CONTENTS

18.1 INTRODUCTION

Dopamine (DA), serotonin (5-hydroxytryptamine, 5-HT), and norepinephrine (NE) are important neurotransmitters in the human brain. These neurotransmitters activate postsynaptic and presynaptic receptors, and their concentration is regulated by active reuptake into presynaptic terminals by transporters.

DA and 5-HT receptors are found in multiple subtypes that are divided into subclasses based on structural and pharmacological similarities. The DA and 5-HT receptors are all putative seven transmembrane (TM) G protein-coupled receptors (GPCRs) except for the 5-HT$_3$ receptor, which is a ligand-gated ion channel regulating the permeability of sodium and potassium ions. Five subtypes of DA receptors are known and grouped into the D$_1$-like receptors (D$_1$ and D$_5$) and the D$_2$-like receptors (D$_2$, D$_3$, and D$_4$), whereas 14 subtypes of 5-HT receptors are known and grouped into seven subclasses, namely 5-HT$_1$ (5-HT$_{1A}$, 5-HT$_{1B}$, 5-HT$_{1D}$, 5-HT$_{1E}$, and 5-HT$_{1F}$), 5-HT$_2$ (5-HT$_{2A}$, 5-HT$_{2B}$, and 5-HT$_{2C}$), 5-HT$_3$, 5-HT$_4$, 5-HT$_5$ (5-HT$_{5A}$ and 5-HT$_{5B}$), 5-HT$_6$, and 5-HT$_7$. In addition, a variety of polymorphic and splice variants (functional and nonfunctional) have been described for subtypes of both DA and 5-HT receptors.

Transporters for DA (DAT), 5-HT (SERT), and NE (NET) belong to the same family, the so-called solute carrier 6 (SLC6) gene family of ion-coupled plasma membrane cotransporters. These transporters are able to transport DA, 5-HT, and/or NE from the synapse and into the cell using the sodium gradient. They are not specific for their substrates, and NET is, for example, important for the transport/clearance of DA in the cortex. This also fits with the fact that the highest homology among the cloned human transporters is found between DAT and NET. Recently, a high-resolution crystal structure of a bacterial homolog (LeuT$_{Aa}$) of these transporters was published revealing a dimeric protein with each of the protomers being a 12 TM spanning protein in a unique fold. Thus, this x-ray structure has given a new insight into the structure of mammalian transporters, and a SERT homology model will be discussed as an example of the application of computational methods in drug design (Section 18.3.1.4). The structure and function of transporters are discussed in more detail in Chapter 14.

Selective ligands have been described for many of these receptor subtypes and transporters and in the following text, we have chosen to focus on ligands that have shown potential as antipsychotic or antidepressant drugs or which have been important in the discovery of these ligands. Antipsychotic drugs that are used in the treatment of schizophrenia will be discussed as an example of ligands for DA and 5-HT receptors (Section 18.2), whereas antidepressant drugs that are used for the treatment of depression and anxiety will be discussed as examples of ligands for transporters (Section 18.3).

18.2 RECEPTOR LIGANDS

18.2.1 Antipsychotic Drugs

Antipsychotic drugs are primarily used to treat schizophrenia. Schizophrenia is distinguished from other psychotic disorders based on a characteristic cluster of symptoms, where the positive symptoms appear to reflect an excess or distortion of normal function (i.e., delusion, hallucinations, disorganized thinking, disorganized behavior, and catatonia), whereas the negative symptoms appear to reflect a diminution or loss of normal functions (i.e., affective flattening, poverty of speech, and an inability to initiate and persist in goal-directed activities). The cognitive symptoms (i.e., impairment of memory, executive function, and attention) have in recent years attracted more and more attention, and recently much research is directed toward understanding the role of these symptoms.

The antipsychotic drugs are divided into the classical and the atypical antipsychotic drugs. The classical antipsychotic drugs were discovered in the 1950s with chlorpromazine (**18.1**, Figure 18.1) as the first prominent example, whereas the atypical antipsychotic drugs were introduced into the treatment of schizophrenia during the 1990s. It is believed that the antipsychotic drugs exert their effect on positive symptoms by reducing DA hyperactivity in limbic areas of the brain.

The term classical antipsychotic drug is linked to compounds that show effect in the treatment of positive symptoms at similar doses that induce extrapyramidal symptoms (EPS, i.e., Parkinsonian symptoms, dystonia, akathisia, and tardive dyskinesia). It is believed that EPS is caused by the blockade of DA activity in striatal areas of the brain. The classical antipsychotic drugs are without effect on negative and cognitive symptoms, and these drugs may even worsen these symptoms. It has been argued that the deterioration of negative and cognitive symptoms by classical antipsychotic drugs may be a consequence of their EPS, and the separation of the antipsychotic effect and EPS is the foremost important property of the atypical antipsychotic drugs.

Thus, the term atypical antipsychotic drug is linked to a diverse group of drugs having antipsychotic effect at doses not giving EPS. However, all drugs from this group have their own compound specific limitations, such as a strong tendency to increase weight for some of the compounds, whereas others have a tendency to prolong the QT interval (total duration of cardiac ventricular electrical activity) in the surface electrocardiogram. In the following text, the classical as well as atypical antipsychotic drugs will be discussed with focus on their discovery, including structural considerations and pharmacological profile of key compounds.

18.2.1.1 Classical Antipsychotic Drugs

Chlorpromazine was discovered in the beginning of the 1950s, and the structure of chlorpromazine with its phenothiazine backbone was an excellent lead for medicinal chemists. Thus, the modification of chlorpromazine without changing the phenothiazine backbone led to a number of drugs such as perphenazine (**18.2**) and fluphenazine (**18.3**) (Figure 18.1). Medicinal chemists also replaced the phenothiazine backbone with other tricyclic structures, and these modifications led to other classes of classical antipsychotic drugs such as the thioxanthenes and the 6-7-6 tricyclics. Lundbeck in

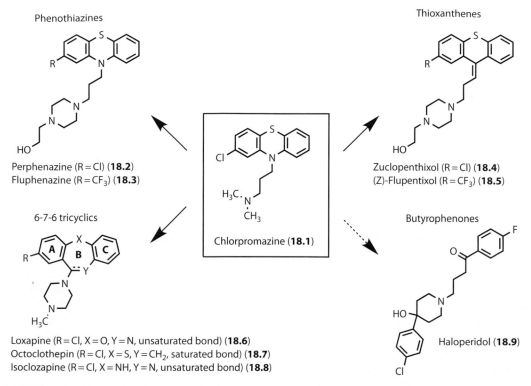

FIGURE 18.1 Classical antipsychotic drugs.

Denmark investigated in particular the thioxanthene backbone, and this work resulted in drugs such as zuclopenthixol (**18.4**) and (Z)-flupentixol (**18.5**) (Figure 18.1). The 6-7-6 tricyclic backbone has also led to a number of classical antipsychotic drugs such as loxapine (**18.6**), octoclothepin (**18.7**), and isoclozapine (**18.8**) (Figure 18.1). The R group found in all of these compounds is called the "neuroleptic substituent," and this substituent increases the D_2 affinity/antagonism relative to unsubstituted molecules and is essential for potent neuroleptic effect.

In the late 1950s, researchers at Janssen discovered an entirely new class of classical antipsychotic drugs without a tricyclic structure, namely the butyrophenones. Haloperidol (**18.9**, Figure 18.1) is the most prominent representative of this class of compounds, and today haloperidol is considered the archetypical classical antipsychotic drug for both preclinical experiments and clinical trials.

The classical antipsychotic drugs were all discovered using *in vivo* animal models, as the current knowledge about receptor multiplicity and *in vitro* receptor-binding techniques were unknown at that time. However, many of the *in vivo* models, which were used at that time as predictive for antipsychotic effect, is today considered more predictive of various side effects, e.g., EPS, and in hindsight it was difficult to find new antipsychotic drugs without the potential to induce EPS with the models available at that time. Thus, the development of new animal models modeling key properties of putative new antipsychotics is essential to the progress toward novel pharmacotherapies of schizophrenia. The ventral tegmental area (VTA) and the substantia nigra pars compacta (SNC) model is a good example of such a "model breakthrough" (see the following text).

An examination of the classical antipsychotic drugs by today's range of receptor-binding techniques and other more advanced biochemical methods has revealed that these drugs are postsynaptic D_2 receptor antagonists, and it is believed that this accounts for both their antipsychotic effect as well as their potential to induce EPS. However, these drugs also target several other receptors, which may contribute to both their antipsychotic effect, and to their side effect profile.

18.2.1.2 Atypical Antipsychotic Drugs

Isoclozapine (**18.8**, Figure 18.1), which has the "neuroleptic chloro substituent" in benzene ring **A**, is a classical antipsychotic drug. On the contrary, clozapine (**18.10**, Figure 18.2), which has the chloro substituent in benzene ring **C** (Figure 18.1), has revolutionized the pharmacotherapy of schizophrenia. Thus, clozapine was the first antipsychotic drug that was effective in the treatment of positive symptoms of schizophrenia and free of EPS, but unfortunately clozapine can cause fatal agranulocytosis in a small percentage (1%–2%) of individuals, and much effort has been directed toward the identification of new antipsychotics with a clozapine-like clinical profile but without the risk of causing agranulocytosis.

This search has resulted in a number of atypical antipsychotics such as olanzapine (**18.11**), quetiapine (**18.12**), risperidone (**18.13**), ziprasidone (**18.14**), aripiprazole (**18.15**), and sertindole (**18.16**) (Figure 18.2). The structure of these compounds reveals that olanzapine and quetiapine were obtained by structural modification of clozapine, whereas risperidone and ziprasidone were obtained from the butyrophenones. Aripiprazole and sertindole are quite different in chemical structure, and the discovery of sertindole will be discussed in more detail in the following text.

In vitro binding data for selected DA, 5-HT, and NE receptors as well as data from the catalepsy model (*in vivo* rat model predictive of EPS in humans) are shown for haloperidol and key atypical antipsychotics (Table 18.1). All the compounds display mixed receptor profiles with affinity for even more receptors and sites than included in the table (data not shown). The tendency is that classical antipsychotics display high affinity for D_2 receptors relative to 5-HT_{2A} receptors, whereas the atypical antipsychotics display an increased affinity for 5-HT_{2A} receptors relative to D_2. The relative ratio of D_2 versus 5-HT_{2A} receptor affinity has been suggested as a reason for atypicals not giving rise to EPS at therapeutic doses. A number of other factors may influence both the antipsychotic potential and the propensity to induce EPS, such as the affinity and efficacy at some of the other receptors, but also *in vivo* preference for limbic versus striatal regions of the brain might explain these differences. Interestingly, aripiprazole is a partial D_2 receptor agonist, whereas the other antipsychotics are D_2 receptor antagonist. Partial D_2 receptor agonists are envisaged to stabilize a dysfunctioning

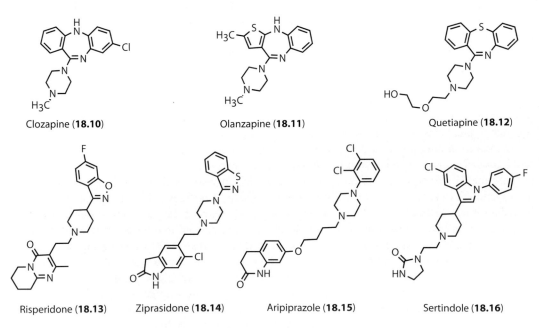

FIGURE 18.2 Atypical antipsychotic drugs.

TABLE 18.1

Receptor Profile and EPS Potential of Antipsychotic Drugs

Compounds	Receptor Binding K_i (nM)							*In Vivo* ED$_{50}$ (μmol/kg)
	D_1^a	D_2^a	D_3^b	D_4^b	5-HT$_{2A}^a$	5-HT$_{2C}^a$	α_1^a	Catalepsy Max., sc^a
Classical antipsychotic drug								
Haloperidol (**18.9**)	15	0.82	1.1	2.8	28	1500	7.3	0.34
Atypical antipsychotic drugs								
Risperidone (**18.13**)	21	0.44	14	7.1	0.39	6.4	0.69	17
Olanzapine (**18.11**)	10	2.1	71	32	1.9	2.8	7.3	37
Quetiapine (**18.12**)	390	69	1100	2400	82	1500	4.5	>80
Ziprasidone (**18.14**)	9.5	2.8	n.t.	73	0.25	0.55	1.9	>48
Sertindole (**18.16**)	12	0.45	2.0	17	0.20	0.51	1.4	>91
Clozapine (**18.10**)	53	36	310	30	4.0	5.0	3.7	120

Sources: Data from^a Arnt, J. and Skarsfeldt, T. *Neuropsychopharmacology*, 18, 63, 1998;^{b} Lundbeck Screening Database, H. Lundbeck A/S, Valby, Denmark.

Note: n.t; not tested.

DA system, inhibiting transmission in synapses with high tonus, and increasing function in those with low activity. This profile might explain why aripiprazole does not induce EPS.

18.2.1.2.1   Discovery of Sertindole

In the 1970s, Lundbeck had successfully marketed a number of classical antipsychotic drugs, and the medicinal chemistry program at Lundbeck was still aimed at finding new antipsychotic drugs based on the phenothiazine or thioxanthene template. However, in 1975, the first compounds were synthesized in a project directed toward the identification of nonsteroidal anti-inflammatory drugs (NSAIDs), and fortunately the compounds were also examined in *in vivo* models predictive of antipsychotic and antidepressant action. It was found that the two *trans*-1-piperazino-3-phenylindanes **18.17** and **18.18** (Figure 18.3) were relatively potent in the methyl phenidate-induced hyperactivity model (predictive of antipsychotic effect/mechanistic model for D_2 antagonism), but at the same time about a factor of 10 weaker in the catalepsy model (predictive of EPS). Thus, these compounds were seen as prototypes of a new class of antipsychotic drugs with an improved side effect profile with respect to EPS.

In 1980, the *trans*-racemate tefludazine (**18.19**, Figure 18.3) was selected from this series as a development agent with potential antipsychotic effect. Tefludazine displayed a similar ratio in the methyl phenidate hyperactivity versus the catalepsy model as the two lead compounds, but

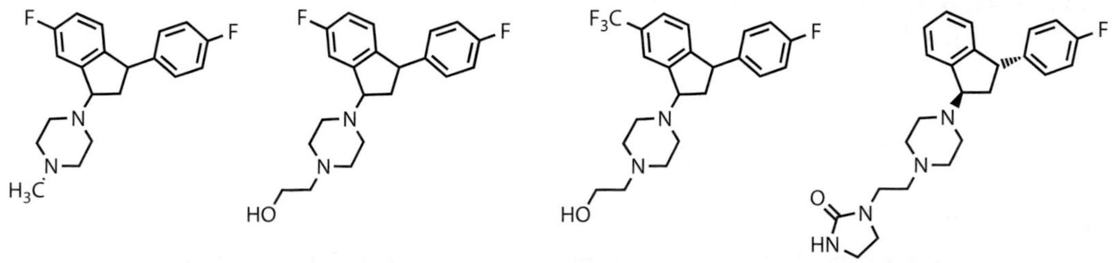

| **18.17** | **18.18** | Tefludazine (**18.19**) | Irindalone (**18.20**) |

FIGURE 18.3   Selected *trans*-1-piperazino-3-phenylindanes.

tefludazine was at least 100 times more potent in these *in vivo* models. It was also shown that tefludazine was an extremely potent and long-acting *in vivo* 5-HT$_2$ receptor antagonist (quipazine model). Thus, tefludazine was a mixed D$_2$ and 5-HT$_2$ receptor compound.

During the 1980s, an electrophysiological *in vivo* model for the evaluation of limbic (linked to positive symptoms) versus striatal (linked to EPS) selectivity was introduced at Lundbeck. The model was a chronic one where rats were treated with a compound for 3 weeks before the number of active dopamine neurons were counted in the VTA/SNC from where neurons project to limbic and striatal areas, respectively. In this so-called VTA/SNC model treatment with classical antipsychotic drugs such as chlorpromazine and haloperidol led to complete inhibition of neurons in both VTA and SNC by equal doses, whereas clozapine selectively inactivated the dopamine neurons in the VTA. These results corresponded to clinical data with regards to antipsychotic effect and EPS, and the model became of key importance at Lundbeck as it had the potential to predict the therapeutic window between antipsychotic effect and EPS of putative new antipsychotic drugs. It was subsequently shown that tefludazine displayed some selectivity in this model predicting tefludazine to be atypical, but unfortunately, the development of tefludazine was discontinued in Phase I due to toxicological findings in dogs.

It was also discovered that the removal of the "neuroleptic substituent" in the indane benzene ring (i.e., the trifluoromethyl group in **18.19**), reduced the D$_2$ receptor antagonism, whereas the 5-HT$_2$ receptor antagonism was retained. Concurrent replacement of the hydroxyethyl side chain with the more bulky 1-ethyl-2-imidazolidinone side chain, resulted in irindalone (**18.20**, Figure 18.3), which was a very potent and selective 5-HT$_2$ antagonist. Irindalone was developed as a potential antihypertensive drug, but the development was discontinued in Phase II in 1989 because of market considerations. Irindalone was, in contrast to tefludazine, developed as the pure (1*R*,3*S*)-enantiomer. This configuration of the 1-piperazino-3-phenylindanes is generally associated with receptor antagonistic properties, whereas the (1*S*,3*R*)- and the (1*R*,3*R*)-enantiomers are NET/DAT reuptake inhibitors.

The piperazinoindanes are chiral molecules, which at that time complicated many stages of drug discovery and development processes. Therefore, the corresponding piperazino-, tetrahydropyridino-, and piperidino-indoles were designed, and it was discovered that the piperidinoindole moiety bioisosterically substituted for the *trans*-piperazinoindane with respect to D$_2$ and 5-HT$_2$ antagonism. One of the compounds synthesized in this series was sertindole (**18.16**, Figure 18.2), which incorporates structural elements from both tefludazine (neuroleptic substituent) and irindalone (imidazolidinone side chain). Despite high affinity for both D$_2$ and 5-HT$_2$ receptors, sertindole displayed an *in vivo* profile of a selective 5-HT$_2$ receptor antagonist. Therefore, it was surprising that sertindole in the VTA/SNC model displayed a more than 100-fold selectivity for inhibition of dopamine neurons in the VTA as compared to the SNC. Sertindole was subsequently pushed through development and marketed in 1996 for the treatment of schizophrenia. Sertindole was temporarily withdrawn from the market in 1998 because of uncertainties regarding the relation between QT prolongation and the ability to induce potentially fatal cardiac arrhythmias in humans. However, the suspension of sertindole was lifted in 2002, and in January 2006, Estonia became the first country to reintroduce it. Sertindole is currently being introduced in several European, Asian, Latin American and the USA. Interestingly, recent preclinical evidence suggests sertindole being effective in the treatment of cognitive impairment in schizophrenia, with superiority over other antipsychotics such as risperidone, olanzapine, and clozapine. The effect has putatively been linked to potent 5-HT$_6$ receptor antagonism combined with lack of antimuscarinic activity. It is currently being investigated whether the preclinical finding can be confirmed in clinical trials.

18.3 TRANSPORTER LIGANDS

18.3.1 ANTIDEPRESSANT DRUGS

Antidepressant drugs represent ligands that target DAT, SERT, and NET to various degrees, and these include first generation antidepressants (i.e., tricyclic antidepressants, TCAs), selective serotonin reuptake inhibitors (SSRIs), combined serotonin and norepinephrine reuptake inhibitors (SNRIs), and the more recently introduced allosteric serotonin reuptake inhibitor (ASRI), escitalopram.

The SSRIs have been highly successful in the treatment of depression due to their high safety in use, and a number of new indications (e.g., panic disorder, obsessive compulsive disorder, and social phobia) have been registered for many of these drugs in addition to major depression. However, there are still major unmet needs in the treatment of depression, and since the inhibition of 5-HT reuptake ensures a certain degree of antidepressant activity, there has been a large interest in combining 5-HT reuptake inhibition with additional pharmacological effects. Some of these have resulted in marketed antidepressants, such as the SNRIs. In the following text, key events related to the discovery of antidepressants and in particular citalopram and escitalopram at Lundbeck will be discussed, including the use of pharmacophore and homology models.

18.3.1.1   First Generation Drugs

The pharmacotherapy of depression started in the late 1950s with the introduction of the two drugs iproniazid (**18.21**) and imipramine (**18.23**, Figure 18.4). Iproniazid was originally an antituberculosis drug, but it was noticed that the drug had an antidepressant effect. It was subsequently discovered that iproniazid was an unselective, irreversible inhibitor of the enzymes MAO-A and MAO-B, which deaminate the monoamines NE, DA, and 5-HT. Structural modifications of the tricyclic antipsychotic drugs with chlorpromazine (**18.1**, Figure 18.1) as a prototype led to the 6-7-6 tricyclic compound imipramine that was found to block the transporters for NE and 5-HT. These mechanisms led to an increase in the concentrations of NE and 5-HT in the synapse, which in turn led to the so-called amine hypothesis of depression, stating that there is a decreased availability of these neurotransmitters in depression.

Although the discovery of these two classes of drugs was of major therapeutic importance, it quickly turned out that both types had fatal side effects. Treatment with MAO inhibitors could induce a hypertensive crisis because of a fatal interaction with foodstuffs containing tyramine such as cheese. Dietary restrictions during treatment with MAO inhibitors were therefore required. Reversible MAO-A inhibitors (such as moclobemide (**18.22**)) have later been developed, but such drugs are still not completely devoid of the "cheese-effect" because the tyramine potentiation is inherent to blockade of MAO-A in the periphery. MAO inhibitors are therefore only used to a lesser extent in antidepressant therapy.

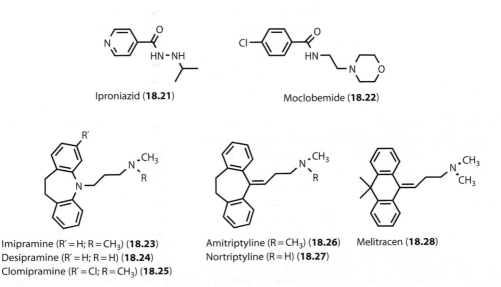

Iproniazid (**18.21**)   Moclobemide (**18.22**)

Imipramine (R′ = H; R = CH₃) (**18.23**)   Amitriptyline (R = CH₃) (**18.26**)   Melitracen (**18.28**)
Desipramine (R′ = H; R = H) (**18.24**)   Nortriptyline (R = H) (**18.27**)
Clomipramine (R′ = Cl; R = CH₃) (**18.25**)

FIGURE 18.4   Antidepressant drugs from MAO-inhibitor and tricyclic classes.

A major problem with the tricyclic antidepressants such as imipramine (**18.23**), desipramine (**18.24**), amitriptyline (**18.26**), nortriptyline (**18.27**), and melitracen (**18.28**) is that, due to their fundamental tricyclic structures, in addition to their blockade of SERT and/or NET, they also block a number of postsynaptic receptors notably for acetylcholine, histamine, and NE. Therefore, they may induce a number of anticholinergic, antihistaminergic, and cardiovascular side effects, such as dryness of the mouth, constipation, confusion, dizziness, sedation, orthostatic hypotension, tachycardia, and/or arrhythmia. Moreover, they are potentially lethal in overdose. So even if these drugs represented a major therapeutic breakthrough, it became clear that there was an inevitable need for better and safer drugs.

18.3.1.2 The Selective Serotonin Reuptake Inhibitors

Nortriptyline (**18.27**) is a relative selective NE reuptake inhibitor, while the corresponding dimethyl derivative, amitriptyline (**18.26**), is a mixed 5-HT/NE reuptake inhibitor with concomitant high affinity for the postsynaptic receptors mentioned earlier. The same is true for the corresponding pair desipramine (**18.24**)/imipramine (**18.23**). Swiss psychiatrist Paul Kielholz coupled these observations to the clinical profiles of these drugs, and Swedish scientist Arvid Carlsson noticed that the tertiary amine drugs, which were mixed 5-HT and NE reuptake inhibitors, were "mood elevating," while the secondary amines, being primarily NE reuptake inhibitors, increased more "drive" in the depressed patients. As the foremost quality of an antidepressant drug should be mood elevation (elevation of drive before mood could induce a suicidal event), Carlsson advocated for the development of selective 5-HT reuptake inhibitors. Consequently, a number of pharmaceutical companies initiated drug discovery programs aiming at design of such drugs in the early 1970s.

18.3.1.2.1 Discovery of Citalopram

In the mid-1960s, chemists at Lundbeck were looking for more potent derivatives of the tricyclic compounds amitriptyline, nortriptyline, and melitracen, which the company had developed and marketed previously. The trifluoromethyl group had in other in-house projects proved to increase potency in thioxanthene derivatives with antipsychotic activity (see Figure 18.1), and it was therefore decided to attempt to synthesize the 2-CF$_3$ derivative (**18.30**) of melitracen (Figure 18.5). The precursor molecule **18.29** was readily synthesized, but attempts to ring-close it in a manner corresponding to the existing melitracen method, using concentrated sulfuric acid, failed. However, another product was formed, which through meticulous structural elucidation proved to be the bicyclic phthalane (or dihydroisobenzofuran) derivative **18.31**. Fortunately, this compound was examined in models for antidepressant activity and was very surprisingly found to be a selective NET inhibitor. Some derivatives were synthesized, among them two compounds that later got the International Nonproprietary Name (INN) names talopram (**18.32**) and talsupram (**18.33**). These compounds are still among the most selective NE reuptake inhibitors (SNIs) ever synthesized (Figure 18.5 and Table 18.2).

Both talopram and talsupram were investigated for antidepressant effect in clinical trials but were stopped in Phase II for various reasons, among which an activating profile in accordance with their potent NE reuptake inhibition. A project was therefore started in the beginning of 1971 with the aim of discovering an SSRI from the talopram structure.

It may not be obvious to use an SNI as template structure for an SSRI. However, in the first series synthesized, two compounds (**18.35** and **18.36**, Table 18.2) without the dimethylation of the phthalane ring showed a tendency for increased 5-HT reuptake, and in accordance with the structure–activity relationship (SAR) studies mentioned earlier for tricyclics, the N,N-dimethyl derivative **18.36** was the more potent. Therefore, compound **18.36** became a template structure for further structural investigation.

In this phase of the project, test models for measuring neuronal reuptake were not available, so 5-HT reuptake inhibition was measured as inhibition of tritiated 5-HT into rabbit blood platelets, while inhibition of NE reuptake was measured *ex vivo* as inhibition of tritiated NE into the heart of the mouse (Table 18.2). Although these models were not directly comparable, they were acceptable for the discovery of selective compounds.

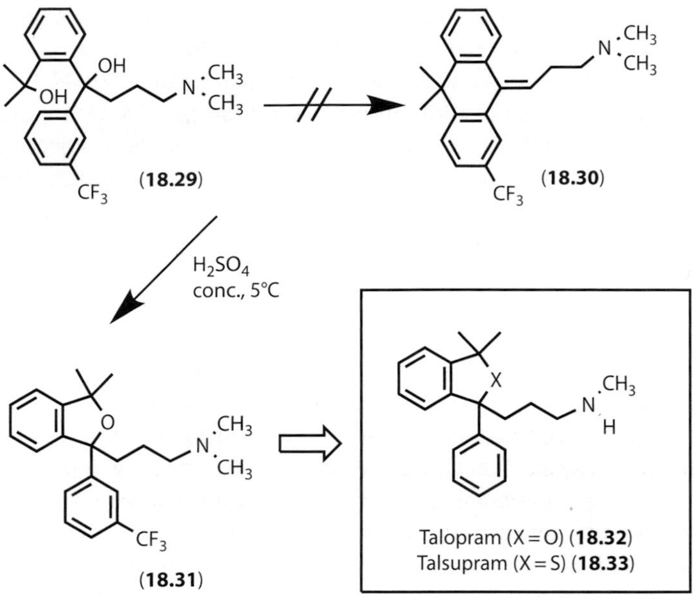

FIGURE 18.5   Discovery of phenylphthalane antidepressants.

TABLE 18.2
5-HT and NE Reuptake Inhibition of Selected Talopram Derivatives

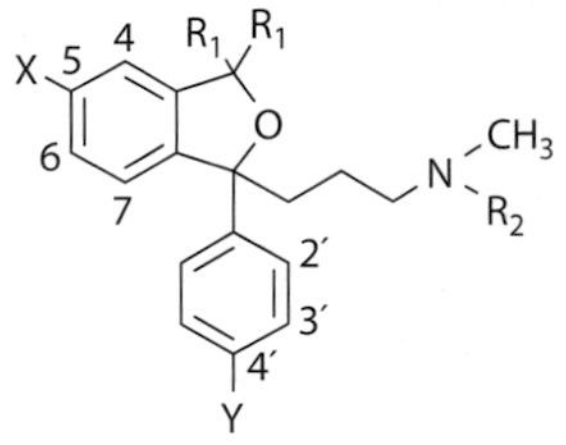

Compound	R_1	R_2	X	Y	5-HT Reuptake (*In Vitro*) Rabbit Blood pl. IC_{50} (nM)	NE Reuptake (*In Vivo*) Mouse Heart ED_{50} (μmol/kg)
Talopram (**18.32**)	CH_3	H	H	H	3,400	2.2
(**18.34**)	CH_3	CH_3	H	H	53,000	5
(**18.35**)	H	H	H	H	1,300	43
(**18.36**)	H	CH_3	H	H	600	66
(**18.37**)	H	CH_3	H	Cl	110	170
(**18.38**)	H	CH_3	Cl	H	220	>200
(**18.39**)	H	CH_3	Cl	Cl	24	>80
(**18.40**)	H	CH_3	H	Br	310	n.t.
(**18.41**)	H	CH_3	H	CN	54	23
(**18.42**)	H	CH_3	CN	Cl	10	>80
Citalopram (**18.43**)	H	CH_3	CN	F	38	>40

Source: Data from Lundbeck Screening Database, H. Lundbeck, Valby, Denmark.

The introduction of a chloro substituent into the template structure **18.36** further increased 5-HT reuptake and decreased NE reuptake inhibition (**18.36** and **18.38**), in accordance with observations by Carlsson that halogen substituents in both zimelidine (**18.47**, Figure 18.6) (see the following text) derivatives and in of imipramine (clomipramine, **18.25**, Figure 18.4) increased 5-HT reuptake. Indeed, the dichloro derivative **18.39** proved to be a selective 5-HT reuptake inhibitor. So the goal of obtaining an SSRI from an SNI was achieved very fast (in 1971), when less than 50 compounds had been synthesized.

The SAR were further explored, and it was established that high activity was generally found in 5,4′-disubstituted compounds where both substituents were halogen or other electron-withdrawing groups. Cyano-substituted compounds were obtained by the reaction of the bromo precursors (e.g., **18.40**) with CuCN. One of the cyano-substituted compounds was (**18.43**), later known as citalopram (INN name). The compound was synthesized for the first time in August 1972. The cyano group could be metabolically labile, but it was subsequently shown not to be the case neither in animals nor in humans. Citalopram displayed the best overall preclinical profile within this series and was consequently selected for development. The 5-cyano substituent in citalopram also proved to be chemically stable in a surprising manner; for example, it does not react with Grignard reagents, which has led to a new and patentable process for its production.

Citalopram was launched in Denmark in 1989, and it has since been registered worldwide. Citalopram is a racemate, having an asymmetric carbon at the 1-position. When it was synthesized in 1972, classical resolution via diastereomeric salts was the only realistic alternative for separation of the enantiomers. However, it is generally difficult to make salts of citalopram, and eventually direct resolution was given up. Finally, an intermediate was resolved in this way, and the resolved intermediate could then be transformed into the pure (*S*)- and (*R*)-enantiomers of citalopram. Subsequent testing showed that all the 5-HT reuptake inhibition resided in the (*S*)-enantiomer. The high stereospecificity was later rationalized in the SSRI pharmacophore model discussed in the following text. The (*S*)-enantiomer (INN name escitalopram) has subsequently emerged as a new drug defining a new group of antidepressants, namely the ASRIs. The discovery and its implications are discussed in the following text.

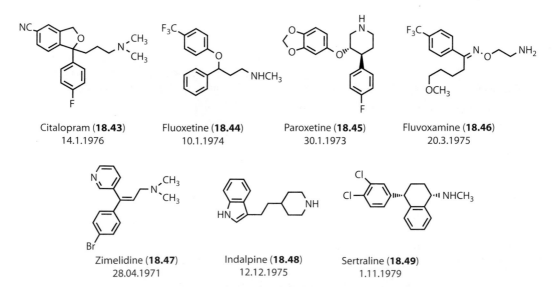

Citalopram (**18.43**)	Fluoxetine (**18.44**)	Paroxetine (**18.45**)	Fluvoxamine (**18.46**)
14.1.1976	10.1.1974	30.1.1973	20.3.1975

Zimelidine (**18.47**)	Indalpine (**18.48**)	Sertraline (**18.49**)
28.04.1971	12.12.1975	1.11.1979

FIGURE 18.6 Selective serotonin reuptake inhibitors (SSRIs).

18.3.1.2.2   Other Selective Serotonin Reuptake Inhibitors

In Figure 18.6, the seven SSRIs that have reached the market, with the priority dates of the first patent application indicated are shown. However, the two first compounds on the market were both withdrawn due to serious, although rare, side effects. Zimelidine (**18.47**) was found to induce an influenza-like symptom in 1%–2% of the patients, which in rare cases (1/10,000) resulted in the so-called Guillain-Barré syndrome. The drug was withdrawn in 1983 after 1½ years on the market. Indalpine (**18.48**) induced agranulocytosis in one of 20,000 patients and was withdrawn in 1984.

As it appears from Figure 18.6, all the marketed SSRIs (except sertraline) were discovered in the first half of the 1970s, meaning that the companies lacked sufficient information regarding the structural classes their competitors were developing. Accordingly, rather diverse structures were developed. However, they were all selective 5-HT reuptake inhibitors (Table 18.3), although their selectivity ratios vary significantly, citalopram/escitalopram being the most selective compounds. In general, the SSRIs have low affinity for receptors for DA, NE, 5-HT, and other neurotransmitters, although exceptions exist. With regard to interaction with cytochrome P450 enzymes there are vital differences, e.g., paroxetine and fluoxetine having significant affinity for CYP2D6.

TABLE 18.3
The Effect of SSRIs, Talopram, and Talsupram on the Inhibition of Reuptake of 5-HT, NE, and DA

Compound	Uptake Inhibition IC$_{50}$ (nM)			Ratio	
	5-HT	**NE**	**DA**	**NE/5-HT**	**DA/5-HT**
Citalopram (**18.43**)	3.9	*6100*	*40,000*	1560	10,300
Escitalopram (*S*) (**18.43**)	2.1	2500	65,000	1200	31,000
R-citalopram (*R*) (**18.43**)	275	6900	54,000	25	200
Indalpine (**18.48**)	2.1	2100	1,200	1000	570
Sertraline (**18.49**)	*0.19*	*160*	*48*	840	250
Paroxetine (**18.45**)	*0.29*	*81*	*5,100*	280	17,600
Fluvoxamine (**18.46**)	*3.8*	*620*	*42,000*	160	11,000
Zimeldine (**18.47**)	56	3100	26,000	55	460
Fluoxetine (**18.44**)	*6.8*	*370*	*5,000*	54	740
Talopram (**18.32**)	1400	2.5	44,000	0.0017	0.00006[a]
Talsupram (**18.33**)	770	0.79	9,300	0.0010	0.00008[a]

Sources: Data in italics are from Hyttel, J. *Int. Clin. Psychopharmacol.* 9(Suppl. 1), 19, 1994; Remaining are from Lundbeck Screening Database, H. Lundbeck A/S, Valby, Denmark.

[a] NE/DA.

18.3.1.3   Discovery of Escitalopram—An Allosteric Serotonin Reuptake Inhibitor

The 5-HT reuptake inhibition of citalopram resides in the (*S*)-enantiomer (escitalopram) whereas the (*R*)-enantiomer is about 100 times less potent. Escitalopram was launched as a single-enantiomer drug in 2002 and is today an effective antidepressant with several advantages as compared to citalopram and other SSRIs. In preclinical studies escitalopram shows greater efficacy and faster onset of action than comparable doses of citalopram. This is attributed to the fact that a number of studies have shown that the (*R*)-enantiomer of citalopram counteracts the activity of the (*S*)-enantiomer. Further in randomized, controlled clinical studies escitalopram shows better efficacy than citalopram, with higher response and remission rates, and faster onset of action.

The existence of an allosteric-binding site on SERT and its possible relevance for the effect of antidepressants has been known since the early 1990s, and among the SSRIs only citalopram and to a lesser extent paroxetine exert an effect via this low-affinity binding site. The dual action on both the allosteric and the primary-binding site results in an increased dissociation half-life of escitalopram from its primary-binding site. The (R)-enantiomer has a three times weaker allosteric effect, and it significantly reduces the association rate for [^{3}H] escitalopram binding to SERT in low (40–80 nM) concentrations. These effects of the (R)-enantiomer may be important in relation to its inhibition of the (S)-enantiomer in citalopram.

18.3.1.4   The SSRI Pharmacophore and SERT Homology Model

Despite very different molecular structures, the SSRIs all bind to SERT. As information about the 3D structure of the transporter is lacking, development of a pharmacophore model was of major interest in the early 1990s. Thus, a pharmacophore model of the SSRIs was developed at Lundbeck based on extensive conformational studies and superimpositions of SSRIs and other reuptake inhibitors (Figure 18.7a). The model operates with three fitting points, namely the centroids of the two aromatic rings and a site point positioned 2.8 Å from the nitrogen atom in the direction of the lone pair. The nitrogen site point mimics a hypothetical hydrogen-binding atom on SERT, most likely the carboxy group of Asp 98 (see also discussion in the following text). The use of a nitrogen site point as fitting point in this model gave a very good superimposition of all key SSRIs (i.e., escitalopram, (S)- and (R)-fluoxetine, (1S,4R)-sertraline, and (3S,4R)-paroxetine), which was not possible when the basic nitrogen atoms were superimposed. Additionally, many SSRIs have aromatic substituents (cyano, trifluoromethyl, chloro, methylendioxo, etc.) that all, in this model, occupy the same volume marked in yellow. Hence, this volume of the transporter is allowed for SSRIs but not for NRIs. On the contrary, the volume marked in white defines a forbidden volume for SSRI ligands. Protrusion of ligands into this volume allows for design of NRIs (Figure 18.7a).

The pharmacophore model has been validated with a number of 5-HT and NE reuptake inhibitors in addition to the compounds in Figure 18.6. Importantly, the model explains the more than 100-fold stereoselectivity of the citalopram enantiomers. Thus, it is possible to find a conformation of the (R)-enantiomer that is superimposable with the proposed bioactive conformation of escitalopram, but the conformational energy penalty is 2.8 kcal/mol, which corresponds closely to a 100-fold affinity difference. The enantiomers of fluoxetine display no stereoselectivity at SERT, and they can be fitted to the model with no differences in conformational energy in accordance with their equipotency.

An x-ray crystal structure of a leucine transporter (LeuT$_{Aa}$), which is a bacterial homolog of SERT, from *Aquifex aeolicus* has offered an opportunity to build a more reliable homology model of SERT as compared to earlier transporter models based on more distantly related proteins. In Figure 18.7b, the secondary structure of the LeuT$_{Aa}$ is illustrated with the primary-binding site for leucine marked with a triangle, and in Figure 18.7c the corresponding homology model of SERT built at Lundbeck is shown. The homology model of SERT is colored and oriented in a similar manner as LeuT$_{Aa}$ in Figure 18.7a and with escitalopram docked into the proposed primary-binding site. In Figure 18.7d a "close up" of the primary-binding site with bound escitalopram is shown. The "close up" is taken from the intracellular side looking up into the transporter. From this illustration it is possible to see the putative interaction points between escitalopram and SERT, notably the hydrogen bond between the basic amine of escitalopram and Asp 98. It is also possible to see the putative hydrophobic interaction between the two aromatic rings of escitalopram and the hydrophobic amino acid Ile 172. Importantly, it has been shown in a number of mutation studies that Asp 98 and Ile 172 are essential for the binding of escitalopram to SERT giving validity to this model.

The nine amino acids (shown in white in Figure 18.7c) that have been shown to be involved in the allosteric binding of escitalopram on SERT are located in TM 10, 11, and 12 near the C-terminal. Very little is known about the interaction of escitalopram with this putative-binding site. However, it is highly likely that the allosteric effect is important for the superior effect of escitalopram as an antidepressant, especially in patients with severe depression.

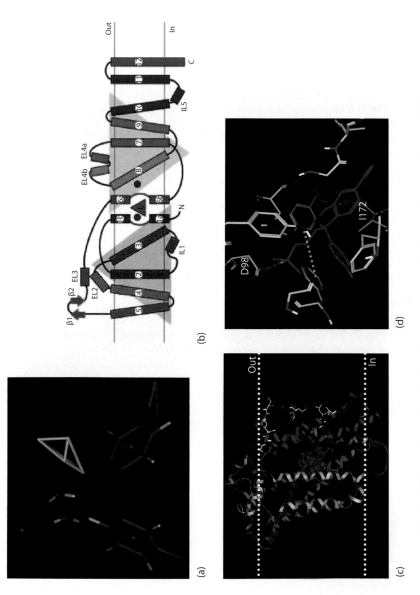

FIGURE 18.7 (a) Pharmacophore model of the 5-HT reuptake site. Purple: Three fitting points; centroids of aromatic rings and site point for interaction with transporter amine. Green: Phenyl rings. Blue: Nitrogen atoms. Yellow: Allowed volume for SSRI substituents. White: Forbidden volume at SERT, allowed at NET. Red: Possible hydrogen bond acceptor site. (Courtesy of Klaus Gundertofte, H. Lundbeck A/S, Copenhagen-Valby, Denmark.) (b) The leucine transporter topology. The position of leucine is depicted as a yellow triangle. (From Yamashita, A. et al. *Nature*, 437, 215, 2005. With permission.) (c) The 5-HT transporter (SERT) homology model. Escitalopram is highlighted in green/blue space filling and docked into the primary-binding site. The transmembrane domains (TMs) involved in the primary-binding site are shown in the same colors as in (b) (TM1: red, TM3: orange, TM6: green, and TM8: blue), and the important amino acids are shown in yellow. The TMs important for allosteric effects are highlighted in purple (TM 10–12), and the corresponding amino acids are shown in white. (Courtesy of Anne Marie Jørgensen, H. Lundbeck A/S, Copenhagen-Valby, Denmark.) (d) A "close up" of the primary-binding site of Figure 18.7c with escitalopram docked into the site. The three fitting points from the pharmacophore model is illustrated with purple spheres, including the site point on Asp 98 involved in the hydrogen bond (yellow dotted line) with escitalopram. (Courtesy of Anne Marie Jørgensen, H. Lundbeck A/S, Copenhagen-Valby, Denmark.)

18.4 CONCLUDING REMARKS

DA, 5-HT, and NE receptors and transporters have shown their relevance as drug targets, and although research toward the pharmacotherapy of schizophrenia and depression nowadays are targeting other targets as well, future antipsychotics and antidepressants may very well interact with these receptors and sites also. Recently, it has been shown that DA signals via several different second messenger systems and as discussed earlier an allosteric site has been identified on SERT. Only the future will tell us whether these new discoveries will result in new and effective pharmacotherapies based on the DA and 5-HT systems.

FURTHER READINGS

Gether, U., Andersen, P. H., Larsson O. M., and Schousboe, A. (2006) Neurotransmitter transporters: Molecular function of important drug targets. *Trends Pharmacol. Sci.*, **27**, 375–383.

Gray, J. A. and Roth, B. L. (2007) The pipeline and future of drug development in schizophrenia. *Mol. Psychiatry*, **12**, 904–922.

Hyttel, J. (1994) Pharmacological characterisation of selective serotonin reuptake inhibitors (SSRIs). *Int. Clin. Psychopharmacol.*, **9**(Suppl. 1), 19–26.

Jørgensen, A. M., Tagmose, L., Jørgensen, A. M. M., Topiol, S., Sabio, M., Gundertofte, K., Bøgesø, K. P., and Peters, G. H. (2007) Homology modeling of the serotonin transporter: Insight into the primary escitalopram-binding site. *ChemMedChem*, **2**, 815–826.

Moltzen, E. K. and Bang-Andersen, B. (2006) Serotonin reuptake inhibitors: The corner stone in treatment of depression for half a century—A medicinal chemistry survey. *Curr. Top. Med. Chem.*, **6**, 1801–1823.

Sánchez, C., Bøgesø, K. P., Ebert, B., Reines, E. H., and Bræstrup, C. (2004) Escitalopram versus citalopram: The surprising role of the *R*-enantiomer. *Psychopharmacology*, **174**, 163–176.

Werkman, T. R., Glennon, J. C., Wadman, W. J., and McCreary, A. C. (2006) Dopamine receptor pharmacology: Interactions with serotonin receptors and significance for the aetiology and treatment of schizophrenia. *CNS Neurol. Disord. Drug Targets*, **5**, 3–23.

Wood, M. and Reavill, C. (2007) Aripiprazole acts as a selective dopamine D_2 receptor partial agonist. *Expert Opin. Invest. Drugs*, **16**, 771–775.

Yamashita, A., Singh, S. K., Kawate, T., Jin, Y., and Gouaux, E. (2005) Crystal structure of a bacterial homologue of Na^+/Cl^--dependent neurotransmitter transporters. *Nature*, **437**, 215–223.

19 Opioid and Cannabinoid Receptors

Rasmus P. Clausen and Harald S. Hansen

CONTENTS

19.1 OPIOID RECEPTORS

Presently she cast a drug into the wine of which they drank to lull all pain and anger and bring forgetfulness of every sorrow.

The Odyssey, Homer (ninth century BC)

The history of opioids and its receptors spans several millennia. The first evidences of uses of the seed pods of *Papaver somniferum* dates back to 4200 BC and numerous findings and descriptions through out the history witness the use of different parts of this plant in food, anesthesia, and ritual purposes. Opium (from *opos*, the Greek word for juice) refers to the liquid that appears on the unripe seed capsule, when it is notched. This liquid contains as much as 16% of morphine, a compound that was isolated already in 1806 as the major active ingredient in opium (Figure 19.1). A few years, later codeine was also isolated. Morphine could now be produced and applied in its pure form for the treatment of pain and as an adjunct to general anesthetics, but it was quickly realized that morphine had the same potential of abuse as opium. In 1898, heroin was synthesized and claimed to be a safer, more efficacious, and nonaddicting opiate as were several other analogues around that time; however, they all proved not to be safer later on. Heroin is an early example of a

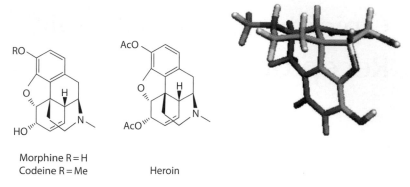

FIGURE 19.1   Chemical structures of morphine, codeine, and heroin. 3D-structure of morphine.

prodrug since the highly potent analgesic properties can be attributed to the rapid metabolism to 6-monoacetylmorphine and morphine, combined with higher blood–brain barrier penetration due to better lipid solubility compared to morphine.

19.1.1   Opioid Receptor Subtypes and Effector Mechanism

The idea that morphine and other opioids caused analgesia by interacting with a specific receptor arose around the 1950s. The observation 40 years earlier that the *N*-allyl analogue of codeine antagonized the respiratory action of morphine was actually an evidence of such a proposal. However, it was first fully realized, when similar *N*-allyl analogue of morphine (nalorphine) was shown to antagonize the analgesic effects of morphine.

Today, it is known that all of the opioid receptors are G-protein coupled receptors (GPCRs) belonging to family A (Figure 19.2) that mediates its effects through Gi/Go proteins. So far, four different opioid receptor subtypes have been cloned sharing more than 60% sequence homology. These are termed μ, κ, and δ receptors (corresponding to MOR, KOR, and DOR, respectively) and an "orphan" receptor termed ORL_1, which was the first orphan GPCR to be cloned.

The different effects mediated by each receptor type (μ–euphoria versus κ–dysphoria; μ–supraspinal analgesia versus ORL_1–supraspinal antagonism of opioid analgesia) in the intact animal are the result of different anatomical localizations and not due to different cellular responses. Each receptor type has been further subdivided into μ_1/μ_2, κ_1/κ_2, and δ_1/δ_2 receptors based on pharmacological and radioligand studies. However, the origin of this subdivision is not genetically based, and it is not known whether it arises from posttranslational modification, cellular localization, or interactions with other proteins; however, it was recently shown that heterodimerization of the receptors could be important for some of these pharmacological differences.

Morphine has the ability to both excite and inhibit single neurons. Opioid inhibition of neuronal excitability occurs largely by the ability of opioid receptors to activate various potassium channels. Another well-established mechanism of action is the inhibition of neurotransmitter release. The observation in 1917 that morphine inhibited the peristaltic reflex in the guinea-pig ileum (giving rise to constipation, one of the side effects of morphine) was 40 years later shown to result from the inhibition of acetylcholine release. Also glutamate, GABA, and glycine release throughout the central nervous system (CNS) can be inhibited by opioid receptor activation. In general, the CNS effects of opioids are inhibitory, but certain CNS effects (such as euphoria) result from excitatory effects (Table 19.1).

19.1.2   Endogenous Opioid Receptor Ligands

It was proposed in the early 1970s, that the physiological role of opioid receptors was not to be target for opium alkaloids, but that endogenous agonists might exist as mediators of the opioid system.

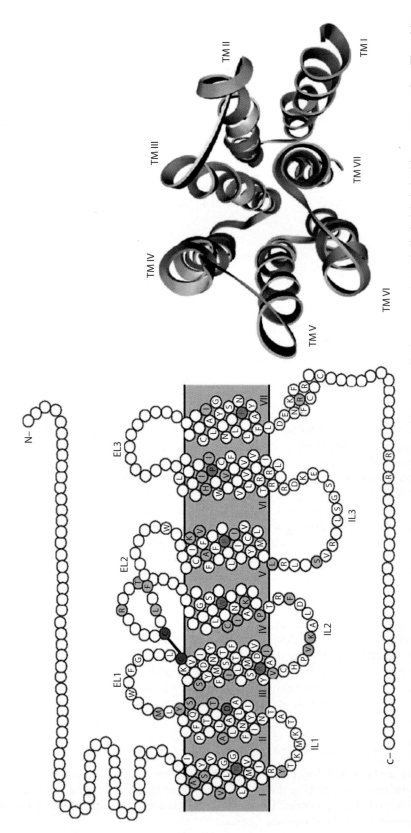

FIGURE 19.2 Structure of opioid receptors. (Left) Serpentine model of the opioid receptor. Each transmembrane helix is labeled with a roman number. The white empty circles represent nonconserved amino acids and white circles with a letter represent identical amino acids among the four opioid receptors. Violet circles represent further identity between the MOR, DOR, and KOR. Green circles highlight the highly conserved fingerprint residues of family A receptors. Yellow circles depict the two conserved cysteines in EL loops 1 and 2, likely forming a disulfide-bridge. IL, intracellular loop; EL, extracellular loop. (Right) Proposed arrangement of the seven transmembrane helices of opioid receptors as viewed from the top (extracellular side). (From Waldhoer, M. et al., *Ann. Rev. Biochem.*, 73, 953, 2004.)

TABLE 19.1
Opioid Receptor Ligands

Receptor	Agonist	Antagonist	Agonist Effect(s)
M	Morphiceptin	Naloxone	Analgesia
	DAGO		Respiratory depression
	Normorphine		Miosis
	Sufentanyl		Reduced gastrointestinal motility
			Nausea
			Vomiting
			Euphoria
Δ	Deltorphin	ICI 154,126	Supraspinal analgesia
	DPDPE	ICI 174,864	
	DADLE		
K	U 50,488	MR2266	Analgesia (spinal level)
			Trifluadom miosis (weak)
			Respiratory depression (weak)
			Dysphoria

Note: DAGO, Tyr-D-Ala-Gly-MePhe-Gly-ol; DPDPE, [D-Pen^2, D-Pen^5]enkephalin; Pen, penicillamine; DADLE, [D-Ala^2, D-Leu^5]enkephalin; deltorphin II, Tyr-D-Ala-Phe-Glu-Val-Val-Gly-NH$_2$; morphiceptin, β-casomorphin-(1–4)-amide or Tyr-Pro-Phe-Pro-NH$_2$.

At that time, there were no hints of what kind of compounds to look for. After 2 years of collecting extracts from pig brain and applying them in a functional bioassay, Kosterlitz and coworkers in 1975 identified two closely related endogenous pentapeptide opioids (Table 19.2).

The amino acid sequences are YGGFM and YGGFL and termed as [Met]- and [Leu]-enkephalin, respectively. Since then, many other peptide opioids of varying lengths have been identified. They are all cleavage products of longer peptides and can be divided into four families based on their precursors. Three of these families all start with the [Met]- and [Leu]-enkephalin. The endogenous opioid peptides have varying affinities for the opioid receptor subtypes; however, none of them are specific for a single subtype, although the neuropeptide nociceptin is the endogenous ligand specific for ORL$_1$. The precursors are often made up of repeating copies of the opioid peptide products.

High affinity opioid receptor peptides (dermorphins and deltorphins) have been isolated from frog skins and are quite unusual in having D-amino acids in the sequence. Also milk-derived casomorphins, hemorphins from hemoglobin, and cytochrophins (fragments of cytochrome B), have low affinity for the opioid receptors.

Besides the endogenous peptides, it has been shown that morphine is present in various tissues and body fluids and SH-SY5Y human neuroblastoma cells are capable of producing morphine. The biosynthetic route is similar to that found in *Papaver somniferum*.

19.1.3 NONENDOGENOUS OPIOID RECEPTOR LIGANDS

The synthetic efforts in the opioid field over the last century have mainly been stimulated by the search for a safer alternative to morphine that maintained the analgesic effects but was devoid of respiratory depression and abuse potential. Different medicinal chemistry approaches have been followed in the development of opioid receptor ligands:

TABLE 19.2
Endogenous Opioid Peptides

Precursor	Opioid Peptide Product	Amino Acid Sequence
Pro-enkephalin	[Met]-enkephalin	**YGGFM**
	[Leu]-enkephalin	**YGGFL**
		YGGFMRF
		YGGFMRGL
	Peptide E	**YGGFM**RRVGRPEWWMDYQKR**YGGFM**
	BAM 22P	**YGGFM**RRVGRPEWWMDYQKRYG
	Metorphamide	**YGGFM**RRVNH$_2$
Pro-opiomelanocortin	β-Endorphin	**YGGFM**TSEKSQTPLVTLFKNAIIKNAYKKGE
Prodynorphin	Dynorphin A	**YGGFL**RRIRPKLKWDNQ
	Dynorphin A(1–8)	**YGGFL**RRI
	Dynorphin B	**YGGFL**RRQFKVVT
	α-Neoendorphin	**YGGFL**RKYPK
	β-Neoendorphin	**YGGFL**RKYP
Pronociceptin/orphanin-FQ	Nociceptin/orphanin-FQ	FGGFTGARKSARKLANQ
	Endomorphin-1	YPWF-NH$_2$
	Endomorphin-2	YPFF-NH$_2$
Prodermorphin and prodeltorphin[a]	Dermorphin	Y(D)AFGYPS-NH$_2$
	Deltorphin	Y(D)MFHLMD-NH$_2$
	Deltorphin I	Y(D)AFDVVG-NH$_2$
	Deltorphin II	Y(D)AFEVVG-NH$_2$

Note: The pentapeptide sequences corresponding to [Met]- and [Leu]-enkephalin contained in other opioid peptides are shown in bold. Note that β-endorphin and most of the opioid peptides derived from proenkephalin contain [Met]-enkephalin at their N-termini, whereas the sequence of [Leu]-enkephalin is present in those peptides derived from prodynorphin.

[a] Dermorphin and deltorphins are derived from multiple precursors and all have a naturally occurring D-amino acid in position 2.

- Chemical modification of morphine and related structures (opiates)
- Simplification of the morphine structure
- Dimerization (bivalent ligands)
- Peptides and peptidomimetics

The early development was focused on the first two approaches. Examples of opiates that display similar affinity to all subtypes are shown in the upper part of Figure 19.3. Introduction of bulky substituents to the morphine structure generally yields antagonists, and naloxone and naltrexone are unselective antagonists. *N*-Allyl analogue nalorphine is an example of a mixed agonist–antagonist. It was originally characterized as an antagonist, but later shown to have antagonist activity at MOR but agonist activity at KOR. Nalorphine was one of the first compounds to be extensively tested in the clinic in combination with morphine to find an ideal agonist–antagonist ratio for maximizing analgesics properties and minimizing adverse effects. Buprenorphine is a potent analgesic and partial agonist at the MOR and antagonist at DOR and KOR. Diprenorphine is reported as an unselective antagonist.

An increasing number of subtype selective ligands has been reported and a few examples are shown in the lower part of Figure 19.3. Compounds that are μ-selective include morphine that is

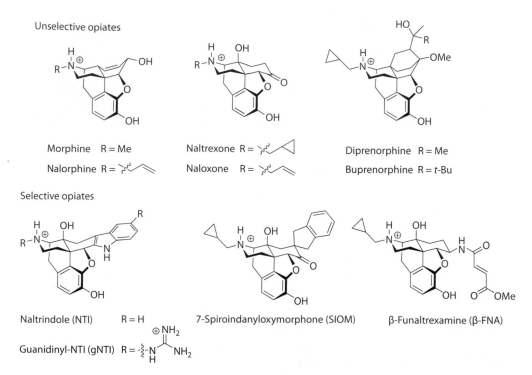

FIGURE 19.3 Chemical structures of classical unselective opiates (except morphine, which is μ-selective) based on the morphine scaffold and chemical structures of selective opiates. Morphine and β-FNA are μ-selective agonist and irreversible antagonist, respectively. SIOM and NTI are examples of δ-selective antagonists, whereas the introduction of charged guanidinium group converts NTI into the κ-selective antagonist gNTI.

an agonist and the irreversible antagonist β-funaltrexamine (β-FNA). 7-Spiroindanyloxymorphone (SIOM) and naltrindole (NTI) are examples of δ-selective antagonists, and guanidinyl-NTI (gNTI) represents a κ-selective antagonist. Interestingly, gNTI was recently shown to have higher affinity toward HEK-293 cells expressing KOR together with DOR or MOR compared to cells expressing only KOR, and the agonist effect of gNTI at the heterodimers KOR/DOR and KOR/MOR could be blocked by antagonists selective against DOR and MOR, respectively. This underlines the importance of heterodimerization and the fact that gNTI is analgesic when injected into the spinal cord but not when injected into the brain, could arise from different heteromeric populations.

The development of selective opiates has followed the "message–address" concept. This states that the amino and the aromatic group in the morphine determine the activity (the "message") of the opiates, whereas the lipophilic region around the allylic alcohol confers selectivity (the "address"). This is demonstrated by the conversion of NTI from a δ-selective antagonist into the κ-selective antagonist gNTI by the introduction of charged guanidinium group. Already in the 1960s, a 3D-pharmacophore model was conceived stating the importance of the spatial placement of the amine, the aromatic group, and the lipophilic region for ligand affinity. The successive breakdown of the morphine structure has led to a number of simpler nonopiate structures obeying this early and simple 3D-pharmacophore model. This breakdown is shown schematically in Figure 19.4 defining a structural classes of opioid receptor ligands developed over the last century.

Examples of these classes are μ-selective agonist fentanyl (piperidine), ethyl-ketocyclazine (benzomorphane), methadone (phenylpropylamine), and meperidine (piperidine) (Figure 19.5). However, other structural classes have appeared, such as δ-selective agonist SNC-80 and κ-selective agonist U50,488, and more recently several new scaffolds come from screening compound libraries of the cloned opioid receptors including heteromeric combinations.

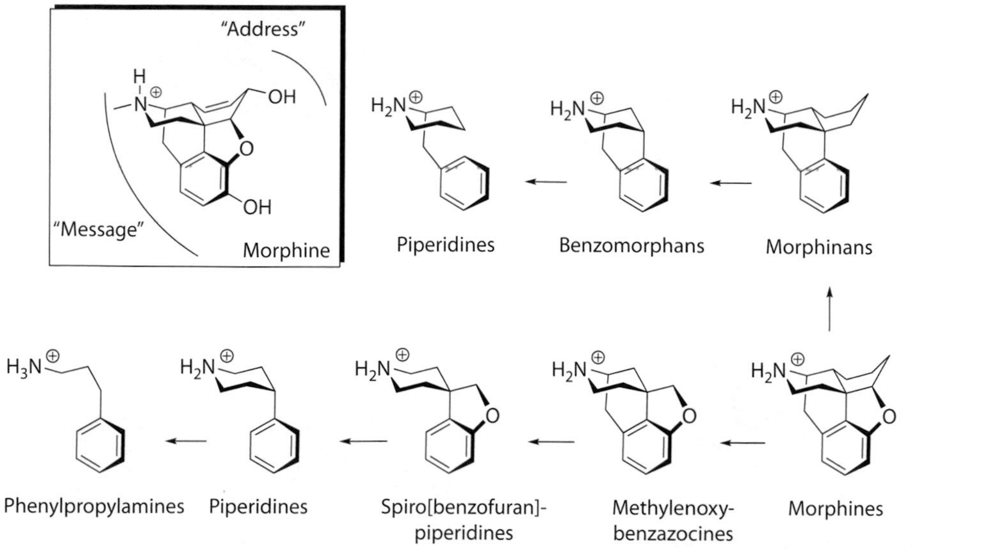

FIGURE 19.4 The message-address concept of the development of opiates is shown schematically in the box. The message region defines the activity of the compounds whereas the address region defines the selectivity of the compounds. The structural development in the progressive simplification of the morphine scaffold via morphinans and benzomorphans to piperidines, but also via benzazocines, spiropiperidines to piperidines and phenylpropylamines.

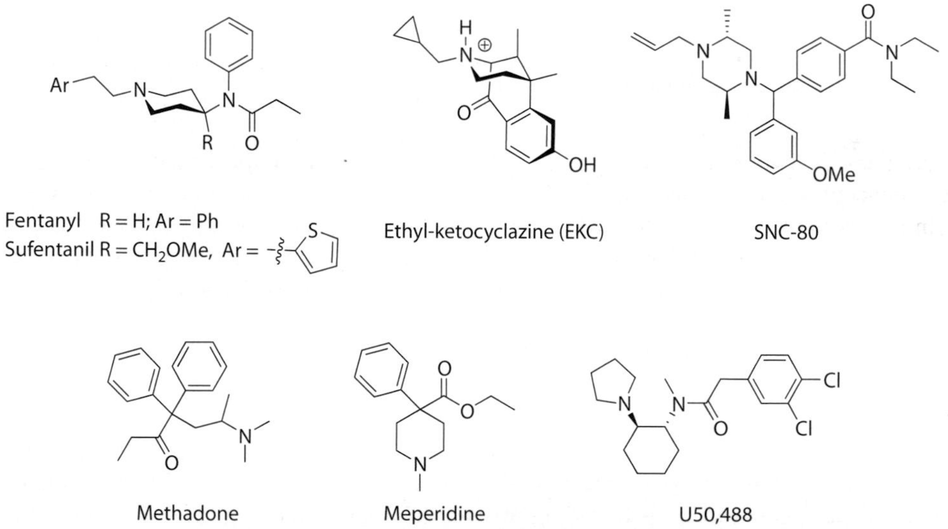

FIGURE 19.5 Examples of different structural classes of opioid receptor ligands.

The dimerization of ligands is a popular strategy in medicinal chemistry to alter the pharmacological properties of a monomeric ligand. This strategy was advanced in the early 1980s by Portoghese and coworkers using opiates. Initially, the idea was to develop such bivalent ligands with a spacer of optimal length that would exhibit a potency that is greater than that derived from the sum of its two monovalent pharmacophores. This would provide an evidence that the receptors existed as dimers. One of the first series where compounds **19.1** (Figure 19.6, n = 0, 2, 4, 6, 8), dimerizing a naltrexone analogue. The optimal spacer length was shown to be n = 4 giving the highest activity.

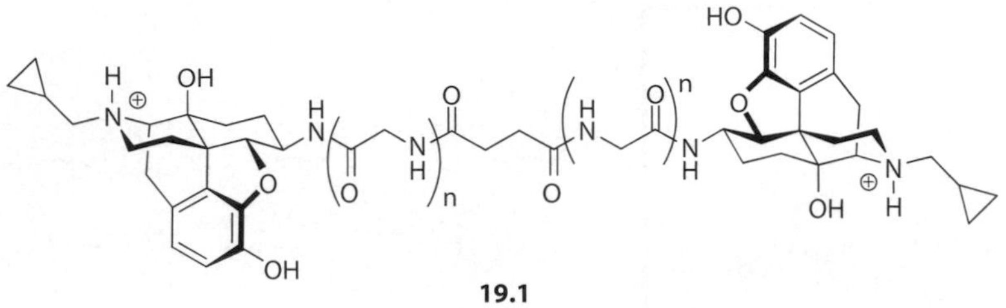

FIGURE 19.6 Examples of dimeric or bivalent opioid receptor ligands.

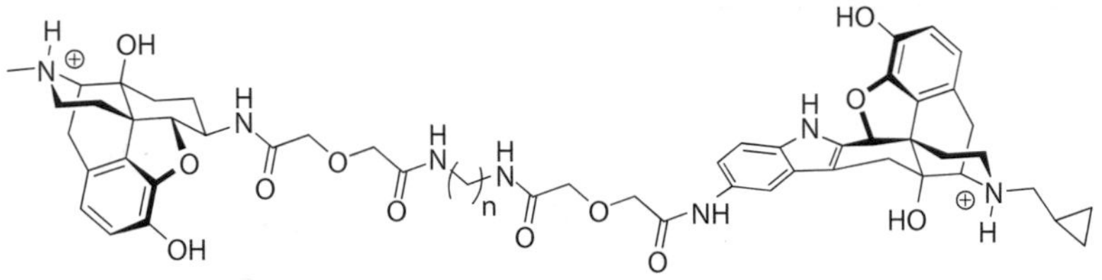

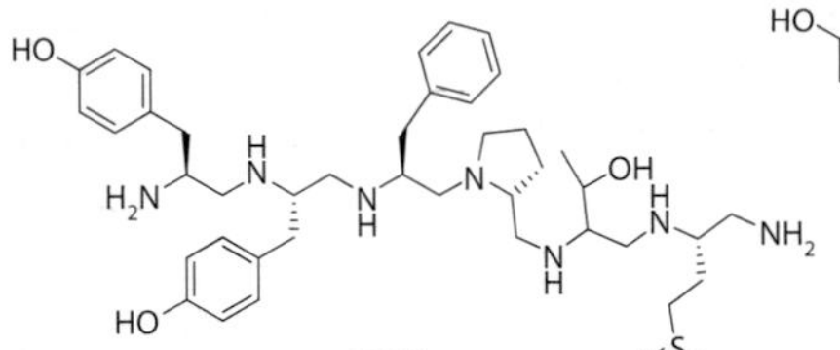

FIGURE 19.7 Examples of peptidomimetic opioid receptor ligands.

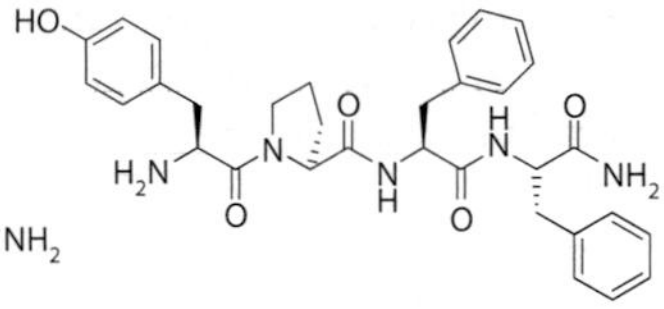

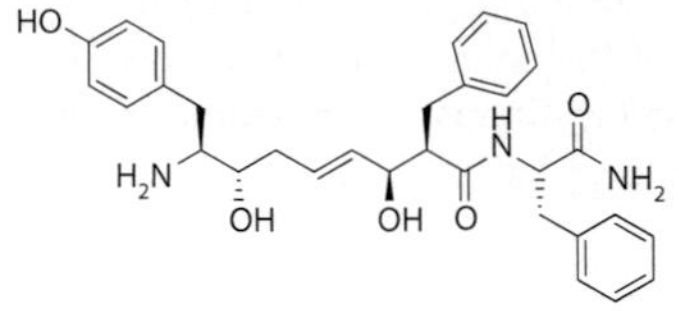

Today there is a substantial evidence to show that G-protein coupled receptors exist as dimers. The concept of making bivalent ligands has been shown to be applicable in many other areas wherein it is possible to modulate other pharmacological properties of a ligand such as degradation, uptake, etc. The concept has also been used to target heterodimeric receptor populations. For example, the dimerization of the analogues of naltrexone and NTI yields a series of heterobivalent ligands **19.2** ($n = 2–7$), where tolerance and dependence are significantly reduced with increasing linker length, whereas agonist potency is increased. It is hypothesized that δ-κ heterodimers are targeted specifically with longer linkers. Also, ligands that selectively target δ-κ heterodimers, which are localized in the spinal cord, have been developed.

The last approach that will be mentioned here is the use of peptides and peptidomimetics. New agonists and antagonists of opioid receptors have been obtained by making large combinatorial libraries of D- and L-amino acids including mix libraries and screening these compounds against MOR, KOR, and DOR. The sequences span from tetra- to decapeptides. In this way, potent and selective peptides have been obtained that differ from the endogenous peptides. Furthermore, the modification of the peptide backbone has yielded potent peptidomimetics. The modifications include minor modifications such as backbone amide alkylation. But examples of more extensive modifications are the use of a polyamine backbone as in compound **19.3**, or compound **19.4**, which is a peptidomimetic analogue of endomorphin-2, a potent agonist of MOR with high selectivity for MOR over DOR and KOR (Figure 19.7).

19.1.4 THERAPEUTIC APPLICATIONS AND PROSPECTS

Although the development of opiates has been spurred primarily by the search for efficient analgesics with few side effects, other clinical applications of opioid receptor agonists and antagonists are known. Agonists are primarily applied as analgesic, anesthetic, antitussive, and in the treatment of diarrhea. Morphine and codeine are mostly used as analgesics. Fentanyl is a very potent analgesic and used in anesthesia. Meperidine is used for acute pain. Methadone is applied to control the withdrawal of heroin from addicts. Antagonists are used for the reversal of some of the effects induced by agonists. Thus, naloxone has been used to reverse coma and respiratory depression of opioid overdose (methadone and heroin). It is also indicated as an adjunct agent to increase blood pressure under septic shock. Naltrexone has been approved as an adjunctive therapy in the treatment of alcohol dependence and the treatment of narcotic addiction to opioids. However, there are also potential indications including obesity, obsessive compulsive disorder, and schizophrenia.

19.2 CANNABINOID RECEPTORS

Different parts of the plant *Cannabis sativa* has for millennia been used for recreational and medicinal purposes, as it can be seen from old Chinese, Assyrian, and Roman literature. However, it was first in 1964 that the active principle causing the psychoactive effects was isolated and found to be Δ^9-tetrahydrocannabinol (THC) (Figure 19.8). Originally, it was thought that THC due to its lipophilicity somehow acted through fluidizing the cellular membranes, but in the early 1990s it was discovered that THC activates two receptors, cannabinoid receptor-1 (CB_1-receptor) and cannabinoid receptor-2 (CB_2-receptor). Cannabinoid effect in rodents is characterized by the so-called tetrad test. In this test, measurement of spontaneous activity, thermal pain sensation, catalepsy, and rectal temperature are made, and compounds with cannabinoid activity should produce hypomotility, analgesia, catalepsy, and hypothermia. Shortly after the discovery of the receptors, two endogenous compounds were identified that could activate these receptors, i.e., anandamide (*N*-arachidonoylethanolamine or arachidonoylethanolamide) and 2-arachidonoylglycerol (2-AG) (Figure 19.9), and they are called endocannabinoids. Both the endocannabinoids are lipids and thus not very water soluble. They may associate with albumin or lipoproteins in the extracellular space and endocannabinoids function in an autocrine and paracrine fashion where they are formed "on demand" and then degraded, i.e., they are not stored in vesicles like neurotransmitters or peptide hormones. Tissue levels of anandamide and 2-AG are usually in the pmol/g and nmol/g tissue, respectively, but it is not clear whether these levels represent the ligand concentration available to the receptors. However, it is generally considered that there is an endogenous tone of endocannabinoid level in most tissues. Endocannabinoids are also found in very small concentrations in plasma where they are thought to represent spill over from the tissues.

19.2.1 ENDOCANNABINOID SYSTEM

The endocannabinoid system comprises the two cannabinoid receptors, the two endocannabinoids and the enzymes that synthesize and degrade the endocannabinoids. Biosynthesis of anandamide is complex, but it is formed from an unusual phospholipid having three fatty acids, *N*-arachidonoyl-phosphatidylethanolamine that again is formed from phosphatidylethanolamine by the acylation of the amino group catalyzed by a poorly known calcium-stimulated *N*-acyltransferase.

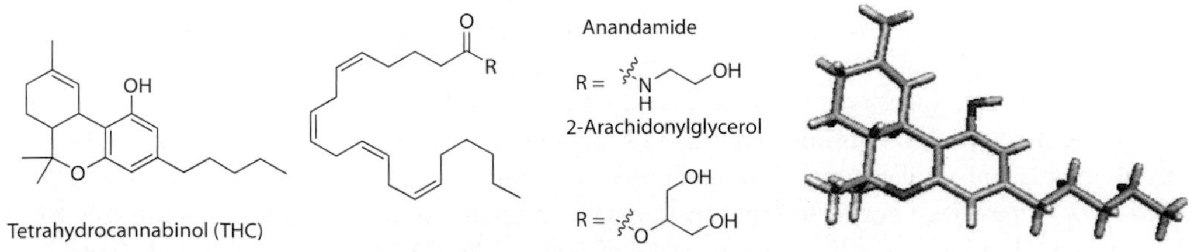

FIGURE 19.8 Plant cannabinoid (THC) and two endocannabinoids.

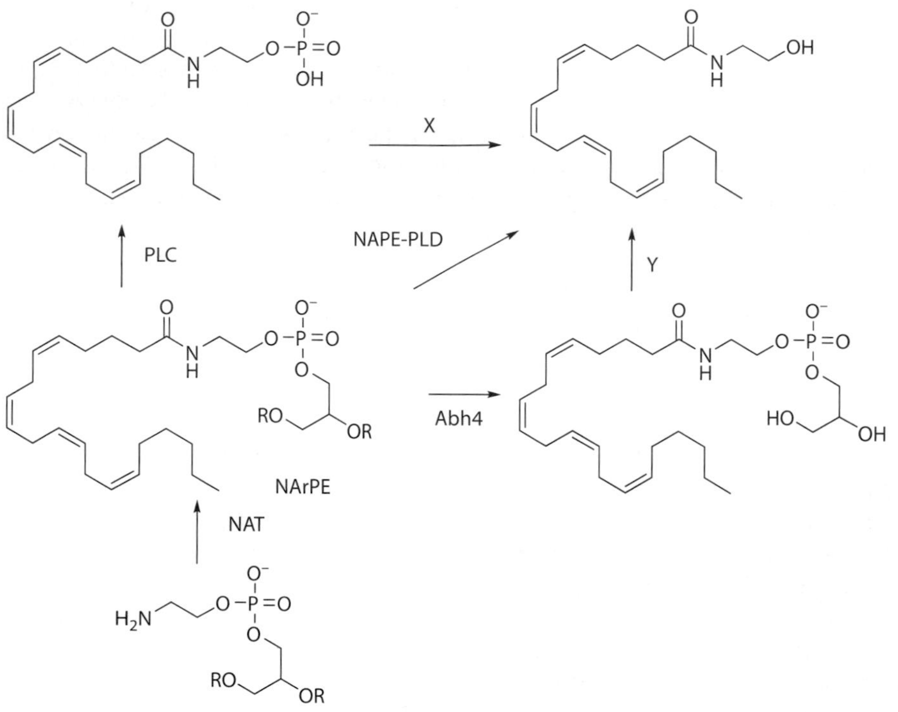

FIGURE 19.9 Biosynthesis of anandamide. The precursor phospholipids (NArPE) is generated from phosphatidylethanolamine by a *N*-acyltransferase (NAT). It can then be hydrolyzed by a phospholipase C (PLC), by *N*-acyl-phosphatidylethanolamine-hydrolyzing phospholipase D (NAPE-PLD), or by alpha-beta-hydrolase 4 (Abh4). Other acylethanolamides may be formed by the same enzymes. (R = fatty acids). The enzymes "X" and "Y" are not well characterized yet.

This enzyme uses acyl groups (e.g., arachidonoyl) from the *Sn*-1 position of phosphatidylcholine as substrate in the acylation process, and it will use whatever fatty acid is present. Thus, a number of *N*-acylphosphatidylethanolamines are always formed having different fatty acids in the *N*-acyl position of which *N*-arachidonoyl is only a minor component and those with palmitic acid, stearic acid, or oleic acid are much more abundant. From this precursor phospholipid, *N*-acyl-phosphatidylethanolamine, a number of different enzyme-catalyzed pathways can result in the generation of acylethanolamides including anandamide that usually amounts to less than 5% of the acylethanolamides (Figure 19.9). Which pathway is most relevant for a particular tissue or a particular physiological/pathophysiological setting is not known at present. The cellular localization of anandamide formation is not known and several of the involved enzymes have not been cloned yet. The different acylethanolamides have a number of more or less specific biological activities, e.g., palmitoylethanolamide is anti-inflammatory and oleoylethanolamide has anorexic and neuroprotective activity that may be mediated via the activation of a transcription factor PPARα, and/or an orphan receptor GPR119. Other acylethanolamides do not bind to the cannabinoid receptors. Anandamide is a partial agonist for the cannabinoid receptors but it can also activate vanilloid receptor and several different ion channels, but it is uncertain to what degree it does this *in vivo*. All acylethanolamides are degraded by a fatty acylethanolamide hydrolase (FAAH) and FAAH-knock out mice have increased levels of ananadmide and other acylethanolamides and increased pain threshold. Acylethanolamides can also be degraded by some other hydrolases. Endogenous levels of anandamide and other acylethanolamides are low and can be increased several fold during tissue injury. It has been suggested that there exists an anandamide transporter responsible for the uptake of anandamide into cells before it is degraded by the FAAH enzyme that is located in the endoplasmic reticulum. However, a transporter protein has not been characterized and the concept of an uptake transporter for a lipophilic molecule is disputed.

2-AG is formed primarily from diacylglycerol, e.g., 1-stearoyl-2-arachidonoyl-glycerol, catalyzed by a Sn-1 specific diacylglycerol lipase and it is degraded by a monoacylglycerol lipase. The precursor, diacylglycerol, is known to be formed during receptor-stimulated turnover of inositol phospholipids where inositol-1,4,5-trisphosphate is also formed. Thus, this diacylglycerol formation occurs in the cell membrane where the diacylglycerol lipase also is located. It is generally accepted that 2-AG is formed in postsynaptic neurons upon activation of neuronal inositol phospholipids whose turnover depends on phospholipase Cβ, whereupon it activates, in a retrograde fashion, the presynaptic CB_1-receptor that subsequently results in the inhibition of neurotransmitter release (Figure 19.10). It is not exactly known how 2-AG travels through the aqueous fluid to reach the presynaptic neuron. This retrograde signaling can decrease the release of glutamate, GABA, acetylcholine and other neurotransmitters. 2-AG is degraded by a monoacylglycerol lipase that is located in the presynaptic neuron. In this way 2-AG and the CB_1-receptors may contribute to homosynaptic plasticity of excitatory synapses and heterosynaptic plasticity between excitatory and inhibitory contacts that is part of the basic mechanism in learning and memory. This retrograde control is also called as the depolarisation-induced suppression of inhibition (DSI) and depolarisation-induced suppression of excitation (DSE) for GABAergic and glutamatergic synapses, respectively.

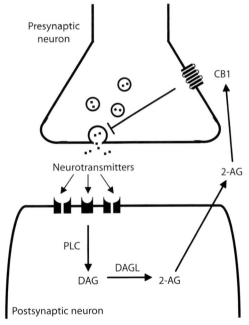

FIGURE 19.10   Synaptic endocannabinoid formation during neurotransmitter release. PLC, phospholipase C; DAG, diacylglycerol; DAGL, diacylglycerol lipase; 2-AG, 2-arachidonoyl glycerol; CB_1, cannabinoid receptor-1.

19.2.2   CANNABINOID RECEPTOR 1

CB_1-receptor belongs to the Class A rhodopsin-like family of GPCRs and it couples through $G_{i/o}$ proteins negatively to adenylate cyclase and positively to mitogen-activated protein kinase. In addition, CB_1-receptor can also couple to ion channels through the same G-proteins, positively to A-type and inwardly rectifying K^+-channels and negatively to N-type and P/Q type Ca^{2+}-channels. CB_1-receptor is primarily localized in the brain where it is particularly abundant in cortex, hippocampus, amygdale, basal ganglia, cerebellum, and the emetic centers of the brain stem. CB_1-receptor can also be found in lower abundance in spleen, tonsils, white blood cells, gastrointestinal tissue, urinary bladder, adrenal gland, heart, lung, and reproductive organs. CB_1-receptor may form homodimeric complexes, and heterodimeric complexes with μ-opioid receptor or dopamine D2 receptor. Anandamide and 2-AG may reach the receptor from the lipid phase of the membrane and not from the aqueous site.

In vitro CB_1-receptor seems to have constitutive activity or it may be under endocannabinoid stimulatory tone. Several antagonists have also been shown to be inverse agonists but it is unclear whether this has any *in vivo* significance.

Knock out mice that are lacking the receptor protein have been generated. They are generally healthy and fertile with no apparent gross anatomical defects. However, they do have a number of abnormalities, e.g., the dysregulation of the hypothalamus–pituitary–adrenal axis suggesting a role of endocannabinoids in modulating neuroendocrine functions. Furthermore, they have a lighter and leaner body phenotype and seem to have higher energy expenditure. In a number of experimental settings these mice do also behave differently, e.g., in studies of alcohol dependence. CB_1-receptor knock out mice do not show hypothermia, hypoalgesia, and hypoactivity in response to THC. There is evidence of splice variation of CB_1-receptor of very low abundance but their biological significance is unclear. Activation of GPR55 can be blocked by the CB receptor antagonist Rimonabant (refer to Figure 19.12), but not by the CB_2-receptor antagonist SR 144528 (refer to Figure 19.13).

19.2.3 Cannabinoid Receptor 2

CB_2-receptor also belongs to the 7TM-receptors and the human CB_2-receptor has 44% homology with human CB_1-receptor. It seems to couple to the same G-proteins and signaling pathways as does CB_1-receptor. However, CB_2-receptor is found primarily in the spleen, immune cells, tonsils, and brain microglial cells where its expression can be induced by transformation of microglia to macrophage-like cells. Furthermore, CB_2-receptor is found in osteoblasts, osteocytes, and osteoclast where it plays a critical role in the maintenance of normal bone mass. CB_2-receptor knock out mice appear healthy and fertile, but they have low-bone mass. In animal models of pain and inflammation, these CB_2-receptor knock out mice have indicated a clear role for this receptor in modulating acute pain, chronic inflammatory pain, postsurgical pain, cancer pain, and pain associated with nerve injury.

19.2.4 Other Cannabinoid Receptors

Since 2-AG and especially anandamide have a number of pharmacological effects that cannot be fully explained by activation of the known cannabinoid receptors, it has been suggested that there may exist more cannabinoid-like receptors. One such receptor may be the recently described GPR55 that is found in the brain and that can potently be activated by both THC, 2-AG, anandamide and the cannabinoid receptor agonist CP 55,940 (refer to Figure 19.12). Another cannabinoid receptor agonist, WIN 55,212-2, can however not activate GPR55. A number of agonists (e.g., CP 55,940, WIN 55,212-2, O-1812) and antagonists (e.g., Rimonabant, AM251, LY 320125) (refer to Figure 19.12) have been developed. The biological significance of GPR55 is at present not known. Especially anandamide can activate a number of other receptors and ion channels as mentioned above and this may add to the confusion regarding the existence of additional cannabinoid receptors.

19.2.5 Therapeutic Use and Potential

THC in capsules (Marinol, Solway Pharmaceutical) is used for treatment of nausea and vomiting that are common side effects of chemotherapy, and for stimulation of appetite in AIDS patients. In Canada, THC in the form of an extract of *cannabis sativa* called Sativex (GW Pharmaceuticals) is provided as a mouth spray for multiple sclerosis patients who can use it to alleviate neuropathic pain and spasticity. Sativex also contains other cannabinoids including cannabidiol that may add to its function. Medicinal cannabis, i.e., marihuana or hashish prescribed by a doctor for increased well being and alleviation of pain, spasticity, or loss of appetite by patients having AIDS, cancer, and multiple sclerosis has been approved in the Netherlands and there is a strong lobby for approval in certain states of United States.

Sanofi-Aventis has a CB$_1$-antagonist (SR141716A, Rimonabant, trade name Acomplia, refer to Figure 19.12) on the European market for the treatment of obesity (body mass index above 30). Large clinical trials have shown that Rimonabant induces a weight loss of ~10% of initial body weight within 1 year. The discontinuation of Rimonabant treatment results in the regain of lost weight. It is not clear how large a fraction of the weight loss effect of rimonabant is mediated by CB$_1$-receptors in the CNS relative to CB$_1$-receptors in the peripheral organs like the liver and adipose tissue. Side effects are increased frequency of nausea and mood disturbances like depression. Rimonabant is not approved in United States. A CB$_1$-receptor antagonist that does not cross the blood–brain barrier may possibly also have beneficial effects on energy metabolism.

Emerging evidence points to a possible participation of the endocannabinoid system in the regulation of the relapsing phenomenon of drug abuse in animal models. CB$_1$-receptor seems to be important in drug as well as cue-induced reinstatement of drug seeking behavior. Stimulation may elicit relapse not only to cannabinoid seeking but also to cocaine, heroin, alcohol, and methamphetamine, and this effect is significantly attenuated in animal experiments by pretreatment with CB$_1$-receptor antagonists.

Potential clinical application involves drugs that can increase or decrease endocannabinoid levels (i.e., FAAH-inhibitors and monoacylglycerol-lipase inhibitors or diacylglycerol-lipase inhibitors, respectively) or serve as agonists/antagonists for the two cannabinoid receptors (Table 19.3). Thus the potential is large but so is also the risk for side effects since the endocannabinoid system appears to be so ubiquitous.

TABLE 19.3
Potential Clinical Applications of the Endocannabinoid System

Clinical Conditions	Therapeutic Target			
	CB$_1$-Agonists	CB$_2$-Agonists	CB$_1$-Antagonists	FAAH-Inhibitors
Alzheimer's disease	X	X		
AIDS and Cancer				
Appetite stimulation and inhibition of nausea and vomiting	#			
Cancer				
Inhibition of growth, angiogenesis, and metastasis	X	X		X
Inflammatory bowel diseases				
Stimulation of gastrointestinal mobility and reduction of inflammation	X	X		
Multiple sclerosis	X			X
Inhibition of tremors and spasticity				
Obesity X			X#	
Weight loss				
Osteoporosis		X#	X	
Inhibition of bone loss				
Parkinson's disease			X	
Pain				
Chronic, inflammatory, and neuropathic	X	X		X

Note: X, based on preclinical data from corresponding animal models; #, based on data from human studies or clinical trial.

19.2.6   FAAH-Inhibitors and Anandamide Uptake Inhibitors

As discussed above it is not clear whether an anandamide transporter exists and several compounds believed to be uptake inhibitors have turned out to be inhibitors of FAAH. URB597 is shown as an example of a experimental FAAH-inhibitor (Figure 19.11).

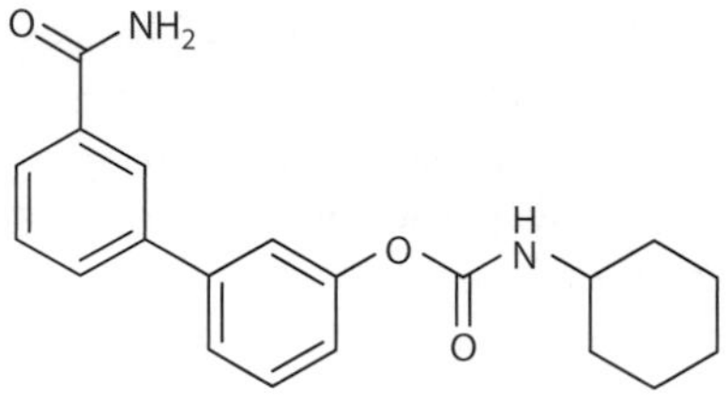

URB597

FIGURE 19.11   Inhibitor of FAAH, the enzyme degrading anandamide and other acylethanolamides.

19.2.7   CB₁-Receptor Agonist/Antagonist

Besides THC, anandamide, and 2-AG, experimentally used synthetic CB$_1$-receptor agonists involves CP55,940 and the aminoalkylindole WIN55,212-2 that target both the cannabinoid receptors. More selective CB$_1$-receptor agonists are arachidonic acid derivatives like O-1812. Selective CB$_1$-receptor antagonist involves rimonabant that are in clinical use and the experimental compounds AM251 and LY 320135 (Figure 19.12).

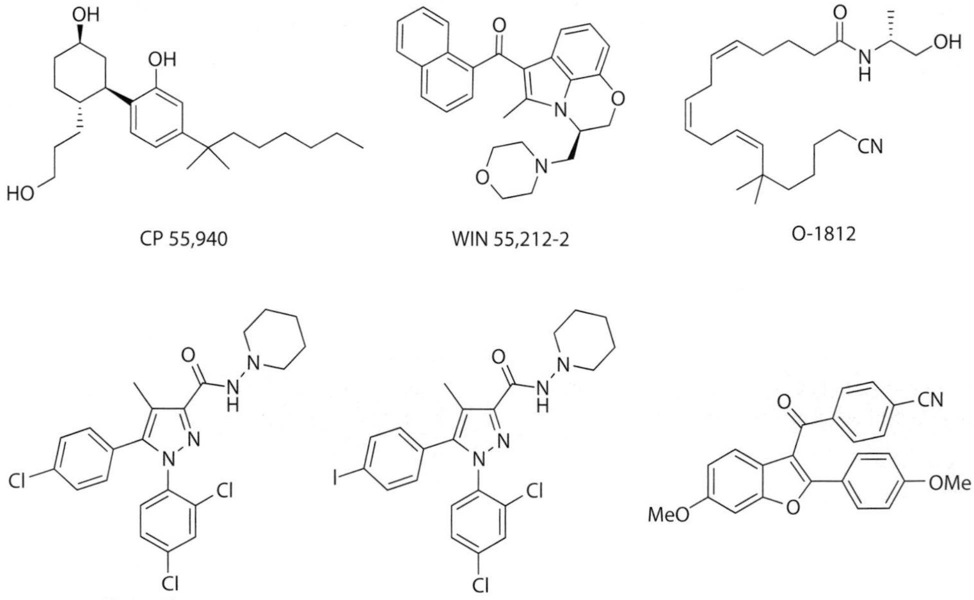

CP 55,940   WIN 55,212-2   O-1812

Rimonabant   AM251   LY 320135

FIGURE 19.12   Agonists and antagonists for CB$_1$-receptor.

19.2.8   CB₂-Receptor Agonist/Antagonist

THC, 2-AG, and to a lesser extent anandamide activates the CB$_2$-receptor. Selective synthetic agonists include JHW 133 and GW 405833 (partial agonist) that are used experimentally (Figure 19.13). SR 144528 is a selective CB$_2$-receptor antagonist.

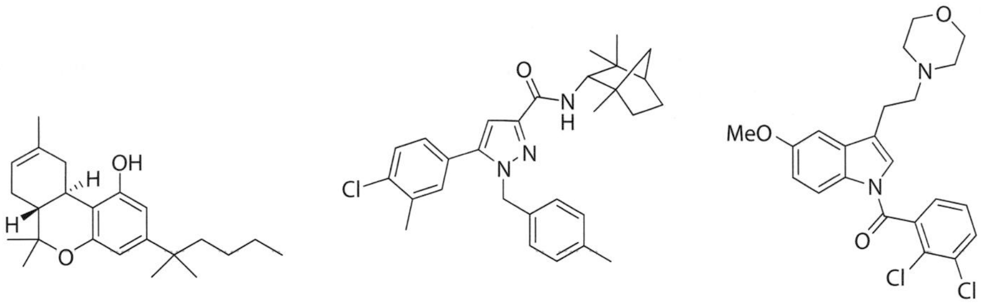

JHW 133 GW 405833 SR 144528

FIGURE 19.13 Agonists and antagonists for CB_2-receptor.

19.2.9 Diacylglycerol Lipase Inhibitors

There are no really specific inhibitors of diacylglycerol lipase, but RHC 80267, tetrahydrolipstatin (Orlistat), and O-3841 have been used in experimental settings (Figure 19.14).

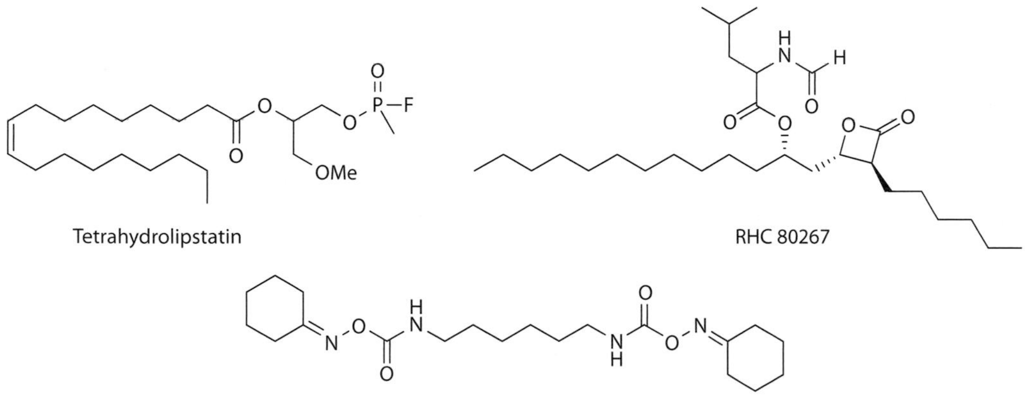

Tetrahydrolipstatin RHC 80267

O-3841

FIGURE 19.14 Inhibitors of diacylglycerol lipase, the enzyme generating 2-arachidonoylglycerol.

19.2.10 Monoacylglycerol Lipase Inhibitors

URB602 (Figure 19.15) is used as an inhibitor of monoacylglycerol lipase in experimental settings, but is not a very specific inhibitor.

As can be seen from the description of the enzyme inhibitors above, this research area is at the very early stage of drug discovery, but if potent and specific inhibitors can be found, there are numerous clinical settings where such enzyme inhibitors may be alternatives to cannabinoid receptor agonists or antagonists.

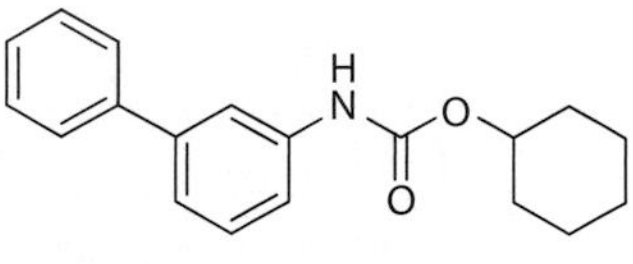

URB602

FIGURE 19.15 Inhibitor of monoacylglycerol lipase, the enzyme degrading 2-arachidonoylglycerol.

FURTHER READINGS

Corbett, A. D., Henderson, G., McKnight, A. T., and Paterson, S. J. (2006) 75 years of opioid research: The exciting but vain quest for the Holy Grail. *Br. J. Pharmacol.* 147, S153–S162.

Di Marzo, V. and Petrosino, S. (2007) Endocannabinoids and the regulation of their levels in health and disease. *Curr. Opin. Lipidol.* 18, 129–140.

Fattore, L., Spano, M. S., Deiana, S., Melis, V., Cossu, G., Fadda, P., and Fratta, W. (2007) An endocannabinoid mechanism in relapse to drug seeking: A review of animal studies and clinical perspectives. *Brain Res. Rev.* 53, 1–16.

Hansen, H. S., Petersen, G., Artmann, A., and Madsen, A. N. (2006) Endocannabinoids. *Eur. J. Lipid Sci. Technol.* 108, 877–889.

Kaine, B. E., Svensson, B., and Ferguson, D. M. (2006) Molecular recognition of opioid receptor ligands. *AAPS J.* 8 (1), Article 15, E126–E137.

Matias, I. and Di Marzo, V. (2007) Endocannabinoids and control of energy balance. *Trends Endocrinol. Metab.* 18, 27–37.

Pacher, P., Batkai, S., and Kunos, G. (2006) The endocannabinoid system as an emerging target of pharmacotherapy. *Pharmacol. Rev.* 58, 389–462.

Pertwee, R. G. (editor) (2005) Cannabinoids. *Handbook of Experimental Pharmacology*, Vol. 168. Springer-Verlag, Berlin, Heidelberg.

Portoghese, P. S. (2001) From models to molecules: Opioid receptor dimers, bivalent ligands and selective opioid receptor probes. J. *Med. Chem.* 44(14), 2260–2269.

Waldhoer, M., Bartlett, S. E., and Whistler, J. L. (2004) Opioid receptors. *Ann. Rev. Biochem.* 73, 953–990.

20 Hypnotics

Bjarke Ebert and Keith A. Wafford

CONTENTS

20.1 INTRODUCTION

Problems with sleep (falling asleep, maintaining sleep, waking up early, or not feeling refreshed after sleep) can occur in some patients as a disease in its own right and in other patients as a comorbidity associated with diseases like depression, stress, and pain. In addition to the impairment of daytime alertness and reduced performance, chronic insomnia has dire consequences for the quality of life. Clinical epidemiological studies have linked poor sleep with depression, increased blood pressure, and type 2 diabetes. The treatment of poor sleep may therefore have more wide ranging effects than purely increase the sleep time during the night. However, when novel hypnotics have been developed over the last decades, focus has not been on the improvement of quality of life or day time function in general, but rather the metrics of sleep. The reason for this is probably, that the regulatory guidelines indicate that hypnotics should be characterized in primary insomniacs, which are insomnia patients devoid of clinical consequences of the impaired sleep. With the absence of daytime consequences pharmaceutical companies, academia, and regulatory authorities in unison have neglected or forgotten that sleep is not about metrics during the night but the quality of life during the day.

Hypnotic drugs used to treat insomnia sufferers are characterized by their effects on induction and maintenance of sleep. These hypnotics—primarily benzodiazepines and benzodiazepine receptor agonists (BzRAs)—are highly effective drugs, which are able to induce and maintain sleep. When it comes to improvement of daytime performance and thereby the quality of life; however, existing hypnotic drugs may not improve or may even worsen these types of parameters.

This chapter will cover the basics of sleep physiology, followed by a description of different treatment modalities from the current benzodiazepines and BzRAs to novel strategies like $5\text{-}HT_{2A}$, $GABA_A$ receptor agonists, melatonergics, and orexinergics, which may advance the treatment of sleep problems.

20.2   SLEEP BASICS

Sleep is a state of the brain, which is shared by most organisms from invertebrates to mammals. The mechanisms underlying the transition from awake to sleep are now understood in some detail, whereas the function of sleep still remains enigmatic. As illustrated in Figure 20.1, the transition from wake to sleep is associated with a change in the activity of different neuronal pathways. Earlier sleep was considered a resting state of the brain—an impression conveyed by the clinical signs of sleep: reduced blood pressure, reduced heart rate, immobility, and reduced sensitivity toward external stimuli. However, measurements of brain activity and imaging studies performed during sleep have clearly shown that sleep is an active state of the brain. Sleep itself is not homogenous, and can be divided into various states representing different patterns of electrical activity. The underlying biology and complex interaction of various neurotransmitter systems in initiating, maintaining, and shaping sleep are now beginning to be better understood, paving the way for more precise and effective pharmacological treatment of sleep disorders.

Like waking, sleep is an active state of the brain. During this heterogeneous and rapidly changing state several restorative functions take place, although the neural substrates of somatic and cognitive restoration remain elusive.

Sleep is generally considered to consist of two substates, rapid eye movement (REM) and nonrapid eye movement (NREM) sleep, which alternate to form a cycle lasting approximately 90 min (Figure 20.2). REM and NREM sleep can clearly be differentiated on the basis of a number of physiological variables including muscle tone, electroencephalographic (EEG), and electromyelographic (EMG) features, and the presence or absence of REMs. Distinct physiological roles for REM and NREM stages have been proposed, but compelling empirical data are scarce.

Real sleep in a living brain is a continuous state without clear transitions. Therefore, a temporal description of the waves and alterations in the amount of both frequencies and amplitudes should most likely be based on an analysis of these waveforms. However, for historical reasons sleep stages are described as either REM or NREM stages 1–4 using visual scoring criteria based, in part, on the quantity and gross type of EEG waveforms per unit time. These are combined together graphically into a hypnogram as shown in Figures 20.2 and 20.3. NREM stages 1 and 2 have been described as

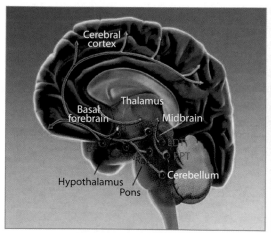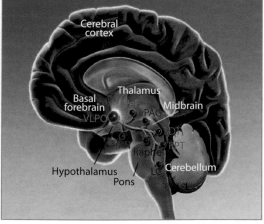

FIGURE 20.1   Neuronal pathways active during wake and sleep. Left: During wake, several arousal systems are active. These include monoaminergic (noradrenalin and serotonin originating in LC, Raphe, and TMN) and orexinergic (originating in LH) systems. Right: During sleep, the inhibitory GABA (VLPO) and melatonergic systems (originating in the pineal gland and projecting to thalamus) take over and initiate and maintain sleep and the transition between different sleep stages. PPT, pedunculopontine nuclei; LDT, laterodorsal tegmental nuclei; LC, locus coeruleus; TMN, tuberomammillary nucleus; vPAG, A10 cell group; LH, lateral hypothalamus; BF, basal forebrain; VLPO, ventrolateral preoptic nucleus; PeF, perifornical neurons.

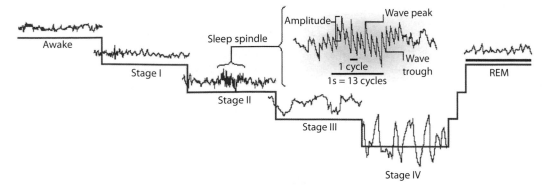

FIGURE 20.2   EEG patterns in humans during wake and different sleep stages. (Data from Pace-Schott, E.F. and Hobson, J.A., *Nat. Rev. Neurosci.*, 3, 591, 2002.)

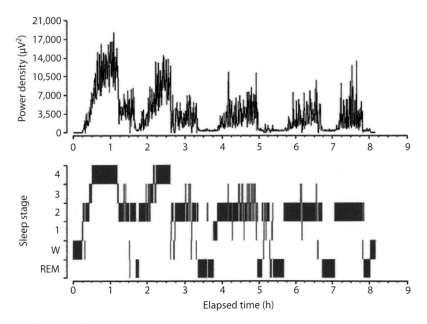

FIGURE 20.3   SWS and SWA. The bottom panel illustrates a hypnogram showing the amount of time spent in the different sleep stages over the course of the night, using traditional visual scoring criteria. Each progression through the five stages (stages 1–4 and REM) constitutes a sleep cycle. The top panel represents a fast Fourier transformation of EEG data. The frequency band selected for power analysis was 0.75–4.5 Hz. This representation illustrates that SWA is present throughout the night, even when the sleeper does not reach stage 3 or 4 according to traditional visual scoring criteria. (From Ebert, B. et al., *Pharmacol. Ther.*, 112, 612, 2006. With permission.)

light sleep, while stages 3 and 4 are often described as deep or slow wave sleep (SWS). The hallmark waveform of SWS consists of rhythmic, low frequency waves (~0.5–4.5 Hz) with large amplitude. The amount of slow waves can be quantified by a Fourier transformation, whereby any pattern can be described by a combination of a series of sinusoidal waves with different frequencies and amplitudes. The numeric amplitude squared is a reflection of the amount of a certain frequency and is termed power. The power spectrum for slow waves (slow wave activity or SWA) during NREM sleep (Figure 20.3 top) is particular strong during stages 3 and 4, but are present throughout the sleep period, although to a much smaller extent.

SWS/SWA in particular may play an important role in somatic and cognitive restoration, including the consolidation of certain forms of procedural and declarative memory. A substantial diminution in the amount of SWS/SWA occurs across the human lifespan. This decline is beginning already in adolescence and middle-aged adults have only 25% of the SWS observed in young adults, whereas the elderly have almost none. While the clinical importance of these phenomena is unknown, it is reasonable to speculate that they may be related to the increase of sleep complaints associated with aging.

20.3   PHARMACOLOGICAL MODULATION OF THE SLEEP SYSTEM

The treatment of sleep disorders currently focuses at inducing and maintaining sleep. This is ensured via the positive modulation of the $GABA_A$ receptor system. Other drugs with sedative side effects (e.g., antihistamines and antidepressants) are also used, but these are often prescribed off-label, and have not been evaluated within this indication with the same rigor as the hypnotics. In addition to this, new pharmacological agents, such as, antagonists at orexin receptors, which are involved in wake promotion, and $5HT_{2A}$ antagonists, which promote SWS are in clinical development. Since most reported studies with these compounds have included patients with primary insomnia, and thus no comorbidity (meaning either other disease or functional consequences of the sleep disturbances), it still remains an open question whether these approaches might constitute an advantage over the BzRAs in this regard. Finally, modification of the circadian rhythm, which in depressed patients and in elderly patients is severely impaired, is being targeted using melatonin and melatonergic agonists.

20.3.1   INDUCTION AND MAINTENANCE OF SLEEP

20.3.1.1   Benzodiazepines and Benzodiazepine Receptor Agonists

Using the benzodiazepine structure as a template, thousands of molecules with similar pharmacological activities in vivo have been developed. Structure–activity studies have identified essential and forbidden areas of the structure and this analysis has led to the development of the so-called non-benzodiazepine hypnotics (hypnotics without the 1,4 benzodiazepine ring structure) (Figure 20.4).

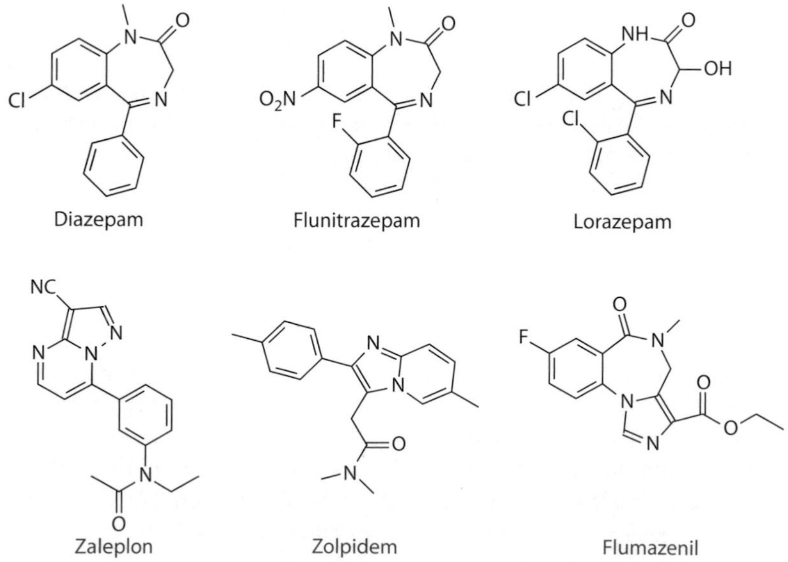

Diazepam	Flunitrazepam	Lorazepam
Zaleplon	Zolpidem	Flumazenil

FIGURE 20.4   Structure of BzRAs and the competitive benzodiazepine receptor antagonist flumazenil.

Since all these compounds are increasing activity at the GABA$_A$ receptor via a positive allosteric interaction with one common binding motif, the term agonist has been applied. It should be noted, however, that the activity of benzodiazepines is dependent on the activation of the receptor by GABA, the ligand that directly gates the ion channel (see Chapter 15). These compounds are best referred to as positive allosteric modulators or BzRAs.

The GABA$_A$ receptor is a pentameric structure composed of five subunits surrounding a central ion channel pore. The receptor subunits comprise of families of related proteins termed α1–6, β1–3, γ1–3, δ, ϵ, π, θ, and ρ. The majority of receptors are composed of 2α, 2β, and 1γ subunits. Other subunits can substitute for γ, such as, δ to form a benzodiazepine insensitive subtype of receptor. The binding site for BzRAs is located at the interface between α and γ subunits in the pentameric assembly of subunits (Figure 20.5). Since the γ subunit of the GABA$_A$ receptor contains a binding motif for a synaptic intercellular protein, γ containing receptors are predominantly synaptically located.

Detailed pharmacological studies have revealed that BzRAs enhance the opening frequency of the activated receptor, thereby allowing more current (more specifically chloride ions) to pass through the receptor-controlled channel within a fixed period of time. On the macroscopic level,

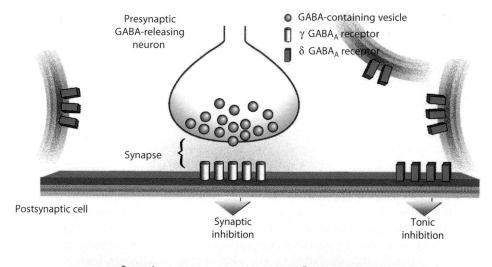

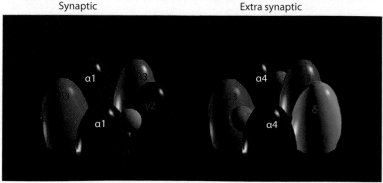

FIGURE 20.5   Top: Localization of GABA$_A$ receptors. γ containing GABA$_A$ receptors are predominantly located in the synapse and determine phasic inhibition. In contrast, δ containing GABA$_A$ receptors are located outside the synapse (extra-synaptically) and contribute to tonic inhibition. Bottom: Synaptic and extra-synaptic receptors are different in subunit composition and pharmacology. Whereas synaptic receptors contain the γ subunit and are sensitive to BzRAs, extra-synaptic receptors contain δ (or only α and β) and are insensitive to modulation by BzRAs. The binding site for BzRAs is located at the interface of α and γ, whereas the GABA binding site is located at the interface between α and β.

this is detected as an enhanced response to a fixed concentration of GABA or a leftward parallel shift of the GABA concentration response curve.

BzRAs can, depending on their ability to shift the GABA dose response curve, be characterized as full or partial agonists. The efficacy (or rather maximum effect) of the BzRA is determined at a concentration, which saturates the allosteric binding site, whereby the GABA concentration is the determining factor.

Variation in α subunit and γ subunit confer a degree of heterogeneity at the benzodiazepine binding site, and it has been possible to develop several binding-affinity based selective BzRAs. It should be noted however that the $\gamma 2$ subunit is by far the most predominant subtype and limits the selectivity based on α-subtype. To date, compounds with modest $\alpha 1$ selectivity and $\alpha 5$ selectivity have been discovered. Those with $\alpha 1$ selectivity appear to be more sedative in nature than nonselective BzRAs. Since the binding site for BzRAs is allosteric in nature, it has also been possible to generate compounds with functional selectivity, behaving as agonists at one subtype but antagonists at another. Compounds with some selectivity for $\alpha 1$ have been developed as hypnotic agents, such as, zolpidem (Ambien) and zaleplon (Sonata).

By the application of molecular biology techniques, it has been possible to identify the amino acid residues essential for the affinity of the BzRAs. The mutation of one single amino acid (Histidine 101 to Arginine) can completely abolish benzodiazepine affinity and this difference in pharmacology can be observed when BzRAs are compared at $\alpha 1$ and $\alpha 1$ H101R containing $GABA_A$ receptors. By introducing the $\alpha 1$ H101R mutation in mice, it has been possible to show that the strong sedative effects of the unselective benzodiazepine diazepam and the $\alpha 1$ selective BzRA zolpidem were strongly reduced. Furthermore, sleep studies with zolpidem in these transgenic mice strongly indicated that the hypnotic effects of zolpidem indeed are mediated primarily via $\alpha 1$ containing $GABA_A$ receptors. These types of studies are obviously very valuable for the characterization of the contribution of different receptor populations to the overall pharmacological consequences of a compound in vivo. However, the pharmacological consequences of a compound are a composite of interactions with potentially several different types of receptors. In order to address the potential for functional heterogeneity in vivo, it is necessary to obtain knowledge on exposure (CNS concentrations and eventually a time dependent profile) and activity (potency and efficacy) at the relevant individual receptor populations. Very seldom are all these data available. However, in a recent study, a series of compounds were systematically characterized at different human $GABA_A$ receptors, expressed in *Xenopus* oocytes, by means of electrophysiology. Since most of these compounds have been characterized clinically and BzRAs freely penetrate the blood–brain barrier (BBB), a very good estimate of both CNS exposure and receptor activation can be obtained.

As illustrated in Figure 20.6, CNS concentrations for these compounds are well above the EC_{50} values at the individual receptor combinations. A compound like indiplon in fact is used clinically at such a high concentration that the functional selectivity is not determined by the subunit dependent potency, but rather by the maximum response at the different receptor types. This means that the in vitro selectivity of these compounds, which is seen at very low concentrations, is misleading when the clinical relevant concentrations are factored in. Consequently, interpretation of in vivo data from man or animal studies, solely based on the in vitro profile may be highly ambiguous.

BzRAs have several serious clinical limitations. Firstly, a fading in response after long-term treatment or tolerance. Secondly, risks for development of dependence especially over long-term treatment, and thirdly, abuse liability. All these aspects can be addressed in preclinical studies, and very consistently, BzRAs (irrespective of in vitro subtype selectivity) have been shown to possess all these risk factors. However, although compounds may show development of tolerance in animal studies, this may not be directly translatable into the clinical situation. Examples of this are the fast acting hypnotics zolpidem and indiplon, which after long-term dosing in animals induces a down regulation in $GABA_A$ receptors, predictable for tolerance development and withdrawal symptoms. However, very few reports on these types of side effects are available in the clinical literature to date, and as a consequence, the American FDA has now removed the restriction of only short term

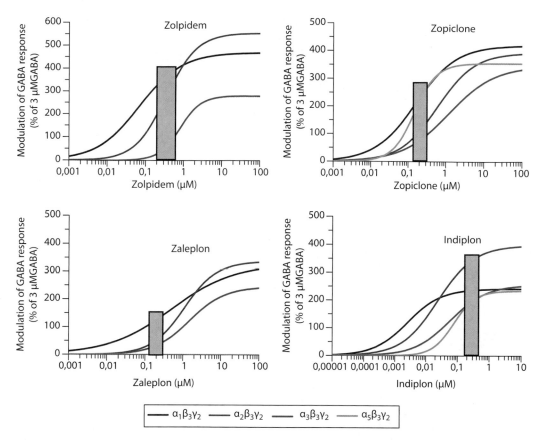

FIGURE 20.6   Modulation of $3\,\mu M$ GABA by a series of BzRAs at therapeutic relevant concentrations. The therapeutic relevant concentration is marked with a blue box. Dose response curves are from in vitro functional experiments in *Xenopus* oocytes. (Adapted from Petroski, R.E. et al., *J. Pharmacol. Exp. Ther.*, 317, 369, 2006.)

use of these BzRAs. The reason for the discrepancy between animal and human data most likely relates to differences in levels and duration of exposure. Most preclinical studies are conducted with the aim of demonstrating a certain pharmacological effect. Therefore most tolerance development studies and receptor down regulation studies are carried out with constant and high exposure, which under normal circumstances may be highly irrelevant from a clinical perspective, but certainly addresses aspects of the mechanisms underlying tolerance development. In contrast, therapeutic exposure with hypnotics usually last for only a few hours per 24h and the peak concentration is selected as a compromise between optimal effectiveness and side effects. The duration of exposure relative to nonexposure is therefore very small and this may allow resensitization of desensitized receptors and prevent significant down regulation.

This should also be borne in mind with the development of sustained release formulations, in that tolerance development may be more likely to occur with this type of therapy, where receptors are exposed to the drug for longer. Since these are relatively new to the market, no clinical data has yet been published on these new formulations.

Preclinical studies have, as indicated above, consistently demonstrated that BzRAs are abusable. Since all hypnotic BzRAs possess a very strong affinity for α1 containing receptors, it has been assumed that this subunit drives both abuse liability and hypnotic effects. If this indeed is the case, BzRA-based compounds will always be associated with this problem. However, as illustrated above, at clinically meaningful concentrations, BzRAs all show significant activities at α2 and α3 containing receptors. Abuse liability studies conducted by Professor Ator at Johns Hopkins School of medicine, have indicated that the level of α2 modulation may be a primary determinant of self-administration, suggesting that truly functionally selective α1 modulators, may be devoid of abuse potential.

Currently no compounds with this particular pharmacological profile have been reported, but since several large pharmaceutical companies have conducted BzRA projects over the last 2–3 decades, these compounds may already have been synthesized.

An alternative to the full BzRAs are partial BzRAs, which in insomnia-related indications are sufficiently efficacious to induce the desired induction and maintenance of sleep. As an example, EVT-201 is active throughout the night and at the same time devoid of residual effects as measured the next morning. The clinical advantage of partial positive allosteric modulators over full BzRAs should in principle be a reduced risk of tolerance development, and this may be demonstrated in preclinical studies. However, it will be important to establish these benefits in the clinical setting.

20.3.2 MODULATING SLOW WAVE SLEEP AND SLOW WAVE ACTIVITY

20.3.2.1 GABA$_A$ Receptor Agonists

Since the GABA receptor system is the major inhibitory neurotransmitter system in the CNS, it is hardly surprising that GABA$_A$ receptor agonists and positive allosteric modulators are sedative. However, sleep is not just a period of unconsciousness. Sleep is a very dynamic process with a large degree of ongoing neuronal general and unspecific dampening of neuronal activity. Therefore, sleep-enhancing agents will act somewhat differently to anesthetics, which produce unconsciousness combined with insensitivity to external stimuli and often combined with analgesia. Hence anesthetics like propofol and etomidate do not induce sleep, but rather nonrousable unconsciousness. While these are very different processes, it is now clear that anesthetics act at least in part through sleep-inducing pathways, and do not just produce a general dampening effect throughout the CNS. Further work will further differentiate these processes.

Quite surprisingly, GABA$_A$ receptor agonists appear to affect sleep in a very particular manner. Studies by Lancel and coworkers and later by Winsky-Sommerer and Tobler have demonstrated that muscimol and gaboxadol (Figure 20.7) specifically modulate sleep stages in a manner, which is highly dependent on particular receptor populations. Gaboxadol and muscimol which both penetrates the BBB are functionally selective for the extra-synaptically located δ-subunit containing GABA$_A$ receptors (non-γ containing receptors, which are insensitive to modulation by BzRAs; see Figure 20.5). This functional selectivity for extra-synaptic receptors is in fact shared by all GABA$_A$ receptor agonists. However, since gaboxadol and muscimol readily penetrates the BBB, only these compounds have been characterized using in vivo studies.

Binding affinity using a radiolabeled agonist for the GABA binding site demonstrates little in the way of subtype selectivity for a variety of agonists and subtypes. This is understandable since the agonist binding site, which is located at the interface between α and β subunits remains conserved across all receptor subunits characterized so far. The functional consequences of receptor activation, however, are highly dependent on the receptor subunit composition. As illustrated in Figure 20.8, the potency and relative maximum response of gaboxadol cover the ranges from low

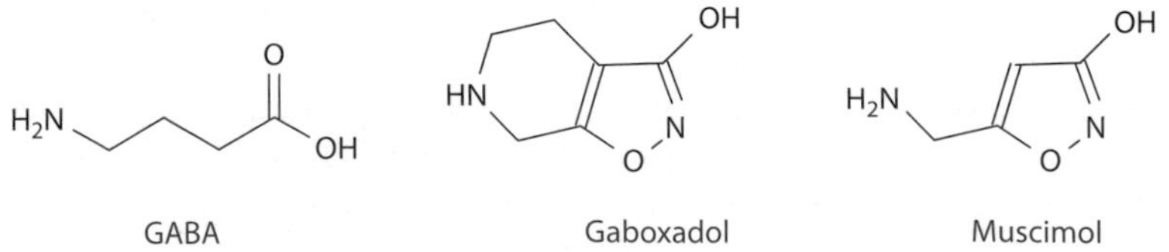

FIGURE 20.7   Structures of GABA, gaboxadol, and muscimol.

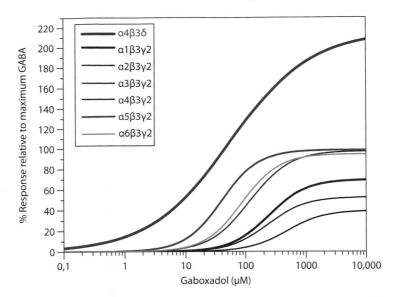

FIGURE 20.8 Dose response curves for gaboxadol at different $GABA_A$ receptor populations expressed in *Xenopus* oocytes. The therapeutic relevant concentration is 1–2 micromolar. (Data from Storustovu, S. and Ebert, B., *J. Pharmacol. Exp. Ther.*, 316, 1351, 2006.)

efficacy and low potency to high potency and high efficacy. For example at non-γ-subunit containing $GABA_A$ receptors, gaboxadol elicits a larger maximal response than GABA. $\alpha4\delta$ containing GABA receptors are located extra-synaptically and are present at neurons projecting to areas contributing to the modulation of sleep. One of these important areas is the ventrobasal region of the thalamus, which seems to play a key role for the hypnotic activity of gaboxadol. In normal animals, gaboxadol enhances the amount of SWS and, in particular, the SWA as recorded using EEG technology. Furthermore, in mice lacking the δ subunit, Gaboxadol dose not enhance SWA, illustrating the contribution of this receptor population to the physiological effects of gaboxadol. Since the relative in vitro profile of gaboxadol is shared by other $GABA_A$ receptor agonists, including partial agonists, a novel CNS penetrant $GABA_A$ receptor agonist most likely will share the hypnotic effects of gaboxadol.

Gaboxadol was in clinical development for insomnia but was discontinued due to variable effects on sleep induction in the final studies. However, since the clinical development program only included primary insomniacs, little is known about the clinical consequences of enhanced SWS and SWA. The hypnotic profile of gaboxadol indicates that $GABA_A$ receptor agonists may play a role in diseases where reduced SWS or enhanced and early onset of REM sleep is a core part of the disease. These conditions include psychiatric diseases (e.g., depression, anxiety, and bipolar disorder), neurological diseases (Alzheimer's disease and Parkinson's disease), stress and chronic pain syndromes.

20.3.2.2   5-HT_{2A} Antagonists

As speculated above, compounds with positive effects on SWS may have a particular value in a number of diseases. Therefore other strategies for enhancing SWS have been attempted. A very promising approach relates to the inhibition of a subpopulation of serotonergic receptors: 5-HT_{2A} receptors. Compounds with antagonist properties at these receptors were initially developed in the psychosis area, but at lower doses these drugs demonstrated a selective and short-term enhancement of the SWS/SWA during the night. Examples of these compounds are ritanserin, ketanserin, and eplivanserin (Figure 20.9). Interestingly, none of these compounds have yet demonstrated effects on

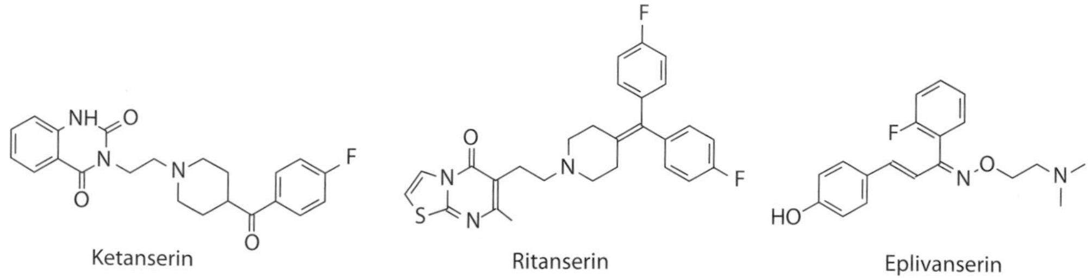

FIGURE 20.9 Structure of the 5-HT$_{2A}$ antagonists ketanserin, ritanserin, and eplivanserin.

the classical objective endpoints: induction and maintenance in clinical studies carried out to date. Therefore in order to have the compounds approved, a clinically meaningful consequence of the enhanced SWS has to be demonstrated. This means that effects on daytime performance must be available at the time of registration. If these compounds are approved for the treatment of insomnia, this may change the possibility for treating insomnia associated with depression or anxiety.

20.3.3 OREXINERGICS

Under normal circumstances, a reciprocal inhibition between the wake promoting orexinergic system and the GABAergic sleep-enhancing system exists. Preclinical experiments have demonstrated that the endogenous peptide orexin is wake promoting. Recent interest from a number of pharmaceutical companies have resulted in the development of selective orexin receptor antagonists acting at OR-1 or OR-2 receptors for the promotion of sleep. The role played by OR-1 and OR-2 receptors in sleep is still not completely established. Preclinical data have demonstrated the hypnotic effects of orexinergic antagonists GSK649868 and ACT-078573 (Figure 20.10). Indeed one nonselective antagonist (ACT-078573) has been shown to promote REM and non-REM sleep in healthy human volunteers. However, one important question is whether hyperfunction of the orexinergic system is involved in the pathophysiology of insomnia and whether orexin-based therapies have any unwanted side effects.

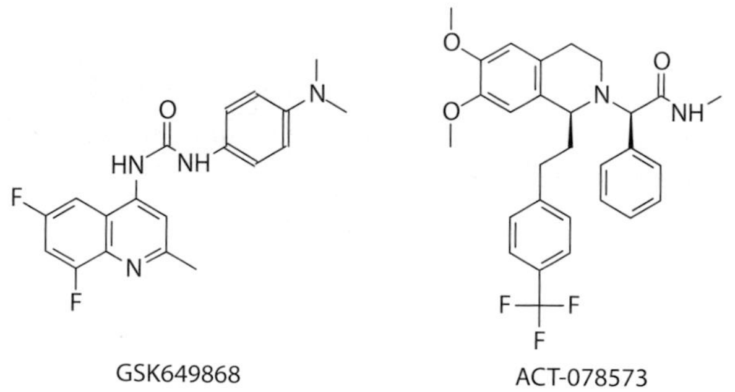

GSK649868   ACT-078573

FIGURE 20.10 Structures of two orexinergic antagonists.

20.3.4 MELATONIN AND MELATONERGIC AGONISTS

Melatonin (Figure 20.11) is an endogenous hormone, which in response to darkness is secreted from the pineal gland and subsequently activates the G-protein coupled melatonin receptors (MT1–3). Activation of MT1 and 2 leads to a release of GABA in the hypothalamus and this contributes to the entrainment of the circadian cycle. Melatonin is therefore a compound, which may

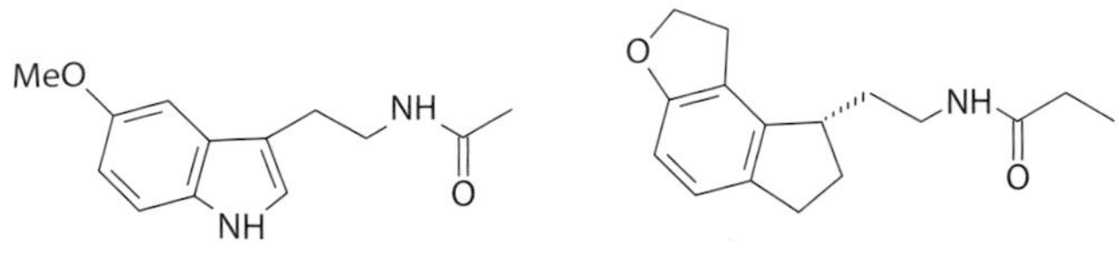

FIGURE 20.11   Structures of melatonin and the melatonergic agonist ramelteon.

bring a person with a disrupted sleep pattern in synchrony with their normal circadian rhythm. Several studies have provided contradictory results for melatonin; a situation that has led to a discussion about the value of melatonin as a drug. Data with a modified release formulation of melatonin has consistently demonstrated that in patients with a dysfunctional melatonin system (e.g., elderly), subjective sleep parameters and subjective daytime quality of life were significantly enhanced after 4 weeks of treatment. This formulation of melatonin was approved for primary insomnia in elderly and may reflect a change in the attitude of the European authorities, such that subjective parameters are sufficient for obtaining regulatory approval. In addition to melatonin, one melatonergic agonist is currently approved for insomnia in the U.S. Ramelteon (Figure 20.11) has been demonstrated to robustly induce sleep (using polysomnography–EEG measurements) in primary insomniacs. However, the effects in terms of minutes faster asleep are smaller than those observed for BzRAs. This does not necessarily predict a weaker effect on daytime performance. A correlation between reduction in time to sleep and subjective (or objective) daytime performance has never been established. The really exciting aspect of these compounds is therefore not whether they may induce or maintain sleep, but whether they will have positive consequences for the quality of life during the day.

20.4   CONCLUDING REMARKS

The development of hypnotics has for years been limited by our lack of insight into the mechanisms underlying sleep and how these translate into daytime function. With the ongoing integration of electrophysiology, molecular biology, imaging techniques, and cognitive research, focus in insomnia is moving from sleep induction and maintenance to effects of sleep on cognition and other types of daytime performance. The acceptance of insomnia as a chronic disease of its own and not as a symptom of other diseases stresses the need for novel types of hypnotic drugs. The coming years may therefore open up for hypnotic compounds focusing entirely on cognitive or psychiatric consequences of insomnia. The desired receptor profile of such compounds still remains to be established and this together with the medicinal chemistry challenges will be a challenge in the coming decade.

FURTHER READINGS

Akerstedt, T., Billiard, M., Bonnet, M., Ficca, G., Garma, L., Mariotti, M., Salzarulo, P., and Schultz, H. 2002. Awakening from sleep. *Sleep Med. Rev.* 6: 267–286.

Curry, D.T., Eisenstein, R.D., and Walsh, J.K. 2006. Pharmacologic management of insomnia: Past, present, and future. *Psychiatr. Clin. North Am.* 29: 871–893.

Ebert, B., Wafford, K.A., and Deacon, S. 2006. Treating insomnia: Current and investigational pharmacological approaches. *Pharmacol. Ther.* 112: 612–629

Kryger, M., Roth T., and Dement W.C., eds. *Principles and Practice of Sleep Medicine.* Philadelphia: W.B. Saunders, 2005.

NIH State-of-the-Science Conference Statement on manifestations and management of chronic insomnia in adults. 2005. *NIH Consens. State Sci. Statements* 22: 1–30.

Nofzinger, E.A. 2004. What can neuroimaging findings tell us about sleep disorders? *Sleep Med.* 5 (Suppl 1): S16–S22.

Pace-Schott, E.F. and Hobson, J.A., 2002. The neurobiology of sleep: Genetics, cellular physiology and sub-cortical networks. *Nat. Rev. Neurosci.* 3: 591–605.

Petroski, R.E., Pomeroy, J.E., Das, R., Bowman, H., Yang, W., Chen, A.P., and Foster, A.C. 2006. Indiplon is a high-affinity positive allosteric modulator with selectivity for alpha1 subunit-containing GABA$_A$ receptors. *J. Pharmacol. Exp. Ther.* 317: 369–377.

Storustovu, S. and Ebert, B. 2006. Pharmacological characterization of agonists at delta containing GABA$_A$ receptors: Functional selectivity for extra synaptic receptors is dependent on absence of gamma. *J. Pharmacol. Exp. Ther.* 316: 1351–1359.

21 Neglected Diseases

Søren B. Christensen

CONTENTS

21.1 INTRODUCTION

Infections including parasitic diseases account for approximately one-third of the worldwide disease burden but only 5% of the disease burden in high-income countries. Pharmaceutical companies assume that only a marginal profit can be made from drugs against diseases in low-income countries and consequently less than 2% of new chemical entities marketed in the last 30 years have been anti-infective drugs. The majority of the few marketed anti-infective agents were antiretroviral drugs, the development of which benefited from a serious political commitment from high-income countries for finding drugs against acquired immune deficiency syndrome (AIDS).

The limited interest for development of drugs against infectious diseases like malaria, African trypanosomiasis, Chagas disease, schistosomiasis, leishmaniasis, and tuberculosis has led to the use of the term neglected diseases, even though infectious diseases worldwide are responsible for a heavier burden than cardiovascular or central nervous system (CNS) diseases.

Medicine for Malaria Venture is a nonprofit organization founded in 1999, which has set up a public private partnership (PPP) in order to develop drugs according to the highest international standards against diseases of the developing countries. The organization includes academic institutions like Yale University, the University of Oxford, the University of California, San Francisco, private companies like F. Hoffmann-La Roche, Novartis Pharma, Korea Shin Poong Pharm, Holleykin Pharmaceutical Company (China), and international organizations like the World Health Organization (WHO).

In the past, the major inspiration for development of drugs against infectious diseases has come from natural products (see Chapter 6).

21.2 INFECTIONS CAUSED BY HELMINTHIC PARASITES

A parasite is an organism that lives in or on and takes its nourishment from another organism. A parasite cannot live independently. A helminth is a multicellular parasitic worm. In general, a helminth is visible to the naked eye in its adult stages. Parasites might have more hosts. The smaller host is generally called the vector. Many neglected diseases are caused by parasites. In contrast to bacteria, which are prokaryotes, parasites are eukaryotes.

21.2.1 SCHISTOSOMIASIS

Schistosomiasis (bilharziasis) is caused by infection with flatworms belonging to the genus *Schistosomas*, *S. mansoni* is found in South America and Africa, *S. haematopium* found throughout Africa, in particular in Egypt, and *S. japonicum* is confined to the Far East. Approximately 200 million people in more than 70 developing countries suffer from the diseases, 20 million suffer severe consequences, such as colonic polyposis with bloody diarrhoea (*S. mansoni*), splenomegami, and portal haematemesis with vomiting of blood (*S. japonicum* and *S. mansoni*), cystitis, and ureteritis, which might lead to bladder cancer (*S. haematopium*), and CNS lesions. The diseases are estimated to cause 280,000 deaths each year. Three safe, effective drugs, praziquantel, oxamniquine, and metrifonate, are now available for schistosomiasis and are included in the WHO model list of essential drugs. WHO has defined essential drugs as "those drugs that satisfy the health care needs of the majority of the population; they should therefore be available at all times in adequate amounts and in appropriate dosage forms, at a price the community can afford."

21.2.2 FILARIASIS

Filariasis is caused by parasites belonging to the order Filaroidea. *Wuchereria bancrofti* and *Brugia malayi* both cause lymphatic filariasis (elephantiasis), *Loa loa* causes fugitive swelling, in particular, around the eyes, and *Onchocerca volvulus* causes onchocerciasis (river blindness). All the diseases are caused by helminthic worms transmitted through bites of insects belonging to the order Diptera. The symptoms pertaining to the diseases are caused by the presence of parasites restricting the flow of lymph fluid. Approximately, 120 million people are infected with the parasites and 40 millions are severely disabled. Onchocerciasis is estimated to have infected 17.7 millions people, of which 500,000 have visual impairment and 270,000 are blinded. The disease is limited to the vicinity of rivers where the vector, blackflies of the genus *Similium*, is endemic. Unfortunately the burden of the disease often forces the population to leave these areas uninhabited. The infection can be treated with ivermectin (**21.1**) (Figure 21.1). Ivermectin, a dihydro derivative of avermectin B_{1a}, acts by opening invertebrate specific glutamate-gated chloride ion channels in the nerve end and muscles

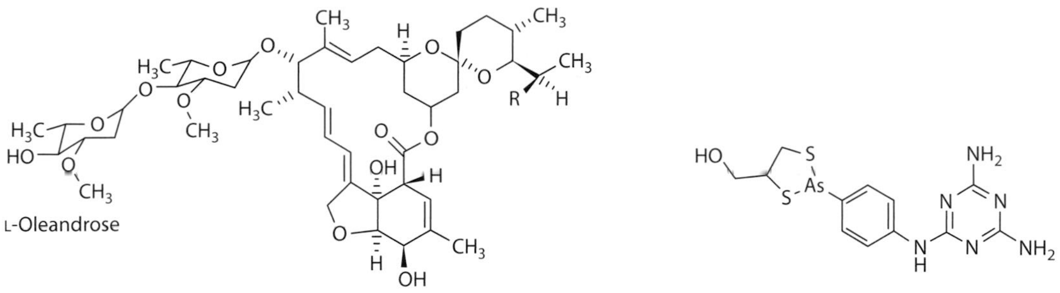

L-Oleandrose

21.1 R = –C$_2$H$_5$ (80%) or –CH$_3$ (20%)   **21.2**

FIGURE 21.1   Ivermectin (**21.1**) is a mixture containing at least 80% of the analog in which R = C$_2$H$_5$, and not more than 20% of the analog in which R = CH$_3$. Configuration of melarsopol (**21.2**).

of the parasites. This leads to death of microfilariae, the first larval stage. The drug does not cause immediate death of the adult parasite but reduces the worm's life span.

21.3   INFECTIONS CAUSED BY PROTOZOAN PARASITES OTHER THAN *PLASMODIUM*

Protozoan parasites are single-celled organisms, which have an animal-like nutrition (they cannot perform photosynthesis). The life cycle of protozoan parasites involves two hosts; the smaller of which typically is named the vector. Important genera are *Plasmodium*, *Trypanosoma*, and *Leishmania*.

21.3.1   TRYPANOSOMIASIS

Two major tropical diseases, American trypanosomiasis (Chagas disease) and African trypanoso-miasis (sleeping sickness) are caused by *T. cruzi* and subspecies of *T. brucei*, respectively. African trypanosomiasis is spread with tsetse flies (*Glossina* species). If untreated, the disease may be lethal since the parasites enter the CNS causing coma (explaining the name sleeping disease) and death. Approximately 48,000 persons are estimated to die from the disease each year. The only drug avail-able for treatment of the disease in a late stage is the arsenical drug melarsoprol (**21.2**) (Figure 21.1), developed more than 50 years ago. It is known to cause a range of side effects including convulsions, fever, loss of consciousness, rashes, nausea, and vomiting. It is fatal in a significant fraction of cases. Early stages of the disease are treated with pentamidin and suramin.

21.3.2   LEISHMANIASIS

Leishmaniasis is caused by parasites of the genus *Leishmania*. The diseases vary from simple self-healing skin ulcers (cutaneous leishmaniasis), severe disfiguring of nose, throat, and mouth cavities (mucocutaneous leishmaniasis) to life-threatening infections (visceral leishmaniasis). Visceral leishmaniais can be fatal if untreated. Approximately 12 million humans are infected with leishma-niasis and it is estimated that 59,000 die each year. The parasites nourish in the macrophages. The life cycle is illustrated in Figure 21.2.

Until recently no orally active drug was known for leishmaniasis but the treatment was based on amphotericin B (**21.3**) (Figure 21.3), pentamidine or antimony containing drugs like sodium stibo-gluconate and meglumine antimonate. Liposomal formulations of amphotericin B have increased the efficiency of the drug. Application of a drug in vesicles as liposomes will target the drug

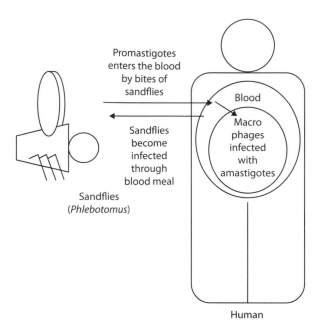

FIGURE 21.2 Life cycle of leishmania parasites: by taking a blood meal sandflies, belonging to the genus *Phlebotomus*, introduce promastigotes into the blood. The promastigotes are phagocytized by macrophages and converted into amastigotes in the blood cells. The amastigotes multiply in the macrophages. A sandfly feeding on the infected person will ingest parasites and thereby conclude the cycle.

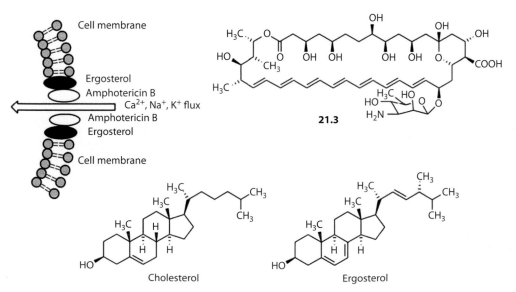

FIGURE 21.3 Mechanism of action of amphotericin B (**21.3**). The polyene region interacts with the double bonds of ergosterol, which is found in the cell membrane of parasites. Mammalian cells contain cholesterol, in which the presence of only one double bond causes the formation of a weaker complex with amphotericin B. The orientation of the amphotericin B–ergosterol complexes creates an ion channel, through which an unregulated flux of small inorganic ion passes. Inability to control the concentration of inorganic ions eventually kills a cell.

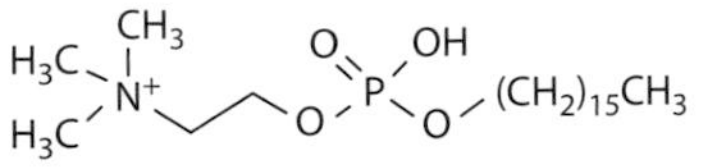

21.4

FIGURE 21.4 Configuration of miltefosin (**21.4**).

against cells performing phagocytosis such as the macrophages. Since the macrophages host the parasites some selectivity in activity is obtained. The main mechanism of action of amphotericin B is based on the amphiphilic nature of the molecule consisting of a lipophilic heptaene region and a hydrophilic polyol region. The polyene region complexes with steroids in the parasite's membrane. The hydrophilic polyol region form an ion channel permeable to small ions (Figure 21.3). Some selectivity is obtained because the drug has higher affinity for the double bonds of ergosterol dominating in the cell membrane of the parasites than for cholesterol in the membrane of mammalian cell.

Serendipitously it was discovered that the cancer drug miltefosine (**21.4**) (Figure 21.4) is an orally active drug against visceral leishmaniasis. Growth inhibition of leishmania parasites induced by miltefosine is correlated with a change in the phosphatidylcholine to the phosphatidylethanolamine ratio in the parasite's membrane. The selectivity might reside on different ways of formation of phosphatidylcholine in vertebrates and in leishmania parasites.

21.4 MALARIA

Malaria is a leading cause of morbidity and mortality in the tropical world; some 300–500 million of the world population are infected with malaria parasites, presenting 120 million clinical cases each year. It is estimated that between 1.5 and 2.7 million persons die from malaria each year and that 1 million of those are African children younger than 5 years. Among the more than 100 species of *Plasmodium* parasites, only four can infect humans: *P. falciparum* (causing malignant tertian malaria), *P. malariae* (quartan malaria), *P. ovale* (ovale tertian malaria), and *P. vivax* (benign tertian malaria). *P. falciparum* is responsible for the majority of deaths.

The life cycle of the malaria parasite encompassing several stages is depicted in Figure 21.5. A bite from an infected female mosquito belonging to the genus *Anopheles* introduces malarial parasites in the sporozoite stage into human with the salvia, which contains agents that prevent clotting of the blood. The sporozoites grow and multiply in the liver for about 5–15 days depending on the species. During this period, the patient has no symptoms. After having multiplied in the liver, the parasites enter the bloodstream as merozoites and invade the red blood cells (the erythrocytes). In the erythrocytes, the parasites proliferate and emerge as merozoites in a synchronous manner in about 48 h (tertian malaria) or 72 h (quartan malaria). This results in the clinical symptoms of the disease, namely, chills with rising temperatures, followed by fever and intense sweating. In addition, there might be severe headache, fatigue, dizziness, nausea, lack of appetite, and vomiting. Since the sporozoites catabolize the hemoglobin of the erythrocytes, a heavy infection will also induce anemia. After eruption, some merozoites reinvade erythrocytes and complete a new erythrocytic cycle. In the erythrocytes, some parasites change into gametocytes. After entering a mosquito stomach, blood meal gametocytes undergo another cycle in the mosquitoes. Merozoites will be digested in the stomach. The falciparum parasites cause the erythrocytes to adhere to the walls of capillary vessels resulting in reduced blood flow to organs. Reduced blood flow to brain contributes to cerebral malaria, which can be fatal. Because of the symptoms, some persons chronically infected with malaria, such as, the majority of Africans perform poorly. Studies suggest that national income in some African countries was suppressed by much as 18% because of malaria.

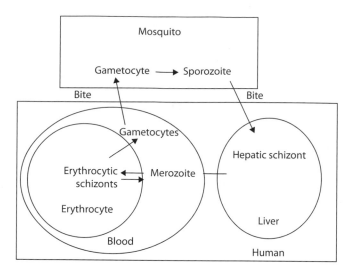

FIGURE 21.5 A mosquito belonging to the genus *Anopheles* pumps salvia into dermis of humans through the proboscis during feeding. If the mosquito is infected with malarial parasites, the salvia also will contain sporozoites, which will infect liver cells. In the liver, the parasites will develop into merozoites, which will be released into the blood by rupture of the liver cells. In the red blood cells, the merozoites will proliferate. At certain intervals, the red blood cells will rupture to release merozoites and male and female gametocytes. If a mosquito takes a blood meal on an infected human, the intraerythrocytic schizonts will be digested but the extraerythrocytic gametocytes will undergo a sexual proliferation in the mosquito enabling the mosquito to infect a new human.

Today the major burden of malaria is restricted to the tropical world: India, South East Asia, SubSaharan Africa, and Central and South America, but the endemic area of malaria besides the tropics also encompasses the subtropics and the major part of the temperate zones. During the 1950s and 1960s, a combined use of dichlorodiphenyltrichloroethane (DDT) for control of the vector mosquitoes and chloroquine (**21.7**) (refer to Figure 21.8) for the control of the parasites almost eradicated malaria from the Indian subcontinent. Resistance of the parasites toward chloroquine and of the mosquitoes toward DDT and the environmental consequences of extended use of DDT led to discontinuation of the project and return of the malaria burden.

21.5 DRUGS AGAINST MALARIA

In the absence of vaccines, malaria therapy relies on small molecule drugs. A number of antibiotics are used successfully either individually or more common in combination with other drugs. It may be surprising that antibiotics display considerable activity against the eukaryotic malarial parasite. This contradiction can be explained by the presence of two essential organelles in the parasites, namely, the mitochondria and the apicoplasts (Figure 21.6).

The apicoplast, probably, is a remnant of endosymbiotic cyanobacteria, which in plants have developed into the photosynthetic chloroplasts. Even though the apicoplasts do not perform photosynthesis, their metabolic pathways are still essential for the parasites. Both organelles have their own machinery for replication. Most antibiotics used in malaria therapy affect the apicoplasts.

21.5.1 Drugs Targeting Hemozoin Formation

The ultimate diagnosis of malaria is microscopic observation of parasites in the erythrocytes of a thick blood film. The presence of the malaria pigment, hemozoin in the erythrocytes, unequivocally reveals the presence of parasites. Hemozoin is formed from the heme (ferroprotoporphyrin IX, **21.5**) remaining after digestion of the peptide part of hemoglobin (Figure 21.7). The digestion proceeds in

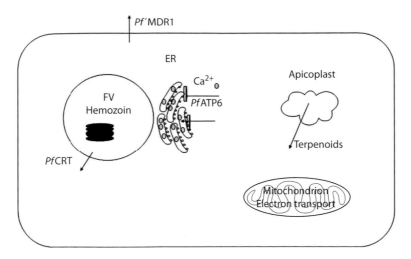

FIGURE 21.6 *Plasmodium* parasite: The Ca^{2+}-ATPase (*Pf*ATP6) pumps Ca^{2+} from the cytosol into the endoplasmic reticulum (ER) maintaining a low cytosolic Ca^{2+} concentration and a high concentration inside the ER. The single circle outside the ER represents the low cytosolic Ca^{2+} concentration in contrast to the several order of magnitudes higher ER concentration (many circles). Artemisinin and its analogs block this pump. Blockage of the pump leads to a prolonged high cytosolic Ca^{2+} concentration, which is lethal for the cell. Hemozoin is accumulated inside the food vacuole (FV). Blockage of hemozoin formation by, e.g., chloroquine is lethal to the cell. Terpenoids are formed in the apicoplasts by the nonmevalonate pathway. Blockage of this pathway by, e.g., fosmidomycin is lethal to the cell. *De novo* pyrimidine synthesis includes reduction of dihyroorotate into orotate, a reaction that is dependent on the electron flow in the mitochondria. Atovaquone blocks the electron flow. By removing chloroquine from the FV, the pump *Pf*CRT makes the parasite resistant toward chloroquine. The pump *Pf*MDR1 removes a number of drugs from the cytosol and induces resistance.

the food vacuole of the parasite. The food vacuole is characterized by a pH between 5.0 and 5.5. The parasites use at least three types of proteases for the catabolism of hemoglobin. Since nude heme is toxic to all kind of cells the parasites have to detoxify it by converting it into an insoluble complex.

The heme detoxification process is concluded by precipitation of microcrystalline hemozoin (**21.6**) (Figure 21.7), which by precipitation loses the effect on the biological system. The mechanism of action behind a series of antimalarial drugs consists in blocking the formation of hemozoin by association with hematin. Two criteria have to be fulfilled for drugs that act by preventing detoxification of heme: (1) the drug must accumulate in the food vacuole of the parasite, and (2) the drug must bind to hemozoin. For a drug to be active in the human patient, several other factors including absorption, metabolism, and distribution need to be taken into account. Thus association with hematin is a necessary, but not sufficient requirement, for an antimalarial drug targeting the hemazoin formation. Criterion 1 might be fulfilled by introduction of a basic aliphatic amine into the molecule, thus taking advantage of the low pH of the food vacuole.

21.5.1.1   4-Aminoquinolines

The 4-aminoquinolines are characterized by the presence of an amino group in the 4-position of the quinoline nucleus.

It is claimed that the drug, which with no comparison has saved most human lives, is chloroquine (**21.7**) (Figure 21.8). For more than 40 years, chloroquine was the first-line therapeutic and prophylactic agent for malaria. In the last three decades, however, chloroquine resistant *P. falciparum* and *P. vivax* strains have developed, but the drug is still believed to be efficient toward infections by *P. ovale* and *P. malariae*. The resonance interaction between the electron pair of the exocyclic amino group and the quinoline nitrogen atom (Figure 21.8) will give the protonated 4- and 2-aminoquinolines higher pK_a values (8.1 and 10.2) than other aminoquinolines.

21.6

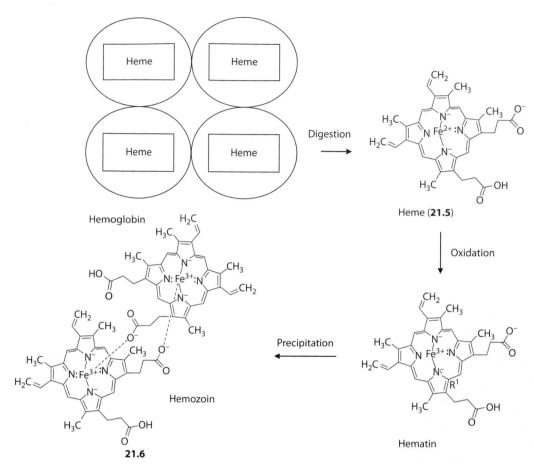

FIGURE 21.7 Digestion of hemoglobin, which is a tetramer consisting of four protein strings and four heme (**21.5**) molecules, leads to liberation of the heme molecules. Oxidation of iron(II) to iron(III) converts heme into hematin. Ionic interactions between the propanoate side chain and the iron(III) yield the poorly soluble β-hematin = hemozoin (**21.6**), which precipitates (shown as a dimer, but in the cell hemozoin will precipitate as a polymer).

The quinoline nucleus Chloroquine **21.7**

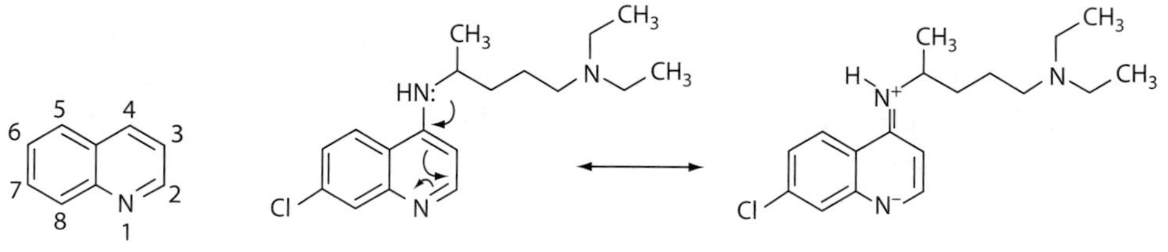

FIGURE 21.8 The quinoline nucleus and configurations of chloroquine (**21.7**) and hydroxychloroquine and resonance structures revealing the delocalization of the electrons in the electron-rich pyridine ring. Protonation of the two basic amino groups ensures accumulation in the food vacuole.

The lipophilicity of the neutral molecule enables it to cross the membranes of the erythrocytes and the parasites. Having entered the food vacuole with a pH ~ 5.2, the two amino groups will become protonated preventing chloroquine from leaving the vacuole by passive penetration. Furthermore, the protonation of quinoline nitrogen enables a cation–π interaction between the quinoline and the hemozoin (Figure 21.9).

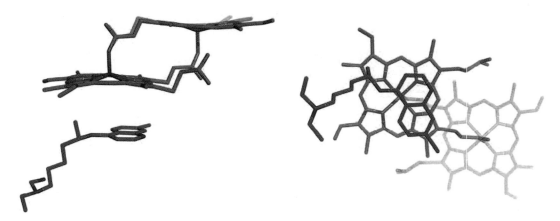

FIGURE 21.9 Two orthogonal views of a suggested binding mode of chloroquine (green) to hemozoin (yellow). Hetero-atoms are colored red, blue, and orange for oxygen, nitrogen, and iron, respectively. (Figure prepared by Dr. F. S. Jørgensen, University of Copenhagen, Denmark.)

The complexation between chloroquine and hemozoin further contributes to the concentration of choroquine in the food vacuole. Actually, the concentration of chloroquine, in sensitive parasites, is four orders of magnitude higher inside the vacuole than outside. The presented model also confirms the finding that variations in the side chain, only to a minor extent, influence the strength of the association. Notice the absence of stereocenters in the porphyrin nucleus. The achirality of this target explains that the two enantiomers of chloroquine have the same binding affinity. In agreement with the target being heme, chloroquine affects the erythrocytic stages of the parasite only, the only stages in which hemozoin is formed.

21.5.2 DRUGS TARGETING A Ca^{2+} PUMP OF *PLASMODIUM* PARASITES

Whereas European physicians did not get access to an efficient antimalarial drug until the sixteenth century, Chinese authors described the effect of *qing hao* and *cao hao* against intermittent fever two thousand years ago. Since the Chinese way of describing symptoms of diseases is very different from the Western terminology, intermittent fever cannot in a simple way be translated into a well-known term. However, intermittent fever could include the fever caused by malaria. *Quing hao* has later been identified as *Artemisia apiacea* (Asteraceae) and *cao hao* as *A. annua* (sweet wormwood). In the 1970s, Chinese scientists isolated artemisinin (**21.8**) (Figure 21.10) from both these species and showed that the compound was a potent antimalarial agent. Artemisinin is an irregular sesquiterpene lactone containing an endoperoxide bridge. Synthetic analogs, in which the peroxide bridge has been removed, show no activity toward *Plasmodium* parasites. The amount of artemisinin that may be extracted from the wormwood varies between 0.01% and 0.8% of dry weight, a serious limitation for the commercialization of the drug.

Originally, it was suggested that artemisinin by reaction with the iron (II) ion of heme generates a radical, which reacted with the heme skeleton and thereby prevented this from precipitation (Scheme 21.1). Although this idea still is favored in some laboratories, later studies indicate that artemisinin inhibits a plasmodial intracellular calcium pump (*Pf*ATP6). All cells maintain a low cytosolic Ca^{2+} concentration by removing Ca^{2+} from the cytosol with a pump sitting in the membrane of the endoplasmic reticulum. Blocking this pump yields a high cytosolic Ca^{2+} concentration, which eventually leads to cell death (Figure 21.6). A crucial difference between the mammalian pump, the sarco-/endoplasmic calcium ATPase (SERCA), and the *Pf*ATP6 pump is the presence of Glu255 in SERCA, whereas a Leu is located in the equivalent position 263 in *Pf*ATP6.

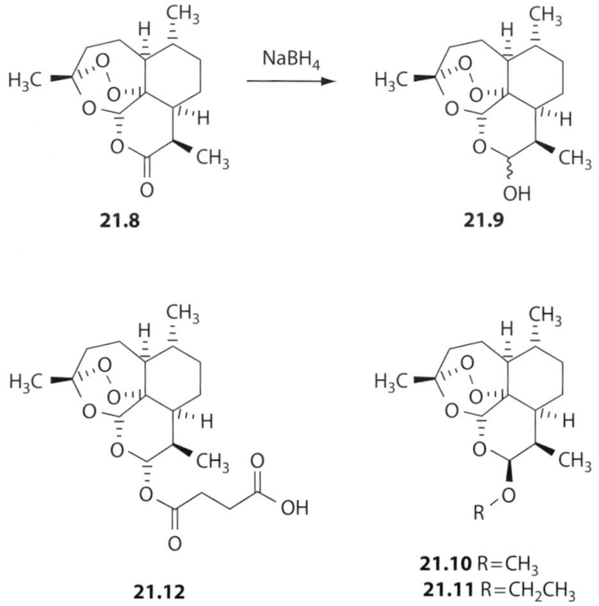

FIGURE 21.10 Treatment of artemisinin (**21.8**) with sodium borohydride selectively reduces the ester to the semiacetal dihydroartemisinin (**21.9**). The semiacetal can be converted into the lipid-soluble artemether (**21.10**) or arteether (**21.11**). Monoesterification with dicarboxylic acids results in the formation of water-soluble esters of the hemiacetal, e.g., artesunate (**21.12**).

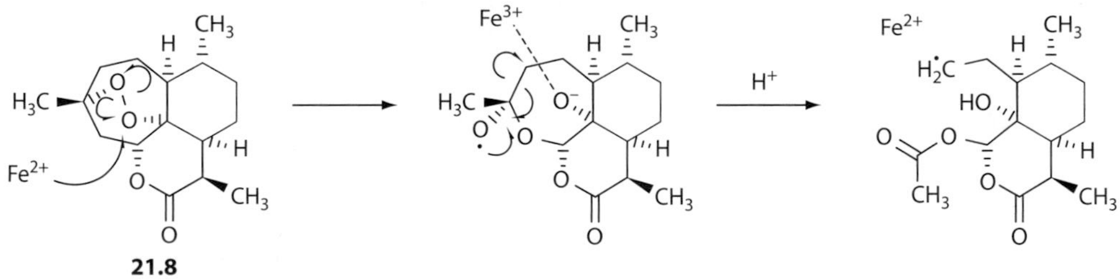

SCHEME 21.1 Possible reductive scission of the peroxide bond of artemisinin to give a reactive carbon centerd radical.

Both pumps are inhibited by the sesquiterpene lactone thapsigargin, but *Pf*ATP6 only by artemisinin and analogs. A mutation of Leu263 into Glu in *Pf*ATP6 preserves thapsigargin but abolish artemisinin sensitivity. Very recent results, however, question the effect of artemisinin on *Pf*ATP6.

The poor solubility of artemisin allows only oral or rectal administration of the drug. Since oral administration is impossible for severely sick patients, lipid-soluble derivatives (artemether [**21.10**], and arteether [**21.11**]) and water-soluble derivatives (e.g., artesunate [**21.12**]) have been developed by selective reduction of the ester carbonyl group followed by ether or ester formation (refer to Figure 21.10).

To avoid recrudescense (reappearance of a disease after it has been quiescent), artemisinin and artemisinin derivatives are preferentially given in combination therapy. Some examples of combination therapies are artesunate in combination with chlorproguanil and dapsone (**21.34**) (refer to Figure 21.18), artesunate in combination with mefloquine (**21.27**) (refer to Figure 21.15), and artesunate in combination with sulphadoxine (**21.33**) (refer to Figure 21.18) and pyrimethamine (**21.31**) (refer to Figure 21.16).

21.5.3 Drugs Targeting Deoxyxylulosephosphate-Reductoisomerase

In plants, the biosynthesis of isopentenyl diphosphate (**21.16**), the precursor of all terpenoids, follows two independent pathways: (1) the mevalonate pathway and (2) the 1-deoxy-D-xylulose-5-phosphate pathway (the nonmevalonate—or the Röhmer pathway). In plants, the mevalonate pathway occurs in the cytosolic compartment, but the deoxyxylulose pathway takes place in the plastids, which are analogous to apicoplasts. A search in the genome of *Plasmodium* parasites revealed the presence of genes encoding the enzymes of the deoxyxylulose pathway including the genes encoding 1-deoxy-D-xylulose 5-phosphate reductoisomerase (DOXP-reductoisomerase). This enzyme catalyses a crucial step in the deoxyxylulose pathway (Scheme 21.2): the reductive rearrangement of 1-deoxy-D-xylulose 5-phosphate (**21.13**) into 2-C-methyl-D-erythritol 4-phosphate (**21.15**).

Fosmidomycin (**21.17**) (Scheme 21.2), an antibiotic and herbicidal agent isolated from cultures of *Streptomyces lavendulae*, efficiently inhibits the plants' carotenoid and phytol synthesis. Fosmidomycin is a structural analog to 2-methylerythrose 4-phosphate (**21.14**), a never isolated compound but likely to be an intermediate in the conversion of (**21.13**) into (**21.15**). Combination therapy using fosmidomycin and clindamycin has revealed high antimalarial activity and only mild gastrointestinal side effects. The drug, however, is still in development.

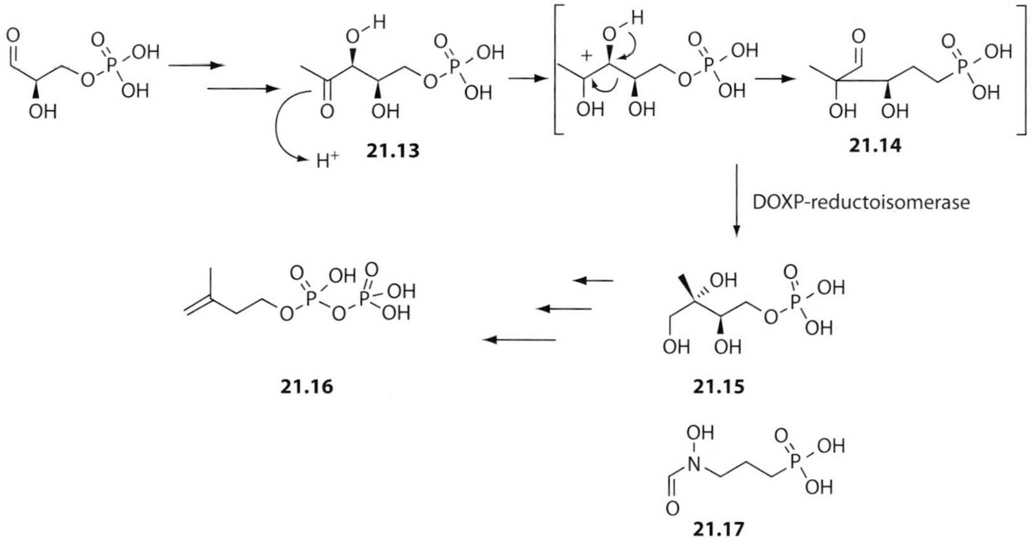

SCHEME 21.2 Reductive rearrangement of 1-deoxy-D-xylulose 5-phosphate (**21.13**) to 2-C-methyl-D-erythritol 4-phosphate (**21.15**) in the deoxyxylulose pathway via the never isolated 2-methylerythrose 4-phosphate (**21.14**). Isopentenyl diphoshate (**21.16**) is the precursor for terpenoids. Configuration of the DOXP-reductoisomerase inhibitor, fosmidomycin (**21.17**).

21.5.4 Drugs Targeting Mitochondrial Functions

In contrast to the hosts, the mitochondrial electron transport of *Plasmodium* parasites is not coupled to the synthesis of ATP. An important function of the mitochondria is to maintain an electron transport needed for nucleotide synthesis. Parasites are dependent on *de novo* synthesis of the nucleotides. The mitochondrial cytochrome bc_1 complex is a part of the electron transport. The enzyme consists of a cytochrome and a Rieske protein bound to an iron sulfur subunit. In the ubiquinol binding pocket, two electrons are transferred from ubiquinol via the subunit to cytochrome c heme iron (Scheme 21.3).

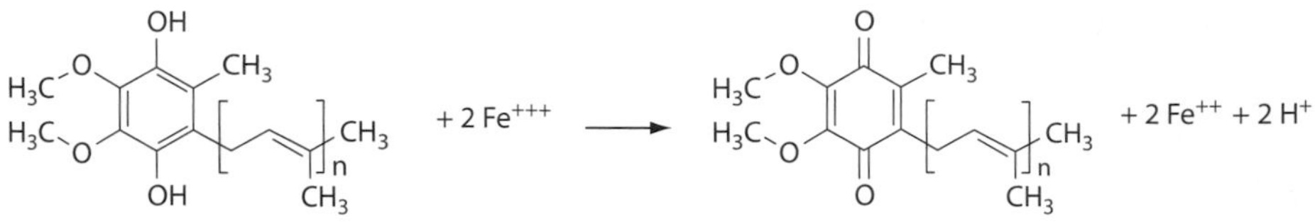

SCHEME 21.3 Oxidation of ubiquinol in the bc_1 complex.

21.5.4.1 Naphtoquinones

The antimalarial naphtoquinones are developed from naturally occurring naphtoquinones such as, lapachol (**21.18**) (Figure 21.11). The problem of fast metabolism, however, prevented the clinical use. Among the several hundreds of napthoquinones synthesized and tested atovaquone (**21.18**) was finally selected for use. Atovaquone is assumed to bind to the ubiquinol oxidation pocket of the parasite and thereby prevent the electron transfer. Model studies performed on the yeast bc_1 complex suggest that a hydrogen bond between the hydroxyl group of atovaquone and nitrogen of His181 of yeast Rieske-protein and a hydrogen bond between Glu272 of bc_1 complex via a water molecule and one of the carbonyls of atovaquone stabilize the complex and thereby prevent transfer of the electrons to the iron–sulfur complex. Replacement of Leu275 with the more bulky Phe275 as found in bovine bc_1 prevents the binding of atovaquone in the pocket (Figure 21.12). Similar atovaquone only possesses a poor affinity for human cytochrome bc_1.

Rapid development of resistance and a high rate of recrudescence necessitated the use of combination therapy. Proguanil (**21.29**) (refer to Figure 21.16)–atovaquone combination (Malarone®) is at the present an effective therapy for multidrug resistant falciparum malaria. Unfortunately, the high costs of this treatment limit its use.

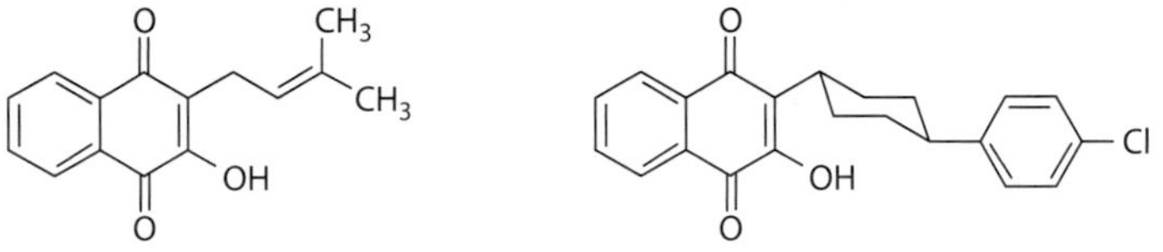

FIGURE 21.11 Configurations of lapachol (**21.18**) and atovaquone (**21.19**).

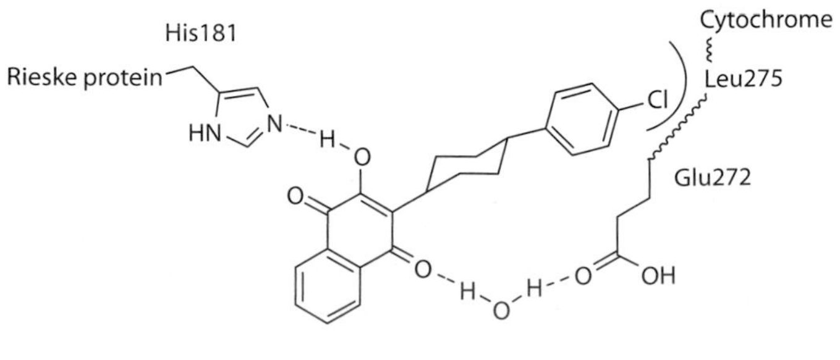

FIGURE 21.12 Suggested binding of atovaquone to the ubiquinol binding site.

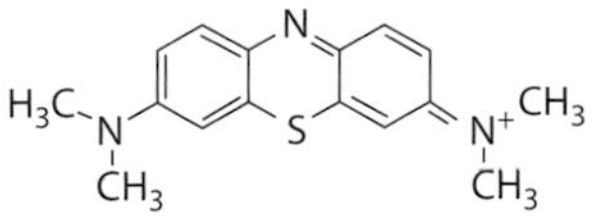

FIGURE 21.13 Configurations of methylene blue (**21.20**), primaquine (**21.21**), and tafenoquine (**21.22**).

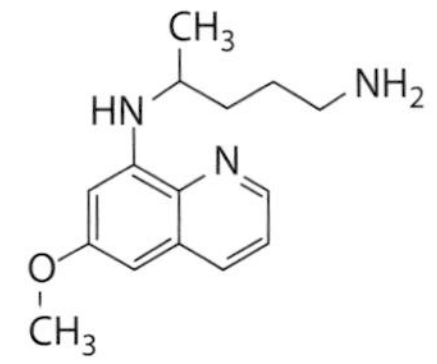

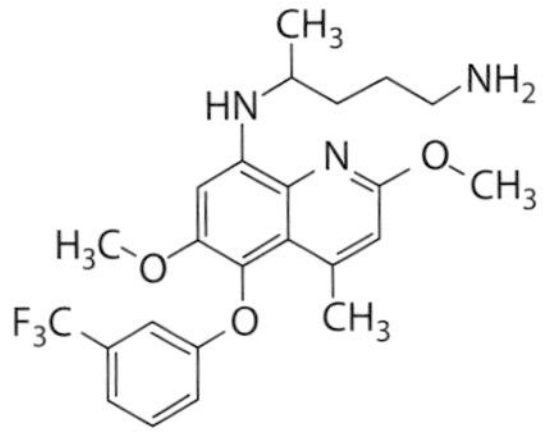

21.5.5 DRUGS WITH NONESTABLISHED TARGETS

21.5.5.1 8-Aminoquinolines

Approximately 100 years ago, Paul Ehrlich (1854–1915) noticed a selective uptake and staining of tissues with dyes such as methylene blue (**21.20**) (Figure 21.13). Based on the pioneering idea, at that time, that this selective staining was caused by selective receptors for the dyes, he discovered that methylene blue had antimalarial activity. Elaborating of this idea led to the development of the 8-aminoquinolines, among which primaquine (**21.21**) is the more important.

Primaquine remains the only drug approved for the cure of vivax malaria. The 8-aminoquinolines possess activity toward all stages of the parasite, including the hypnozoites in the liver and the gametocytes in the blood. Killing of hypnozoites prevents relapse, which is caused by the activation of hypnozoites resting in the liver. Relapse is pronounced for vivax malaria. Killing of gametocytes prevents transmission. The ability to affect all stages reveals that the mechanism of action of the 8-aminoquinolines must differ from that of the 4-aminoquinolines, which only affect parasites digesting hemoglobin (Section 21.5.1.1). Drawbacks of primaquine include a narrow therapeutic window, a short half-life (4–6h), which requires repeated administration for 14 days to achieve a cure, and hemolysis and methemoglobin formation. The latter side effect is particularly pronounced in patients with an inborn deficiency of glucose-6-phosphate dehydrogenase, a genetic abnormality common in areas where malaria is endemic. Structure–activity relationships (SARs) have revealed that an appropriate substitution in the 2-position improved efficacy and decreased general systemic toxicity, a methyl group in the 4-position improved not only the therapeutic activity but also toxicity, and that a phenoxy group in the 5-position decreased toxicity and maintained activity. The studies led to synthesis of tafenoquine (**21.22**). The substituent in the 5-position provides tafenoquine with a half-life of 2–3 weeks. A long half-life is essential for the development of a single-dose oral cure for malaria.

The mechanism of action of the 8-aimonquinolines has not yet been established. It is suggested that the compounds might affect the calcium homeostasis, affect the mitochondria by causing oxidative stress, or act by a combination of these effects.

21.5.5.2 4-Quinolinemethanols

The quinolinemethanols might be considered as methanol substituted with 4-quinoline and an aliphatic substituent (Figure 21.14).

21.5.5.2.1 The Cinchona Alkaloids

The oldest representatives for the quinolinemethanols are (–)-quinine (**21.23**), (–)-cinchonidine (**21.24**), (+)-quinidine (**21.25**), and (+)-cinchonine (**21.26**). All four alkaloids are isolated from the bark of trees belonging to the genus *Cinchona* (Rubiaceae), often referred to as fever trees. Until isolation of quinine in large scale in 1820, the crude bark was the only efficient drug in Europe for

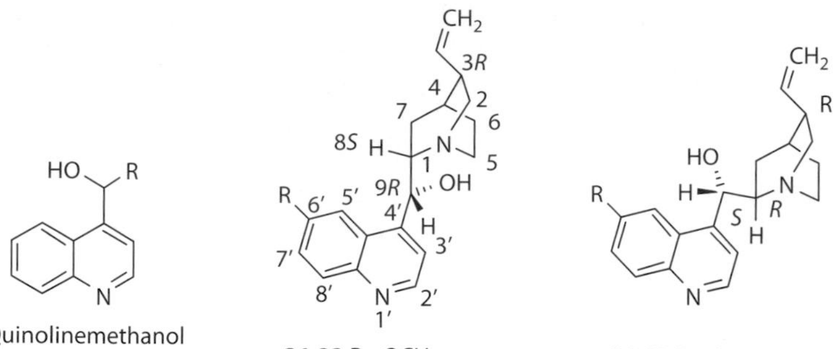

4-Quinolinemethanol

21.23 R=OCH_3
21.24 R=H

21.25 R=OCH_3
21.26 R=H

FIGURE 21.14 4-Quinoline methanols. Absolute configurations of (–)-quinine (**21.23**), (–)-cinchonidine (**21.24**), (+)-quinidine (**21.25**), and (+)-cinchonine (**21.26**).

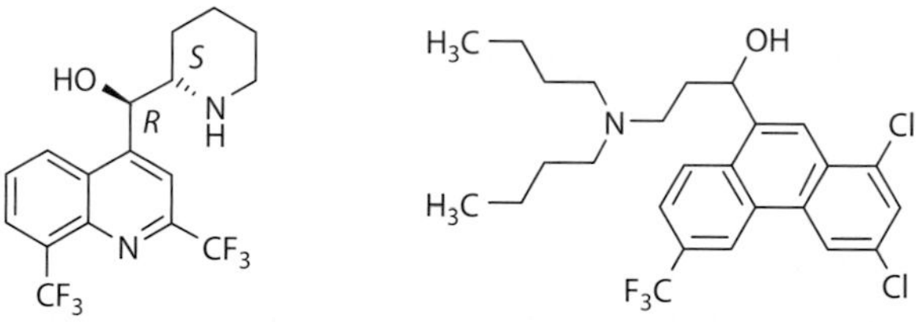

21.27

21.28

FIGURE 21.15 Absolute configuration of (+)-mefloquine (**21.27**) and configuration of halofantrine (**21.28**).

treatment of malaria. The narrow therapeutic window of the cinchona alkaloids in combination with different concentrations of the alkaloids in the bark depending on species and time of harvesting made access to the homogenous compounds as a major therapeutic improvement. In spite of the several severe side effects including cardiac arrhythmia, insulin release causing hypoglycemia, and peripheral vasodilatation the cinchona alkaloids are still used for treatment of multiresistant malaria, most frequently in combination with tetracyclines. The treatment, however, requires careful supervision.

The mechanism of action of the cinchona alkaloids is unknown, but the target might be situated in the cytosol of the parasite. Clinically quinidine and cichonine (both 8*R*,9*S*) are two- to threefold more active than quinine and cinchonidine (8*S*,9*R*). Quinidine and cinchonine, however, should be used only in the case of shortage of quinine because of a narrow therapeutic window. The importance of the stereochemistry at C8 and C9 is illustrated by the lack of activity of 9-epiquinine (8*S*,9*S*).

21.5.5.2.2 Mefloquine

Mefloquine (**21.27**) (Figure 21.15) was selected among 300 quinolinemethanols prepared in order to develop agents that are effective against chloroquine resistant agents. Mefloquine is marketed as a 1:1 mixture of the two diastereomeric racemic pairs. The four stereoisomers show similar ability to inhibit hemazoin formation in vitro. The pharmacokinetic, however, is different for the two since the half-life of the (–)-isomer (enantiomer of **21.27**) is significantly longer than that of the (+)-isomer of **21.27**.

The 2,8-bis-trifluoromethyl arrangement proved to be the most active of the series. Compounds that possessed 2-aryl groups were found to have augmented antimalarial activity, but at the same time unacceptable phytotoxic side effects. Mefloquine has been marketed as Lariam, a drug that has serious hallucinogenic side effects in some patients. Resistance against mefloquine has led to the use of combination therapy using mefloquine and arteminisin derivatives.

21.5.5.3   Phenanthrenemethanols

The dibutylaminopropyl groups of halofantrine (**21.28**) (Figure 21.15) were found to give optimal antimalarial effect in the 9-phenanthrene system. Even though the evidence for the mechanism of action for this compound is less convincing than that for chloroquine the findings that mefloquine only affects erythrocytic stages of the parasite and that some studies show association with hemazoin support the suggestion that halofantrine acts by preventing detoxification of hemazoin. Like quinine halofantrine can induce cardiac arrhythmias.

21.5.6   DRUGS TARGETING FOLATE SYNTHESIS

Tetrahydrofolic acid is an important coenzyme in parasites as well as their hosts. The coenzyme is involved in the biosynthesis of thymine, pyrine nucleotide, and several amino acid syntheses. Malaria parasites are dependent on *de novo* folate synthesis (Scheme 21.4) whereas mammalian cells take up fully formed folic acid as vitamin B_9. Consequently, dihydropteroate synthase is absent in humans. In the mammalian as well as in parasitic cells, the precursors [folate or dihydrofolate (**21.32**), respectively] have to be reduced to the enzymatically active tetrahydrofolate, a reaction that is catalyzed by dihydrofolate reductase (DHFR). DHFR and thymidylate synthase are separate enzymes in mammalians, whereas they are covalently linked to one bifunctional enzyme (DHFR-TS) in protozoan parasites. The binding site of dihydrofolate in DHFR-TS is sufficiently different from the binding site in the human DHFR to allow selectivity. The binding site of DHFR-TS inhibitors like cycloguanil (**21.30**) (Figure 21.16) and pyrimethamine (**21.31**) and the enzyme is illustrated in Figure 21.17 using pyrimethamine as an example. Proguanil (**21.29**) will metabolically be converted into cycloguanil in the liver. The negatively charged carboxylate of Asp54 of the enzyme binds to the positively charged amino group of pyrimethamine. The 4-amino group forms hydrogen bonds with the backbone carbonyl groups of Ile14 and Ile164. The coenzyme of DHFR, NADPH, is oriented through a hydrogen bond to Ser108.

Sulfadoxine (**21.33**) (Figure 21.18) and dapsone (**21.34**) act as antimetabolites of *para*-aminobenzoic acid (**21.35**) (Scheme 21.4), which is a building block in the dihydrofolate synthesis. An antimetabolite is an agent, which prevents the incorporation of a structural related endogenic metabolite.

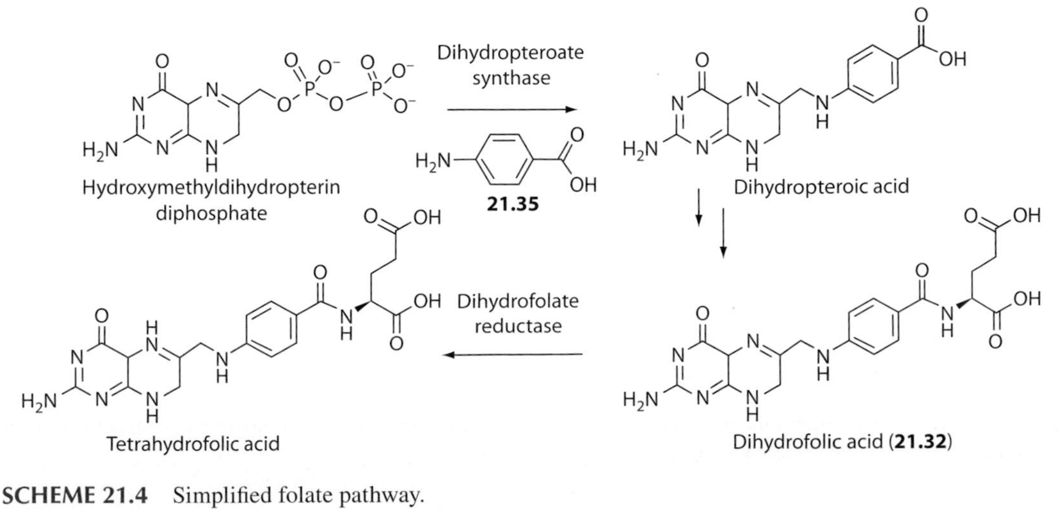

SCHEME 21.4   Simplified folate pathway.

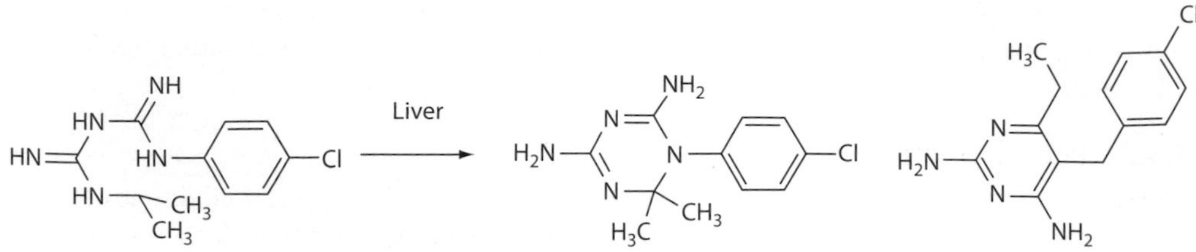

21.29 **21.30** **21.31**

FIGURE 21.16 Metabolic conversion of proguanil (**21.29**) into cycloguanil (**21.30**) and configurations of pyrimethamine (**21.31**).

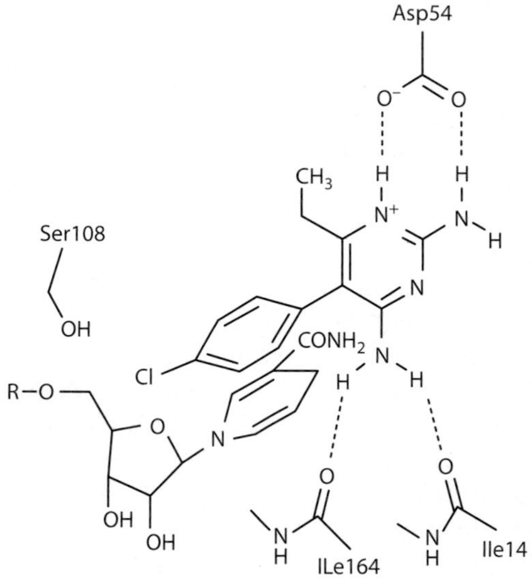

FIGURE 21.17 Binding of pyrimethamine (**21.31**, Figure 21.16) to the dihydrofolate binding site of dihydrofolate reductase (DHFR). The binding is stabilized through hydrogen bondings between the carboxylate of Asp54 and the positively charged NH group of the 2-amino group and the nitrogen of the pyrimidine ring, of hydrogen bonds between 4-amino group and the backbone carbonyl groups of Ile14 and Ile164. In addition, a charge-transfer interaction between the chlorophenyl residue and the dihydropyridine ring of the NADPH and finally a hydrogen bond between the Ser108 and one of the hydrogen acceptors at the NADPH molecule stabilize the complex. (*R* = the remaining part of the NADPH molecule.)

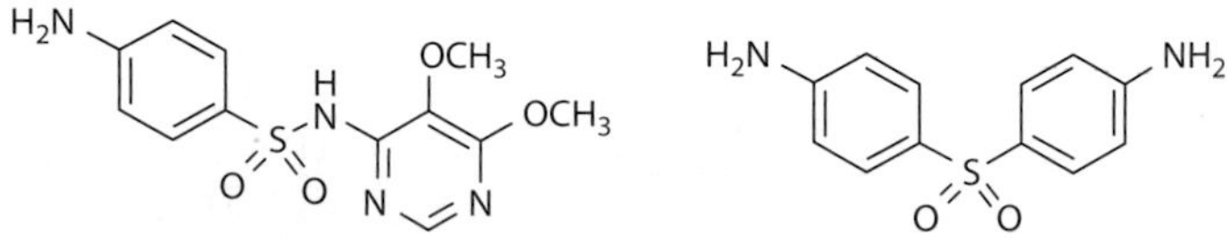

21.33 **21.34**

FIGURE 21.18 Configurations of sulfadoxine (**21.33**) and dapsone (**21.34**). Compare these structures with *para*-aminobenzoic acid (**21.35** Scheme 21.4).

21.6   RESISTANCE

Treatment of malaria frequently fails because of the development of resistance. In fact, regular mutations of the parasites force continued development of new drugs. Resistance develops in different ways e.g. mutations cause changes in the target proteins preventing interaction with the drug, or a transport system develops, which decreases the concentration of the drug at the target site.

21.6.1   Chloroquine Resistance

Chloroquine resistance has been correlated to a mutation in a wild-type food vacuolar membrane protein termed *P. falciparum* chloroquine resistance transporter (*Pf*CRT). A mutation replacing Lys76 with Thr enables the protein, in an energy-dependent manner, to transport chloroquine out of the food vacuole thus, decreasing the chloroquine concentration to below the pharmacologically active concentration. Other 4-aminoquinolines might also be substrates for the transporter explaining cross resistance. Chloroquine analogs with a modified side chain are developed in order to make analogs, which are not substrates for the transporter.

A number of other hypotheses for resistance including an increased value of the pH in the food vacuole or prevention of association between chloroquine and heme cannot be excluded.

21.6.2   4-Quinolinemethanol Resistance

The membrane transport *P*-glycoprotein pump, *Pf*MDR1, which is an analog of the mammalian ABC multidrug-transporter, has a central role in the resistance development of *P. falciparum* parasites. An increased number of *Pf*MDR1 transporters facilitate removal of the drug from the putative target in the cytosol.

21.6.3   Antifolate Resistance

Resistance toward antifolate drugs is caused by mutations that alter the active site resulting in different binding affinities for different drugs. The resistance conferring mutations occur in a stepwise sequential fashion with a higher level of resistance occurring in the presence of multiple mutations. The decreased affinity for the drug often is followed with a decreased activity for the natural substrate, suggesting the parasites containing mutated forms of the enzyme might be selected against in the absence of drug. Mutation of Ser108 into Asn causes steric interaction between the Asn side chain and the chlorophenyl group of pyrimethamine and thereby reduces the affinity of pyrimethamine to *Pf*DHFR (Figure 21.17). This mutation, however, only causes a moderate loss of susceptibility to cycloguanil (**21.30**), in which the side chain is shorter. Additional replacement of Asn51 into Ile results in a higher pyrimethamine resistance but only a moderate effect of cycloguanil. On the contrary, replacement of Ser108 into Thr coupled with Ala16 into Val confers resistance to cycloguanil but only modest loss of susceptibility to pyrimethamine.

21.7   CONCLUDING REMARKS

A series of examples of drugs targeting biological systems present only in parasites has been given. In principle, addressing targets not present in the host but essential for the survival of the parasite should give a therapy without side effects. Unfortunately, very few drugs are truly selective. Quinine, as an example, has targets in the parasite cytosol, but does also cause a number of effects in the patient such as insulin release, inducing severe hypoglycemia, and cardiac arrhythmia. Another serious problem in the treatment of parasitic diseases is the development of resistance, which is addressed by giving a combination of drugs. Artemisin and derivatives of artemisinin are given in combination with, e.g., proguanil. A further advantage of combination therapy is prevention of recrudescence, which in particular is a problem after treatment with artemisinin.

In this chapter, no attempt has been made to include all of the drugs or putative targets.

FURTHER READINGS

Azzouz, S., Maache, M., Garcia, R. G., and Osuma, A. 2005. Leishmanicidal activity of edelfosine, miltefosine and ilmofosine. *Pharmacology & Toxicology* 96: 60–65.

Cook, G. C. and Zumla, A. I. (eds.). 2003. *Manson's Tropical Diseases*, 21st edn. London: Elsevier Science.

Croft, S. L., Barrett, M. P., and Urbina, J. A. 2005. Chemotherapy of trypanosomiasis and leishmaniasis. *Trends in Parasitology* 21: 508–512.

Egan, T. J. 2004. Haemozoin formation as a target for the rational design of new antimalarials. *Drug Design Reviews—Online* 1: 93–110.

Rosenthal, P. J. (ed.). 2001. *Antimalarial Chemotherapy. Mechanism of Action, Resistance, and New Directions in Drug Discovery*. Totowa, NJ: Humana Press Inc.

Schlitzer, M. 2007. Malaria chemotherapeutics. *ChemMedChem* 2: 944–986.

Willcox, M., Bodeker, G. and Rasoanaivo, P. (ed.). 2004. *Traditional Medicinal Plants and Malaria*. Boca Raton, FL: CRC Press.

22 Immunomodulating Agents

Ulla G. Sidelmann

CONTENTS

22.1 INTRODUCTION

The purpose of the immune system is to combat infectious diseases caused by bacteria and viruses; however, evidence suggests that the immune system plays a central role in protecting the body against cancer and in combating cancer that has already developed. When the immune system is weakened, it can be more easily overwhelmed by cancerous cells. Cancer may occur when the immune system can no longer defend against invading tumor cells. There is evidence in many cancer patients that rebuilding the immune system slows down the growth and spread of tumors.

The immune system can also lead to pathological consequences for the individual. The first example is the normal attack of a healthy immune response on a transplant leading to transplant rejection. The second example is when tolerance to self break downs and self tissues are attacked by the immune system leading to autoimmune disease. The third example is when the immune system responds dramatically to otherwise harmless antigens leading to allergy or hypersensitivity.

Immunomodulating agents can be separated into agents that suppress or block the immune system when it is reacting against self antigens, hyperreacting or in relation to transplantation (autoimmune diseases, allergy/asthma, inflammation, and transplantation), agents that stimulate or activates the immune system to work harder or smarter (virus infections and cancer), and finally the removal of unwanted cellular subtypes of the immune system by depletion via specific surface antigens (auto-immune disease and cancer).

This chapter will give an overview of the key cellular and molecular drivers of the immune system. The following molecular approaches to treat diseases will then be covered: general immune suppression, depletion of B cells, modulation of T cells, and finally cytokines. Focus will be on autoimmune diseases, inflammation, and cancer.

22.2 BRIEF INTRODUCTION TO THE IMMUNE SYSTEM

The immune system has evolved to fight antigens invading the human body. Antigens are nonself entities, e.g., parts of bacteria, virus, parasites, foreign materials such as splinters and host generated threats such as cancer. An immune response can be divided into two phases: (1) the recognition of the antigen and (2) the elimination of the antigen.

There are two types of immune responses, namely, an innate response and an adaptive response both capable of distinguishing between self and nonself antigens but with different mechanism and specificity. The innate and the adaptive immune responses are interdependent and elements of one response enhance the other. The innate response is nonspecific and is designed to recognize molecular patterns common to a wide variety of pathogens. On second encounter of a pathogen, the response is exactly the same as the first. The response involves physical, chemical, and molecular barriers that distinguish and exclude the antigens. The adaptive immune response involves highly specific recognition and effector actions involving a variety of cells in the body. Unique antigens from the foreign entity will be specifically recognized by receptors expressed on B and T cells and an effector action against the entity is initiated. On second attack by the same entity, a faster and much stronger response called a memory adaptive response is induced, which will prevent disease from occurring the second time.

22.2.1 CELLS OF THE IMMUNE SYSTEM

The cells mediating the immune response are the leukocytes or white blood cells. The innate immune response is constituted by (1) granulocytes, that by phagocytosis, can engulf and destroy pathogens; (2) monocytes that can differentiate into macrophages; and (3) dendritic cells (DCs) that are also capable of phagocytosis and secretion of cytokines and growth factors that activate T cells in the adaptive immune system. Macrophages and DCs can present antigens to the T and B cells and fall into the category of antigen presenting cells (APCs).

B cells and T cells are lymphocytes, and are mediators of the adaptive immune response. Although mature lymphocytes look alike, they are diverse in their functions. B cells are produced and mature in the bone marrow, whereas the precursors of T cells leave the bone marrow and mature in the thymus (which accounts for their designation). The specificity of binding is defined by their respective receptors for antigens, the B cell receptor (BCR) and the T cell receptor (TCR), respectively.

Both BCRs and TCRs are integral membrane proteins, present in many identical copies, and exposed at the cell surface. They are present before the cell encounters an antigen and characterized by a unique binding site. A portion of the antigen called an epitope binds to the binding site, through noncovalent interactions. Successful binding of the antigen receptor to the epitope (accompanied by additional signals) will result in stimulation of the cell to enter the cell cycle. Repeated mitosis leads to the development of a clone of cells bearing the same antigen receptor; thus, a clone of cells of identical specificity. BCRs and TCRs are different with respect to their structure, the genes that encode them and the type of epitope they bind.

22.2.1.1   B Cells

When BCRs bind soluble antigens, the bound antigen molecules are engulfed by the B cell by receptor-mediated endocytosis. The antigen is digested into fragments, which are then displayed at the cell surface inside a histocompatibility molecule (MHC). Helper T cells specific for this structure (with a complementary TCR) bind to the B cell and secrete lymphokines. Lymphokines are messenger proteins secreted by lymphocytes that effect their own activity and/or that of other immune cells. Examples of lymphokines are interleukins and cytokines. Lymphokines stimulate the B cell to enter the cell cycle and develop into a clone of cells with identical BCRs. These cells switch from synthesizing their BCRs as integral membrane proteins to a soluble version called antibodies (Abs) and to differentiate into plasma cells that secrete these antibodies (Figure 22.1).

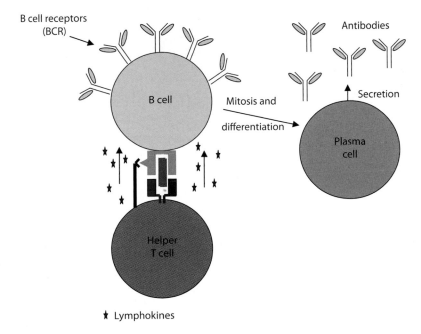

FIGURE 22.1   Schematic representation of an antigen displayed at the cell surface of a B cell inside a histocompatibility molecule and Helper T cells specific for this structure (with a complementary TCR) binding the B cell and secreting lymphokines. Differentiation of the B cell into an antibody secreting plasma cell.

22.2.1.2   T Cells

The TCR binds a bimolecular complex displayed at the surface of an APC. This complex consists of a fragment of an antigen lying within the groove of a histocompatibility molecule constituted by two α-helixes exposed on the surface (Figure 22.2). Most of the T cells in the body belong to one of two subsets. These are distinguished by the presence, on their surface, of one or the other of two glycoproteins, namely, CD4 and CD8. Which of these molecules is present determines what types of cells the T cell can bind to.

- CD8$^+$ T cells bind epitopes that are part of class I histocompatibility molecules (MHC class I). Almost all the cells of the body express class I molecules.
- CD4$^+$ T cells bind epitopes that are part of class II histocompatibility molecules (MHC class II). Only specialized APCs express class II molecules (DCs, phagocytic cells like macrophages and B cells).

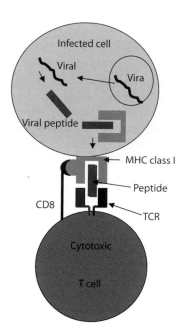

FIGURE 22.2   Schematic representation of an antigen (peptide) displayed at the cell surface of an APC in the class I histocompatibility molecules leading to the activation of a cytotoxic T cell.

The best understood CD8$^+$ T cells are cytotoxic T-lymphocytes (CTLs). They secrete molecules that destroy the cell to which they are bound. In general, the role of the CD8$^+$ T cells is to monitor all the cells of the body after they have been primed by APCs in the thymus or in the lymph nodes, ready to destroy any cell that express foreign antigen fragments in their class I molecules.

CD4$^+$ T cells bind an epitope consisting of an antigen fragment lying in the groove of a class II histocompatibility molecule. CD4$^+$ T cells are essential for both the cell-mediated and antibody-mediated (humoral branch) branches of the immune system.

22.2.1.2.1   Cell-Mediated Immunity

CD4$^+$ cells bind to antigen presented by APCs like phagocytic macrophages and DCs. The T cells then release lymphokines that attract other cells to the area. The result is an inflammation where immune cells and molecules accumulate and attempt to protect from and destroy the antigenic material.

22.2.1.2.2   Antibody-Mediated Immunity

These CD4$^+$ cells, called helper T cells, bind to antigen presented by B cells. The result is the development of clones of plasma cells secreting antibodies against the antigenic material (as described above).

22.2.1.2.3   Building the T Cell Repertoire

After the naive T cell (N) encounters an antigen in the context of an APC (1) it becomes activated and begins to proliferate (divide) into many clones or daughter cells. (2) Some of the T cell clones will differentiate into effector T cells (E) that will perform the function of that cell (e.g., produce cytokines in the case of helper T cells or invoke cell killing in the case of cytotoxic T cells). (3) Some of the cells will form memory T cells (M) that will survive in an inactive state in the host for a long period of time until they reencounter the same antigen and reactivate (Figure 22.3).

T helper (Th) 17 cells represent a newly identified subset of CD4$^+$ T cells that protects against extra cellular microbes, but are responsible for autoimmune disorders in mice. However, their

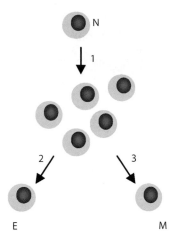

FIGURE 22.3 Building the T cell repertoire.

properties in humans are only partially known. Th17 cells, some of which produce both interleukin (IL)-17 and interferon gamma (IFNγ) have the ability to help B cells, have low cytotoxicity and poor susceptibility to regulation by autologous regulatory T cells.

Regulatory T cells (Tregs) is a cell population of the T cell repertoire that is recognized by specific cell surface markers, most importantly FOXP3, and is reported to have regulatory activity. They are found both in the CD4$^+$ and CD8$^+$ T cell population. Tregs specifically suppress the action of other cells in the immune system and maintain homeostasis of immune system and tolerance to self. Thereby, they prevent the establishment of an immune response. The interest in regulatory T cells has increased due to the evidence gathered from experimental mouse models demonstrating that the immunosuppressive potential of these cells can be used therapeutically to treat autoimmune diseases and facilitate transplantation tolerance or the regulatory T cells can be specifically eliminated to potentiate cancer immunotherapy.

22.2.2 ANTIBODIES

Antibodies are glycoproteins that are built of subunits containing two identical light chains (L chains), and two identical heavy chains (H chains), which are at least twice as long as light chains. The first ~100 amino acids at the N-terminal of both H and L chains vary greatly from antibody to antibody. These are the variable (V) regions. The amino acid sequence variability in the V regions is especially pronounced in three hypervariable regions. The structure of antibodies brings the three hypervariable regions of both the L and the H chains together. Together they construct the antigen binding site against which the epitope fits. For this reason, the hypervariable regions are also called complementarity determining regions (CDRs) (Figure 22.4).

Antibody molecules have two functions, to recognize and bind to an epitope on an antigen and trigger a useful response to the antigen. The V regions are responsible for epitope recognition and the C regions are responsible for triggering recruitment of other cells and molecules to destroy and dispose of pathogens to which the antibody is bound. The Fc part of the C region recognizes Fc receptors that are specialized receptors present on immune effector cells. The Fc part also facilitates active transport of the antibodies and initiation of the complement cascade.

22.2.3 NATURAL KILLER CELLS

Natural killer cells (NK cells) are large granular, non-T- and non-B-lymphocytes that are able to kill diseased cells. They are part of the innate immune system and help to fight viruses and other

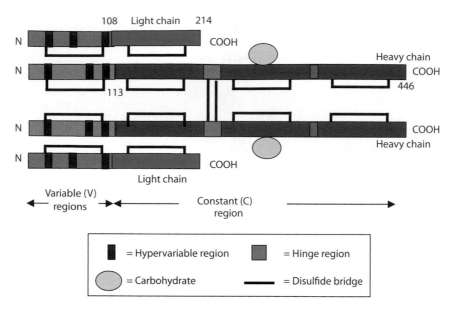

FIGURE 22.4   This image represents the polypeptide chain structure of a molecule of IgG. The numbers indicate the number of amino acids. In the actual molecule, the chains are folded so that each cysteine is brought close to the partner with which it forms a disulfide (S–S) bridge.

intracellular pathogens. NK cells are actors in antibody dependant cell-mediated cytotoxicity (ADCC). ADCC is the killing of antibody coated target cells by immune cells with FC receptors that recognize the C-region of the bound antibody. NK cells have the Fc receptor CD16a or FCγIIIa on their surface to facilitate the recognition.

22.3   DISEASES OF THE IMMUNE SYSTEM

Malfunction of the immune system will result in pathological consequences such as transplant rejection, autoimmune disease, or allergy. These situations are accompanied by inflammation that leads to tissue damage. Individuals may also be immunodeficient (lacking a component of the immune system) and have an increased susceptibility to infections and cancer; one such example is HIV (decrease in the number of CD4+ T cells).

22.3.1   AUTOIMMUNE DISEASES

Adaptive immune responses are sometimes elicited by antigens that are not associated with infectious agents and this causes serious disease and tissue damage. Autoimmunity is the failure of an organism to recognize its own constituent parts as "self," which results in an immune response against its own cells and tissues. Any disease that results from such an immune response is termed as an autoimmune disease. Prominent examples include insulin-dependent diabetes mellitus type 1 (IDDM), multiple sclerosis (MS), and rheumatoid arthritis (RA).

In IDDM, the immune system attacks the beta cells in the Islets of Langerhans of the pancreas, destroying them, or damaging them sufficiently to reduce and eventually eliminate insulin production leading to hyperglycemia. The autoimmune attack may be triggered by reaction to an infection.

In MS, the inflammatory responses are launched in the absence of a pathogen. It seems that certain T cells by mistake recognize the insulating sheaths around nerves (myelin) as a foreign invader resulting in the formation of auto antibodies against the myelin sheaths leading to destruction of the insulating layer surrounding neurons in the brain and spinal cord. When the myelin sheath is destroyed, nerve messages are sent less efficiently. Patches of scar tissue, called plaques, form over the affected areas, further disrupting nerve communication. The symptoms of MS occur when the brain and spinal cord nerves no longer communicate properly with other parts of the body. MS causes a wide variety of symptoms and can affect vision, balance, strength, sensation, coordination, and bodily functions.

RA is a chronic autoimmune disease that causes inflammation and deformity of the joints. In RA, the underlying event that promotes RA in a person is unknown. Given the known genetic factors involved in RA, some researchers have suggested that an outside event occurs that triggers the disease cycle in a person with a particular genetic profile. The body's normal response to such an organism is to produce cells that can attack and kill the organism, protecting the body from the foreign invader. In an autoimmune disease like RA, this immune cycle spins out of control. The body produces misdirected immune cells, which accidentally identify parts of the person's body as foreign. These immune cells then produce a variety of chemicals that injure and destroy parts of the body. Autoantibodies (antibodies against self protein) are observed in 80% of the patients.

IDMM and MS are examples of organ-specific autoimmune diseases as only one organ is affected whereas RA is an example of a systemic autoimmune disease involving many tissues.

22.3.2 INFLAMMATION

An inflammation is a manifestation of the immune system's response to invading organisms or substances. At the site of the infection, there are a number of physiological changes that take place to assist the destruction of the invaders. These include

- Increased blood flow to the area to maximize the number of leukocytes that can get to the infection site.
- A thinning of the cells in local blood capillary walls (endothelial cells) to allow the leukocytes to squeeze through.
- An increase in local temperature that has an antibiotic effect.
- A large number of immune system signaling molecules (chemokines) are released by leukocytes to coordinate the immune response and to call more leukocytes to the site.

Once the invader has been dealt with, the body terminates the immune response by killing off the leukocytes in the locality. This is done by depriving them of nutrients (necrosis) and by apoptosis.

22.3.3 CANCER IMMUNOLOGY

Recently, we have learned that the immune system may play a central role in protecting the body against cancer and in combating cancer that has already developed. This latter role is not well understood, but there is evidence that in many cancer patients the immune system slows down the growth and spread of tumors. On the other hand, when the immune system is weakened by old age or environmental factors, it can be more easily overwhelmed by cancerous cells.

One immediate goal of research in cancer immunology is the development of methods to harness and enhance the body's natural tendency to defend itself against malignant tumors. Immunotherapy represents a new and powerful weapon in the arsenal of anticancer treatments. Immunotherapies involving certain cytokines or antibodies that recognize cancer cells specifically and deplete the cells via ADCC have now become part of the standard cancer treatment.

22.4   IMMUNOSUPPRESSIVE AGENTS

The proliferative nature of the immune response may be controlled with immunosuppressive drugs. These drugs work by inhibiting the division of cells and, therefore, also suppress nonimmune cells leading to side effects such as anemia, neurotxicity, hepatotoxicity, nephrotoxicity, and diabetes. In addition the same immune system that we are suppressing in order to avoid graft rejection in transplantation or reactivity against auto antigens in autoimmune diseases is responsible for pathogen infections and tumor surveillance; thus, the immunosupressed patient is liable to come down with opportunistic infections and malignancies.

The most used immunosuppressive drugs are azathioprine, cyclosporine A, leflunomide, cyclophosphamide, glucocorticoids, and methotrexate. The mechanism of action of these drugs on the immune system is briefly described in the following.

Azathioprine is rapidly hydrolyzed in the blood to 6-mercaptopurine (Figure 22.5). In this form (as a purine analog), it incorporates into the DNA, inhibiting nucleotide synthesis by causing feedback inhibition in the early stages of purine metabolism. This ultimately prevents mitosis and proliferation of rapidly dividing cells, such as activated B- and T-lymphocytes. Through this action, Azathioprine is able to block most T cell functions and inhibit primary antibody synthesis.

Cyclosporine A is a small (11 amino acids) fungal (nonribosomal) cyclic peptide that is a calcineurine inhibitor (Figure 22.6). Cyclosporine works by binding to a protein found in the cytosol: cyclophilin. This complex inhibits calcineurin and ultimately leading to the inhibition of IL-2 production and secretion. The interaction between IL-2 and the IL-2 receptor is crucial in the activation and differentiation of B and T cells. Cyclosporine is therefore a highly effective immunosuppressant. The clinical introduction of cyclosporine has significantly increased graft survival and significantly reduced the occurrence of acute rejection in transplant patients.

Leflunomide (Figure 22.5) interferes with an enzyme called dihydroorotate dehydrogenase, an enzyme involved in the de novo pyrimidine synthesis. Thus leflunomide inhibits the synthesis of pyrimidines and thereby inhibits lymphocyte proliferation, and reduces adhesion molecules that allow the immune cells to home in to the area of inflammation. As a result the immune process is slowed. The drug is developed for RA and is also used in combination with methotrexate.

Cyclophosphamide is an inactive cyclic phosphamide ester of mechlorethamine. It is transformed via hepatic and intracellular enzymes to active alkylating metabolites, 4-hydroxycyclophophosphamide, aldophosphamide, acrolein, and phosphoramide mustard (Figure 22.7). Cyclophosphamide causes the prevention of cell division primarily by cross-linking DNA strands. It is therefore referred

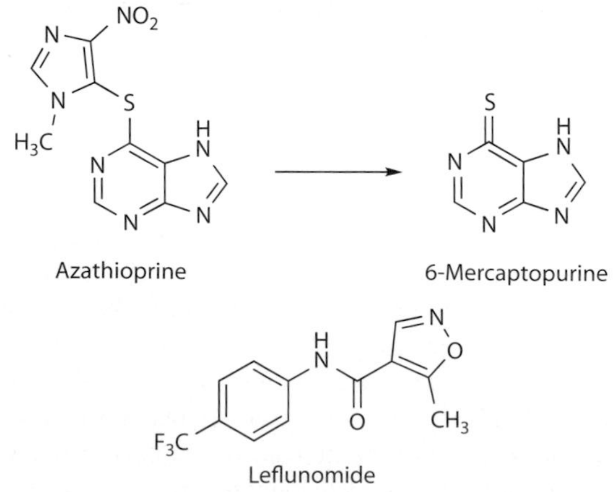

FIGURE 22.5   Azathioprine is an immunosuppressant and it is a prodrug, converted in the body to active metabolites 6-mercaptopurine and 6-thioinosinic acid, which is a purine synthesis inhibitor. Leflunomide is used in moderate to severe RA and psoriatic arthritis and is a pyrimidine synthesis inhibitor.

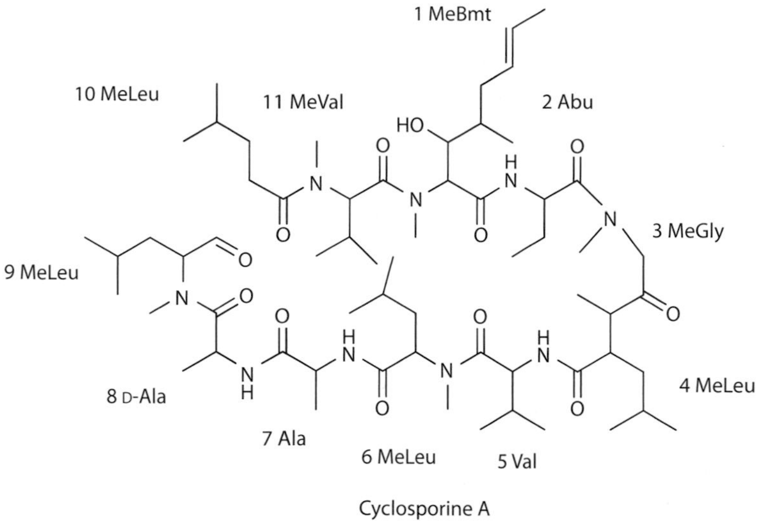

FIGURE 22.6 Structure of cyclosporine A, which is a cyclic nonribosomal peptide of 11 amino acids (undecapeptide) produced by the fungus *Tolypocladium inflatum Gams*, and contains unnatural amino acids, including D-amino acids.

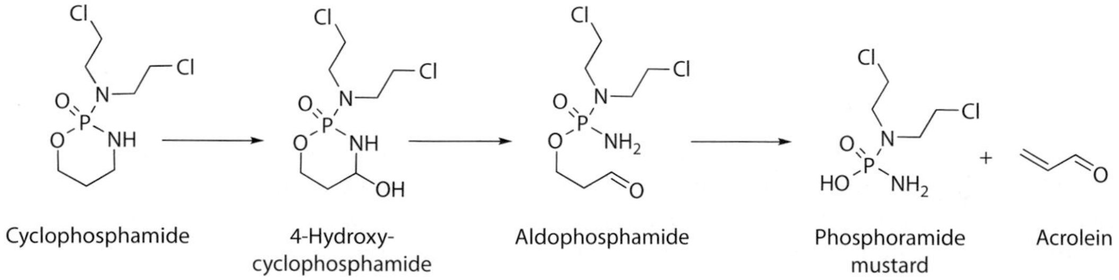

FIGURE 22.7 Cyclophosphamide is a nitrogen mustard alkylating agent, which is used to treat various types of cancer and some autoimmune disorders. It is a prodrug and is converted in the liver to active forms such as phosphoramide mustard that have chemotherapeutic activity.

to as a cytotoxic drug. Unfortunately, normal cells also are affected, and this results in serious side effects. It is used in cancer treatment and to treat severe cases of RA and other autoimmune diseases.

Glucocorticoids such as prednisone (Figure 22.8) works principally to block T cell and APC derived cytokine and cytokine-receptor expression. The major elements blocked are IL-1 and

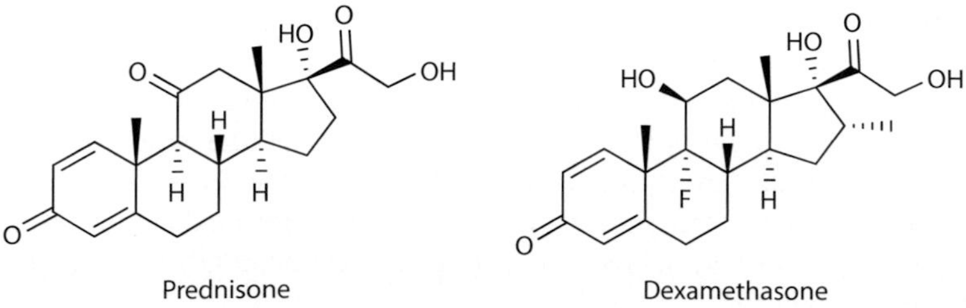

FIGURE 22.8 Glucocorticoids are steroid hormones characterized by an ability to bind with the glucocorticoid receptor. Synthetic derivatives of the natural corticoids, including prednisone and dexamthasone, which are used as effective immunosuppressants.

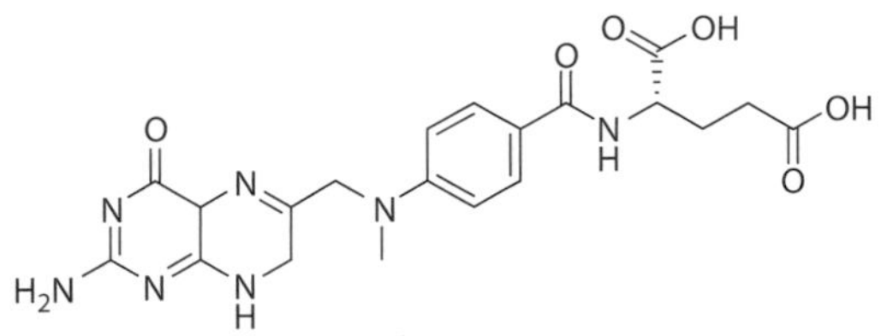

Methotrexate

FIGURE 22.9   Structure of methotrexate, which is a structural analog of folic acid, and is used in the treatment of severe autoimmune and inflammatory diseases.

IL-6. Secondary effects of corticosteroids include the blocking of IL-2, IFN-γ, and tumor necrosis factor alpha (TNF-α). These elements, notably IL-1, are essential for lymphocyte and APC communication. A decrease in production of these cytokines effectively obstructs an APC's capacity to activate antigen-specific lymphocytes. Glucocorticoids have a hydrophobic structure that allows them to easily diffuse into cells and bind to specific cytoplasmic receptors. The resulting complexes progress to the nucleus, where they are able to inhibit the transcription of the cytokine genes. Corticosteroids are also able to inhibit cytokine production in macrophages. This subsequently inhibits the macrophage phagocytosis and chemotaxis properties. Corticosteroids are potent nonspecific anti-inflammatory agents–administration of corticosteroids results in an acute reduction of circulating lymphocytes and monocytes.

Methotrexate (Figure 22.9), a structural analog of folic acid, is involved in the first line treatment of severe autoimmune and inflammatory diseases. It is classified as an antimetabolite drug, which means it is capable of blocking the metabolism of cells. As a result of this effect, it is used in treating diseases associated with abnormally rapid cell growth, such as cancer, RA, and psoriasis. Methotrexate and its active metabolites compete for the folate binding site of the enzyme dihydrofolate reductase (DHFR). Folic acid is reduced to tetrahydrofolic acid by DHFR for DNA synthesis and cellular replication to occur (see Chapter 20). Competitive inhibition of the enzyme leads to blockage of tetrahydrofolate synthesis, depletion of nucleotide precursors, and inhibition of DNA, RNA, and protein synthesis. Methotrexate also inhibits thymidylate synthase and the transport of reduced folates into the cell. Methotrexate is believed to diminish inflammation by diminishing cytokine production. It has been shown that it has direct effect on T cell function in vitro and in vivo. Due to the nonspecific effects on cell proliferation methotrexate treatment is accompanied by serious side effects and more specific alternatives are required. The new immunomodulating biologics (see below) may offer alternative possibilities for less toxic treatments.

22.5   IMMUNOMODULATING BIOLOGICS

22.5.1   Recombinant Protein and Engineered Proteins

Biologics cover peptides or proteins that are used as therapeutic modalities for medical treatment. Immunomodulating biologics are proteins such as cytokines, interferons (IFNs), or interleukins or it could be monoclonal antibodies (mAbs) blocking their action. It has so far not been successful to make small molecules or peptide mimetics of the large extra cellular protein–protein interactions through which these molecules exert their action and we therefore see a lot of new biologics coming to the market in this area. Many proteins that may be used for medical treatment are normally expressed at very low concentrations; however, through recombinant DNA technology a large quantity of proteins can be produced.

Protein engineering can be used to stabilize proteins and improve in vivo half lives either by introduction of mutations or by introducing chemical modifications such as PEGylation, the latter meaning that a poly ethylene glycol (PEG) moiety is specifically attached to the protein such that it does not interfere with the protein interactions (see also Chapter 4). The resulting molecule has an increased size and renal clearance of the molecule is decreased, thereby extending the half-life of the molecule.

22.5.2   MONOCLONAL ANTIBODIES

Monoclonal antibodies (mAbs) are artificial antibodies against a particular target (the "antigen") and are produced in the laboratory. The original method involved hybridoma cells (a fusion of two different types of cells) that acted as factories for antibody production. A major advance in this field was the ability to convert these antibodies, which originally were made from mouse hybridomas, to "humanized" antibodies that more closely resemble our natural antibodies. mAbs have been widely used in scientific studies of cancer, as well as in cancer diagnosis. They are now used as a very successful molecular format in treatment of several diseases where there is a high unmet medical need and where targets are the so-called nondruggable, i.e., it has not been successful to make small molecules against the target. The format is highly attractive due to the relatively long in vivo half-life of the antibodies and the fact that we have high levels of antibodies circulating so the body does not see them as foreign entities and the risk of immunogenic effects is therefore reduced. The primary drawback of using antibodies is that so far have to be delivered as injectibles either by the intravenous or subcutaneous administration.

22.5.3   CYTOKINES

Cytokines are relatively small proteins that play a major role in modulating the immune system. Among the large number of cytokines are the IFNs, TNF, and the interleukins. Cytokines are manufactured and released by cells of the immune system and perform a variety of functions including cell activation, inflammation, tissue breakdown, and repair as well as cell death. Various different cytokines send different complex signals to other cells including: calling immune system cells to the site of the infection, telling endothelial (blood vessel "lining") cells to let these cells through and telling immune system cells to activate themselves.

Using the body's own immunomodulators is becoming an exciting possibility to target inefficient or misdirected immune responses that result in diseases. The potential benefits in terms of treatment of human diseases are enormous and still largely unexplored. Thus, using cytokines and their antagonists as therapeutic agents is an emerging and growing area of research.

22.5.3.1   Interferons

IFNs belong to the cytokine protein family. They are produced by white blood cells in the body (or in the laboratory) in response to infection, inflammation, or stimulation. They have been used as a treatment for certain viral diseases, including hepatitis B and C as well as MS. IFNs can be divided into three groups, IFN-α, IFN-β, and IFN-γ, respectively. Sub-variants of each group have been developed as therapeutic agents in the form of recombinant proteins for injection.

IFNs regulate cell function in the immune system by blocking cell growth and differentiation and stimulating monocytes and macrophages. In MS, IFN-β works indirectly on the central nervous system (CNS) by reducing the inflammatory reaction. The exact mechanism is unknown but it is believed that IFN-β can improve the activity of regulatory T cells, reduce the production of proinflammatory cytokines such as, TNF-α, IL-6, IL-1, and IL-8. They have been shown to downregulate antigen presentation and the mobility of the activated T cells in the CNS.

IFN-α was one of the first cytokines to show an antitumor effect, and it is able to slow tumor growth directly, as well as help to activate the immune system. IFN-α has been approved by the FDA and is now commonly used for the treatment of a number of cancers, including multiple myeloma, chronic myelogenous leukemia, hairy cell leukemia, and malignant melanoma.

Some of the problems with these cytokines, including many of the IFNs and interleukins, are their side effects, which include flu-like syndromes when given at a high dose.

22.5.3.2 Tumor Necrosis Factor

TNF-α is a proinflammatory cytokine that is produced by macrophages, NK cells as well as B and T cells. Its action is to promote local inflammation, and to activate endothelial cells. TNF-α, in soluble or membrane form, binds to two types of receptors that will cause the activation of the target cell.

TNF-α plays a predominant part in the inflammatory process of RA and is highly up regulated in the synovial fluid of the joints in RA patients. TNF-α causes release of intra-articular metalloproteases, which destroy bone cartilage, and via activation of transcription factors (NF-kB), TNF-α causes the production of proinflammatory and immunomodulating cytokines.

The implication of TNF-α in the inflammatory process and in the destruction of bone cartilage (RA) and in inflammatory intestinal lesions (Crohn's disease) has enabled the development of two approaches to treatment using TNF-α inhibitor agents, two mAbs and a soluble receptor, which are used in combination with methotrexate.

There are two mAbs directed at TNF-α, one chimeric, Infliximab (Remicade®); the other entirely humanized, Adalimumab (Humira®). A soluble receptor of TNF-α, Etanercept (Enbrel®), which limits the biological activity of TNF-α, acts by binding to it and preventing it from interacting with its receptors. The TNF-α inhibitors are successfully used for treatment of RA, psoriasis, and Chron's disease.

22.5.3.3 Interleukins

The role of interleukins is to mediate and control the immunologic and inflammatory response (Table 22.1). The list of known ILs is still increasing most of which have only been discovered in the last few years. Their role within the immune system is only beginning to be understood and they are just starting to be utilized in the treatment of a wide variety of diseases including cancer, AIDS, and autoimmune diseases. The following table briefly describes the role of each of the interleukins where the action is well understood.

TABLE 22.1
The Role of Interleukins

Interleukin	Secreting Cells	Action
IL-1	Macrophages	Stimulates T cells to secrete interleukin-2 and activate the inflammatory response. It also causes the hypothalamus to increase the body temperature.
IL-2	Helper T cells	Causes activated T- and B cells to proliferate themselves. It also induces antibody synthesis.
IL-3	T cells	Causes other leukocytes to be proliferated—it does this by making certain types of stem cell in the bone marrow to differentiate and grow.
IL-4	Helper T cells	Causes T- and B cells to grow. It is also a factor in the production of IgE antibodies.
IL-5	Helper T cells	Stimulates B cells, and eosinophils. It causes B cells that produce IgA antibodies to proliferate
IL-6	T cells and macrophages	Works in combination with alpha interferon to induce B cell differentiation. It also causes the production of acute phase proteins in the liver and stimulates T cells and other leukocytes.
IL-7	Stromal cells	Causes lymphoid stem cells to differentiate into progenitor T and B cells.

TABLE 22.1 (continued)
The Role of Interleukins

Interleukin	Secreting Cells	Action
IL-8	Macrophages and endothelial cells	IL-8 is "sticky" for T cells and neutrophils and helps to bring them to the site of an inflammation.
IL-9		Induces growth in Helper T cells.
IL-10	T cells, B cells, monocytes, and macrophages	Acts to inhibit some aspects of the immune system while stimulating others. It represses the production of other cytokines within the immune system, especially INF-γ, TNF-α, IL-1, and IL-6. It inhibits antigen presentation but activates B cells.
IL-11		Causes plasmacytoma cells to proliferate.
IL-12	Macrophages and DCs	Causes T cells and NK cells to proliferate. Promotes Th17 lineage.
IL-13	T cells	Promotes B cell differentiation but inhibits inflammatory cytokine production.
IL-14	DCs and T cells	Enhances memory B cell production and proliferation.
IL-15	Monocytes and macrophages	Enhances T cell proliferation in the blood and NK cell activation.
IL-16	T cells	Acts as a chemo attractant and adhesion molecule and activator for T cells. Plays a part in both asthma and autoimmune diseases.
IL-17	T cells	Activates neutrophils.
IL-18	Leukocytes and nonleukocytes	Stimulates the release of Th1 cytokines.
IL-19	Monocytes	May be involved in regulation of proinflammatory cytokines.
IL-20	Unknown	May regulate inflammation in the skin.
IL-21	Activated T cells	Stimulates the proliferation of activated T cells.
IL22	T cells and mast cells	Production of acute phase proteins, increases the number of basophils and platelets.
IL-23	Activated DCs	Acts on memory CD4+ T cells to support their differentiation. IL-23 sustains differentiated Th17 cells.
IL-24	T cells	Can promote induction of apoptosis in cancer cells.
IL-25	Not known	Cytokine production.
IL-26	Activated memory T cells	May act as autocrine growth factor.
IL-27	Activated monocytes, macrophages, and DCs	Expression of IL12RRE2 making T cells responsive to IL-12.
IL-28,29	Various cells	Induce an antiviral state in infected cells.

IL-1Ra, an antagonist of the IL-1 receptor, is a natural inhibitor of IL-1, a proinflammatory cytokine that is involved in inflammation and joint destruction. It has been shown that mice deficient in IL-1Ra develop a type of inflammatory joint disease similar to RA. In the animal arthritis models, IL-1Ra improves the clinical symptoms and slows bone and joint destruction. Anakinra (Kineret®) is a recombinant nonglycosylated IL-1Ra. It is administered subcutaneously in combination with methotrexate. This product received regulatory approval for the treatment of active RA insufficiently controlled by methotrexate.

ILs with antitumor activity includes IL-2 (Figure 22.10). IL-2 is frequently used to treat kidney cancer and melanoma. IL-2 is the major growth and differentiation factor of immunocompetent killer cells, including the CTLs, NK cells, and monocytes. In metastatic renal cell cancer and melanoma, IL-2-based treatments have induced therapeutic responses. Efficacy of IL-2 as a single agent has been reported in 15%–25% of cases in these diseases and could possibly be increased by the addition of other agents such as IFN or chemotherapy. There are many side effects associated with IL-2 treatment; a serious, but very uncommon side effect of IL-2 in high doses is "capillary leak syndrome." Capillary leak syndrome is a potentially serious disease in which fluids within the vascular system (veins and capillaries) leak into the tissue outside the bloodstream. Due to the serious side effects, low dose IL-2 treatment is now applied.

FIGURE 22.10 IL-2 is a soluble protein of 133 amino acids. It is characterized by four α-helixes that are bundled together and belong to the structural family of cytokines called the four α-helix bundle family. Other interleukins in this subfamily are IL-3, IL-4, IL-5, IL-6, IL-7, IL-9, IL-11, IL-12, IL-13, IL-15, IL-21, and IL-23.

Basiliximab (Simulect®) is a chimeric (70% human and 30% murine) mAb utilized in the prevention of acute organ rejection. This mAb has specificity and high affinity for the subunit of the IL-2 receptor (IL-2Ra, also known as CD25 or Tac) preventing IL-2 from binding to the receptor on the surface of activated T cells. By acting as an IL-2 antagonist, basiliximab inhibits IL-2-mediated activation and proliferation of T cells, the critical step in the cascade of cellular immune response of allograft rejection. Therefore, basiliximab has a long half-life of ~7–12 days and saturates the IL-2 receptor for up to 59 days. Other ILs or IL receptor antagonists in clinical development are IL-6, IL-10, IL-12/23, IL-15, IL 20, and IL-21.

22.5.4 B Cell Depletion

As a therapy for cancer, mAbs can be injected into patients to seek out the cancer cells, potentially leading to disruption of cancer cell activities or to enhancement of the immune response against the cancer. This strategy has been of great interest since the original invention of mAbs in the 1970s. After many years of clinical testing, researchers have proven that improved mAbs can be used effectively to help treat certain cancers. An antibody called rituximab (Rituxan®) can be useful in the treatment of leukemias and other new mAbs are undergoing active testing.

Rituximab is a chimeric mAbs directed against the CD20 antigen specific to B-lymphocytes. Once the antibody recognizes the surface antigen, the Fc part of the antibody recognizes Fc receptors on the NK cells and induces ADCC. It has been used for a number of years in the treatment of B cell lymphomas. Various data now suggest an important role for B-lymphocytes in the inflammatory cascade of RA that causes the destruction of cartilage and erosion of bone. Rituximab may intervene by destroying the B cells that produce auto antigens (rheumatoid factor). Systemic lupus erythematosus (SLE) is a disease that is driven by B cells that produce antibodies directed against self-antigens to form immune complexes that deposit in the tissues and instigate an inflammatory process. Rituximab has shown very promising clinical results in the treatment of SLE. Antibodies against other surface antigens are being pursued for B cell depletion therapy; these include CD19, CD22, CD32, CD38, and CD138.

22.5.5 TARGETING T CELLS

Agents that can interfere with T cell functions open up an important new avenue to the treatment of various autoimmune diseases as well as transplantation. There are several mechanisms that can be exploited: depletion of the T cells, interfering with TCR-mediated signaling and targeting costimulatory signaling to induce anergy (immune unresponsiveness), and finally targeting several cytokines indirectly modulates the T cell function (see Section 22.5.3). Below are examples of mAbs developed for depletion of T cells as well as mAbs interfering with the TCR described.

Anti-CD3 (Muromomab-CD3) is the first type of murine mAb directed against the epsilon chain of the CD3 molecule (an integral part of the TCR complex). It thereby inactivates T-cell function blocking both naive T cells and CTLs. This results in rapid depletion of T cells from circulation and cytokine release. This antibody is used to treat acute graft rejection in transplantation. Several severe adverse effects as a result of Muromonab-CD3 are thought to be a product of the cytokine release (also known as cytokine storm). Humanized versions of this antibody are being pursued that are also engineered in the Fc part of the molecule to be nondepleting and to decrease the cytokine storm. The antibodies are believed to induce tolerance by up regulation of regulatory T cells. These antibodies are being investigated in clinical trials for treatment of Type I diabetes and other autoimmune and inflammatory diseases such as RA, psoriasis, and inflammatory bowel disease.

Alemtuzumab (Campath®) is a humanized anti-CD52 mAb that has emerged as an effective lymphocyte-depleting agent for organ transplantation, RA, MS, and vasculitis. Most of the depleted T cells return to near normal in 3 months after dosing. There is a risk that anti-CD52 treatment elicits autoimmune diseases; this is believed to be due to depletion of the regulatory T cells.

Abatacept (Orencia®) is a fusion protein of CTLA-4 (a surface protein of CTLs) and the Fc fragment of the human immunoglobulin IgG. It acts at the level of the T-lymphocytes where it blocks activation by interference with the path of costimulation between the APC and the T-lymphocyte. Abatacept thus modulates the activity of the B cells, DCs, macrophages, and other cells implicated in the inflammatory cascade and it is currently being developed for treatment of RA.

22.6 PROSPECTS OF IMMUNE MODULATING AGENTS

The immune system is highly complex and new immunological mechanisms and how they relate to human disease is still being identified. In this chapter, focus has been on the more established therapeutic approaches, general immune suppression that is still the first line treatment in autoimmune disease, depleting antibodies against cell surface ligands, the messenger proteins of the immune system—the cytokines and finally the targeting of the T cells. There are several other molecular avenues that can be explored, the first inhibitory antibody to the complement system has just reached the market; some are in clinical development including inhibitory antibodies to the innate immune system and blocking of the Th17 pathway. We expect to see many more immunological pathways to be explored over the years to come as we learn more about this highly fascinating biological system with plenty of therapeutic opportunities.

FURTHER READINGS

E. S. L. Chan and B. N. Cronstein, Molecular action of methotrexate in inflammatory diseases, *Arthritis Research*, 4, 266–274, 2001.

J. C. W. Edwards, B cell targeting in rheumatoid arthritis and other autoimmune disease, *Nature Reviews Immunology*, 6, 394–403, 2006.

M. Feldmann and R. N. Maini, Anti TNFα therapy of rheumatoid arthritis: What have we learned? *Annual Review of Immunology*, 19, 163–196, 2001.

C. Janeway, P. Travers, M. Walport, M. Shlomchik, *Immuno Biology*, 6th edition, Earland Science Publishing, New York, 848, 2005.

E. H. Liu, R. M. Siegel, D. M. Harlan, and J. J. O'Shea, T cell-directed therapies: Lessons learned and future prospects, *Nature Immunology*, 8, 25–30, 2007.

T. W. Mak and M. E. Saunders, The immune response, *Basic and Clinical Principles*, Elsevier Academic Press, San Diego, CA, 1194, 2006.

23 Anticancer Agents

Fredrik Björkling and Lars H. Jensen

CONTENTS

23.1 THE DISEASE

Cancer is the second most common cause of death in the United States, and the rest of the western world, exceeded only by heart diseases. Thirty years ago, 50% of the Americans diagnosed with cancer died due to the disease within 5 years; today's 5-year survival rate is 65%. The FDA has approved 43 new cancer drugs over the last 10 years, compared with 27 in the previous 10 years. Table 23.1 gives an overview of some of the anticancer agents approved today indicating their primary target/mechanism of action and the indications they are used for. Despite this improvement in anticancer therapy, more and more people are being diagnosed with cancer; the primary reason for this being the prolongation of mean life span as a result of more effective treatments against deadly infections and cardiovascular diseases. The need for effective anticancer treatments is therefore still urgent. In this chapter, we will give an overview of the molecular and cellular alterations associated with the development of cancer, and the challenges to be overcome for the effective treatment of this disease. Finally, we will focus on the development and characterization of some anticancer agents used in today's praxis.

23.1.1 THE HALLMARKS OF MALIGNANT CANCER

Metastatic cancer, responsible for 90% of all human cancer deaths, is characterized by a number of molecular, cellular, and morphologic characteristics that has been described as the six hallmarks, or acquired capabilities, of cancer. These hallmarks relate to different levels of dysregulation required

TABLE 23.1

Examples of Currently Used Anticancer Agents with Indication of (1) Their Site of Interaction, (2) Their Interactions and Compound Classes, (3) Their Name/Indications/Approval Year, and (4) Their Specific Target/Mode of Interaction

Site of Interaction	Interaction/Compound Class	Examples of Launched and Experimental Drugs, Trade Name (Generic Drug Name, FDA Approval Year, Indication)	Specific Targets/Mode of Action
Nucleus	*Nonspecific DNA break*	Cytoxan, Neosar (cyclophosphamide, 1959, broad; ovary, lung breast)	DNA cross linking
Tumor cell DNA	Nitrogen mustards	Temodar (temozolomide, 1999, anaplastic astrocytoma)	DNA methylation
	Nitrosoureas	Blenoxane (bleomycin, 1973, testicular, ovarian)	DNA fragmentation
	Triazenes	Platinol (cisplatin, 1978, testicular)	DNA cross-linking
	Platinum compounds	Eloxatin (oxaliplatin, 2002, colorectal)	DNA cross-linking
Nucleus	*DNA-related proteins*	Adriamycin, Rubex (doxorubicin, 1974, ovary)	Topo II-induced breaks
Tumor cell DNA	Epipodophyllotoxines	Idamycin (idarubicin, 1990, AML)	Topo II-induced breaks
	Topo I inhibitors	Camptosar (irinotecan, 1996, colon)	Topo I-induced breaks
	Antimetabolites	Hycamtin (topotecan, 1996, ovary)	Topo I-induced breaks
	Antifolates	Vepesid (Etoposide, VP16, 1983, lung, testicular)	Topo II-induced breaks
	Anthracyclines	Cerubidine (daunorubicin, daunomycin, 1987, ovary)	Topo II-induced breaks
		Methotrexate (methotrexate, 1953, osteosarcoma)	Dihydrofolic reductase—DHFR
		Adrucil (fluorouracil, 5-FU, 1962, broad colon, rectum, breast)	Thymidylate synthase—TS
		FUDR (floxuridine, 1970, metastatic gastrointestinal)	TS
		Xeloda (capecitabine, 2001, breast, colon)	TS
		Cytosar-U (cytarabine, 1969, leukemia)	DNA polymerase
		Gemzar (gemcitabine, 1996, pancreas, lung)	DNA polymerase
		Alimta (pemetrexed disodium, 2004, mesothelioma, lung)	GRAFT, TS, and others
Nucleus	*Hormone receptor*	Nolvadex (tamoxifen, 1977, breast)	Estrogen-receptor antagonist
Nuclear receptors	Retinoids	Vesanoid (ATRA, 1995, APL)	Retinoic acid receptor
	Anti hormones	Femara (letrozole, 1997, breast)	Aromatase/CYP-19 inhibitor
	Steroids	Aromasin (exemestane, 1999, breast)	Arometase/CYP-19 inhibitor

Target	Class	Example (agent, year, indications)	Mechanism
Microsomes	*Oxidoreductases* Small molecules		
Nucleus Histones, DNA	*Epigenetics* Hydroxamic acids	Zolinza (vorinostat, 2006, CTCL)	Histone deacetylases
		Belinostat (pxd101, phase II, various indications)	Histone deacetylases
	Cyclic peptides	(valporic acid, phase II, various indications)	Histone deacetylases
	Benzamide	Decogen (decitabine, 2006, MDS)	DNA methyl transferases
Cytoplasm Acetylated proteins	Short chain fatty acids		
Cell-membrane *Membrane receptors*	MoAb	Rituxan (rituximab, 1997, non-Hodgkin's lymphoma)	CD20 antagonist
		Tarceva (erlotinib, 2004, lung, pancreas)	EGFR antagonist
Membrane-bound tumor proteins	Small molecule antagonists	Iressa (gefitinib, 2003, lung)	EGFR antagonist
		Herceptin (trastuzumab, 1998, breast)	Her-2 antagonist
		Sutent (sunitinib maleate, 2006, gastrointestinal, renal)	Multiple tyrosine kinase inhibitor
		Nexavar (sorafenib, 2005, renal)	Receptor tyrosine kinase inhibitor
Cytoplasm	*Signal transduction kinase inhibitors*	Gleevec (imatinib mesylate, CML, gastorintestinal, 2001)	Bcr/alb and c-kit
	Small molecule antagonists	Nexavar (sorafenib, 2005, renal)	MAPK kinase pathway
	Protein degradation Proteasome inhibitor	Velcade (bortezomib, 2006, multiple myeloma)	Proteasome inhibitor
	Tubulin modulator Taxanes and vinca alkaloids	Oncovin (vincristine, 1963, broad leukemia, brain, breast)	Hyperstabilization of microtubules
		Taxol (paclitaxel, 1992, Kaposi's sarcoma, breast, ovary)	Hyperstabilization of microtubules
		Taxotere (docetaxel, 1996, breast, lung)	
Endothelial cells	*Endothelial receptors of VEGF, bFGF, and TGF-α* Monoclonal antibodies	Avastin (bevacizumab, 2006, lung, colon)	Vascular endothelial growth factor receptor (VEGFR) antagonist
Cell membrane	Small molecules antagonists		
Cells of the immune system	*Lymphocytes, macrophages, and dendritic cells* Interferons	Roferon A (interferon alfa 2a, 1986, leukemias)	Stimulates the ability of immune cells to attack cancer cells
	Interleukin2		
	Cancer vaccines (DNA- or protein-based)		

for cancer cells to develop into and maintain malignancy. These include (1) self-sufficiency in growth signals, (2) insensitivity to antigrowth signals, (3) evading apoptosis, (4) limitless replicative potential, (5) sustained angiogenesis, and (6) tissue invasion and metastasis.

23.1.2 Acquired Capabilities of Cancer Cells

1. Self-sufficiency in growth signals can be achieved directly by the production of growth-promoting signaling molecules by the cancer cell itself. This phenomenon is known as autocrine stimulation, thus creating a positive growth feed back loop. The production of platelet derived growth factor (PDGF), transforming growth factor α (TGF-α) by cancer cells are examples. Another way of achieving self-sufficiency in growth signals is to constitutively activate downstream signal transduction pathways normally activated by progrowth signals. The SOS-Ras-Raf-MAPK signal (Figure 23.1) transduction pathway is a good example, and the Ras proteins are indeed altered in 25% of all cancers leading them to produce mitogenic signals in the absence of progrowth signals (Figure 23.1).

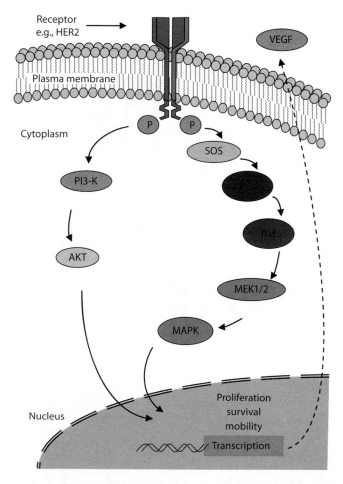

FIGURE 23.1 Cartoon depicting two signal transduction cascades often deregulated in cancer, namely, the PI3K-AKT and the SOS-Ras-Raf-Mek cascades. One way of activating these pathways is by signaling through the HER2 transmembrane tyrosine kinase receptor.

2. At the G1 cell cycle checkpoint, also termed the restriction point, normal cells detect the composition of growth/antigrowth signals including cytokines and nutrients in their environment, and then decide whether or not to enter another round of cell division. One key effector halting the cell cycle at G1 is the retinoblastoma protein (pRb). Hypophosphorylated pRb blocks cell proliferation by sequestering E2f family transcription factors, thereby inhibiting the binding of these to DNA so that the genes required for entry into the S phase of the cell cycle are not expressed. Deletion of pRb is frequent in cancers and thus represents one of the most common means by which cancer cells obtains resistance to antigrowth signals.

3. Evading apoptosis, or programmed cell death, is also a key acquired capacity of cancer cells. In normal cells, apoptosis targets cells for self-destruction when their DNA is damaged beyond repair, or when the cells do not receive the proper prosurvival signals from their environment. A key family of regulatory proteins involved in regulating apoptosis is the Bcl-2 family containing proapoptotic (Bax, Bak, Bid, and Bim) as well as antiapoptotic (Bcl-2, Bcl-XL, and Bcl-W) regulatory proteins. The Bcl-2 protein itself is frequently unregulated in cancer cells leading directly to resistance toward drug-induced apoptosis. Furthermore, deletion or mutational inactivation of the tumor suppressor gene p53 represents another way of evading apoptosis, because p53 normally induces up regulation of proapoptotic Bax protein upon DNA damage. p53 is inactivated in 50% of all human tumors.

4. Normal cells have the capacity to divide only a finite number of times, typically 60–70 doublings before they die due to senescence, a cell fate different from apoptosis. This is because the telomeres, DNA sequences at the end of the chromosome arms, shorten a little at each mitotic cell division. Loss of telomeres results in DNA recombination and end to end fusions of chromosomes leading to cell senescence. More than 90% of human cancers avoid telomere shortening by expressing the catalytic subunit of telomerase hTERT, thereby avoiding senescence. The remaining use DNA recombination for this purpose.

5. Angiogenesis is the process of developing new blood vessels. In order to grow beyond the size of a few $100\,\mu m$, tumors must develop neovascularization through the process of angiogenesis. However, the tumor cells do not directly participate in neovascularization, which is mediated only by endothelial cells. Instead tumor cells produce the factors required for endothelial cells to perform angiogenesis in that they secrete growth factors like vascular endothelial growth factor (VEGF) or fibroblast growth factor (FGF1/2), which can be produced by different mechanisms as shown for VEGF in Figure 23.1. Angiogenesis is attractive in anticancer therapy because its inhibition is expected to have no implications for normal cells in the body regardless of their proliferation rate.

6. Primary tumor growth is seldom the direct cause of death in cancer patients. Instead, the spreading of tumors known as metastasis is the most frequent cause of death (90%). Invasion is penetration of tumor tissue through basal membranes allowing a tumor to grow out from its original site into the surrounding tissue; whereas metastasis is the spreading of cancer cells to other sites in the body, and the subsequent establishment of secondary tumors in other tissues. Both processes involve changing the interaction of cancer cells with the extracellular environment/matrix.

23.1.3 Enabling Characteristics of Cancer Cells

In addition to these hallmarks, cancer cells display a number of additional alterations enabling cancer growth. One of the most important alterations is genome instability. The number of individual mutations required to induce all six hallmarks of cancer would not normally accumulate in a single cell if not for genome instability. Genome instability is caused by mutations of DNA repair/ checkpoint surveillance mechanisms that normally results in cell cycle arrest until damaged DNA has been repaired, or in case DNA cannot be efficiently repaired results in elimination of the cell via apoptosis. The genome guarding protein p53 as well as proteins sensing DNA damage such as ATR, ATM, and their downstream signal kinases Chk1 and Chk2 play important roles in such checkpoints, and their loss often precedes genome instability.

One of the main obstacles in cancer therapy is tumor heterogeneity that is closely related to genome instability. Within one tumor several different subpopulations of tumor stem cells often exist, due to accelerated chromosomal aberrations driven by genome instability resulting in chromosomal deletions, duplications and translocations. Consequently, while first line chemotherapy does often lead to good responses and tumor regression, tumor heterogeneity makes it very difficult to kill all tumor cells within a tumor. Consequently, some cancer cells may be inherently resistant toward a given drug due to mutation of its primary target/receptor or due to dysregulation of downstream signal transduction cascades. Such cells will often continue to divide and will eventually overgrow the drug sensitive cells resulting in a new tumor that is now resistant. As a result in modern anticancer therapy, combinations of several (five or more) different drugs targeting different receptors/pathways are often used.

23.1.4 ANTICANCER AGENTS

Despite intensive research aimed at understanding the molecular pathology of cancer, a great deal of anticancer agents currently in clinical use were discovered and even entered the clinic before their exact mechanism of action was clarified. These drugs were often discovered in cellular screens of extracts from natural sources, or in in vivo screens using a leukemic P388 mouse model. The drugs discovered typically inhibit DNA synthesis (antimetabolites), damage DNA (DNA alkylating agents and topoisomerase poisons), or inhibit the function of the mitotic microtubule-based spindle apparatus (taxanes) (Table 23.1). The reason for these agents still being in clinical use relates to the fact that they are often highly effective although they have toxic properties toward normal fast proliferating cells as the intestinal and gut lining hair follicles, and the bone marrow cells, leading to the well-known effects of classical chemotherapy including nausea, vomiting, hair loss, and myleosuppression. The cytotoxics stands in contrast to the so-called targeted therapies that are developed in a totally different fashion by applying knowledge concerning the structure of a primary target with molecular in silico screening as well as high-throughput compound library screening, or by designing protein-based medications, which in the case of cancer treatment are often in the form of cell surface tyrosine kinase antagonistic antibodies. Examples of targeted therapeutic anticancer agents are kinase inhibitors (Gleevec, Iressa), proteasome inhibitors (Velcade), histone deacetylase (HDAC) inhibitors (Zolinza, belinostat), and antibodies against cell surface receptors (Herceptin) (Table 23.1).

23.2 ANTICANCER AGENTS CURRENTLY USED

Reviewing the multitude of anticancer agents in current use is an overwhelming task and not within the scope of this chapter. Instead, we will describe the development and mechanism of action of three classic cytotoxics. Two antimetabolites Xeloda and Alimta inhibiting the production of precursors for DNA synthesis in the cell, as well as an agent targeting the structural protein β-tubulin (Paclitaxel) involved in microtubule functioning. We will likewise review the development and mechanism of three new classes of anticancer therapeutics, the HDAC inhibitors exemplified by Zolinza, the kinase inhibitors exemplified by the BCR-ABL tyrosine kinase inhibitor Gleevec, and finally the HER2/Neu tyrosine kinase antagonist monoclonal antibody Herceptin (Table 23.1).

23.2.1 XELODA

The classic antineoblastic agent 5-Fluorouracil, 5-FU (Figure 23.2A), is a fluorinated analog of uracil. 5-FU exhibits its main activity via two known biochemical mechanisms originating from the 5-FU metabolites 5-fluoro-2′-deoxyuridine monophosphate (FdUMP) and 5-fluorouridine triphosphate (FUTP) (Figure 23.2A). FdUMP and the folate cofactor N^5,N^{10}-methylene-tetrahydrofolate bind to thymidylate synthase (TS) to form a covalently bound ternary complex. This binding inhibits the formation of thymidylate from 2′-deoxyuridylate. Since thymidylate is a necessary precursor of thymidine triphosphate, which is essential for the synthesis of DNA, its inhibition causes inhibition of cancer cell division. Second, nuclear transcriptional enzymes can mistakenly incorporate

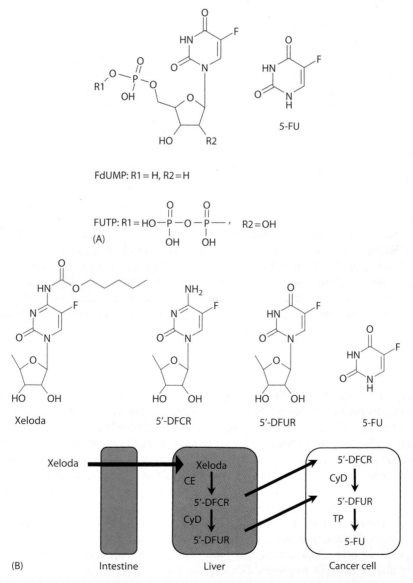

FIGURE 23.2 (A) Chemical structures of 5-FU and its bioactive metabolites. (B) Bioreactive pathways leading to the generation of 5-FU from Xeloda. The relevant enzymes and their compartments are shown.

FUTP in place of uridine triphosphate (UTP) during the synthesis of RNA. This metabolic error can interfere with RNA processing and protein synthesis and consequently with cell growth. It has been suggested that while inhibition of TS is primarily responsible for the anticancer activity of 5-FU, its effect on RNA synthesis may be the main cause of its toxicity.

Even though 5-FU is an efficient drug that has been used for many years in the treatment of solid tumors, such as breast and colorectal cancers, there has been a wish to develop an orally available fluoropyrimidine with improved efficacy and safety profile. The goal was to design derivatives of 5-FU that could specifically be converted to the parent drug (5-FU) by enzymes preferentially located in tumor tissue. Several attempts have been made to make orally active prodrugs of 5-FU, the most advanced version being a 5′-deoxy-5-fluorouridine (5′-DFUR) derivative (Figure 23.2B), which is transformed by pyrimidine nucleoside phosphorylase (PyNPase) enzymes, which are preferentially found not only in tumor tissue but unfortunately also in the intestine. Consequently, even though there is some selective targeting of cancer cell with this compound, 5′-DFUR also releases 5-FU in the intestine causing dose limiting toxicological effects there.

To get around the unwanted effects in the intestine, another strategy was developed. The chemical starting point for 5-FU prodrugs was the 5′-deoxy-5-fluorocytidine (5′-DFCR) described above (Figure 23.2B), which was known to be effectively transformed to 5-FU by cytidine deaminase (CyD), particularly in tumor tissue where it is highly expressed. This in combination with a low activity of the same enzyme in human bone marrow cells indicated that selective killing of tumor cells could be obtained. The strategy was now to find a fluoropyrimidine carbamate that was stable in the intestinal tract but efficiently hydrolyzed to 5′-DFCR by carboxylesterase (CE) located in the liver, where 5′-DFCR could be further transformed to 5′-deoxy-5-fluorouridine (5′-DFUR), and finally to 5-FU by thymidine phosphorylase (TP) (Figure 23.2B).

Among the many prodrugs prepared to achieve this activity pattern, through extensively testing for (1) selectivity for hepatic CE, (2) oral bioavailability, and (3) activity in human cancer xenograft models in vivo, Capecitabine (N^4-pentyloxycarbonyl-5′-deoxy-5-fluorocytidine, Xeloda™), developed and marketed by Roche, was found to have the most favorable characteristics resulting in substantially higher 5-FU concentrations within tumors than observed in plasma and in normal tissue (muscle) (Figure 23.2B). Additionally, the tumor 5-FU levels were much higher than those that could be achieved by the intraperitoneal administration of 5-FU at equitoxic doses. This tumor selective delivery of 5-FU ensures a greater efficacy and a more favorable safety profile than can be obtained by other fluoropyrimidines. Xeloda is now approved in the United States, Canada, and other countries for the treatment of metastatic breast cancer.

23.2.2 ALIMTA

The discovery of Alimta has its chemical origin in the early findings of the antimetabolites aminopteridines and thereafter methotrexate and both inhibit folate metabolism (Figure 23.3C). The impressive anticancer effects found for methotrexate validated folate antimetabolites early on as antiproliferative agents. For decades, researchers have worked on the task to find inhibitors of folate-dependent enzymes such as TS, dihydrofolate reductase (DHFR), and glycinamide ribonucleotide formyltransferase (GARFT), which take part in the folic acid activation (Figure 23.3A). The active form of folate is the reduced form tetrahydrofolate (Figure 23.3B), which plays an important role in the biochemical pathways to donate one carbon unit in the form of methyl, methylene, or formyl groups. These metabolic reactions are essential for the formation of DNA, RNA, ATP, and the catabolism of certain amino acids. Consequently, inhibiting this metabolic pathway abrogates cancer cell proliferation because cancer cells have high demands for ATP, and because they require high levels of nucleic acid precursors for DNA synthesis.

The pathway leading to the formation of tetrahydrofolate (THF) begins when folate (F) is reduced to dihydrofolate (DHF), which is then reduced to THF, DHFR catalyses both steps (Figure 23.4). Methylene tetrahydrofolate (CH_2THF) is formed from tetrahydrofolate by the addition of

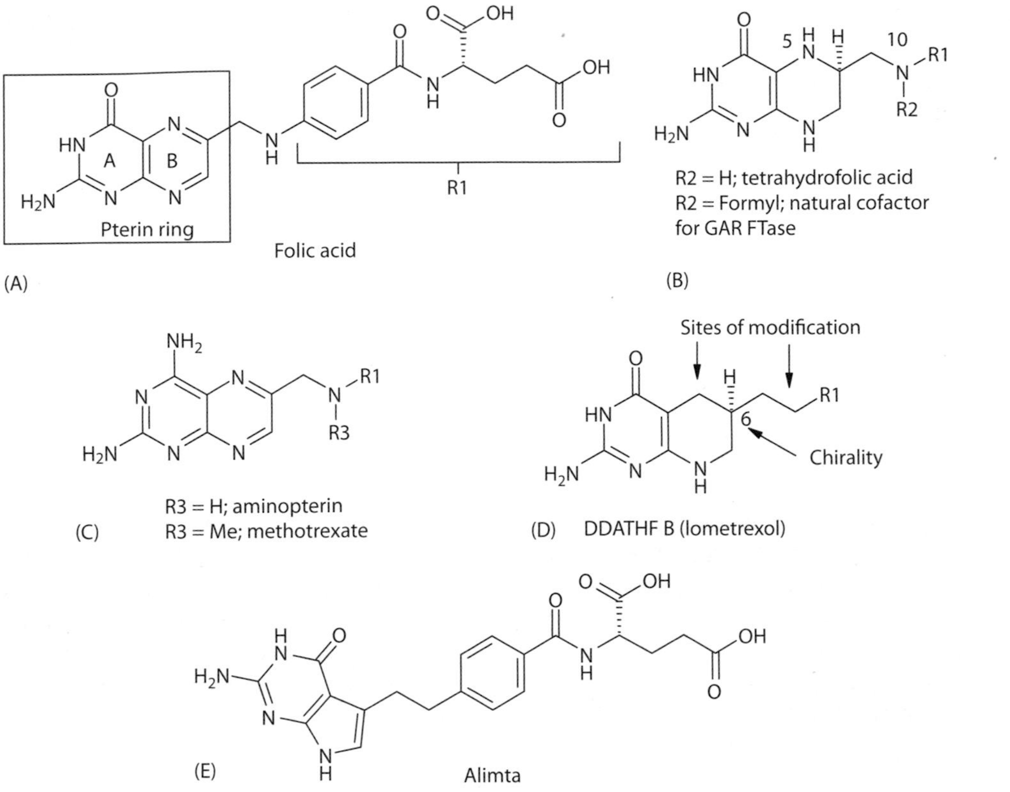

FIGURE 23.3 Chemical structures of various antimetabolites preceding and founding the development of Alimta.

methylene groups forming N^5, N^{10}-methylene tetrahydrofolate from one of the three carbon donors: formaldehyde, serine, or glycine (Figure 23.4). The key reaction is the TS-catalyzed methylation of deoxyuridine monophosphate (dUMP) to generate thymidylate (dTMP), which is needed for DNA synthesis. Methyl tetrahydrofolate (CH$_3$THF) is formed from methylene tetrahydrofolate by reduction of the methylene group and formyl tetrahydrofolate (CHOTHF, folinic acid) results from the oxidation of the same precursor (Figure 23.4).

Inspired by the active pterine structures (Figure 23.3A), many modifications were made in this ring system including modifications in ring A, such as substitution of NH$_2$ with methyl or hydrogen as well as exchange of the fused ring B for a fused phenyl ring. This resulted in compounds having high biological activity as TS inhibitors with concomitant antiproliferative activity. Some of these analogs were indeed taken into early clinical testing, but were stopped due to pharmacokinetic or toxicological problems.

An important new class of potent folate antimetabolites that are active as antitumor agents are represented by 5,10-dideaza-5,6,7,8-tetrahydrofolic acid (DDATHF, Lometrexol) in which the two nitrogens in positions 5 and 10 is exchanged for carbon and the B ring is reduced and as such mimic the structure of THF (Figure 23.3D). The target enzyme for DDATHF was shown to be glycinamide ribonucleotide formyltransferase (GARFT) (Figure 23.4) catalyzing the first folate cofactor-dependent formyl transfer step in the de novo purine biosynthetic pathway instead of the DHFR enzyme, which was the target for earlier folate inhibitors described above.

The two diastereomeric forms of DDATHF were separated and their biological activity examined. Interestingly, they did not show any significant difference in activity and further work was therefore undertaken to remove this chiral center so as to obtain a stereochemically pure compound.

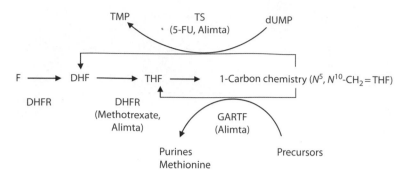

FIGURE 23.4 GARTF is one of the many enzymes involved in the biosynthesis of purines.

After exploring many options such as opening of the ring B, the best candidate was found to be Alimta, developed and marketed by Eli Lilly (Figure 23.3E), which contains a fused pyrrole ring in place of ring B. While early studies with Alimta showed that its primary target was TS, more recent studies have demonstrated that following intracellular polyglutamation of Alimta the product affects folate metabolism dramatically by inhibiting several folate-dependent enzymes: TS, DHFR, GRAFT (Figure 23.4) in addition to aminoimidazole ribonucleotide formyltransferase (AICARFT), and C-1 tetrahydrofolate synthetase (C1-S). This broad mechanism of action on folate activation may be responsible for the efficacy of Alimta. Today Alimta is used in pleural mesothelioma and as a single agent for the treatment of patients with locally advanced or metastatic nonsmall lung cancer.

23.2.3 Taxol

Another example of a successful anticancer drug whose development goes far back in time is taxol. Back around 1960, the National Cancer Institute (NCI) launched a program of screening compounds from plant extracts with the aim of identifying molecules of biomedical interest. Taxol was discovered in 1963. This complex polyoxygenated diterpenoid was isolated from the pacific yew, *Taxus brevifolia*, and later found in several other species of Taxus including *Taxus wallichiana*, the Himalayan yew. By the early 1970s, the structure of taxol was solved (Figure 23.5A). Yet, another decade had to pass before the molecular mechanism of the compound was elucidated, which is stabilization of microtubules with concomitant cell cycle arrest at the G2M cell cycle phase. The binding of taxol to polymerized β-tubulin is depicted in Figure 23.5B. Later on availability problems associated with the limited supply of the Pacific yew was solved by applying a semisynthetic route starting with another natural product isolated from the English yew, *Taxus baccata*, (10-deacetylbaccatin-III) avoiding the exhaustive complete synthesis of taxol. This was achieved in the early 1980s, and accomplished by acetylating 10-deacetylbaccatin-III and attaching a side chain to it. Taxol was finally approved by the FDA at the end of 1992. Today, taxol (Paclitaxel) developed and marketed by Bristol–Meyer Squibb is an effective drug in the treatment of breast-, ovarian-, and lung cancer.

A total of more than 300 taxanes have now been made and extensive structure–activity relationship (SAR) analyses have been carried out. The most important findings are summarized here (Figure 23.5A). In the lower part of the molecule, the activity is reduced by the removal of C1 hydroxy, C4 acetyl, 4,5,20 oxetane ring, and in the C2 benzoyl only limited substitutions are allowed. In the upper part of the molecule derivatization of C7-hydroxyl or change in its stereochemistry has no significant effect on activity. Also, reduction of a C9-ketone slightly increases the activity, and both C10 hydroxyls and acetates retain activity. The C-13 side chain is essentially required for

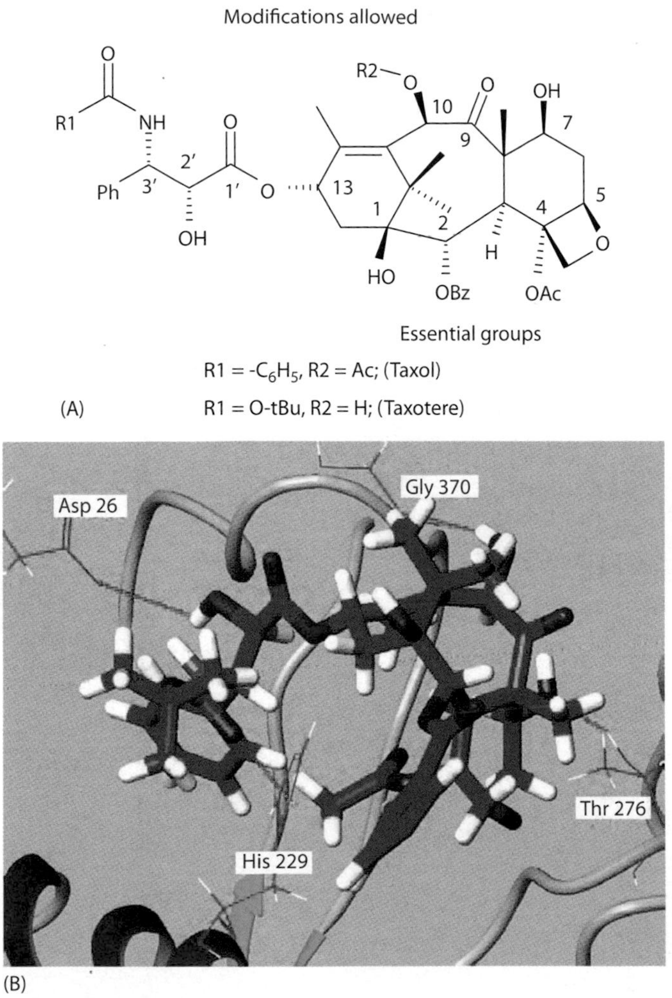

FIGURE 23.5 (A) Chemical structures of the taxanes with associated SAR information; (B) crystal structure of taxol bound to β-tubulin. (Computer model courtesy of Dr. S. Vadlamudi, Topotarget, U.K.)

activity, even though some specific alteration is found to be useful for improving the parent compound. For example, can the C2'-hydroxyl group be used as attachment point for a prodrug ester that results in a compound with in vivo activity, but not in vitro potency. Limited modifications of C2' and C3' are allowed, and the stereochemistry at these positions is important for high activity, with preference for the natural 2'R,3'S isomer. Based on the SAR for Taxol, many derivatives have been prepared in particular with the aim of identifying new compounds with increased solubility, since this is a major problem for taxol having a solubility <0.3 μg/mL. Consequently, taxol needs to be administered with a solubilizing carrier such as polyethoxylated castor oil. One example of an approved analogue is docetaxel (Figure 23.5A), which shows potent anticancer activity and better solubility properties.

23.2.4 Zolinza

HDAC inhibitors together with inhibitors of DNA methyl transferase are drugs interfering with epigenetic gene regulation. In cells, DNA exists in the form of chromatin where the DNA double strand helix is wrapped around core histone complexes, which is a protein octamer consisting of two of each the following histones; H2A, H2B, H3, and H4. The basic repeating unit of eukaryotic DNA is termed the nucleosome, and consists of 146 bp of DNA wrapped 1.65 left-handed turns around a core histone particle. The N-terminal tails of the core histones protrude away from the core histone particle as schematically presented in Figure 23.6A. These tails, especially those of histone H3 and H4, are subjected to a number of posttranslational modifications (PTMs), including arginine and lysine methylation, serine phosphorylation, and lysine acetylation amongst others.

The pattern of histone PTMs together with the pattern of DNA methylation is termed the epigenetic makeup or epigenetic code of the cell. One important form of histone PTM is lysine ε-*N*-acetylation, which is a highly dynamic process, and the result of the opposite activities of histone acetyl transferases (HATs) and histone deacetyl transferases HDACs (Figure 23.6A). HDAC

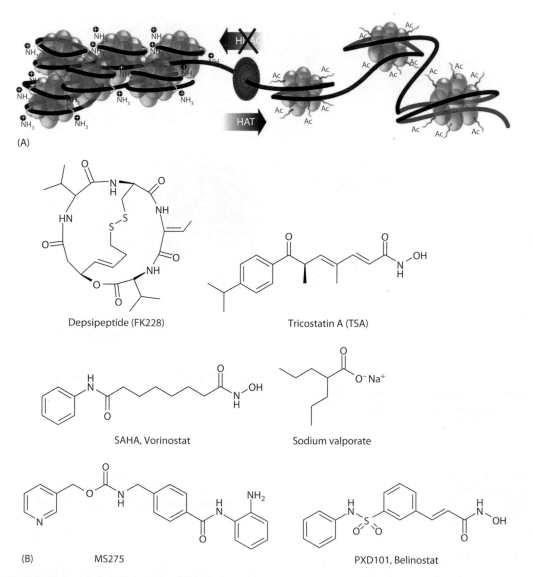

FIGURE 23.6 (A) Schematic representation of core histone complexes on DNA, showing the effect of opposing HAT and HDAC activities on the acetylation status of the *N*-terminal tails of core histones. (B) Chemical structures of some HDAC inhibitors in clinical development.

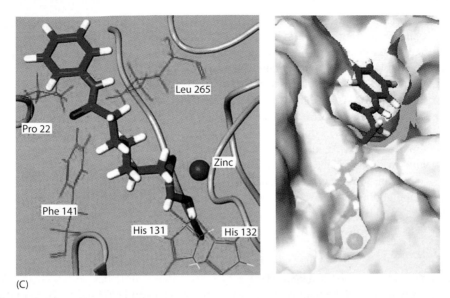

(C)

FIGURE 23.6 (continued) (C) Crystal structure of SAHA complexed with HDAC-like protein. (A) Close-up of interactions with the active site Zn^{2+} and two adjacent binding histidines. (B) Hydrophobic cavity illustrated with the protein surface added. (Computer model courtesy of Dr. S. Vadlamudi, Topotarget, U.K.)

inhibition results in the generation of hyper-acetylated histone tails. These changes in histone acetylation levels (and changes in histone PTMs in general) are accompanied by changes in the cellular gene expression profile. This is because the pattern of histone acetylation marks (and the PTM pattern as a whole) at any given gene affects the transcriptional state of that gene by determining which multiprotein complexes are recruited in its promotor region, in this way regulating the activity of the RNA polymerase II machinery in a local fashion.

Two types of HDACs exists. One uses a coordinated zinc atom to cleave the ε-N lysine bound, while the other deacetylates histones by transferring the lysine ε-N group to NAD^+. Only compounds inhibiting the activity of the zinc dependant family will be addressed here. Several structural and chemical classes of HDAC inhibitors exist, and many are derived from or inspired from natural products (Figure 23.6B). The most advanced HDAC inhibitor is the hydroxamic acid Zolinza, developed by Merck & Co, which is inspired by the natural product antifungicide HDAC inhibitor Tricostatin A (Figure 23.6B). Zolinza comprises (1) a hydrophobic moiety designed to bind to the surface of the zinc dependent HDACs, (2) a carbon linker designed to penetrate through a hydrophobic cavity in the HDAC enzyme, at the bottom of which the coordinated zinc atom is situated, and (3) a hydroxamic acid group at the end of the linker, designed to chelate this active site zinc atom in order to abrogate enzymatic activity. This compound is now approved in the United States by the FDA to treat cutaneous T-cell lymphoma (CTCL). Models of Zolinza bound to zinc-dependent HDAC enzyme is depicted in Figure 23.6C. Another hydroxamic acid in late clinical development (phase III) is belinostat developed by TopoTarget A/S (Figure 23.6B). Other classes of HDAC inhibitors in clinical development is the cyclic peptide depsipeptid (Gloucester Pharmaceuticals) and the benzamide MS275 (Schering AG) (Figure 23.6B). Valporic acid (Figure 23.6B), a known antiepileptic, has recently been found to have weak HDAC inhibitory activity. This compound is now in clinical testing for the treatment of basal cell carcinoma (skin cancer) and familial adenomatous polyposis (an intestinal predisposition to cancer) (TopoTarget A/S).

Some of the genes whose expression most consistently changes across almost all cells examined as a consequence of HDAC inhibition is the p21[Waf1/Cip1] cyclin dependant kinase inhibitor (induced) and various cyclines (repressed). These effects lead to arrest of the cell cycle. Interestingly, HDAC induced cell cycle arrest is often followed by apoptosis in cancer cells while normal cells display a reversible cell cycle arrest. This difference, that is at present poorly understood, is believed to play a role in the cancer selectivity of HDAC inhibitors. In addition, by reprogramming gene expression HDAC inhibitors can cause redifferentiation of leukemic cells, underscoring their utility in the treatment of hematological malignancies. During recent years, the HDACs have been found to have targets other than histones. These include p53, α-tubulin, and many transcription factors. Some of the cellular effects of HDAC inhibition are related to such nonhistone targets.

Although HDAC inhibitors have been shown to have anticancer activity on their own, their main strength will likely be as part of drug combinations. Preclinical studies and emerging clinical studies as well, point toward HDAC inhibitors being capable of increasing the efficacy and broaden the therapeutic window of several classes of unrelated anticancer compounds such as the topoisomerase I poisons (irinotecan, topotecan), the topoisomerase II poisons (etoposide, doxorubicin daunorubicin), the antimetabolite (5-FU), retinoic acids, the proteasome inhibitor Velcade, and the kinase inhibitor Gleevec and the others as well.

23.2.5 GLEEVEC

Imatinib or Gleevec (see also Chapter 11) is a tyrosine kinase inhibitor developed by Novartis, which has greatly improved the treatment of chronic myeloid leukemia (CML). CML is a rather rare condition (prevalence is 1:100,000), which in about 90% of all cases is caused by a specific chromosomal translocation t(29;22) (q34;q11) also referred to as the Philadelphia chromosome. Through the creation of this new chromosomal breakpoint, this translocation creates a unique fusion protein BCR-ABL with oncogenic properties caused by its constitutive tyrosine kinase activity, which activates a number of cellular pathways including JAK/STAT and Ras-Raf-Mek-MAPK leading to the achievement of proliferation and antiapoptotic signaling as depicted in Figure 23.7A.

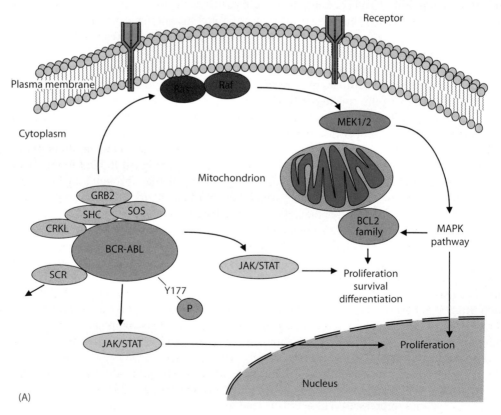

(A)

FIGURE 23.7 (A) Cartoon highlighting pathways affected by the constitutive tyrosine kinase activity of the BCR-ABL onco protein and its cellular consequences.

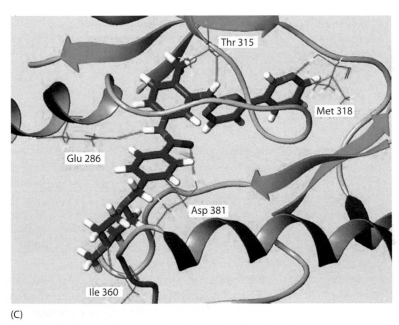

Comp./activity IC50 (μM)	v-Abl-K	PKCα	PKCβ
A	0.4	1.2	23
B	0.4	72	>500
C	0.038	>100	>100

(B)

(C)

FIGURE 23.7 (conitnued) (B) Chemical structure of Gleevec and associated kinase inhibitors. Table shows the effect of SAR variations on the activity toward BCR-ABL and protein kinase C, α, and β isoforms. (C) Crystal structure of the catalytic domain of ABL tyrosine kinase complexed with Gleevec. Hydrogen bonds with interacting amino acids are indicated. (Computer model courtesy of Dr. S. Vadlamudi, Topotarget, U.K.)

The presence of the BCR-ABL protein exclusively in CML cells (cancer specificity) combined with the cellular dependence on this sole protein for survival of transformed cells (also referred to as oncogene addiction) represents an unique situation where targeted cancer therapy is relatively easily achieved. Thus, the development of Gleevec is one of the best examples of a modern targeted anti-cancer therapeutics, and its design builds on a clear rational for intervention (in this case oncogene addiction), combined with detailed knowledge of the three-dimensional structure of the primary molecular target (the BCR-ABL onco-protein). The successful launching of Gleevec has provided a great deal of inspiration and effort into the further development of small molecule kinase inhibitors as anticancer drugs. The finding that Gleevec also inhibits c-KIT has promoted its recent use in the treatment of colorectal cancers.

The molecular starting point for the medicinal chemistry leading to the development of Gleevec was a phenylaminopyrimidine derivative (the blue core of compound A/B as contained in the box seen in Figure 23.7B) found in a screen for protein kinase C (PKC) inhibitors. This compound showed good lead-like properties and could be further optimized with straightforward chemistry to improve activity, selectivity as well as drug like properties. Thus, the addition of a 3′-pyridyl group at the 3′-position of the pyrimidine gave compounds with superior activity as PKC inhibitors in cellular assays. It was further found that derivatives with an amide group in the phenyl ring provided inhibitory activity against tyrosine kinases, such as the BCR-ALB kinase (Figure 23.7B, compound A). Subsequent SAR studies suggested that a substitution at position 6 of the diaminophenyl ring abolished PKC inhibitory activity while retaining activity for BCR-ABL kinase, which was confirmed by the introduction of a methyl group in this position (Figure 23.7B, compound B). Further modification was, however, needed due to the low solubility and bioavailability found for these compounds. Introduction of a highly polar side chain (an N-methylpiperazine) gave a marked improvement of both solubility and oral bioavailability. To avoid the mutagenic potential of aniline moieties, a spacer was introduced between the phenyl ring and the nitrogen atom that now afforded the best compound in the series, Gleevec (Figure 23.7B, compound C). Computational chemistry docking and x-ray crystallography shows that Gleevec binds at the ATP-binding site of an inactive form of the kinase. The high specificity of Gleevec is explained by this unusual binding as well as some strong interactions of the N-methylpiperazine group with the ABL kinase backbone (Figure 23.7C).

23.2.6 Herceptin

In the previous sections, we have looked at small molecule inhibitors of various cancer pathways and targets. A totally different approach is to use natures' own building blocks and principles to target cancer calls selectively for destruction (see also Chapter 22). One of the best examples is the humanized antibody Trastuzumab or Herceptin developed by Genentec targeting the HER2/Neu receptor that is often over expressed in breast cancers. The development of antibody-based anticancer therapies goes back to the 1950s and the first experiments used polyclonal antisera, but their use was hampered by the inherent problems with the polyclonal origin of the antibodies. This problem was alleviated by the 1970s when the application of hybridoma cell culture allowed for the preparation of monoclonal antibodies. At the beginning of clinical experiments using such antibodies, fully murine, rabbit, or rat antibodies were investigated but their applications were often associated with severe clinical problems as a result of anaphylaxis due to the patient's immune response toward these forcing molecules. Today these problems have been effectively solved by the use of so-called humanized antibodies in which the invariable regions have been replaced with the homologous human sequences. The antibodies can also be "armed" with radioisotopes or cellular toxins that are then effectively targeted to the surface of their target cells, or even aimed with activators of prodrugs, leading to high concentrations of the active substances in targeted cells.

The HER2 receptor also called Neu or ErbB2 is a member of the epidermal growth factor receptor (EGFR) family of transmembrane tyrosine kinases. It is composed of an extracellular domain, a hydrophilic transmembrane domain, and an intracellular domain harboring its tyrosine kinase activity which when activated provides the upstream signal for activating cellular signaling pathways responsible for cell proliferation such as the Ras-MAPK and PI3K pathways (Figure 23.1). In contrast to the other EGFR receptor members, HER2 has no identified binding ligand, but its tyrosine kinase activity can be activated upon dimerization of two HER2 receptors and also upon heterodimerization of HER2 with other EGFR family members. HER2 is often overexpressed in breast cancers, which can be caused by gene duplication as well as transcriptional upregulation. This in turn leads to hyper activation of its downstream pathways mentioned above, accelerating, and sustaining tumor growth (Figure 23.1). Therefore, there is a clear rational for targeting the HER2/Neu receptor in breast cancers in cases where it is overexpressed (20%–30% of metastatic breast cancers). Large clinical trails were necessary to establish that the HER2/Neu receptor has to be overexpressed to high levels in order for Herceptin to provide a therapeutic benefit. This finding is now routinely used in diagnostics and in the classification of patients into groups amenable to Herceptin treatment, which significantly benefits this well-defined subpopulation. Today Herceptin is often part of chemotherapeutic regimens also containing topoisomerase II targeting anthracyclines, or microtubule targeting paclitaxel in patient populations with highly elevated levels of HER2 expression, although Herceptin has also been applied as monotherapy, with encouraging results.

23.3 CONCLUDING REMARKS

In this chapter, we have briefly touched on the molecular and cellular background underlying the development of cancer. Furthermore, we have focused on some anticancer agents in current clinical use. The six anticancer agents reviewed represent a continuum in cancer drug development beginning with the antimetabolites used in cancer therapy for more than half a century (Alimta and Xeloda) over therapy targeting the structural function of the cancer cell (Taxol) as well as its epigenetics (Zolinza) to novel highly targeted small molecule (Gleevec) and antibody-based (Herceptin) therapies.

This continuum not only represents a developmental time scale, but also reflects today's clinical practice, attesting to the fact that in the field of cancer treatment old drugs continue to be used alongside much newer ones. This is because the older anticancer drugs, the cytotoxics, are generally highly effective although they show their well-known side effects. On the other hand, while being more cancer specific, the new targeted therapies are often hampered by their rather restricted applications (with regard to indication) or lack of efficacy when administered as single agents. It is therefore often advantageous to combine the classic cytotoxics with new targeted molecules. Here, numerous preclinical and clinical research projects are beginning to indicate the directions toward effective combination regimens including both old and new drugs. We believe these approaches will be further developed in coming years and several new principles, e.g. in gene therapy, cancer immune therapy, and stem cell research will find their way to the patient, possibly even as individually tailored personalized medications, but certainly also as traditional add on to existing therapy.

ABBREVIATIONS

Topo	topoisomerase
GRAFT	glycinamide ribonucleotide formyltransferase
DHFR	dihydrofolic reductase
TS	thymidylate synthase
MoAb	monoclonal antibody
EGFR	epidermal growth factor receptor
VEGF	vascular endothelial growth factor
VEGFR	vascular endothelial growth factor receptor
FGF	fibroblast growth factor
TGF	transforming growth factor
CD20	nonglycosylated phosphoprotein expressed on the surface of all mature B-cells
MAPK	mitogen activated protein kinase
AML	acute myeloid leukemia
APL	acute promyelocytic leukemia
CTCL	cutaneous T-Cell lymphoma
MDS	myelodysplastic syndromes
ATAR	all *trans*-retinoic acid

FURTHER READINGS

Altmann KH and Gertsch J (2007) Anticancer drugs from nature–natural products as a unique source of new microtubule-stabilizing agents. *Nat Prod Rep* 24:327–357.

Costi MP, Ferrari S, Venturelli A, Calo S, Tondi D, and Barlocco D (2005) Thymidylate synthase structure, function and implication in drug discovery. *Curr Med Chem* 12:2241–2258.

DeVita VT, Jr, Hellman S, and Rosenberg SA (2005) *Cancer: Princples and Practice of Oncology*, 7th edn. Lipincott Williams & Wilkins ISBN 0-781-74450-4.

Hanahan D and Weinberg RA (2000) The hallmarks of cancer. *Cell* 100:57–70.

Hudis CA (2007) Trastuzumab–mechanism of action and use in clinical practice. *N Engl J Med* 357:39–51.

Kastan MB and Bartek J (2004) Cell-cycle checkpoints and cancer. *Nature* 432:316–323.

Weisberg E, Manley PW, Cowan-Jacob SW, Hochhaus A, and Griffin JD (2007) Second generation inhibitors of bcr-abl for the treatment of imatinib-resistant chronic myeloid leukaemia. *Nat Rev Cancer* 7:345–356.

Xu WS, Parmigiani RB, and Marks PA (2007) Histone deacetylase inhibitors: Molecular mechanisms of action. *Oncogene* 26:5541–5552.

24 Antiviral Drugs

Erik De Clercq

CONTENTS

24.1 INTRODUCTION

The antiviral drug armamentarium, at the end of 2006, comprises more than 40 compounds that have been officially approved for clinical use. Most of the approved drugs date from the last 5 to 10 years, and at least half of them are used for the treatment of human immunodeficiency virus (HIV) infections. The other antiviral drugs that are currently available are primarily used for the treatment of hepatitis B virus (HBV), herpesvirus (herpes simplex virus [HSV], varicella-zoster virus [VZV], and cytomegalovirus [CMV]), influenza virus, respiratory syncytial virus (RSV), and hepatitis C virus (HCV) infections.

24.2 ANTI-HIV COMPOUNDS

24.2.1 Nucleoside Reverse Transcriptase Inhibitors (NRTIs)

24.2.1.1 Zidovudine

Structure (Figure 24.1): 3′-Azido-2′,3′-dideoxythymidine, azidothymidine (AZT), ZDV, Retrovir®.

Activity spectrum: HIV (types 1 and 2).

Mechanism of action: Targeted at the reverse transcriptase (RT) of HIV, acts as a chain terminator in the RT reaction, following intracellular phosphorylation of AZT 5′-triphosphate, and, after removal of the diphosphate group, incorporation of AZT 5′-monophosphate at the 3′-end of the viral DNA chain (Figure 24.2).

Principal indication(s): HIV infection, in combination with other anti-HIV agents (such as lamivudine and abacavir).

Administered: Orally at 600 mg/day (two 300 mg tablets daily).

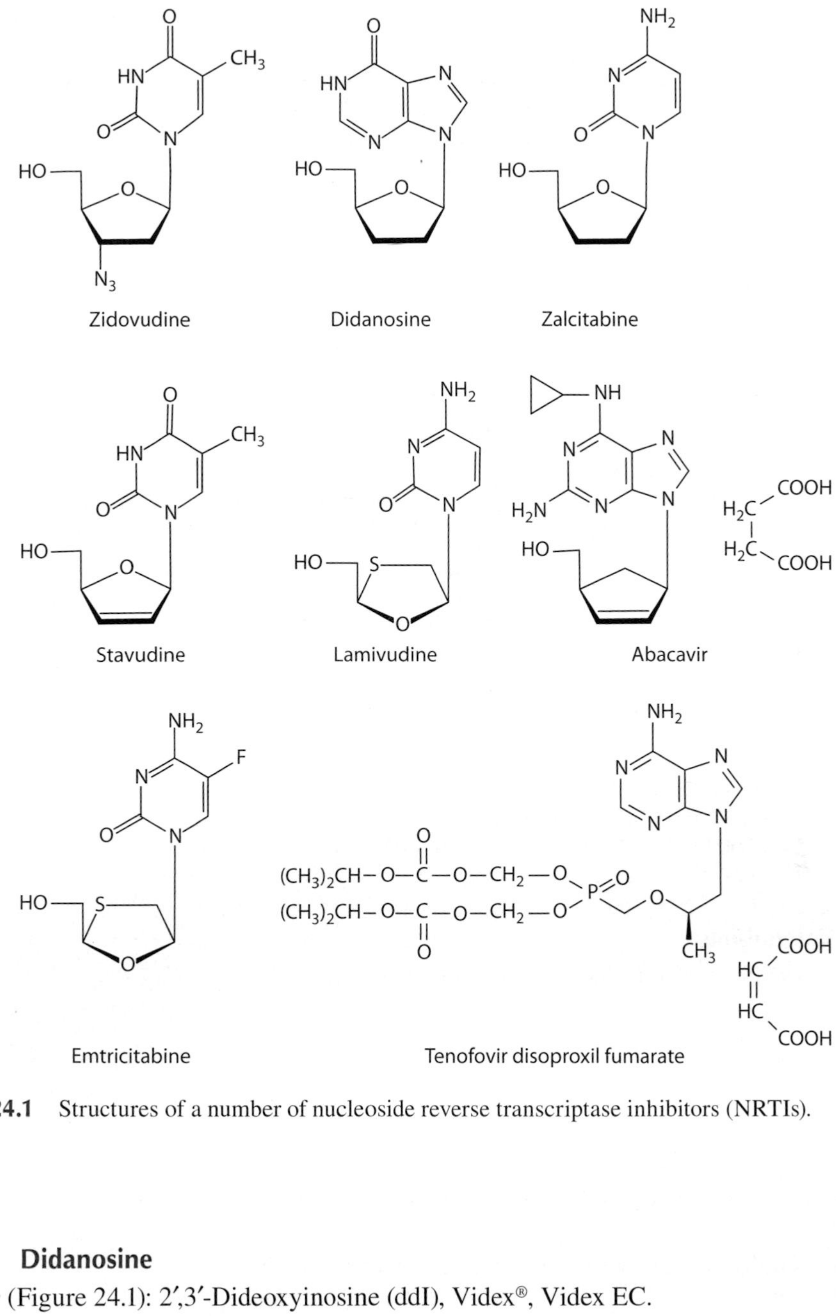

FIGURE 24.1 Structures of a number of nucleoside reverse transcriptase inhibitors (NRTIs).

24.2.1.2 Didanosine

Structure (Figure 24.1): 2′,3′-Dideoxyinosine (ddI), Videx®, Videx EC.

Activity spectrum: HIV (types 1 and 2).

Mechanism of action: Targeted at HIV RT, acts as chain terminator, following intracellular phosphorylation to 2′,3′-dideoxyadenosine (ddA) 5′-triphosphate, and, after removal of the diphosphate group, incorporation of ddA 5′-monophosphate at the 3′-end of the viral DNA chain.

Principal indication(s): HIV infection, especially advanced HIV disease, in combination with other anti-HIV agents.

Administered: Orally at 400 mg/day (Videx: two 100 mg tablets twice a day or two 200 mg tablets once a day; Videx EC: one 400 mg capsule once a day).

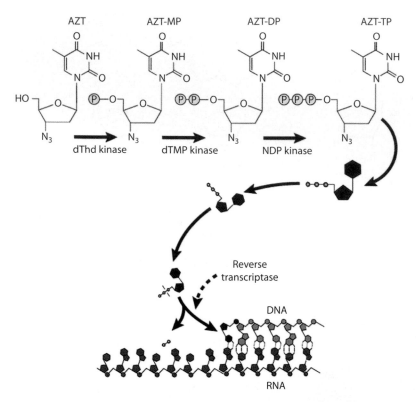

FIGURE 24.2 Mechanism of action of azidothymidine (AZT). AZT needs to be phosphorylated, in three steps, to the triphosphate form before it can interfere with the RT reaction. (After De Clercq, E., *Nat. Rev. Drug Discov.*, 1, 13, 2002.)

24.2.1.3 Zalcitabine

Structure (Figure 24.1): 2′,3′-Dideoxycytidine (ddC), Hivid.

Activity spectrum: HIV (types 1 and 2).

Mechanism of action: Targeted at HIV RT, acts as chain terminator, following intracellular phosphorylation to ddC 5′-triphosphate, and, after removal of the diphosphate group, incorporation of ddC 5′-monophosphate at the 3′-end of the viral DNA chain.

Principal indication(s): HIV infection, especially in adult patients with advanced HIV disease that is intolerant or unresponsive to zidovudine, in combination with other anti-HIV agents (not didanosine).

Administered: Orally at 2.25 mg/day (one 0.75 mg tablet every 8 h).

24.2.1.4 Stavudine

Structure (Figure 24.1): 2′,3′-Didehydro-2′,3′-dideoxythymidine (d4T), Zerit®.

Activity spectrum: HIV (types 1 and 2).

Mechanism of action: Targeted at HIV RT, acts as chain terminator, following intracellular phosphorylation to d4T 5′-triphosphate, and, after removal of the diphosphate group, incorporation of d4T 5′-monophosphate at the 3′-end of the viral DNA chain.

Principal indication(s): HIV infection, especially advanced HIV disease, in combination with other anti-HIV agents.

Administered: Orally at 80 mg/day (one 40 mg capsule every 12 h).

24.2.1.5 Lamivudine

Structure (Figure 24.1): (−)-β-L-3′-Thia-2′,3′-dideoxycytidine (3TC), Epivir® (for HIV), Zeffix® (for HBV).

Activity spectrum: HIV (types 1 and 2) and HBV.

Mechanism of action: Targeted at HIV RT and HBV RT, acts as chain terminator, following intracellular phosphorylation to 3TC 5′-triphosphate, and, after removal of the diphosphate group, incorporation of 3TC 5′-monophosphate at the 3′-end of the viral DNA chain.

Principal indication(s): HIV and HBV infections: for HIV infection, in combination with other anti-HIV agents (such as zidovudine and abacavir).

Administered: Orally at 300 mg/day (one 150 mg tablet twice a day, or one 300 mg tablet once a day). In the treatment of HIV infections, lamivudine can be combined with zidovudine (Combivir®), or with zidovudine and abacavir (Trizivir®). Combivir tablets contain 300 mg zidovudine and 150 mg lamivudine per tablet and are administered orally (two tablets daily). Trizivir tablets contain 300 mg zidovudine, 150 mg lamivudine, and 300 mg abacavir per tablet and are administered orally (two tablets daily). Lamivudine is administered orally at 100 mg/day in the treatment of HBV infections.

24.2.1.6 Abacavir

Structure (Figure 24.1): (1S,4R)-4-[2-Amino-6-(cyclopropylamino)-9H-purin-9-yl]-2-cyclopentene-1-methanol succinate (ABC), Ziagen®.

Activity spectrum: HIV (types 1 and 2).

Mechanism of action: Targeted at HIV RT, acts as chain terminator, following intracellular phosphorylation and conversion (deamination) to the 5′-triphosphate of the corresponding guanosine analog (carbovir), and, after removal of the diphosphate group, incorporation of carbovir 5′-monophosphate at the 3′-end of the viral DNA chain.

Principal indication(s): HIV infection, in combination with other anti-HIV agents such as zidovudine and lamivudine (see above).

Administered: Orally at 600 mg/day (two 300 mg tablets daily).

24.2.1.7 Emtricitabine

Structure (Figure 24.1): (−)-β-L-3′-Thia-2′,3′-dideoxy-5-fluorocytidine, [(−)-FTC], Emtriva®.

Activity spectrum: HIV and HBV.

Mechanism of action: Similar to that of 3TC.

Principal indication(s): HIV infections, where it is used in combination with tenofovir disoproxil fumarate (TDF) as a single tablet to be taken orally once daily.

Administered: Orally as a once-daily 200 mg capsule containing only emtricitabine or as a once-daily tablet containing 200 mg emtricitabine and 300 mg TDF (Truvada®).

24.2.2 NUCLEOTIDE REVERSE TRANSCRIPTASE INHIBITORS (NtRTIs)

24.2.2.1 Tenofovir Disoproxil

Structure (Figure 24.1): Fumarate salt of bis(isopropoxycarbonyloxymethyl) ester of (R)-9-(2-phosphonylmethoxypropyl)adenine, or bis(POC)PMPA, Viread®.

Activity spectrum: HIV (types 1 and 2) and various other retroviruses, and HBV.

Mechanism of action: Serves as oral prodrug of tenofovir (PMPA) that is targeted at HIV RT (and HBV RT), and acts as chain terminator, following intracellular phosphorylation to the diphosphate

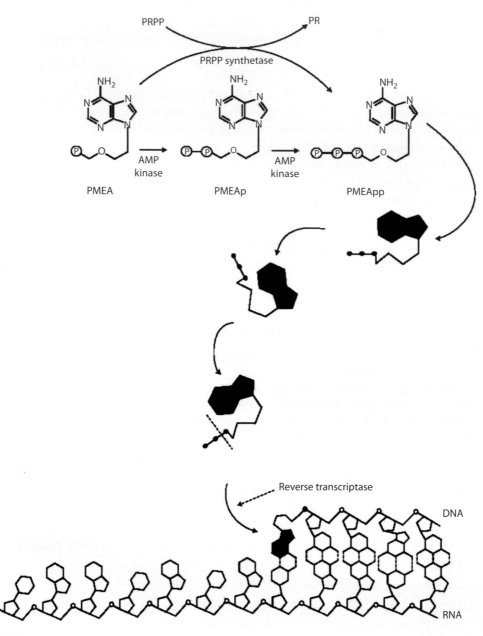

FIGURE 24.3 Mechanism of antiviral action of tenofovir (PMPA) is similar to that of adefovir (PMEA). PMEA and PMPA need to be phosphorylated, in one or two steps, to the diphosphate form before they interfere, as chain terminators, with the RT reaction. (After De Clercq, E., *Expert Rev. Anti-infect. Ther.*, 1, 21, 2003a.)

form, and, after the removal of the diphosphate group, incorporation at the 3′-end of the viral DNA chain (Figure 24.3).

Principal indication(s): HIV infection as such or in combination with emtricitabine (see above), or in combination with emtricitabine and efavirenz (see below).

Administered: Orally as a once-daily 300 mg tablet (Viread) or a 300 mg TDF/200 mg emtricitabine tablet (Truvada) or a 300 mg TDF/200 mg emtricitabine/600 mg efavirenz tablet (Atripla®).

24.2.3 Nonnucleoside Reverse Transcriptase Inhibitors (NNRTIs)

24.2.3.1 Nevirapine

Structure (Figure 24.4): 11-Cyclopropyl-5,11-dihydro-4-methyl-6*H*-dipyrido[3,2-b:2′,3′-f][1,4] diazepin-6-one, Viramune®.

Activity spectrum: HIV type 1.

Mechanism of action: Targeted at an allosteric "pocket," nonsubstrate binding site of the HIV-1 RT (as illustrated in Figure 24.5 for Etravirine, an NNRTI currently in clinical development).

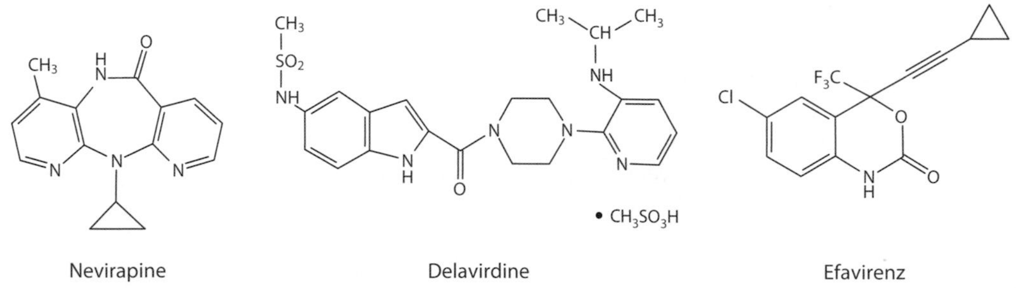

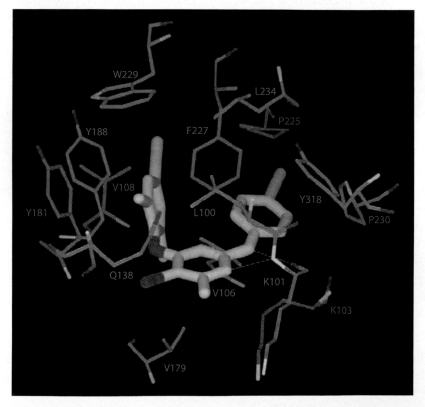

FIGURE 24.4 Structures of a number of nonnucleoside RT inhibitors (NNRTIs).

FIGURE 24.5 Etravirine (TMC125) positioned in NNRTI-binding site of HIV-1 RT. (After Pauwels, R., *Curr. Opin. Pharmacol.*, 4, 437, 2004.)

Principal indication(s): HIV-1 infection, in combination with other anti-HIV agents, particularly NRTIs.

Administered: Orally at 200 mg/day for the first 14 days (one 200 mg tablet per day), then 400 mg/day (two daily 200 mg tablets).

24.2.3.2 Delavirdine

Structure (Figure 24.4): 1-(5-Methanesulfonamido-1*H*-indol-2-yl-carbonyl)-4-[3-(1-methylethyl-amino)pyridinyl)piperazine monomethane sulfonate, Rescriptor®.

Activity spectrum: HIV type 1.

Mechanism of action: Similar to that of nevirapine.

Principal indication(s): HIV-1 infection, in combination with other anti-HIV agents (primarily NRTIs).

Administered: Orally at 1200 mg/day (two 200 mg tablets three times a day).

24.2.3.3 Efavirenz

Structure (Figure 24.4): (−)6-Chloro-4-cyclopropylethynyl-4-trifluoromethyl-1,4-dihydro-2*H*-3,1-benzoxazin-2-one, Sustiva®, Stocrin®.

Activity spectrum: HIV type 1.

Mechanism of action: Similar to that of nevirapine.

Principal indication(s): HIV-1 infection, in combination with other anti-HIV agents, particularly TDF and emtricitabine (see above).

Administered: Orally once-daily at 600 mg/day, as 600 mg tablet, preferably at bedtime to improve tolerability of CNS side effects, or, in combination with TDF and emtricitabine (Atripla) (see above).

24.2.4 PROTEASE INHIBITORS (PIs)

24.2.4.1 Saquinavir

Structure (Figure 24.6): *cis-N-tert*-Butyl-decahydro-2-[2(*R*)-hydroxy-4-phenyl-3(*S*)-[[*N*-2-quinolylcarbonyl-L-asparaginyl]-amino]butyl]-(4a*S*-8a*S*)-isoquinoline-3(*S*)-carboxamide methane sulfonate, hard gel capsules, Invirase®, also available as soft gelatin capsules (Fortovase®).

Activity spectrum: HIV (types 1 and 2).

Mechanism of action: Transition-state, hydroxyethylene-based, peptidomimetic inhibitor of HIV protease (as illustrated for Darunavir (see below) in Figure 24.7).

Principal indication(s): HIV infection, in combination with other anti-HIV agents (i.e., NRTIs) or ritonavir as "booster."

Administered: Orally at 3600 mg/day (six 200 mg soft gelatin capsules three times a day (Fortovase)) or 1800 mg/day (three 200 mg hard gel capsules three times a day (Invirase)), to be taken with a meal or up to 2 h after a full meal. If boosted with ritonavir, 1000 mg saquinavir/100 mg ritonavir twice daily.

24.2.4.2 Ritonavir

Structure (Figure 24.6): [5*S*-(5*R*,8*R*,10*R*,11*R*)]-10-Hydroxy-2-methyl-5-(1-methylethyl)-1-[2-(methylethyl)-4-thiazolyl]-3,6-dioxo-8,11-bis(phenylmethyl)-2,4,7,12-tetraazatridecan-13-oic acid 5-thiazolylmethyl ester, Norvir®.

Activity spectrum: HIV (types 1 and 2).

Mechanism of action: Peptidomimetic inhibitor of HIV protease.

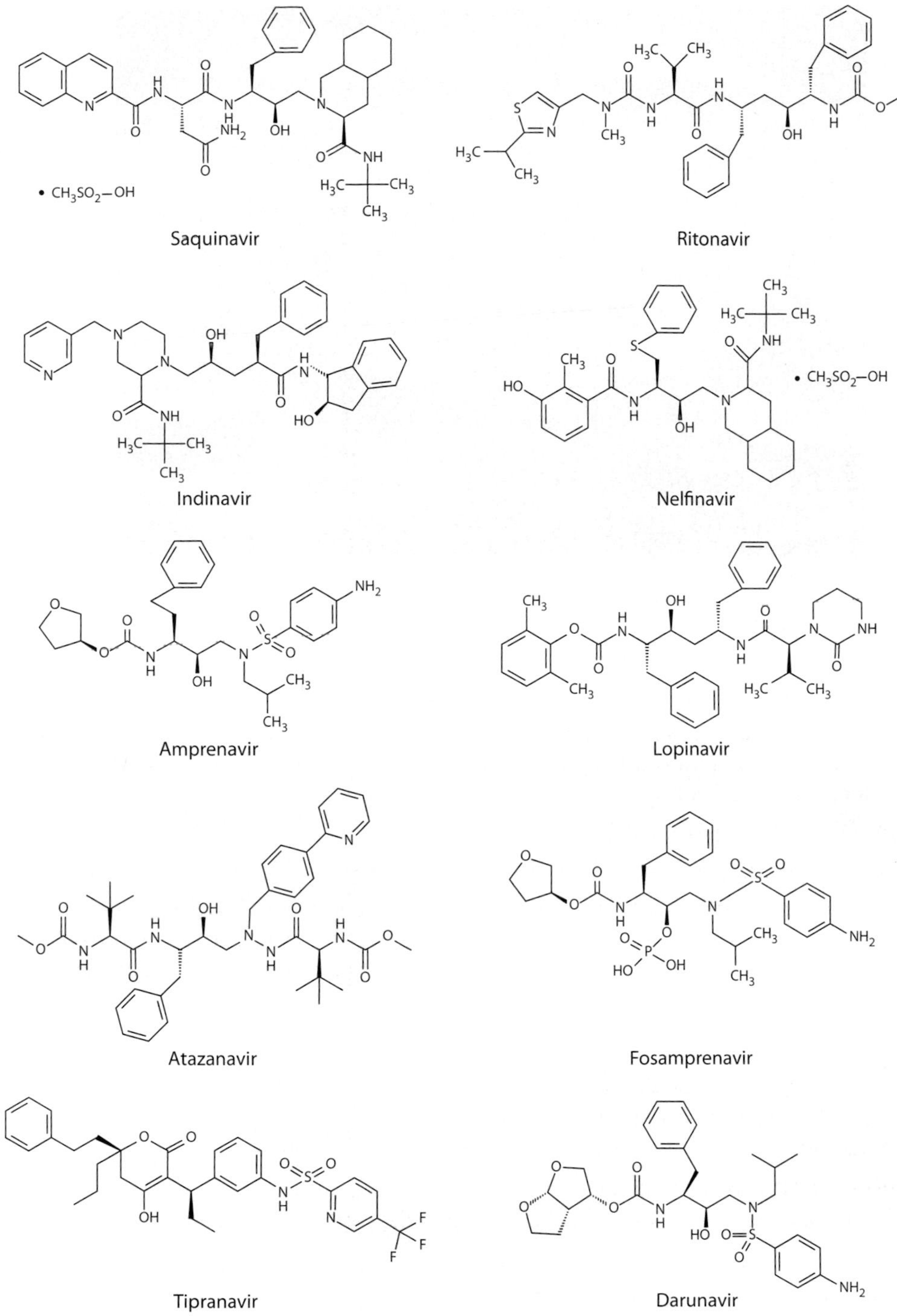

FIGURE 24.6 Structures of a number of HIV protease inhibitors (PIs).

Principal indication(s): HIV infection, in combination with other anti-HIV agents (particularly PIs to boost their anti-HIV activity).

Administered: Orally at 1200 mg/day (six 100 mg capsules, twice a day to be taken with food), or if used as "booster" for other PIs, at 100 mg twice daily.

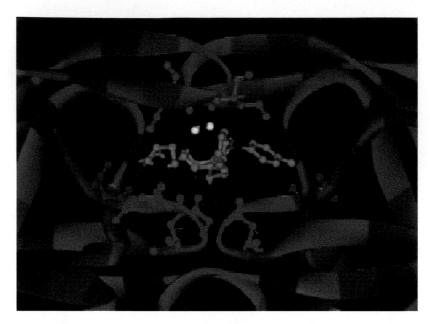

FIGURE 24.7 HIV protease structure with Darunavir (TMC114) in the active site. (After Pauwels, R., *Antiviral Res.*, 71, 77, 2006.)

24.2.4.3 Indinavir

Structure (Figure 24.6): [(1*S*,2*R*,5(*S*)-2,3,5-Trideoxy-*N*-(2,3-dihydro-2-hydroxy-1*H*-inden-1-yl)-5-[2-[[(1,1-dimethylethyl)amino]carbonyl]-4-pyridinylmethyl)-1-piperazinyl]-2-(phenylmethyl-D-erythro)pentonamide, Crixivan®.

Activity spectrum: HIV (types 1 and 2).

Mechanism of action: Peptidomimetic inhibitor of HIV protease.

Principal indication(s): HIV infection, in combination with other anti-HIV agents (i.e., NRTIs). *Administered*: Orally at 2400 mg/day (two 400 mg capsules every 8 h to be taken on empty stomach), plus hydration (at least 1.5 L liquid daily). If "boosted" with ritonavir, 800 mg indinavir/100 mg ritonavir twice daily.

24.2.4.4 Nelfinavir

Structure (Figure 24.6): [3*S*-(3*R*,4a*R*,8a*R*,2'*S*)]-2-[2'-Hydroxy-3'-phenylthiomethyl-4'-aza-5'-oxo-5'-[2'-methyl-3'-hydroxyphenyl)-pentyl]-3-(*N*-(*tert*-butyl)-carboxamide)-decahydro isoquinoline methane sulfonate, Viracept®.

Activity spectrum: HIV (types 1 and 2).

Mechanism of action: Peptidomimetic inhibitor of HIV protease.

Principal indication(s): HIV infection, in combination with other anti-HIV agents (i.e., NRTIs).

Administered: Orally at 2250 mg/day (three 250 mg tablets three times a day) or 2500 mg/day (five 250 mg tablets twice a day), to be taken with a meal.

24.2.4.5 Amprenavir

Structure(Figure 24.6):(3*S*)-Tetrahydro-3-furyl-*N*-[(*S*,2*R*)-3-(4-amino-*N*-isobutylbenzene-sulfonamido)-1-benzyl-2-hydroxypropyl]carbamate, Agenerase®, Prozei®.

Activity spectrum: HIV (types 1 and 2).

Mechanism of action: Peptidomimetic inhibitor of HIV protease.

Principal indication(s): HIV infection, in combination with other anti-HIV agents (i.e., NRTIs).

Administered: Orally at 2400 mg/day (eight 150 mg capsules twice a day, to be taken with or without food, but not with a high-fat meal). If "boosted" with ritonavir, 600 mg amprenavir/100 mg ritonavir twice daily.

24.2.4.6 Lopinavir

Structure (Figure 24.6): *N*-(4(*S*)-(2-(2,6-Dimethylphenoxy)-acetylamino)-3(*S*)-hydroxy-5-phenyl-1(*S*)-benzylpentyl)-3-methyl-2(*S*)-(2-oxo(1,3-diazaperhydroinyl)butanamine, combined with ritonavir at 4:1 ratio; ABT-378/r, Kaletra®.

Activity spectrum: HIV (types 1 and 2).

Mechanism of action: Peptidomimetic inhibitor of HIV protease.

Principal indication(s): HIV infection, in combination with other anti-HIV agents (i.e., NRTIs).

Administered: Orally, as Kaletra, at 1000 mg/day (three 166.6 mg capsules twice a day; each capsule containing 133.3 mg lopinavir + 33.3 mg ritonavir), to be taken with food.

24.2.4.7 Atazanavir

Structure (Figure 24.6): 1-[4-(Pyridin-2-yl)phenyl]-5(*S*)-2,5-bis-{[*N*-(methoxycarbonyl)-L-*tert*-leucinyl]amino}-4(*S*)-hydroxy-6-phenyl-2-azahexane, CGP 73547, BMS-232632, Reyataz®.

Activity spectrum: HIV (types 1 and 2).

Mechanism of action: Peptidomimetic inhibitor of HIV protease.

Principal indication(s): HIV infection, in combination with other anti-HIV agents (i.e., NRTIs).

Administered: Orally at 400 mg/day (once daily two 200 mg capsules), to be taken with food.

24.2.4.8 Fosamprenavir

Structure (Figure 24.6): (3*S*)-Tetrahydrofuran-3-yl (1*S*,2*R*)-3-[[(4-aminophenyl) sulfonyl](isobutyl) amino]-1-benzyl-2-(phosphonooxy) propylcarbamate monocalcium salt, Lexiva®.

Activity spectrum: HIV (types 1 and 2).

Mechanism of action: Peptidomimetic inhibitor of HIV protease.

Principal indication(s): HIV infection, in combination with other anti-HIV agents (i.e., NRTIs).

Administered: Orally at 1400 mg (two 700 mg tablets) twice daily without ritonavir; or 1400 mg (two 700 mg tablets) once daily with ritonavir 200 mg once daily; or 700 mg twice daily plus ritonavir 100 mg twice daily.

24.2.4.9 Tipranavir

Structure (Figure 24.6): 2-Pyridinesulfonamide, *N*-[3-[(1*R*)-1-[(6*R*)-5,6-dihydro-4-hydroxy-2-oxo-6-(2-phenylethyl)-6-propyl-2*H*-pyran-3-yl]propyl]phenyl]-5-(trifluoromethyl), Aptivus®.

Activity spectrum: HIV (types 1 and 2).

Mechanism of action: Nonpeptidomimetic inhibitor of HIV protease.

Principal indication(s): HIV infection, in combination with other anti-HIV agents (i.e., NRTIs).

Administered: Orally as 250 mg capsules, boosted by ritonavir: 500 mg tipranavir/200 mg ritonavir twice daily.

24.2.4.10 Darunavir

Structure (Figure 24.6): [(1*S*,2*R*)-3-[[(4-Aminophenyl)sulfonyl](2-methylpropyl)amino]-2-hydroxy-1-(phenylmethyl)propyl]-carbamic acid (3*R*,3a*S*,6a*R*)-hexahydrofuro[2,3-*b*]furan-3-yl ester mono-ethanolate, Prezista®.

Activity spectrum: HIV (types 1 and 2).

Mechanism of action: Peptidomimetic inhibitor of HIV protease.

Principal indication(s): HIV infection, in combination with other anti-HIV agents (i.e., NRTIs).

Administered: Orally, as two 300 mg tablets twice daily, together with ritonavir 100 mg twice daily and with food.

24.2.5 VIRAL ENTRY INHIBITORS

24.2.5.1 Enfuvirtide

Structure (Figure 24.8): 36-Amino acid peptide, corresponding to amino acid residues 643–678 of the viral glycoprotein precursor gp160 (or amino acid residues 127–162 of the viral glycoprotein gp41), DP-178, pentafuside, T-20, Fuzeon®.

Activity spectrum: HIV-1.

Mechanism of action: Inhibits virus-cell fusion, through a coil–coil interaction with its homologous region in gp41 (Figure 24.9).

Principal indication(s): HIV infection, in combination with other anti-HIV agents; generally used for salvage therapy in highly drug-experienced patients.

Administered: By subcutaneous injection, twice daily at a dose of 90 mg.

24.3 ANTI-HBV COMPOUNDS

24.3.1 LAMIVUDINE

Lamivudine (Figure 24.1) is used in the treatment of both HIV and HBV infections; for the former at an oral dose of 300 mg/day, for the latter at an oral dose of 100 mg/day.

24.3.2 ADEFOVIR DIPIVOXIL

Structure (Figure 24.10): Bis(pivaloyloxymethyl)ester of 9-(2-phosphonylmethoxyethyl)adenine, or bis(POM)PMEA, Hepsera®.

Activity spectrum: HBV, HIV, and other retroviruses, and, to a lesser extent, also herpesviruses (HSV, CMV, etc.).

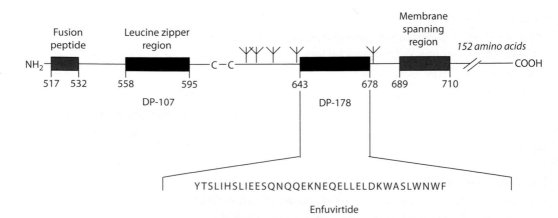

FIGURE 24.8   Structure of enfuvirtide, a 36-amino acid fragment of the viral glycoprotein precursor gp 160.

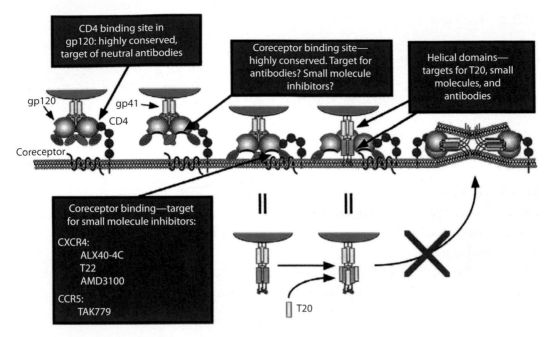

FIGURE 24.9 Targets of opportunity: inhibition of HIV-1 entry. Binding of CD4 to gp 120 leads to exposure of the highly conserved coreceptor binding site in gp 120. This domain and the CD4 binding region in gp 120 are potential targets for neutralizing antibodies. Binding of gp 120 to coreceptor can be inhibited by a variety of small-molecule inhibitors, four of which have been described to date. Coreceptor binding is believed to trigger additional conformational changes in gp41 including the formation of a triple-stranded coiled-coil with insertion of the fusion peptide into the cell membrane. The transition from the triple-stranded coiled-coil to the six helix bundle conformation is the proximal cause of membrane fusion and can be inhibited by T20. T20 is a peptide based on the second helical domain of gp41 that, by binding to the first helical domain, blocks formation of the six helix bundle. (After Doms, R.W., *Virology*, 276, 229, 2000.)

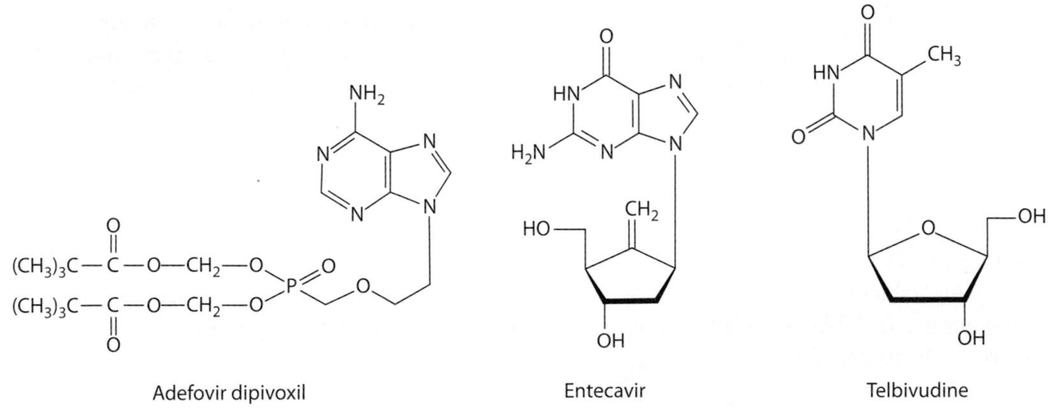

Adefovir dipivoxil	Entecavir	Telbivudine

FIGURE 24.10 Structures of three antihepatitis B virus (anti-HBV) compounds.

Mechanism of action: Serves as oral prodrug of adefovir (PMEA) that is targeted at HBV RT (and HIV RT), and acts as a chain terminator, following intracellular phosphorylation to the diphosphate form, and incorporation at the 3′-end of the viral DNA chain (Figure 24.3).

Principal indication(s): Chronic HBV infection, where it has proven successful in the treatment of both HBeAg-positive and HBeAg-negative patients whether resistant to lamivudine or not.

Administered: Orally as a single dose of 10 mg/day.

24.3.3 ENTECAVIR

Structure (Figure 24.10): 2-Amino-1,9-dihydro-9-[(*1S,3R,4S*)-4-hydroxy-3-(hydroxymethyl)-2-methylenecyclopentyl]-6*H*-purin-6-one, monohydrate, Baraclude®.

Activity spectrum: HBV.

Mechanism of action: Targeted at HBV RT, following intracellular phosphorylation to its triphosphate.

Principal indication(s): Chronic hepatitis B (either HBeAg-positive or -negative patients).

Administered: Orally as 0.5 mg tablet once daily (1 mg once daily for lamivudine-resistant).

24.3.4 TELBIVUDINE

Structure (Figure 24.10): 1-((2*S*,4*R*,5*S*)-4-Hydroxy-5-hydroxymethyltetrahydrofuran-2-y1)-5-methyl-1*H*-pyrimidine-2,4-dione, or 1-(2-deoxy-β-L-ribofuranosyl)-5-methyluracil, Tyzeka®.

Activity spectrum: HBV.

Mechanism of action: Targeted at HBV RT, following intracellular phosphorylation to its triphosphate.

Principal indication(s): Chronic hepatitis B (HBeAg-positive or -negative).

Administered: Orally as a 600 mg tablet once daily.

24.4 ANTIHERPESVIRUS COMPOUNDS

24.4.1 HSV AND VZV INHIBITORS

24.4.1.1 Acyclovir

Structure (Figure 24.11): 9-(2-Hydroxyethoxymethyl)guanine, acycloguanosine (ACG), acyclovir, aciclovir (ACV), Zovirax®.

Activity spectrum: HSV (types 1 and 2) and VZV.

Mechanism of action: Targeted at the viral DNA polymerase, acts as a chain terminator, following intracellular phosphorylation to ACV triphosphate and incorporation of ACV monophosphate at the 3′-end of the viral DNA chain (Figure 24.12). The first phosphorylation step is catalyzed by the virus-encoded thymidine kinase (TK), which explains the specificity of acyclovir for HSV-1, HSV-2, and VZV.

Principal indication(s): Mucosal, cutaneous, and systemic HSV-1 and HSV-2 infections (including herpetic keratitis, herpetic encephalitis, genital herpes, neonatal herpes, and herpes labialis) and VZV infections (including varicella and herpes zoster).

Administered: Orally at 1000 mg/day (five 200 mg tablets [genital herpes]) or 4000 mg/day (four times five 200 mg tablets [herpes zoster]), topically as a 3% ophthalmic cream (herpetic keratitis) or 5% cream (herpes labialis), or intravenously at 30 mg/kg/day (herpetic encephalitis, and other severe HSV or VZV infections).

24.4.1.2 Valaciclovir

Structure (Figure 24.11): L-Valine ester of acyclovir (VACV), Zelitrex®, Valtrex®.

Activity spectrum: As for acyclovir.

Mechanism of action: Serves as oral prodrug of acyclovir, and then acts as described for acyclovir.

Principal indication(s): HSV and VZV infections that can be approached by oral therapy (i.e., genital herpes, herpes zoster). Also used in the prophylaxis of CMV infections in transplant recipients.

Administered: Orally at 1000 mg/day (two 500 mg tablets [genital herpes]) up to 3000 mg/day (three times two 500 mg tablets [herpes zoster]).

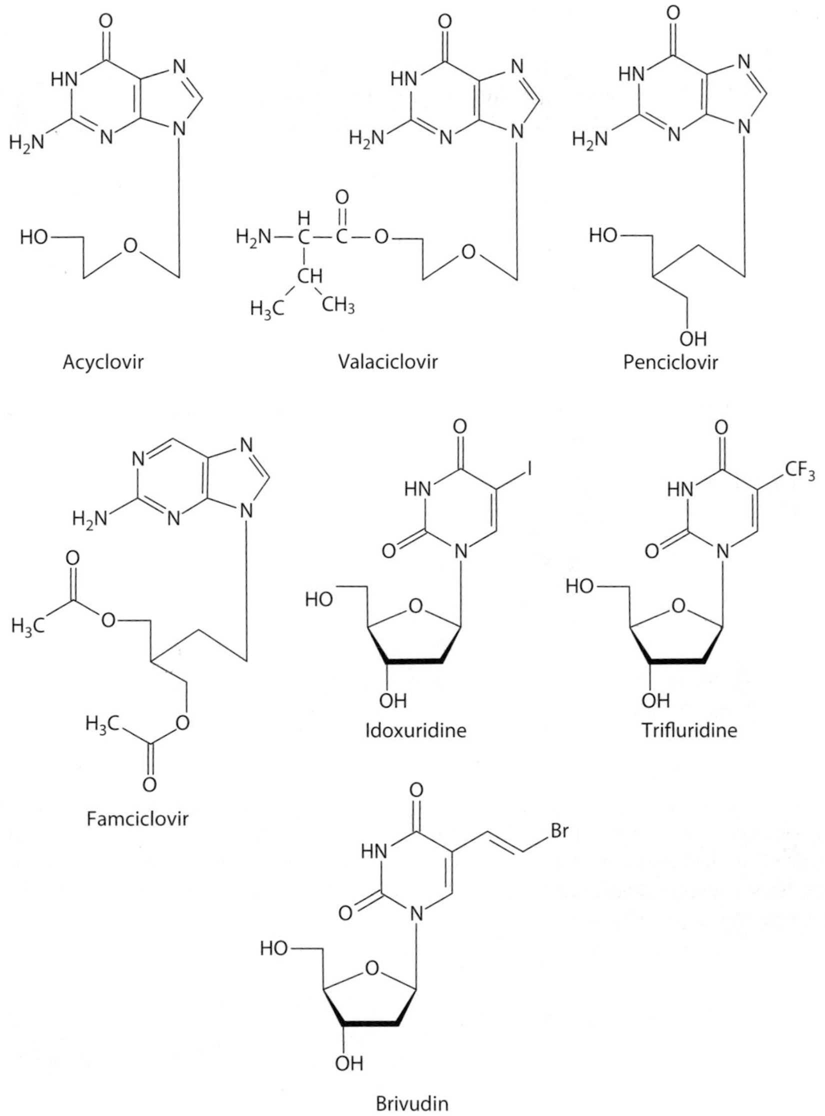

FIGURE 24.11   Structures of a number of antiherpes simplex virus (anti-HSV) and antivaricella-zoster virus (anti-VZV) compounds.

24.4.1.3   Penciclovir

Structure (Figure 24.11): 9-(4-Hydroxy-3-hydroxymethyl-but-1-yl)guanine (PCV), Denavir®, Vectavir®.

Activity spectrum: HSV 1, HSV-2, and VZV.

Mechanism of action: Essentially similar to that of acyclovir.

Principal indication(s): Mucocutaneous HSV infections, particularly recurrent herpes labialis (cold sores).

Administered: Topically as a 1% cream.

24.4.1.4   Famciclovir

Structure (Figure 24.11): Diacetyl ester of 9-(4-hydroxy-3-hydroxymethyl-but-1-yl)-6-deoxyguanine (FCV), Famvir®.

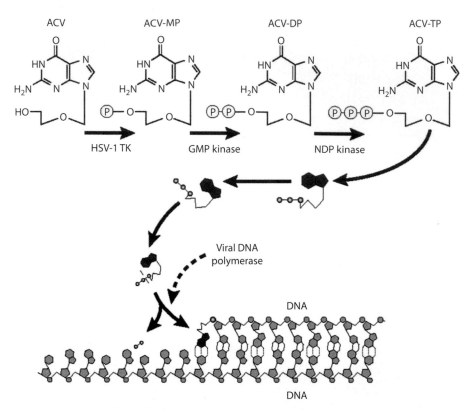

FIGURE 24.12 Mechanism of antiviral action of acyclovir (ACV). ACV targets viral DNA polymerases, such as the herpesvirus (HSV) DNA polymerase. Before it can interact with viral DNA synthesis, it needs to be phosphorylated intracellularly, in three steps, to the triphosphate form. The first phosphorylation step is ensured by the HSV-encoded thymidine kinase (TK), and is therefore confined to virus-infected cells. (After De Clercq, E., *Nat. Rev. Drug Discov.*, 1, 13, 2002.)

Activity spectrum: HSV-1, HSV-2, and VZV.

Mechanism of action: Serves as oral prodrug of penciclovir (to which it is converted by hydrolysis of the two acetyl groups and oxidation at the 6-position), then acts as described for penciclovir.

Principal indication(s): HSV-1, HSV-2, and VZV infections.

Administered: Orally at 750 mg/day (250 mg tablet every 8 h, three times a day), or 1500 mg/day (500 mg every 8 h).

24.4.1.5 Idoxuridine

Structure (Figure 24.11): 5-Iodo-2′-deoxyuridine (IDU, IUdR), Herpid®, Stoxil®, Idoxene®, Virudox®, etc.

Activity spectrum: HSV-1, HSV-2, and VZV.

Mechanism of action: Incorporated into (viral/cellular) DNA, following intracellular phosphorylation to IDU 5′-triphosphate (in virus-infected and uninfected cells).

Principal indication(s): HSV keratitis.

Administered: Topically as eyedrops (0.1%) or ophthalmic cream.

24.4.1.6 Trifluridine

Structure (Figure 24.11): 5-Trifluoromethyl-2′-deoxyuridine, trifluorothymidine (TFT), Viroptic®.

Activity spectrum: HSV-1, HSV-2, and VZV.

Mechanism of action: Inhibits conversion of dUMP to dTMP by thymidylate synthase, following intracellular phosphorylation to TFT 5′-monophosphate.

Principal indication(s): HSV keratitis.

Administered: Topically as eyedrops (1%) or ophthalmic cream.

24.4.1.7 Brivudin

Structure (Figure 24.11): (*E*)-5-(2-Bromovinyl)-2′-deoxyuridine, bromovinyldeoxyuridine (BVDU), Zostex®, Brivirac®, Zerpex®.

Activity spectrum: HSV (type 1), VZV, and some other (veterinarily important) herpesviruses.

Mechanism of action: Targeted at the viral DNA polymerase, can act as competitive inhibitor (with respect to the normal substrate, dTTP) after intracellular phosphorylation to BVDU 5′-triphosphate; can also act as alternate substrate and be incorporated into the viral DNA, thus leading to a reduced integrity and functioning of the viral DNA (Figure 24.13). The first and second phosphorylation steps are catalyzed by the virus-encoded TK (HSV-1 TK, VZV TK), which explains the remarkable specificity of BVDU for these viruses.

Principal indication(s): HSV-1 and VZV infections, particularly herpes zoster, but also HSV-1 keratitis and herpes labialis. Brivudin has been licensed for the treatment of herpes zoster in immunocompetent patients in a number of European countries.

Administered: Orally at 125 mg/day, once-daily (herpes zoster); can also be administered topically, as 0.1%–0.5% eyedrops (herpetic keratitis) or 5% cream (herpes labialis).

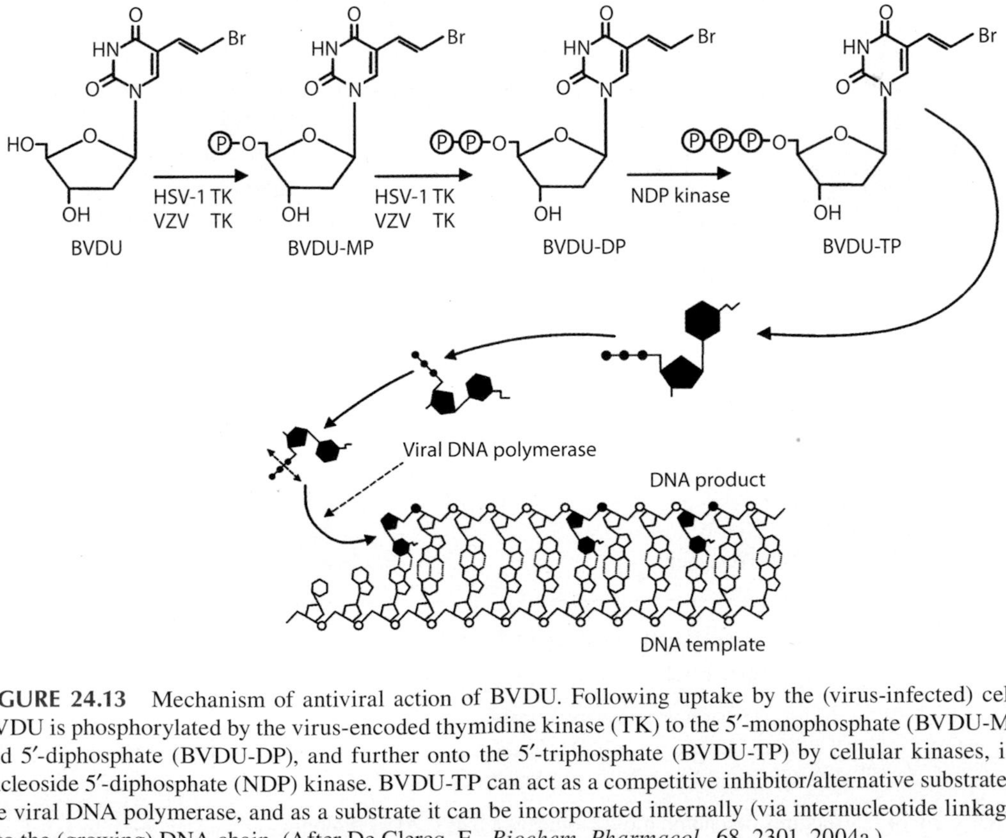

FIGURE 24.13 Mechanism of antiviral action of BVDU. Following uptake by the (virus-infected) cells, BVDU is phosphorylated by the virus-encoded thymidine kinase (TK) to the 5′-monophosphate (BVDU-MP) and 5′-diphosphate (BVDU-DP), and further onto the 5′-triphosphate (BVDU-TP) by cellular kinases, i.e., nucleoside 5′-diphosphate (NDP) kinase. BVDU-TP can act as a competitive inhibitor/alternative substrate of the viral DNA polymerase, and as a substrate it can be incorporated internally (via internucleotide linkages) into the (growing) DNA chain. (After De Clercq, E., *Biochem. Pharmacol.*, 68, 2301, 2004a.)

24.4.2 CMV INHIBITORS

24.4.2.1 Ganciclovir

Structure (Figure 24.14): 9-(1,3-Dihydroxy-2-propoxymethyl)guanine (DHPG), (GCV), Cymevene®, Cytovene®.

Activity spectrum: HSV (types 1 and 2), CMV, and some other herpesviruses.

Mechanism of action: Targeted at the viral DNA polymerase, where it mainly acts as a chain terminator, following intracellular phosphorylation to GCV triphosphate and incorporation of GCV monophosphate at the 3′-end of the viral DNA chain. First phosphorylation step is catalyzed by the HSV-encoded thymidine kinase (TK) or CMV-encoded protein kinase (PK), which explains the specificity of ganciclovir for HSV and CMV, respectively.

Principal indication(s): CMV infections, particularly CMV retinitis in immunocompromised (i.e., AIDS) patients (treatment and prevention).

Administered: Intravenously at 10 mg/kg/day (2 × 5 mg/kg, every 12 h) for induction therapy; orally at 3000 mg/day (three times four 250 mg capsules) for maintenance therapy and for prevention; intraocular (intravitreal) implant (Vitrasert*) of 4.5 mg ganciclovir as localized therapy of CMV retinitis.

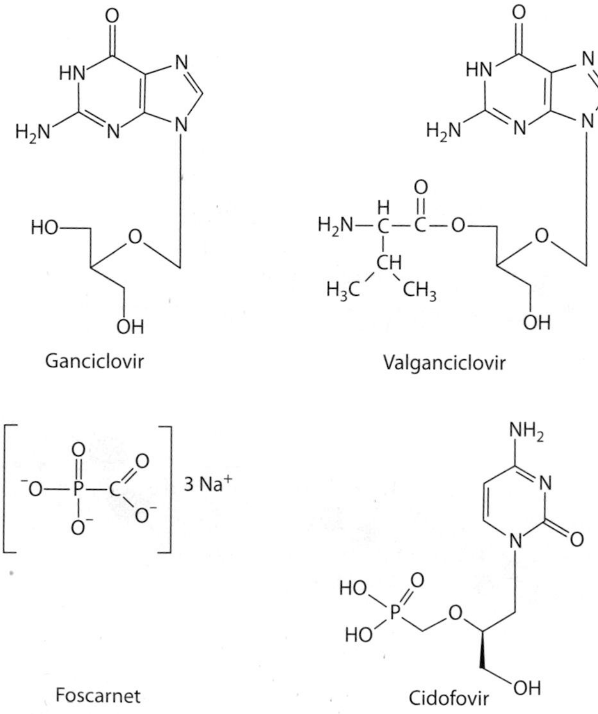

Ganciclovir

Valganciclovir

Foscarnet

Cidofovir

5′-d-[G*C*G*T*T*T*G*C*T*C*T*T*C*T*T*C*T*T*G*C*G]–3′
Sodium salt

* = Racemic phosphorothioate

Fomivirsen

FIGURE 24.14   Structures of a number of cytomegalovirus (CMV) inhibitors.

24.4.2.2 Valganciclovir

Structure (Figure 24.14): L-Valine ester of ganciclovir (VGCV), Valcyte®.

Activity spectrum: As for GCV.

Mechanism of action: Serves as oral prodrug of GCV, and then acts as described for GCV. *Principal indication(s)*: CMV infections. Oral valganciclovir is expected to replace intravenous ganciclovir in both the therapy and prevention of CMV infections.

Administered: Orally at 900 mg/day (two 450 mg tablets daily) for maintenance therapy (900 mg twice daily for induction therapy).

24.4.2.3 Foscarnet

Structure (Figure 24.14): Trisodium phosphonoformate, foscarnet sodium, Foscavir®.

Activity spectrum: Herpesviruses (HSV-1, HSV-2, VZV, CMV, etc.) and also HIV.

Mechanism of action: Pyrophosphate analog, interferes with the binding of the pyrophosphate (diphosphate) to its binding site of the viral DNA polymerase, during the process of DNA polymerization.

Principal indication(s): CMV retinitis in AIDS patients, and mucocutaneous acyclovir-resistant (viral TK-deficient) HSV and VZV infections in immunocompromised patients.

Administered: Intravenously at 180 mg/kg/day (3 × 60 mg/kg, every 8 h) for induction therapy of CMV retinitis; intravenously at 120 mg/kg/day (3 × 40 mg/kg, every 8 h) for maintenance therapy of CMV retinitis and for therapy of acyclovir-resistant mucocutaneous HSV or VZV infections in immunocompromised patients. Dose adjustments for changes in renal function are imperative.

24.4.2.4 Cidofovir

Structure (Figure 24.14): (*S*)-1-(3-Hydroxy-2-phosphonylmethoxypropyl)cytosine (HPMPC), (CDV), Vistide®.

Activity spectrum: Herpesviruses (HSV-1, HSV-2, VZV, CMV, etc.), papilloma-, polyoma-, adeno-, and poxviruses.

Mechanism of action: Targeted at the viral DNA polymerase, acts as a chain terminator, following intracellular phosphorylation to the diphosphate form, and incorporation at the 3′-end of the viral DNA chain (two sequential incorporations needed for chain termination in the case of CMV DNA synthesis) (Figure 24.15).

Principal indications(s): Officially licensed for the treatment of CMV retinitis in AIDS patients. Also shown to be effective in the treatment of acyclovir-resistant (viral TK-deficient) HSV infections, recurrent genital herpes, genital warts, CIN-III (cervical intraepithelial neoplasia grade III), laryngeal and cutaneous papillomatous lesions, molluscum contagiosum lesions, orf lesions, adenovirus infections, and progressive multifocal leukoencephalopathy (PML).

Administered: Intravenously (Vistide) at 5 mg/kg/week during the first 2 weeks, then 5 mg/kg every other week, with sufficient hydration and under cover of probenecid to prevent nephrotoxicity. It can also be administered topically as a 1% gel or cream.

24.4.2.5 Fomivirsen

Structure (Figure 24.14): Antisense oligodeoxynucleotide composed of 21 phosphorothioate-linked nucleosides, ISIS 2922, Vitravene®.

Activity spectrum: CMV.

Mechanism of action: Being complementary in base sequence, it hybridizes with, and thus blocks expression (translation) of, the CMV immediate and early 2 (IE2) mRNA.

Principal indication(s): CMV retinitis (in AIDS patients).

Administered: Intraocularly (intravitreally).

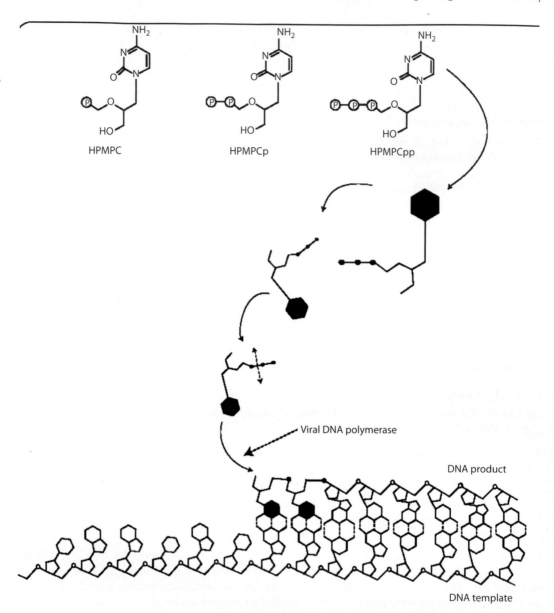

FIGURE 24.15 Mechanism of antiviral action of cidofovir (HPMPC). HPMPC needs to be phosphorylated, in two steps, to the diphosphate form before it interferes, as chain terminator (following two consecutive incorporations in the case of CMV) with the DNA polymerase reaction. (After De Clercq, E., *Expert Rev. Anti-infect. Ther.*, 1, 21, 2003a.)

24.5 ANTI-INFLUENZAVIRUS COMPOUNDS (INCLUDING RIBAVIRIN)

24.5.1 Amantadine

Structure (Figure 24.16): Tricyclo[3.3.1.1.^{3,7}]decane-1-amine hydrochloride, 1-adamantanamine, amantadine HCl, Symmetrel®, Mantadix®, Amantan®, etc.

Activity spectrum: Influenza A virus.

Mechanism of action: Blocks M2 ion channel, and thus prevents the passage of H$^+$ ions that are required for the necessary acidity to allow for the decapsidation (viral uncoating process).

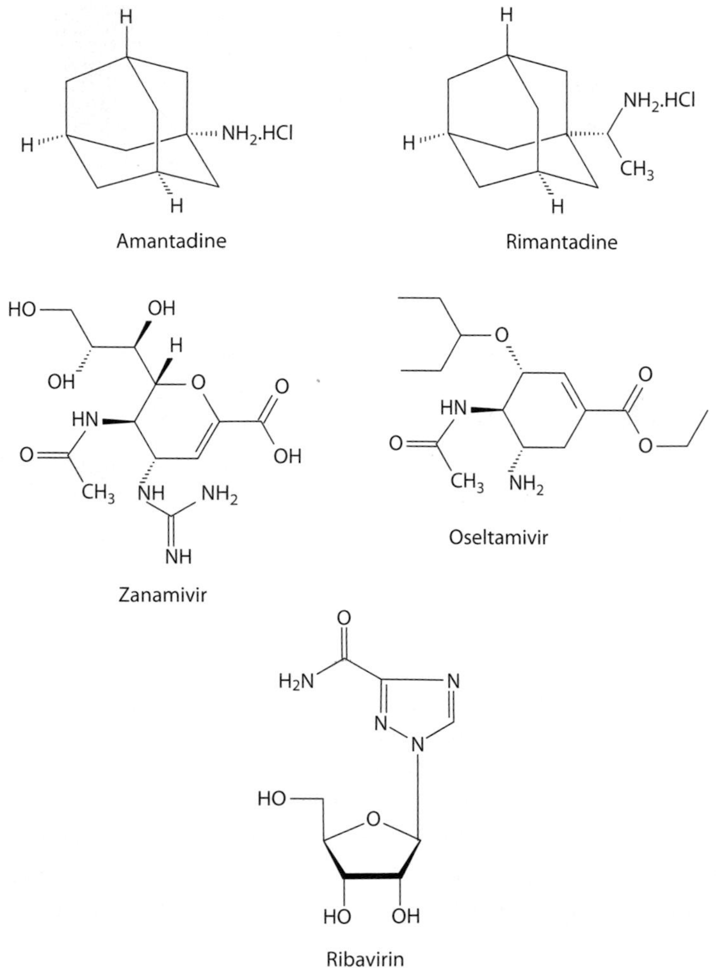

FIGURE 24.16   Structures of a number of anti-influenza-virus compounds.

Principal indication(s): Influenza A virus infections (prevention and early therapy). Also used in the treatment of Parkinson's disease.

Administered: Orally at 200 mg/day (two times a 100 mg capsule).

24.5.2   RIMANTADINE

Structure (Figure 24.16): α-Methyltricyclo[3.3.1.1.^{3,7}]decane-1-methanamine hydrochloride, α-methyl-1-adamantanemethylamine HCl, Flumadine®.

Activity spectrum: Influenza A virus.

Mechanism of action: As for amantadine.

Principal indication(s): Influenza A virus infections (prevention and early therapy).

Administered: Orally at 300 mg/day (two times 150 mg).

24.5.3   ZANAMIVIR

Structure (Figure 24.16): 4-Guanidino-2,4-dideoxy-2,3-didehydro-*N*-acetylneuraminic acid, 5-acetylamino-4-[(aminoiminomethyl)amino]-2,6-anhydro-3,4,5-trideoxy-D-glycero-D-galacto-non-2-enonic acid, CG 167, Relenza®.

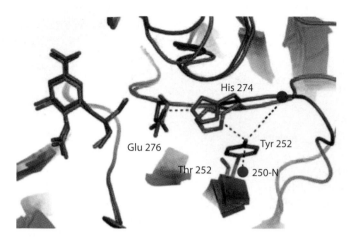

FIGURE 24.17 Locations of oseltamivir-resistance mutations (H274Y) showing that the tyrosine at position 252 is involved in a network of hydrogen bonds in group-1 (H5N1 and H1N1) neuraminidases. (After Russell, R.J. et al., *Nature*, 443, 45, 2006.)

Activity spectrum: Influenza (A and B) virus.

Mechanism of action: Zanamivir and oseltamivir (see below) are *N*-acetylneuraminic (sialic acid) analogues, which are targeted at the influenza viral neuraminidase (sialidase) (Figure 24.17). The viral neuraminidase is responsible for the cleavage of *N*-acetylneuraminic acid present in the influenza virus receptor so that progeny virus particles can be released from the infected cells. The neuraminidase inhibitors zanamivir and oseltamivir keep the virus trapped onto the surface of the cells so that they cannot be released and hence cannot infect other cells.

Principal indication(s): Influenza A and B viral infections (therapy and prevention).

Administered: By (oral) inhalation, at a dosage of 20 mg/day (two times 5 mg, every 12 h) for 5 days. Treatment to be started as early as possible, and certainly within 48 h, after onset of the symptoms.

24.5.4 Oseltamivir

Structure (Figure 24.16): Ethyl ester of (3*R*,4*R*,5*S*)-4-acetamido-5-amino-3-(1-ethylpropoxy)-1-cyclohexane-1-carboxylic acid, GS 4104, Ro 64-0796, Tamiflu®.

Activity spectrum: Influenza (A and B) virus.

Mechanism of action: As for zanamivir.

Principal indication(s): As for zanamivir.

Administered: Orally at 150 mg/day (two times a 75 mg capsule, every 12 h) for 5 days. Treatment should be started as early as possible, and certainly within 48 h, after onset of the symptoms.

24.5.5 Ribavirin

Structure (Figure 24.16): 1-β-D-Ribofuranosyl-1*H*-1,2,4-triazole-3-carboxamide, Virazole®, Virazid®, Viramid®.

Activity spectrum: Various DNA and RNA viruses, in particular orthomyxoviruses (influenza A and B), paramyxoviruses (measles, RSV) and arenaviruses (Lassa, Junin, etc.).

Mechanism of action: The principal target for ribavirin (in its 5′-monophosphate form) is IMP dehydrogenase, that converts IMP to XMP, a key step in the *de novo* biosynthesis of GTP and dGTP. In its 5′-triphosphate form, ribavirin can also interfere with the viral RNA polymerase and in the formation of 5′-capped oligonucleotide primer that is required for transcription of the influenza RNA genome.

Principal indication(s): As a small size droplet aerosol, in the treatment of RSV infections in high-risk infants, and in combination with interferon-α (Intron A®, as in Rebetron®) or pegylated interferon-α (PEG-INTRON® or Pegasys®) in the treatment of HCV infections.

Administered: Orally at doses of 800–1200 mg/day, in the treatment of HCV infections; or by aerosol (solution of 20 mg/mL), which has proved superior to placebo aerosol in the treatment of RSV infections.

24.6 CURRENT STATE OF THE ART

Almost 50 antiviral compounds (not including interferons or immunoglobulins) have momentarily been licensed for the treatment of HIV, HBV, herpesvirus, influenza virus, and/or HCV infections. In the preceding sections these compounds have been discussed from the following viewpoints: chemical structure, activity spectrum, mechanism of action, principal clinical indication(s), route(s) of administration and dosage. Other points that need to be considered before the full clinical potential of any given drug could be appreciated are: (1) duration of treatment, (2) single- versus multiple-drug therapy, (3) pharmacokinetics, (4) drug interactions, (5) toxic side effects, and (6) development of resistance.

As to the duration of treatment, this may vary from a few days (HSV, VZV, influenza virus infections) to several months or years (HIV, HBV, and HCV infections), depending on whether we are dealing with an acute (primary [i.e., influenza] or recurrent [i.e., HSV, VZV]) infection or chronic, persistent (i.e., HIV, HBV, and HCV) infection.

While the short-term treatment (5–7 days) of HSV, VZV, and influenza virus infections, and even the more prolonged treatment of CMV infections, can be based on single-drug therapy, for the long-term treatment of HIV infections a combination of several drugs in a triple-drug cocktail (also referred to as HAART for "highly active antiretroviral therapy") has become the standard procedure. For HIV infection, this triple-drug combination therapy can now be given as a single, once-daily oral pill, i.e., Atripla, containing three active ingredients: TDF, emtricitabine, and efavirenz (De Clercq, 2006b).

Pharmacokinetic parameters to be addressed, when evaluating the therapeutic potential, include bioavailability (upon either topical, oral, or parenteral administration), plasma protein binding affinity, distribution through the organism (penetration into the CNS, when this is needed), metabolism through the liver (i.e., cytochrome P-450 drug-metabolizing enzymes) and elimination through the kidney. Particularly when concocting the multiple-drug combinations for the treatment of HIV infection, possible drug–drug interactions should be taken into account: i.e., some compounds act as P-450 inhibitors and others as P-450 inducers, and this may greatly influence the plasma drug levels achieved, especially in the case of NNRTIs and PIs.

Toxic side effects, both short- and long-term, must be considered when the drugs have to be administered for a prolonged period, as in the treatment of HIV infections. These side effects may seriously compromise compliance (adherence to drug intake), and could, at least in part, be circumvented by reducing the pill burden to, ideally, once-daily dosing.

Finally, resistance development may be an important issue, again for those compounds that have to be taken for a prolonged period, as is generally the case for most of the NRTIs, NNRTIs, and PIs currently used in the treatment of HIV infections. Yet, the nucleoside phosphonate analogues (NtRTIs) tenofovir and adefovir do not readily or rapidly lead to resistance development, even after several years of therapy (as for HIV and HBV infections). Increasing resistance has been noted with HBV against lamivudine after 1 or more years of treatment, but, even if resistant to lamivudine, HBV infections remain amenable to treatment with adefovir dipivoxil. As has been occasionally observed in immunosuppressed patients, HSV may develop resistance to acyclovir, and CMV to ganciclovir, but, if based on ACV TK or CMV PK deficiency, these resistant viruses remain amenable to treatment with foscarnet and/or cidofovir. In immunocompetent patients, treated for an acute or episodic HSV, VZV, or influenza virus infection, short-term therapy is unlikely to engender any drug resistance problems.

24.7   ANTIVIRAL AGENTS IN (PRE)CLINICAL DEVELOPMENT

In addition to the 42 antiviral compounds that by the end of 2006 were available, there are a few more that are under clinical development, and many more that are under preclinical development.

For HIV (De Clercq, 2004), these include the virus adsorption inhibitors (cosalane derivatives, cyanovirin-*N*, cyclotriazadisulfonamide [CADA] derivatives, teicoplanin aglycons); the CXCR4 antagonist AMD070; the CCR5 antagonists maraviroc (UK-427857) and vicriviroc (SCH-D); the NRTIs (±)-2′-deoxy-3′-oxa-4′-thiacytidine (dOTC) (apricitabine); racemic (±)FTC, amdoxovir (diaminopurine dioxolane, DAPD), Reverset (β-D-d4FC), and alovudine (FddThd); the NNRTIs etravirine (TMC125), dapivirine (TMC120), and rilpivirine (TMC278); and the integrase inhibitors Raltegravir (MK-0518) and Elvitegravir (GS-9137). Maraviroc, etravirine, and raltegravir have, in the mean time, been licensed for clinical use in the treatment of HIV infections.

For HBV, four compounds, namely, lamivudine, adefovir dipivoxil, entecavir, and telbivudine have been licensed for medical use and TDF will most likely be the next one and has, in the mean time, been licensed for the treatment of chronic hepatitis B, and for HCV, a variety of compounds targeted at either the viral protease or RNA-dependent RNA polymerase (RdRp) are currently under intensive scrutiny. Also for the herpesviruses (i.e., HSV), new antivirals have been described that target the helicase/primase complex, terminase complex or UL97 protein kinase, and, likewise, new inhibitors are on the horizon for CMV (De Clercq, 2004b) and VZV (De Clercq, 2003c).

24.8   APPRAISAL OF CLINICAL UTILITY

Currently licensed antiviral drugs are particularly focussed on the treatment of HIV, HBV, herpesvirus, influenza virus, and HCV infections, and, so are most of forthcoming antiviral compounds that are in (pre)clinical development.

For the treatment of HIV/AIDS there are now 22 anti(retro)viral drugs available, and to achieve the largest possible benefit, these drugs have to be combined in multiple-drug regimens. Numerous drug combinations could be envisaged. Those that have been generally used consist of two NRTIs, or one NRTI, and one NtRTI (TDF), to which is then added one NNRTI or one PI. Because of the long-term side effects (such as lipodystrophy, diabetes, and cardiovascular disturbances) associated with the PIs that have been longest in use, there is a tendency for starting anti-HIV therapy with PI-sparing regimens.

One such regimen that has proven to be quite efficacious in the treatment of HIV infections, and seems to be well tolerated, is the combination of TDF with (−)FTC (emtricitabine) and efavirenz. Enfuvirtide represents a new dimension in anti-HIV therapy, which could be added onto any (optimized background) regimen, but, because of the costs involved and the fact it has to be administered subcutaneously (twice daily), enfuvirtide should be primarily reserved for salvage therapy.

For the treatment of HBV infections, four compounds, lamivudine, adefovir dipivoxil, entecavir, and telbivudine, besides human interferon, are currently available. TDF is recommended for use in HIV-infected patients who are coinfected with HBV and should soon become available for the treatment of chronic hepatitis B as well. Whether the treatment of HBV infections should be based upon multiple-drug regimens (so as to minimize the emergence of virus-drug resistance), as in the case of HIV/AIDS, needs to be addressed in future studies. A dual-drug regimen that could be envisaged for the treatment of chronic hepatitis B is that of TDF combined with emtricitabine.

The treatment for HSV and VZV infections is since many years fairly well consolidated: it is based on the use of acyclovir, valaciclovir, or famciclovir, and in several European countries, also brivudin (BVDU). Here, there is no need for drug combination therapy, as virus-drug resistance has only rarely proved to be a problem, and, if so (in severely immunocompromised patients), therapy could be switched to, for example, foscarnet or cidofovir. The latter two drugs, which must be administered intravenously, are also used in the treatment of CMV infections in immunosuppressed patients, mostly as a second choice following the use of (val)ganciclovir.

For the therapy and prophylaxis of influenza virus infections the neuraminidase inhibitors zanamivir and oseltamivir have acquired increased momentum (the latter being more practical as it can be administered orally whereas the former has to be administered through (oral) inhalation). These neuraminidase inhibitors may be expected to be effective against new influenza virus types or variants for which no vaccines are available. Finally, for the treatment of HCV infections, the combination of ribavirin with pegylated IFN still stands out as the current choice of treatment, although it provides a durable response only in a portion of the patients (depending on the HCV genotype).

FURTHER READINGS

De Clercq, E. 2002. Strategies in the design of antiviral drugs. *Nat. Rev. Drug Discov.* 1: 13–25.

De Clercq, E. 2003a. Potential of acyclic nucleoside phosphonates in the treatment of DNA virus and retrovirus infections. *Expert Rev. Anti-infect. Ther.* 1: 21–43.

De Clercq, E. 2003b. The bicyclam AMD3100 story. *Nat. Rev. Drug Discov.* 2: 581–587.

De Clercq, E. 2003c. Highly potent and selective inhibition of varicella-zoster virus replication by bicyclic fluro [2,3-d]pyrimidine nucleoside analogues. *Med. Res. Rev.* 23: 253–274.

De Clercq, E. 2004a. Discovery and development of BVDU (brivudin) as a therapeutic for the treatment of herpes zoster. *Biochem. Pharmacol.* 68: 2301–2315.

De Clercq, E. 2004b. Antivirals and antiviral strategies. *Nat. Rev. Microbiol.* 2: 704–720.

De Clercq, E. 2006a. Antiviral agents active against influenza A viruses. *Nat. Rev. Drug Discov.* 5: 1015–1025.

De Clercq, E. 2006b. From adefovir to Atripla™ via tenofovir, Viread™ and Truvada™. *Future Virol.* 1: 709–715.

De Clercq, E. and Holý, A. 2005. Acyclic nucleoside phosphonates: A key class of antiviral drugs. *Nat. Rev. Drug Discov.* 4: 928–940.

*De Clercq, E. 2009. Anti-HIV drugs: 25 compounds approved within 25 years after the discovery of HIV. *Int. I. Antimicrob. Agents,* 33, 307–320.

Doms, R.W. 2000. Beyond receptor expression: The influence of receptor conformation, density, and affinity in HIV-1 infection. *Virology* 276, 229–237.

Gallant, J.E., DeJesus, E., Arribas, J.R. et al. 2006. Tenofovir DF, emtricitabine, and efavirenz vs. zidovudine, lamivudine, and efavirenz for HIV. *N. Engl. J. Med.* 354, 251–260.

Hadziyannis, S.J., Tassopoulos, N.C., Heathcote, E.J. et al. 2005. Long-term therapy with adefovir dipivoxil for HBeAg-negative chronic hepatitis B. *N. Engl. J. Med.* 352: 2673–2681.

Pauwels, R. 2004. New non-nucleoside reverse transcriptase inhibitors (NNRTIs) in development for the treatment of HIV infections. *Curr. Opin. Pharmacol.* 4: 437–446.

Pauwels, R. 2006. Aspects of successful drug discovery and development. *Antiviral Res.* 71, 77–89.

Russell, R.J., Haire, L.F., Stevens, D.J. et al. 2006. The structure of H5N1 avian influenza neuraminidase suggests new opportunities for drug design. *Nature* 443, 45–49.

---

* At present, 25 compounds have been approved for the treatment of HIV infections.

25 Antibiotics

Piet Herdewijn

CONTENTS

25.1 INTRODUCTION

The story of the present-day antibiotics is mainly an old story, as few new entities reached the market during the last decennium. However, the increasing clinical problems encountered with the appearance of resistant strains of bacteria have renewed interest in the field. It is therefore time for a historical overview and for some future perspectives.

Vuillemin first used the word "antibiotic" in 1889 to describe antagonism between living organisms. Waksman formulated the present definition in 1942 as "a chemical substance produced by microorganisms that can inhibit the growth or even destroy other microorganisms." Antibacterial substances found, e.g., in plants were not ranked antibiotics.

The discovery, by Fleming in 1929, of an antibacterial substance produced by a *Penicillium* is well known. It should be mentioned that the production of antibacterial substances by various microorganisms has been described by several authors, e.g., products from various bacteria active against *Anthrax* by Pasteur and Joubert in 1877; pyocyanase from *Pseudomonas aeruginosa* (previous name: *Bac. pyocyaneus)* by Emmerich and Löw in 1899; cultures of *Actinomyces* by Gratia and Dath 1925; and "tyrothricin" by Dubos in 1939. All these products were either too toxic for systemic application or presented a low activity. The treatment of infections by these substances was doubtful and even Fleming wrote in 1941 in a letter to the *Brit. Med. J.* that he had used cultures of *Penicillium* in a few cases of septic wounds, but that, although the results were reasonably good, the trouble of making the product seemed not worthwhile. The discovery of the antibiotic penicillin has been made by H. Florey and his collaborators, Chain, Abraham, and Heatley, who produced the product by fermentation, isolated it, and showed its activity and innocuity in animals and humans. The great effort in England and the United States to produce sufficient quantities of penicillin in the years 1942–1945 is well known. The structure of penicillin was elucidated by a great number of investigators in these two countries.

The discovery of penicillin led the soil microbiologist Selman A. Waksman at the New Jersey Agricultural Experiment Station of Rutgers University, New Brunswick, NJ, to examine his large collection of *Streptomyces* for the production of antibiotics. In 1944, his group announced the discovery of streptomycin. The genus *Streptomyces* has been the most important source of therapeutically useful antibiotics.

Important antibiotics are produced by fungi, which are plant-like, nonphotosynthetic eukaryotes, which grow in filamentous multicellular aggregates, e.g., *Penicillium chrysogenum* (penicillin), *Cephalosporium* (cephalosporins), and *Penicillium griseofulveum* (griseofulvin), which belong to the class of the Fungi imperfecti (Deuteromycetes).

Other important genera are Streptomyces (streptomycin, neomycin, chloramphenicol, tetracyclines, and macrolides), Nocardia (rifamycins and nocardicin), and *Micromonospora* (gentamicin, sisomicin). They belong to the family of the Actinomycetales, which are Gram-positive bacteria that grow in the form of mycelia. Some antibiotics like polymyxin, tyrothicin, and bacitracin are produced by the bacteria of the genus *Bacillus*, which are also prokaryotes. Subsequently, β-lactam antibiotics have been discovered in *Streptomyces*, e.g., cephamycins (7-methoxycephalosporins), thienamycin, and other carbapenems, clavulanic acid, monobactams in *Chromobacterium violaceum, Gluconobacter, Acetobacter,* etc.

25.2 ANTIBIOTICS AFFECTING BACTERIAL CELL WALL FORMATION

25.2.1 β-Lactam Antibiotics

25.2.1.1 General

The structure of penicillin was determined based on a great number of studies performed in England and the United States in the years 1942–1945. After some hesitation concerning the β-lactam-thiazolidine structure, this formula was accepted based on x-ray diffraction studies of the potassium and rubidium salts of penicillin. In the cephalosporins, whose structure was determined in 1961, the thiazolidine cycle is replaced by a hydrothiazine ring. It should be noted that not only the double ring structure but also the configuration at C_3, C_5, and C_6, respectively C_6 and C_7, are essential for activity.

25.2.1.2 Penicillins

The production was and still is performed by fermentation. The original *Penicillium notatum* of Fleming was replaced by *Penicillium chrysogenum*, whose production has been increased more than a 1000-fold by the isolation of mutants. Instead of surface culture, already in 1943 deep fermentation was developed in the United States. Large steel cylinders of 100.000 L and more, which are stirred and aerated are used. Culture media usually contain mineral salts, a carbohydrate source and soy or peanut meal, and/or cornsteep liquor. It was found in 1944 in a Government laboratory in Peoria, that addition of cornsteep liquor increased penicillin production. The use of cornsteep liquor was the origin of the difference between the American penicillin (penicillin G) that has a benzyl side chain and the English penicillin (penicillin F) that has a pentenyl side chain. The difference is due to the presence of phenylethylamine (formed from phenylalanine), which is transformed into phenylacetic acid. Since then, this acid is always added during penicillin G fermentation. On this basis, Behrens et al. in the Lilly laboratories, examined in 1944–1945 several acids as precursors for penicillin production. They obtained 11 new penicillins. All these penicillins were active, but apparently had no interesting new properties. In 1954, Brandl and Margreiter in Austria observed that a new penicillin, which was acid stable, was formed when phenoxyethanol was added to the culture medium. This penicillin that was identified as phenoxymethyl penicillin (penicillin V), which unexpectedly was acid stable, was introduced as oral penicillin. It is produced by adding phenoxyacetic acid to the fermentation medium of *Penicillium chrysogenum*.

The production of new penicillins by this method, however, is very limited because the mold incorporates only monosubstituted acetic acids ($R–CH_2–COOH$). The possibility of preparing other penicillins was vastly increased by the discovery of 6-aminopenicillanic acid (6-APA) by Batchelor, Doyle, and Rolinson in 1959. They could isolate this substance from the fermentation media of *Penicillium chrysogenum* to which no precursor acid was added. The yield was very low but fortunately another method of preparation was found. The Beecham group and other laboratories in the United States and Germany (in 1960) discovered that the side chain of penicillin G could be removed by an enzyme from some bacteria. Many bacteria produce this enzyme, but only certain strains of *Escherichia coli* are used in the industry. It was also found that this enzyme, called penicillin acylase, did not cleave penicillin V. But an enzyme present in other microorganisms like *Fusarium* or *Erwinia* was able to transform penicillin V into 6-APA. Still another practical procedure is the chemical cleavage of penicillin to 6-APA, discovered in 1970 by Weissenburger and Vanderhoeven.

The first semisynthetic penicillins, phenethicillin, and propicillin, obtained by reaction of 6-APA with α-phenoxypropionic or α-phenoxybutyric acid, had a spectrum of activity very similar to that of phenoxymethylpenicillin (Table 25.1). A more important advance was the discovery of ampicillin = α-aminobenzylpenicillin by Doyle et al. in 1962 (Table 25.2). This penicillin had a broader spectrum of activity than the previous products, i.e., it is active against some Gram-negative bacteria like *E. coli*, *Proteus*, *Enterococcus* but not against *Klebsiella*, *Enterobacter*, *Pseudomonas aeruginosa*. The antibacterial spectrum of cyclacillin is similar to that of ampicillin but its activity is lower.

The *p*-hydroxy derivative of ampicillin, amoxicillin, unexpectedly presents a much better oral absorption than ampicillin (1971). Similarly, a better oral absorption occurs also with esters of ampicillin, pivampicillin, talampicillin, and bacampicillin (1970–1975). These prodrugs are hydrolyzed to ampicillin, partially through the action of esterases, partially spontaneously. It also should be noticed that in these penicillins, phenyl glycine or derivative, has the D-configuration because this epimer is more active than the epimer with an L-side chain.

TABLE 25.1
Penicillins

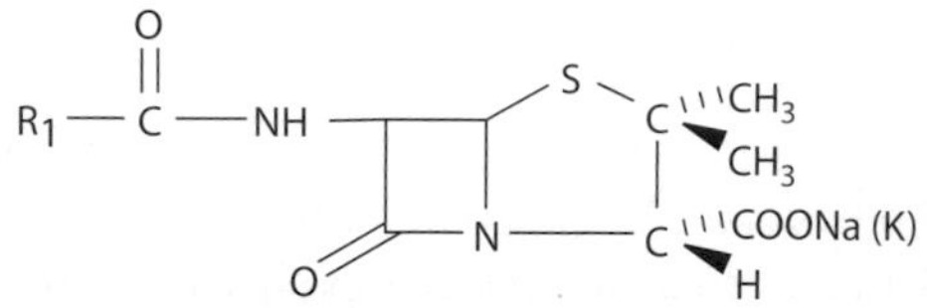

R₁

$C_8H_5-CH_2-$	Benzylpenicillin (Penicillin G)
$CH_3CH_2CH=CH-CH_2-$	Penicillin F
$C_6H_5-O-CH_2-$	Phenoxymethylpenicillin

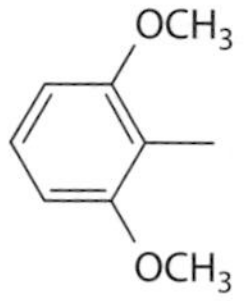

Methicillin

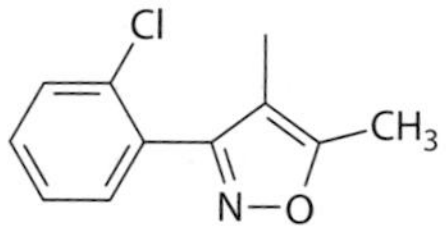

Cloxacillin

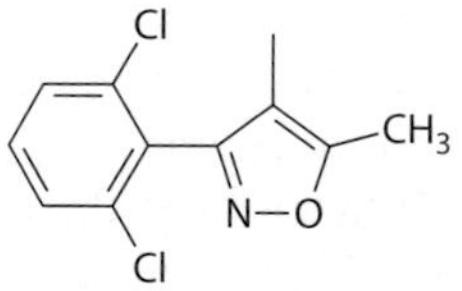

Dicloxacillin

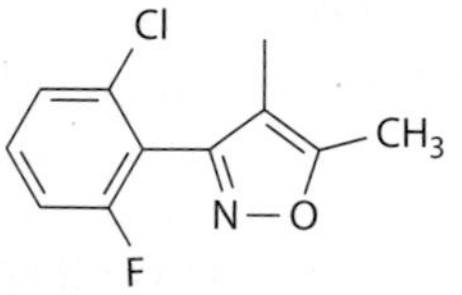

Flucloxacillin

Another important development was the discovery of penicillins resistant to penicillinase. Around 1960, quite a number of staphylococci were resistant to penicillin, because they produced an enzyme, penicillinase, which opens the β-lactam ring. Methicillin was the first penicillin found to be resistant against Staphylococcal penicillinase. More important are the oxacillins discovered by Doyle and Nayler in the Beecham Laboratories (1952–1962), which may also be administrated orally. The introduction of one chlorine (cloxacillin) and two chlorine atoms (dicloxacillin) improves not only the oral absorption but also increases their protein binding. Flucloxacillin seems to be the best compromise for optimum activity and protein binding (Table 25.1).

Carbenicillin (1967), which contains a carboxyl group in the α-position of the side chain, was the first penicillin with antipseudomonal activity. Its activity is rather low (MIC ± 30 μg/mL) and large doses must be injected (500 mg/kg). Ticarcillin is approximately twice as active as carbenicillin against *Pseudomonas aeruginosa*. The acylureido penicillins, mezlocillin and piperacillin, are 8–10 times more active than carbenicillin against *P. aeruginosa*, and also have activity against *K. pneumoniae*, *S. marcescens*, *H. influenzae*, and *Neisseria*. Temocillin, with a 6α-methoxy group,

TABLE 25.2
Penicillins

R₁	R₂	Y	
H	H	H	Ampicillin
OH	H	H	Amoxicillin
H	H	—CH₂—OCOC(CH₃)₃	Picampicillin
H	H	—CH(CH₃)—OCOOC₂H₅	Bacampicillin
H	H	(talampicillin Y group)	Talampicillin
H	(azlocillin R₂ group)	H	Azlocillin
H	(piperacillin R₂ group)	H	Piperacillin

has improved resistance, against β-lactamase. It is quite active against enterococci, but not against *Pseudomonas*.

Mecillinam is not an *N*-acyl derivative of 6-APA but is a semisynthetic 6-amidinopenicillanic acid (Table 25.3). It is very active against some Gram-negative bacteria like *E. coli, Klebsiella*, and *Serratia* but not against *Pseudomonas, Proteus*, and *Haemophilus*. For oral administration, the prodrugs pivmecillinam and bacmecillinam can be used.

25.2.1.3 Cephalosporins

In 1948, Brotzu at the University of Cagliari in Sardinia discovered antibacterial activity in the culture medium of a microorganism, *Cephalosporium acremonium*, collected at the sewage outlet of the town in the sea. This strain was sent to the laboratory of Professor Florey in Oxford. from its culture medium, Abraham et al. isolated in 1951, cephalosporin P, a steroid antibiotic related to fusidic acid, and in 1954 cephalosporin N, later called penicillin N, which is a penicillin with D-α-aminoadipic acid as side chain. When this penicillin was inactivated by acid, still another antibiotic, called cephalosporin C was isolated (Abraham and Newton, 1955). They were able to determine its structure in 1961 and they found that it also had a D-α-aminoadipic acid side chain but a β-lactam-hydrothiazine ring system. The fact that this product was acid stable and resistant to penicillinase indicated that it could be the starting point of a new group of important antibiotics. The side chain, however, was

TABLE 25.3
Penicillins

$$R_1-CO-N(H)-\text{[β-lactam ring with Z, S, C(CH}_3)_2\text{, COOH, H]}$$

Z	R$_1$	
H	phenyl–CH(COONa)–	Carbenicillin
H	thienyl–CH(COONa)–	Ticarcillin
OCH$_3$	thienyl–CH(COONa)–	Temocillin

cyclohexyl(NH$_2$)–CO–NH– [β-lactam ring, S, C(CH$_3$)$_2$, COOH, H]   Cyclacillin

azepane N–CH=N– [β-lactam ring, S, C(CH$_3$)$_2$, COOR$_2$, H]

R$_2$	
$-CH_2-O-CO-C(CH_3)_3$	Pivmecillinam
$-CH(CH_3)-O-COOEt$	Bacmecillinam
$-H$	Mecillinam

responsible for a rather low activity and had to be replaced by other groups. No microorganism was found that could enzymatically remove the side chain, but this transformation could be performed by chemical means. The first two cephalosporins introduced in the clinic, cefalothin (1962) and cefaloridine (1964), have a thienylacetic acid side chain. Of the thousands of cephalosporins that were synthesized, only a few reached the clinic (Table 25.4).

They are often classified in generations. This classification is more or less related to the year of introduction (I:1962–1971, II:1974–1977, III:1976–1980) and their properties. One should not consider that cephalosporins of the third generation are superior in all respects to those of the first one.

Generation I cephalosporins are used mainly in infections with Gram-positive bacteria. The II generation is also active against certain Gram-negative bacteria like *Neisseria* and *Haemophilus*. The introduction of heterocyclic rings in C$_3$ is responsible for the metabolic stability of these cephalosporins. However, some structures like the tetrazolthiomethyl group (in cefamandole, cefotetan) are responsible for a disulfiram effect and prolonged blood coagulation. These side effects influence unfavorably their clinical application.

**TABLE 25.4
Cephalosporins**

R$_1$	R$_2$	
HOOC—CH—(CH$_2$)$_3$— (with NH$_2$)	—CH$_2$—O—C(=O)—CH$_3$	Cephalosporin C
thiophene-CH$_2$—	—CH$_2$—O—C(=O)—CH$_3$	Cefalotin
tetrazole-N—CH$_2$—	—CH$_2$—S—(thiadiazole)—CH$_3$	Cefazolin
thiophene-CH$_2$—	—CH$_2$—N$^+$(pyridine)	Cefaloridine
phenyl-CH(OH)—	—CH$_2$—S—(tetrazole-N-CH$_3$)	Cefamandole
furan-C(=N-OCH$_3$)—	—CH$_2$—O—C(=O)NH$_2$	Cefuroxime
phenyl-CH(OH)—	—CH$_2$—S—(tetrazole-N-CH$_2$-SO$_3$H)	Cefonicid
phenyl(o-CH$_2$NH$_2$)-CH$_2$—	—CH$_2$—S—(tetrazole-N-CH$_2$-COOH)	Ceforanide

In 1971 β-lactams with a methoxy group on C$_7$, cephamycins, were discovered in certain *Streptomyces* strains. This substituent is responsible for a greater resistance to β-lactamase (Table 25.5).

In the Takeda laboratories, it was discovered in 1977 that reaction of a chloracetylcephalosporin with thiourea leads to a product with an aminothiazolylacetic acid side chain, e.g., cefotiam, which had a very good activity especially against Gram-negative bacteria. On the other hand, the presence of an oxime group in nocardicin was the guide for the introduction of this group in cefuroxime, which has a good resistance against β-lactamase. These two observations were combined in the preparation of cefotaxime. A whole series of cephalosporins with an aminothiazol side chain are now available (Table 25.6).

TABLE 25.5
Cephalosporins

R_1	R_2	
HOOC—CH—(CH$_2$)$_3$— with NH$_2$	—CH$_2$—O—C(=O)—NH$_2$	Cephamycin C
thiophene—CH$_2$—	—CH$_2$—O—C(=O)—NH$_2$	Cefoxitin
H$_2$N—CO, NaOOC dithietane —CH—	—CH$_2$—S—tetrazole(N—CH$_3$)	Cefotetan

TABLE 25.6
Cephalosporins

R_3	R_2	
—C(=N—OCH$_3$)—	—CH$_2$—O—C(=O)—CH$_3$	Cefotaxime
—C(=N—O—C(CH$_3$)$_2$—COOH)—	—CH$_2$—N$^+$(pyridine)	Ceftazidime
—C(=N—OCH$_3$)—	—CH$_2$—S—triazinone(N—CH$_3$)	Ceftriaxone
—C(=N—OCH$_3$)—	—CH$_2$—N$^+$(pyrrolidine-CH$_3$)	Cefepim
—C(=N—OCH$_3$)—	—CH$_2$—N$^+$(cyclopentapyridine)	Cefpirom

TABLE 25.7
Cephalosporins

R_3	R_2	
phenyl-CH(NH$_2$)-	$-CH_2-O-C(=O)-CH_3$	Cefaloglycin
phenyl-CH(NH$_2$)-	$-CH_3$	Cefalexin
cyclohexadienyl-CH(NH$_2$)-	$-CH_3$	Cefradine
HO-phenyl-CH(NH$_2$)-	$-CH_3$	Cefadroxil
HO-phenyl-CH(NH$_2$)-	$-CH_2-S-$ (triazole)	Cefatrizine
aminothiazolyl-C(=N-OCH$_2$COOH)-	$-CH=CH_2$	Cefixime
phenyl-CH(NH$_2$)-	$-Cl$	Cefaclor

The blood concentrations obtained after oral administration of the above described cephalosporins are low. Cefaloglycin, the first cephalosporin that gave moderate blood levels after oral administration, has a phenylglycine side chain. This group or a related group is present in several oral cephalosporins. Cefalexin, which is the most widely used product, has an activity that is much lower than that of the injectable compounds. Attempts to increase the activity have not been very successful, but cefaclor and cefixime are somewhat superior (Table 25.7).

25.2.1.4 Nontraditional β-Lactam Antibiotics

25.2.1.4.1 Clavulanic Acid and Sulbactam

Clavulanic acid discovered in the Beecham laboratories in 1976 is a β-lactam produced by the same Actinomycete *Streptomyces clavuligerus* that produces cephamycin C. Its *in vitro* activity is low, but it is a potent inhibitor of many β-lactamases. It is combined with conventional β-lactams like amoxicillin and ticarcillin. It potentiates the actions of these antibiotics against β-lactamase

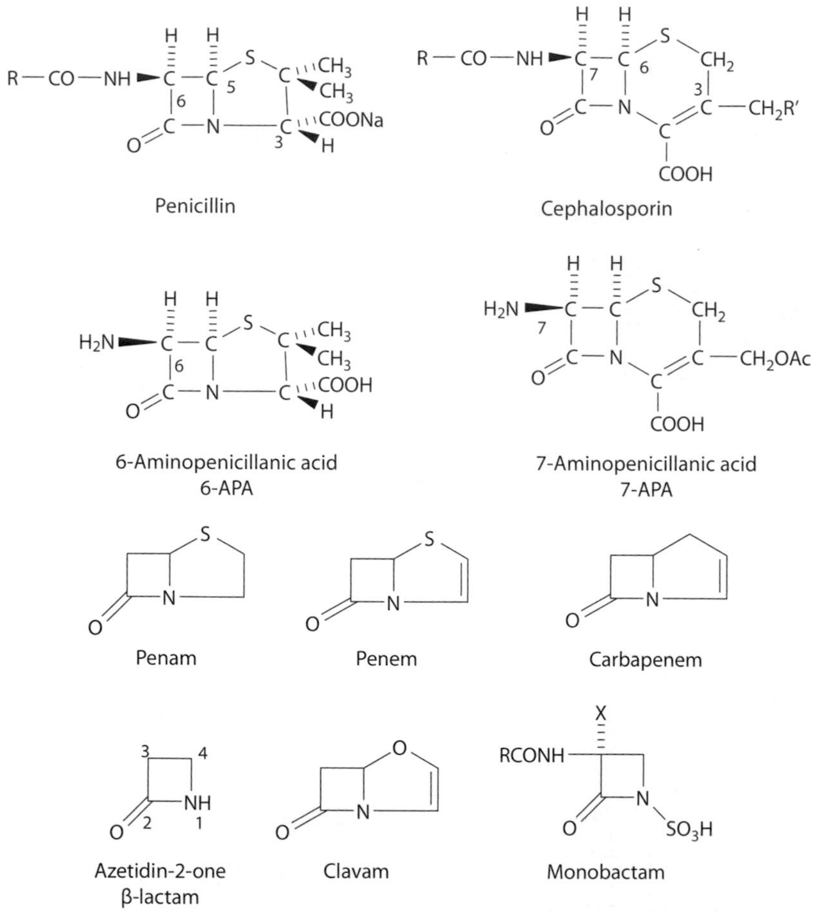

Penicillin

Cephalosporin

6-Aminopenicillanic acid
6-APA

7-Aminopenicillanic acid
7-APA

Penam   Penem   Carbapenem

Azetidin-2-one   Clavam   Monobactam
β-lactam

FIGURE 25.1   β-Lactam structures.

producing bacteria. Penicillanic acid sulfone (sulbactam) and its prodrug pivsulbactam are also inhibitors of β-lactamase and are combined with ampicillin (Figure 25.2).

25.2.1.4.2   Carbapenems

A new β-lactam antibiotic, thienamycin, was obtained in 1978 from cultures of *Streptomyces cattleya* by researchers at Merck Sharp and Dohme. The bicyclic system, with a double bond between C_2 and C_3, is called 2-carbapenem (Figure 25.1). Several related products were discovered, e.g., epithienamycins (isomers in hydroxyethyl side chain). Thienamycin is not stable and is used as the *N*-formimidoyl derivative (imipenem). Thienamycin and imipenem are prepared by total synthesis. Imipenem and meropenem are broad-spectrum antibiotics. Another problem was discovered during clinical studies, i.e., the cleavage of thienamycin by a dehydropeptidase present in the kidney. For that reason, imipenem is associated with an inhibitor of that enzyme, cilastatin. Imipenem exhibits a broad spectrum of activity and is resistant to most β-lactamases (Figure 25.2).

25.2.1.4.3   Penems

In 1977, Woodward described the first synthesis of a penem. This penem, which had a phenacetyl-amido side chain like penicillin, had a rather low activity. The introduction of hydroxyethyl groups in C_6, like in the carbapenems, improved the potency significantly.

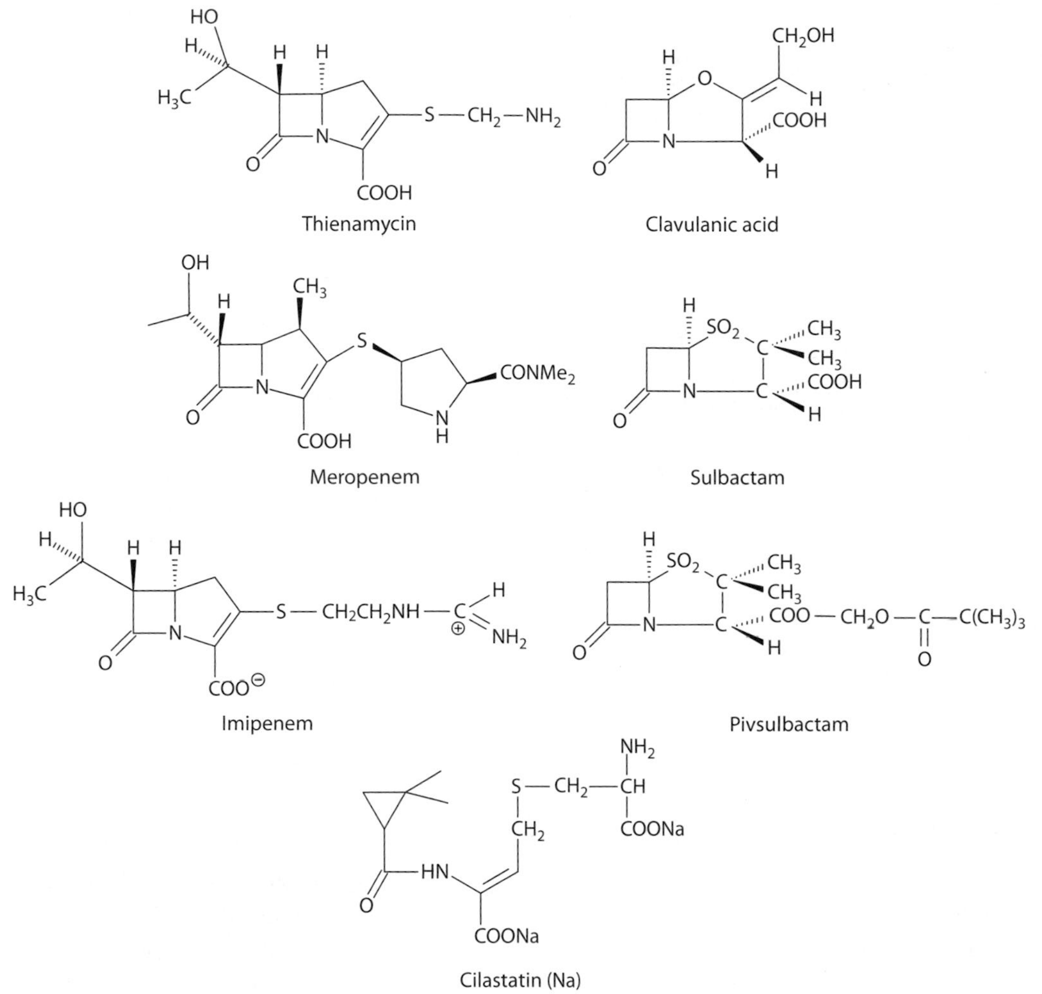

FIGURE 25.2 Nonclassical β-lactams.

25.2.1.4.4 Monobactams

The term monobactam was coined by Sykes et al. (1981) to describe a novel group of bacterially produced monocyclic β-lactams. A product SQ26.180 was isolated by Sykes from *Chromobacterium violaceum* and a related product, sulfazecin was discovered in Japan by Imada et al. (1981) in cultures of *Gluconobacter* sp., and *Pseudomonas* sp. The activity was improved by modification of the amide side chain. The introduction of the cefotaxime side chain gave an interesting new drug, aztreonam. It is very active against Gram-negative bacteria, including *Pseudomonas aeruginosa* (Figure 25.3).

25.2.2 BACITRACIN

Bacitracin is produced by a strain of *Bacillus subtilis* and was isolated in 1945 from the wound of a patient called Tracy. The same product was discovered in Oxford in 1949, in the culture of *Bacillus licheniformis* A5, and was called Ayfivin. Studies by Abraham et al. and by Craig et al. in New York revealed that bacitracin (= ayfivin) was a mixture and it had a peptide structure. The structure of the main component is given in Figure 25.3.

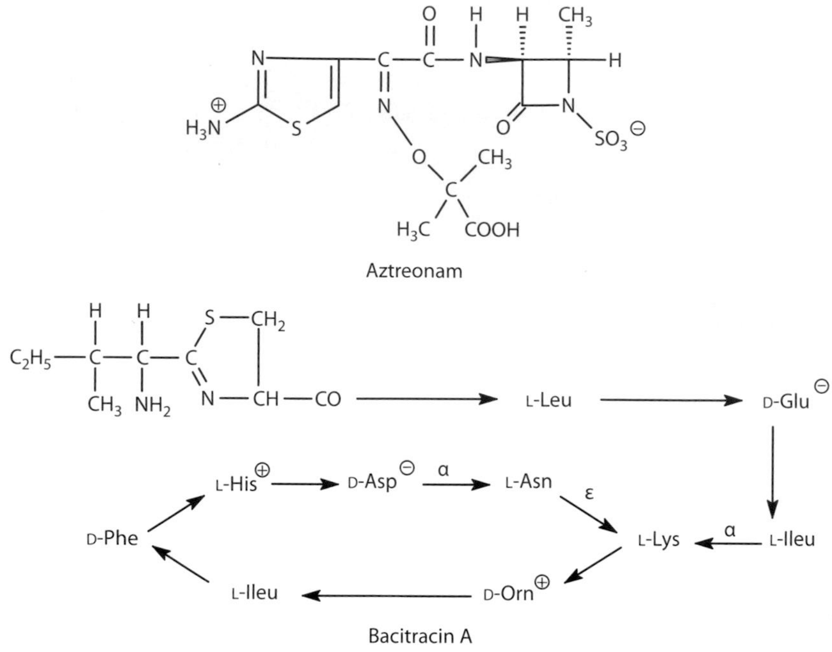

Aztreonam

Bacitracin A

FIGURE 25.3   Aztreonam and bacitracin A.

25.2.3 VANCOMYCIN AND RELATED PRODUCTS

Vancomycin was isolated by Williams in 1956 in the Lilly laboratories from a culture of *Streptomyces orientalis*. Its very complex structure was studied mainly by D.H. Williams in Cambridge.

Vancomycin is active only against Gram-positive microorganisms, mainly staphylococci and streptococci, and also against *Clostridium difficile*. It is a quite toxic product. As intramuscular injections cause pain and necrosis, the administration has to be intravenous. Febrile reactions, thrombophlebitis, ototoxicity, and nephrotoxicity are the other side effects. For these reasons, its use is limited to infections due to resistant staphylococci and streptococci, mainly in patients treated with immunosuppressants. Oral application may be used in pseudomembranous colitis, due to *Clostridium difficile*, which may occur after the administration of antibiotics like lincomycin and clindamycin. Teicoplanin, also named teichomycin, was isolated in 1978 from *Actinoplanes teicho-mycelicus*. Its structure and activity spectrum are related to that of vancomycin.

25.3 ANTIBIOTICS AFFECTING THE CYTOPLASMIC MEMBRANE

25.3.1 ANTIBIOTICS THAT DISORGANIZE MEMBRANE STRUCTURE

25.3.1.1 Tyrothricin

Tyrothricin was isolated from *Bacillus brevis* by R. Dubos in 1939. Because of its toxicity, e.g., lysis of erythrocytes, its application has always been limited to external use like troches for throat infections, nose drops, etc.

It is a mixture of a neutral acetone–ether soluble part (about 20%) called gramicidin and a solvent insoluble part, tyrocidine HCl. Tyrocidine is a cyclic peptide. Gramicidin is a mixture of neutral linear peptides. It is neutral because the *N*-terminal amino group is formylated and because the carboxy group is linked to ethanolamine (Figure 25.4).

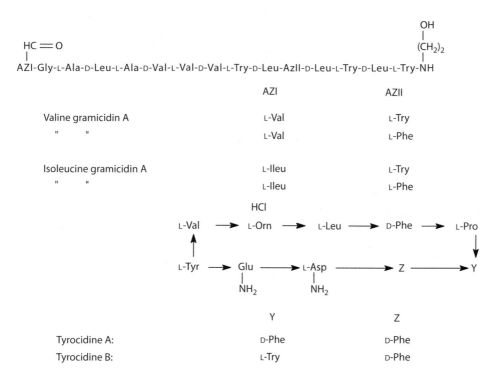

FIGURE 25.4 Tyrothricin components.

25.3.1.2 Polymyxin

Polymyxin was discovered in 1947 almost simultaneously in three different laboratories in the United States and England. It has been shown that *Bacillus polymyxa* and *B. aerosporus* were identical species and an agreement was reached on one name: polymyxin. Several polymyxins A, B, C, D, and E were isolated, but polymyxin B, which seemed to be the least toxic, was introduced in medicine. A peptide antibiotic, colistin, isolated in 1950 in Japan from *B. colistinus* was shown to be identical with polymyxin E. Both polymyxin B and E contain several components differing in the structure of the fatty acid. The most important component is B_1 (or E_1), with some 15%–25% of B_2 (or E_2). Polymyxin and colistin have five free amino groups and are used as the sulfate salt. A sulfomethyl derivative, obtained by reaction with formaldehyde and bisulphite, and which is less toxic, is used for injection. These peptides have been used for the treatment of infections by Gram-negative bacteria like *Pseudomonas*, but this treatment is seldom used now, because less toxic anti-biotics are available. Because these peptides are neurotoxic and nephrotoxic, their use is limited usually to ear- and nose-drops, and treatment of skin infections (Figure 25.5).

25.3.2 ANTIBIOTICS WITH ALTERED MEMBRANE PERMEABILITY

25.3.2.1 Gramicidins

Their structure is given in Figure 25.4.

25.3.2.2 Polyether Antibiotics

An important group of antibiotics with a polyether structure has been described. These antibiotics have a certain activity against Gram-positive bacteria but their main application lies in the treatment of coccidiosis. Monensin, which was isolated in 1967 from *Streptomyces cinnamonensis*, is used as coccidiostat in poultry and as growth promoter in ruminants (cattle and sheep). Other polyethers are lasalocid, salinomycin, and nigericin (Figure 25.6).

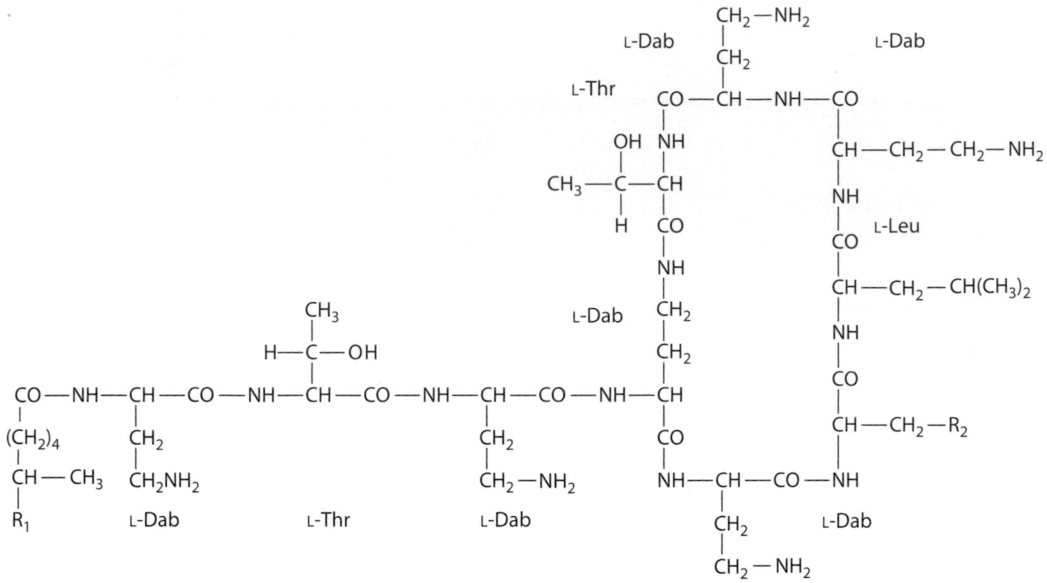

FIGURE 25.5 Peptide antibiotics.

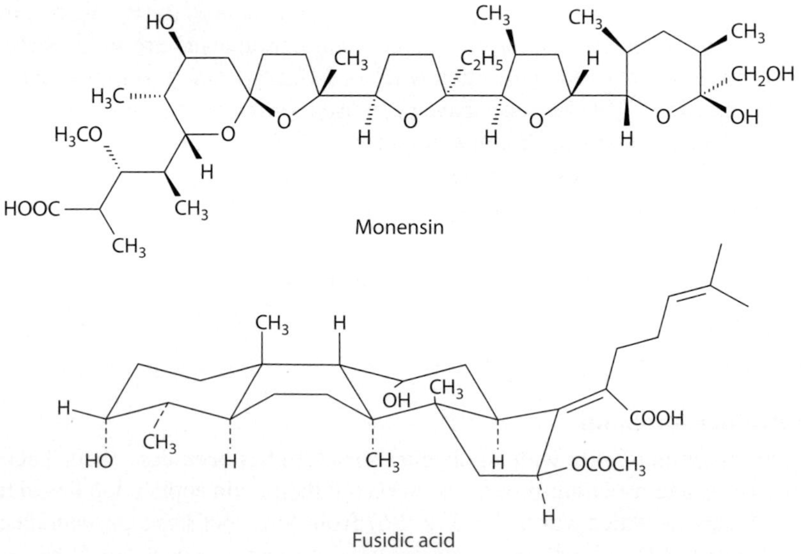

	R_1	R_2
Polymyxin B$_1$	— C$_2$H$_5$	— C$_6$H$_5$ (D-Phe)
Polymyxin B$_2$	— CH$_3$	— C$_6$H$_5$ (D-Phe)
Polymyxin E$_1$	— C$_2$H$_5$	— CH(CH$_3$)$_2$ (D-Leu)
Polymyxin E$_2$	— CH$_3$	— CH(CH$_3$)$_2$ (D-Leu)

Monensin

Fusidic acid

FIGURE 25.6 Monensin, fusidic acid.

25.4 ANTIBIOTICS AFFECTING NUCLEIC ACID SYNTHESIS

25.4.1 ACTINOMYCIN AND ANTHRACYCLINES

One of the first antibiotics discovered by Waksman was actinomycin, isolated from *Streptomyces antibioticus* (Waksman and Woodruff, 1940). Subsequently several related actinomycins were described, and they differed by one or two amino acids in the peptide part. Their structure was elucidated by Brockmann et al. in Germany and A.W. Johnson in England. They have a phenoxazone chromophore, to which are linked two identical peptidyl-lactone rings. The product that is usually used in the clinic is actinomycin D = dactinomycin. It is used for treatment of tumors.

The first anthracycline to be described was rhodomycin (Brockmann and Bauer, 1950). The toxicity of this product was too high even for the treatment of cancer. The first clinically useful product was daunorubicin. It was discovered independently in 1963 at Rhone-Poulenc in France and at Farmitalia. In France it was isolated from *Streptomyces coeruleorubidus* and called rubidomycin and in Italy from *S. peucetius* and called daunomycin. Daunorubicin is a combination of both names. Doxorubicin (also called adriamycin) was obtained in 1969 from *S. peucetius* strain caesius. Both have an anthraquinone chromophore to which the amino sugar daunosamine is attached (Figure 25.7).

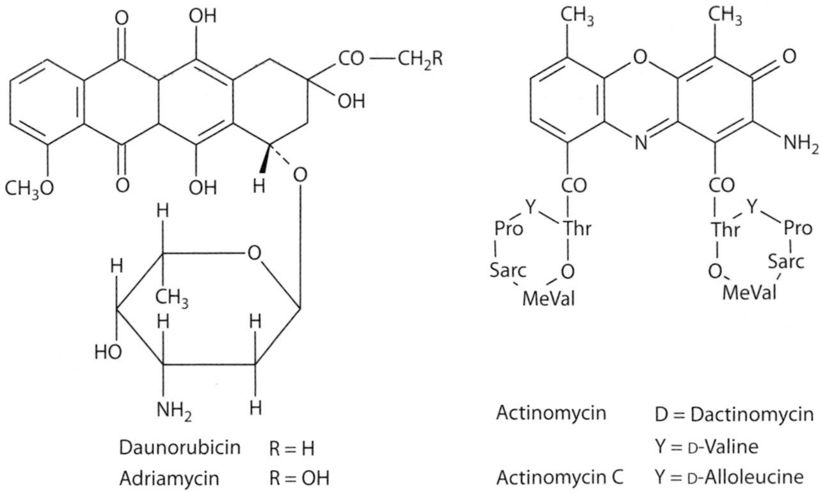

FIGURE 25.7 Anticancer antibiotics.

25.4.2 ANTIBIOTICS THAT INHIBIT RNA POLYMERASE

Rifamycin is an antibiotic that was discovered at Lepetit in Italy in 1957 in a strain of *Streptomyces*, now reclassified as *Nocardia mediterranii*. In 1963, it was discovered that the most important component, rifamycin B, was transformed by oxygenation of the aqueous solution into a more active product, rifamycin O, which on mild reduction gave rifamycin SV (Sensi, Furesz). Rifamycin SV reacts with formaldehyde to form formyl rifamycin, which on condensation with *N*-amino-*N'*-methylpiperazine yields rifampicin. The determination of the structure was performed by Oppolzer and Prelog (1964). Later, rifaximin was introduced whose spectrum is an analogue of rifampicin.

Rifamycin is very active against Gram-positive bacteria and some Gram-negative ones and against *Mycobacterium tuberculosis*. Rifampicin gives a better, more regular absorption and is an excellent drug in the treatment of tuberculosis and leprosy. Rifaximin is used for gastrointestinal infections (Figure 25.8).

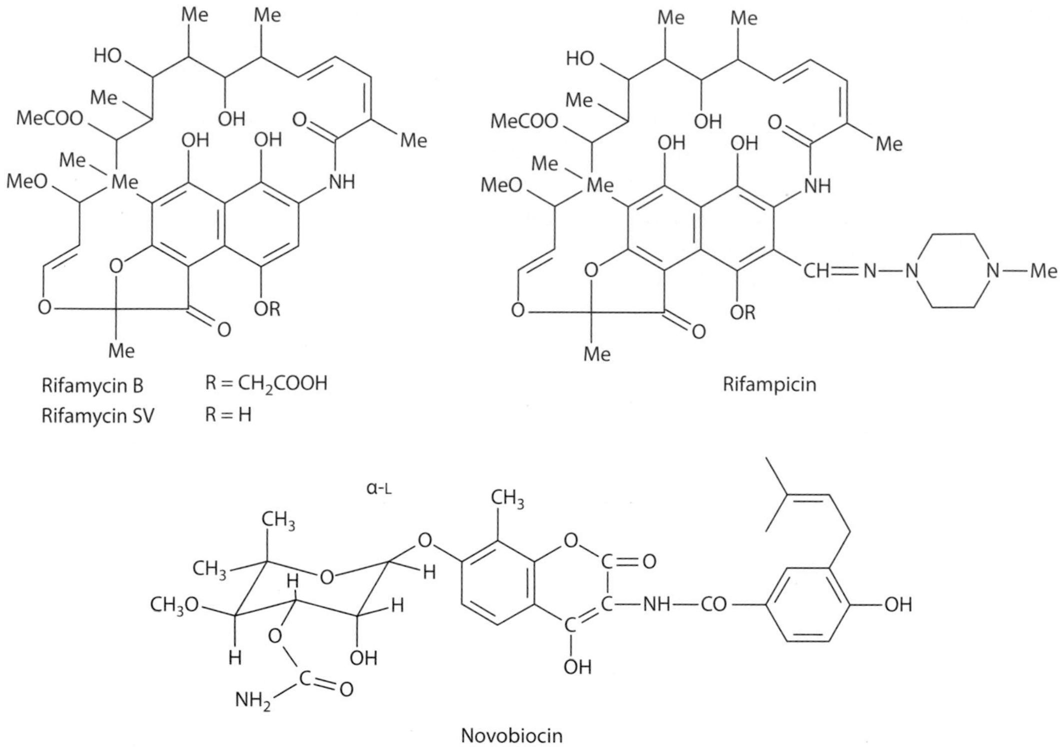

Rifamycin B   R = CH₂COOH
Rifamycin SV   R = H

Rifampicin

Novobiocin

FIGURE 25.8   Rifamycins and novobiocin.

25.4.2.1   Novobiocin

Novobiocin was isolated in 1955, simultaneously in several laboratories in the United States Novobiocin producing microorganisms were *Streptomyces niveus, S. spheroides*. Novobiocin is a coumarin derivative. It is active against Gram-positive bacteria and also *H. influenzae, Neisseria*. It has been used in the treatment of infections caused by resistant staphylococci, but numerous other products are available now.

25.4.2.2   Quinolones

Nalidixic acid was discovered by G. Lesher et al. in the Winthrop laboratories in 1962. Its original trade name Negram® indicates that it is active against Gram-negative bacteria. Subsequently some derivatives were introduced, e.g., oxolinic acid (1967), cinoxacin (1970), norfloxacin (1978). The last product had already a much superior activity both against Gram-positive and Gram-negative bacteria. More recent compounds like ciprofloxacin and ofloxacin have a broad-spectrum activity (Figure 25.9).

25.5   ANTIBIOTICS AFFECTING THE PROTEIN SYNTHESIS

25.5.1   INHIBITORS OF THE 30 S RIBOSOMAL SUBUNIT

25.5.1.1   Aminoglycosides

Streptomycin, discovered in 1944 by Schatz, Bugie, and Waksman in *Streptomyces griseus* was the first aminoglycoside. It was active mainly against Gram-negative bacteria, and for that reason it was a useful complement to penicillin, which was active against Gram-positive bacteria. Streptomycin together with *p*-aminosalicylic acid were the first drugs used in the treatment of tuberculosis.

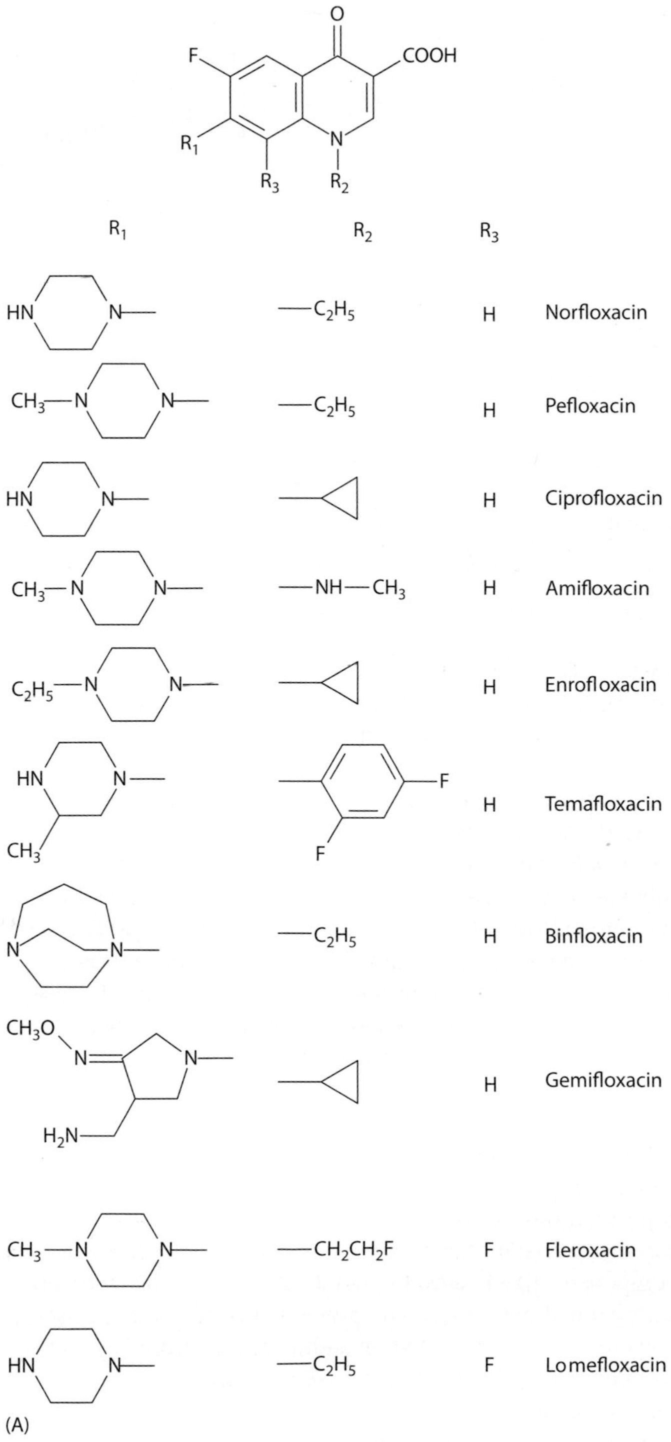

FIGURE 25.9 Quinolones.

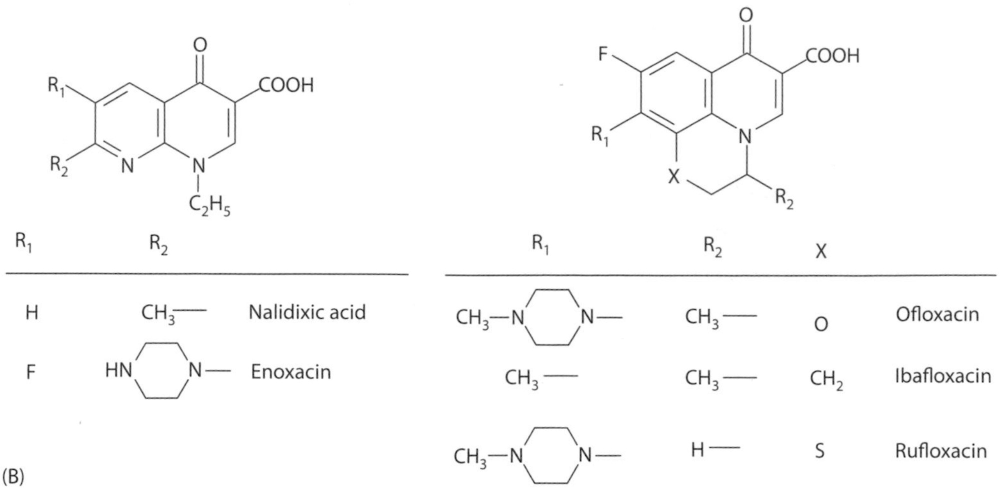

R₁	R₂	
H	CH₃—	Nalidixic acid
F	HN‿N—	Enoxacin

R₁	R₂	X	
CH₃—N‿N—	CH₃—	O	Ofloxacin
CH₃—	CH₃—	CH₂	Ibafloxacin
CH₃—N‿N—	H—	S	Rufloxacin

(B)

FIGURE 25.9 (continued)

Streptomycin has three basic groups, and it is used usually as the sulfate salt. All aminoglycosides have a cyclohexane ring moiety viz. in streptomycin diaminotetrahydroxycyclohexane (streptamine) and in others diaminotrihydroxycyclohexane (deoxystreptamine). In streptomycin, the amino groups of streptamine are replaced by guanidino groups (streptidine). To this ring is attached a pentose containing an aldehyde group (streptose), to which is linked *N*-methyl-L-glucosamine. All aminoglycosides present a certain toxicity: vestibular disturbance, with problems of equilibrium, and cochleotoxicity, which may result in partial and total loss of hearing. Streptomycin, which causes mainly vestibular toxicity, was replaced at a certain time by dihydrostreptomycin (the aldehyde function of the streptose moiety is replaced by an alcohol group). This derivative however had a greater cochleotoxicity, and it is not used anymore in human medicine.

The second aminoglycoside that was introduced in medicine was neomycin. It was isolated in 1949 by Waksman and Lechevalier from the cultures of *Streptomyces fradiae*. Neomycin is a mixture of components B and C. It was demonstrated that dextromycin, fradiomycin, and framycetin were identical to neomycin. The last product is mainly neomycin B. Neomycin presents a good activity against Gram-positive and Gram-negative bacteria but it is very ototoxic. Like other aminoglycosides, it is very slightly absorbed from the digestive tract, so its oral use does not produce this systemic toxicity. Neomycin is useful for the treatment of intestinal infections and for external application (eye drops, eardrops, ointments).

Paromomycin was discovered in 1959 in *Streptomyces rimosus* variant *paromomycinus*. It is also a mixture of two components like neomycin, and it differs from this antibiotic by the replacement of one amino group by a hydroxy group. The product was discovered in several laboratories and catenulin, hydroxymycin, zygomycin, and aminosidin are all paromomycin (Figure 25.10).

Paromomycin sulfate has broad-spectrum antibacterial activity but its use is largely restricted to the treatment of intestinal amebiasis.

Kanamycin was isolated in 1957 in Japan by Umezawa et al. from *Streptomyces kanamyceticus*. Kanamycin (used as sulfate, like most aminoglycosides) has four amino groups. Its ototoxicity and renal toxicity is much lower than that of neomycin and paromomycin. At the time of its introduction, it was applied mainly for the treatment of infections due to penicillin resistant staphylococci. Now it is only used externally (Figure 25.11).

Gentamicin and sisomicin are aminoglycosides produced by *Micromonospora*. Gentamicin was discovered in 1963 in cultures of *Micromonospora purpurea*. The fermentation yields a mixture of

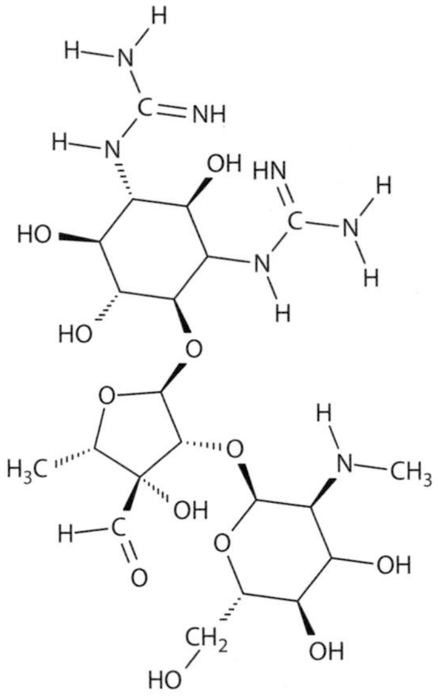

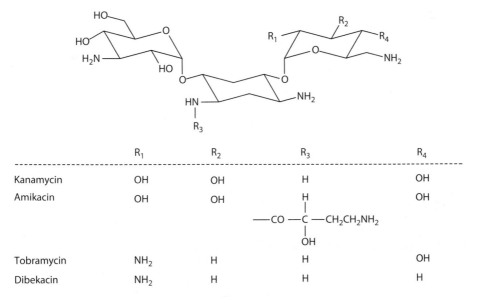

FIGURE 25.10 Streptomycin, neomycin, and paromomycin.

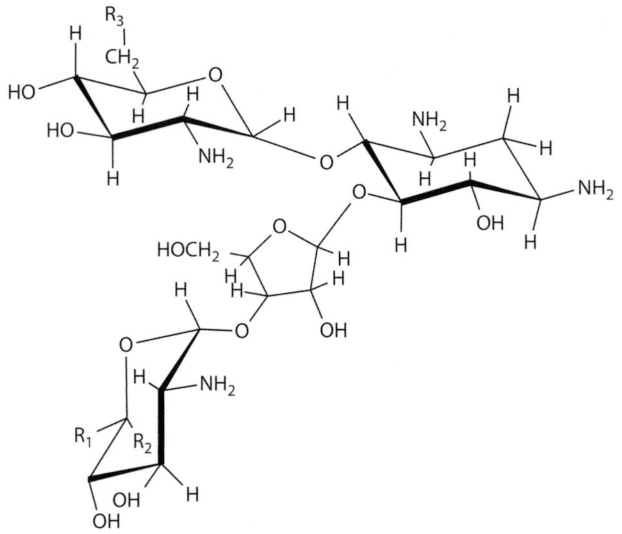

Neomycin B	$R_1 =$	—H	$R_2 =$	—CH_2NH_2	$R_3 =$	—NH_2
Neomycin C	$R_1 =$	—CH_2NH_2	$R_2 =$	—H	$R_3 =$	—NH_2
Paromomycin I	$R_1 =$	—H	$R_2 =$	—CH_2NH_2	$R_3 =$	—OH
Paromomycin II	$R_1 =$	—CH_2NH_2	$R_2 =$	—H	$R_3 =$	—OH

	R_1	R_2	R_3	R_4
Kanamycin	OH	OH	H	OH
Amikacin	OH	OH	H / —CO—C(OH)—$CH_2CH_2NH_2$	OH
Tobramycin	NH_2	H	H	OH
Dibekacin	NH_2	H	H	H

FIGURE 25.11 Kanamycin and related compounds.

gentamicins A, B, and C, from which the C component is isolated. Sisomicin was isolated in 1970 from *Micromonospora inyoensis* and netilmicin is obtained by the ethylation of sisomicin.

Another aminoglycoside that was obtained by partial synthesis is amikacin, where a L-2-hydroxy-4-aminobutyryl group was introduced in the C1 amino group of the deoxystreptamine ring of kanamycin (1972). The increased use of the aminoglycosides for the treatment of infections with Gram-negative bacteria has led to the development of strains resistant to these antibiotics (Figure 25.12).

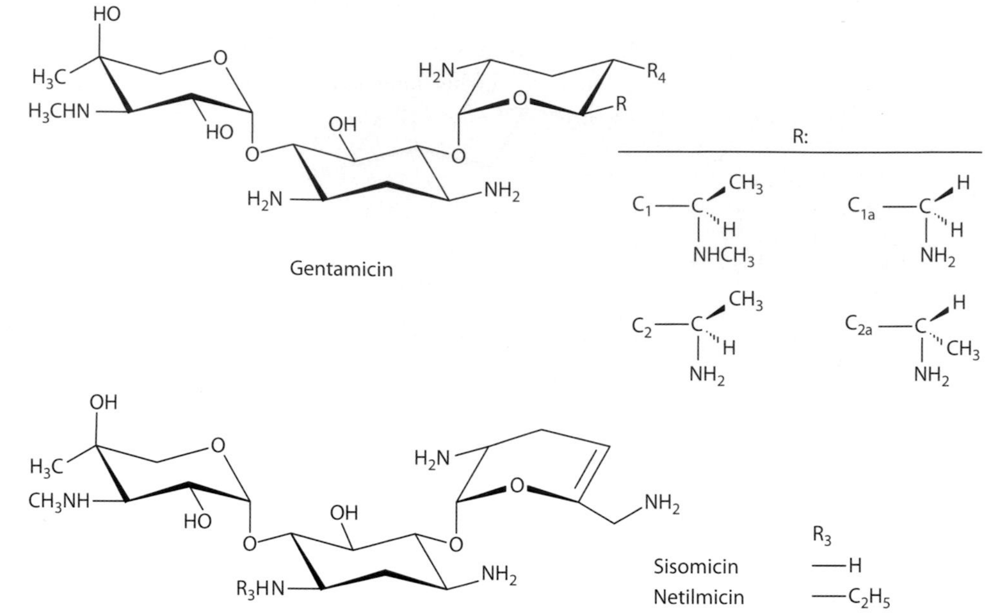

FIGURE 25.12 Gentamicin and derivatives.

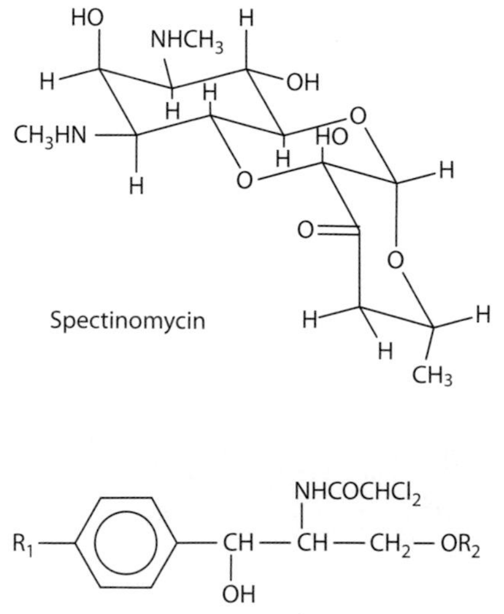

R_1	R_2	
NO_2	—H	Chloramphenicol
NO_2	—$CO(CH_2)_{14}CH_3$	Palmitate
NO_2	—$CO(CH_2)_2COONa$	Hemisuccinate
SO_2CH_3	—H	Thiamphenicol
SO_2CH_3	—$COCH_2NH_2$.HCl	Glycinate

FIGURE 25.13 Spectinomycin, chloramphenicol, and thiamphenicol.

Gentamicin contains fewer groups that may react with these enzymes. The introduction of the 2-hydroxy-4-amino group in amikacin protects many groups against these enzymes, while maintaining the activity of the parent molecule. Tobramycin, which was obtained from *Streptomyces tenebrarius* in 1968, is also less susceptible because one hydroxy group is replaced by hydrogen.

A special product is spectinomycin, which was isolated in 1961 from *Streptomyces spectabilis.* It is a broad-spectrum antibiotic with a moderate activity. It is used for the treatment of gonorrhea (Figure 25.13).

25.5.1.2 Tetracyclines

The first tetracycline, chlortetracycline, was discovered in 1948 in a culture of *Streptomyces aureofaciens*. Oxytetracycline was isolated somewhat later (1950) from *Streptomyces rimosus*. The structure was studied in the laboratories of Lederle (Boothe et al.) and Pfizer (Hochstein et al.) in collaboration with Woodward at Harvard University. The full structure was published in 1952 but complete stereochemistry was obtained later from x-ray diffraction analysis. During these studies, it was found that the removal of chlorine from chlortetracycline by hydrogenolysis led to an active product, tetracycline. In other laboratories, it was shown that tetracycline could be produced by fermentation of a medium poor in chloride using an appropriate strain, e.g., *S. alboniger, S. viridifaciens*, etc.

The tetracyclines derive their name from the tetracyclic ring system, which is octahydronaphthacene. They have three ionizable groups, on C_3, C_4, and the dihydroxy ketone system (C_{10}–C_{12}). The tetracyclines are often used in the form of the hydrochloride, but the tetracycline bases are also used. The C_4. dimethylamino-group may epimerize and these epimers are almost inactive. At pH values between 4 and 7, mixtures of normal and epitetracycline are formed.

In acid medium, the C_6 hydroxy group and C_5 hydrogen are removed in the form of water and anhydrotetracyclines are found. For this reason demeclocyline, which has a C_6 secondary hydroxyl group instead of a tertiary one, is more stable. This product was obtained in 1957 using a mutant strain of *S. aureofaciens*. Chemical manipulations of oxytetracycline led to the production of metacycline. Hydrogenation of metacycline under suitable conditions, gave doxycycline. It should be noted that the C_6 methyl group should have the α-configuration as in oxytetracycline, because the C_6 epimer is less active. Doxycycline is very stable (no C_6–OH group) and has a high lipophilic character. It is more completely absorbed after oral administration. Minocycline was described in 1972 and is prepared by the chemical treatment of 6-deoxy-6-demethyltetracycline.

The tetracyclines are true broad-spectrum antibiotics. They are active against a wide range of Gram-positive and Gram-negative bacteria, spirochetes, mycoplasmas, rickettsiae, and chlamydia.

The *in vitro* activities of the different tetracyclines are very similar. Only the greater activity of minocycline against some Gram-positive bacteria like staphylococci and streptococci should be noted (Figure 25.14).

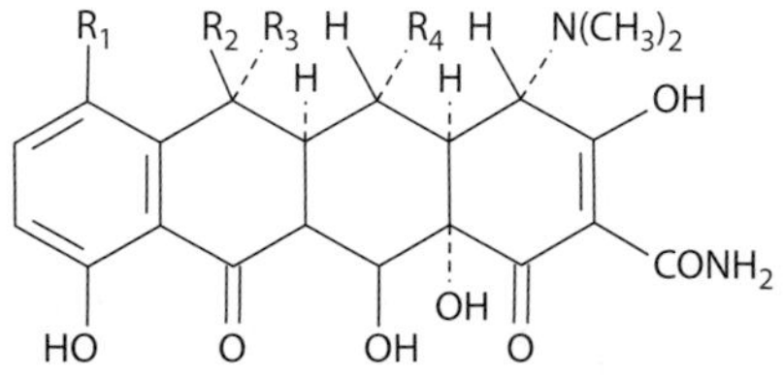

R_1	R_2	R_3	R_4	
H	OH	CH_3	H	Tetracycline
Cl	OH	CH_3	H	Chlortetracycline
H	OH	CH_3	OH	Oxytetracycline
Cl	OH	H	H	Demeclocycline
H	$=CH_2$		OH	Metacycline
H	H	CH_3	OH	Doxycycline
$N(CH_3)_2$	H	H	H	Minocycline

FIGURE 25.14 Tetracyclines.

25.5.2 Inhibitors of the 50 S Ribosomal Subunit

25.5.2.1 Chloramphenicol

The first broad-spectrum antibiotic was chloramphenicol, which was isolated in 1947 from *Streptomyces venezuelae*. Its chemical structure was soon established and in 1949 a synthesis was described. The commercial production always has been by the synthetic route. Of the four possible diastereoisomers, only the R,R isomer is active and is separated during the synthesis.

Many derivatives of chloramphenicol were prepared but only the sulfomethyl analogue, thiamphenicol, has come into clinical use. It is generally less active than chloramphenicol. The glycinate ester is used as a prodrug for injections. Chloramphenicol has a moderate activity against Gram-positive and Gram-negative bacteria. It is not recommended for treatment of these infections, because of the occurrence of serious toxic reactions in the blood (aplastic anemia, thrombocytopenia). It is still used in the treatment of typhus and meningitis caused by *Haemophilus influenzae*.

25.5.2.2 Macrolides

The macrolide antibiotics have in common (a) a large lactone ring (hence the name macrolide); (b) a glycosidically linked aminosugar (sometimes two), and (c) usually a desoxysugar. The lactone ring may contain 12 (macrolides not used in medicine), 14 (erythromycin, oleandomycin), and 16 atoms (leucomycin, spiramycin, tylosin).

The first clinical useful macrolide was erythromycin, isolated from a culture of *Streptomyces erythreus* (1952). The structure was determined by chemical methods (1954–1957) and the stereochemistry and conformation by x-ray diffraction and NMR. Erythromycin is inactivated by acid.

Clarithromycin, where the C_6-hydroxy group is replaced by a methoxy group and roxithromycin, where the ketone group is under the form of an oxime-ether is more acid-stable (Figure 25.15).

The 9-position of erythromycin has been changed more dramatically in dirithromycin and azithromycin. Their spectra are comparable to that of erythromycin itself. Azithromycin has a longer half-life. Recently, telithromycin was introduced as a semisynthetic derivative of erythromycin. It is active a.o. against *S. pneumonia*, β-hemolytic streptococci, *L. pneumophila*, *Chlamydia pneumoniae*, and *Mycoplasma pneumoniae* (Figure 25.16).

Oleandomycin was isolated in 1955 by Sobin et al. from *Streptomyces antibioticus*, It is administered usually as the triacetyl derivative, which gives higher blood levels.

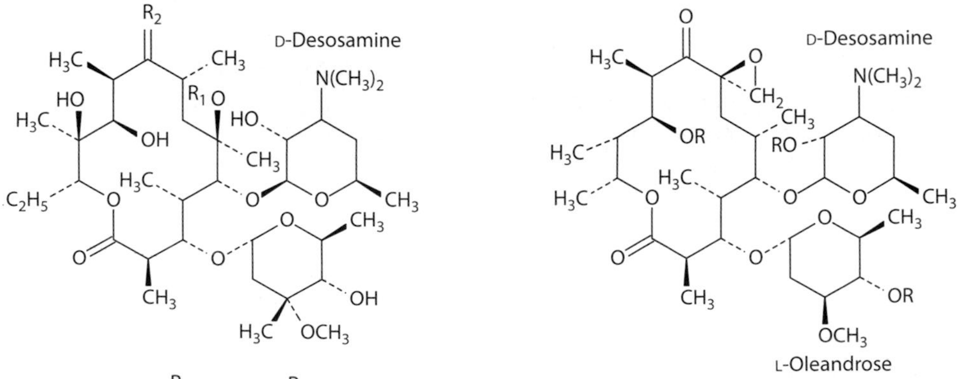

	R_1	R_2
Erythromycin	H	= O
Roxitthromycin	H	= N — O — CH₂CH₂ — OCH₃
Clarithromycin	CH₃	═O

Oleandomycin	R = H
Troleandomycin	R = COCH₃

FIGURE 25.15 Macrolides.

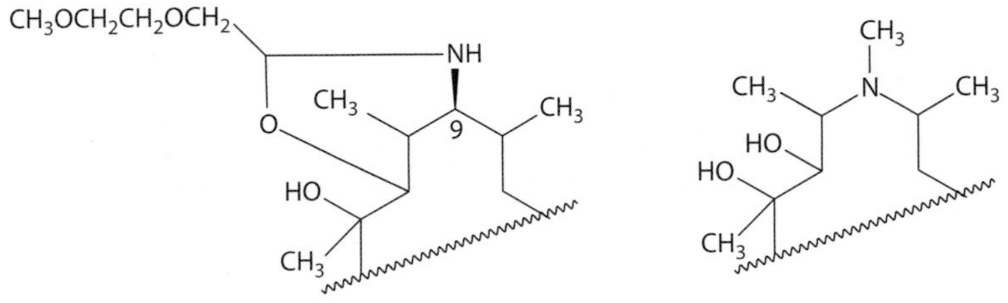

FIGURE 25.16 C_7–C_{12} fragments of dirithromycin and azithromycin.

Erythromycin is very active against Gram-positive bacteria. Oleandomycin has a similar spectrum, but the MIC are generally higher.

Spiramycin and leucomycin are macrolides with a more limited use. Spiramycin was discovered in *S. ambofaciens* (1955). Besides the main component I, it contains some components II (max. 15%) and III (max. 10%).

Another active macrolide is tylosin. Its application is restricted to veterinary medicine.

25.5.2.3 Lincomycin, Clindamycin

Lincomycin is also a basic antibiotic isolated in 1962 from *Streptomyces lincolnensis*. The basic group is in the proline part of the molecule and the sugar moiety contains a methylmercapto group. In 1967, it was shown that replacement of a hydroxy group by chlorine, with inversion of configuration, resulted in a product clindamycin, with improved absorption and higher serum levels. Both antibiotics are active against Gram-positive bacteria, with a spectrum similar to that of erythromycin. Side effects are diarrhea and occasionally serious pseudomembranous colitis, which is caused by an overgrowth of clindamycin-resistant strains of *Clostridium difficile* (Figure 25.17).

25.5.2.4 Fusidic Acid

Fusidic acid (Figure 25.6) was isolated from *Fusidium coccineum* in 1962. It has a unique steroid type (fusidane) structure. Cephalosporin P_1 has a similar structure. Fusidic acid is active against Gram-positive bacteria and Gram-negative cocci. Resistant strains rapidly emerge. Because of this observation, its use is limited.

25.6 CELL WALLS OF BACTERIA

Most prokaryotic cells are surrounded by a cell wall that is responsible for their shape and allows bacteria to live in a hypotonic environment without bursting. In 1884, C. Gram discovered that some bacteria retained crystal violet–iodine complex after washing with alcohol (Gram-positive) and others did not (Gram-negative).

Gram-positive cells are surrounded by a cytoplasmic membrane and a thick cell wall consisting of peptidoglycan to which are linked polyol phosphate polymers called teichoic acids. Gram-negative bacteria have a much thinner cell wall consisting of peptidoglycan and associated proteins, and this cell wall is surrounded by an outer membrane comprising of lipid, lipopolysaccharide, and protein. The osmotic pressure in the cytoplasm of Gram-positive bacteria (±20 atm) is higher than that of in Gram-negative (±5 atm).

Peptidoglycan is an alternating polymer of *N*-acetylglucosamine (NAG) and *N*-acetylmuramic acid (NAM), which are cross-linked by a short peptide bridge, and which is 200–250 Å thick in Gram-positive bacteria. Peptidoglycan forms an enormous bag shaped molecule, also called "murein sacculus" (Latin murus: wall), which surrounds the entire cell.

NAM is the 3-*O*-D-lactylether of *N*-acetylglucosamine. NAG and NAM are linked by $\beta(1–4)$ glycosidic bonds and form a linear structure as in cellulose (glucose units) and in chitin

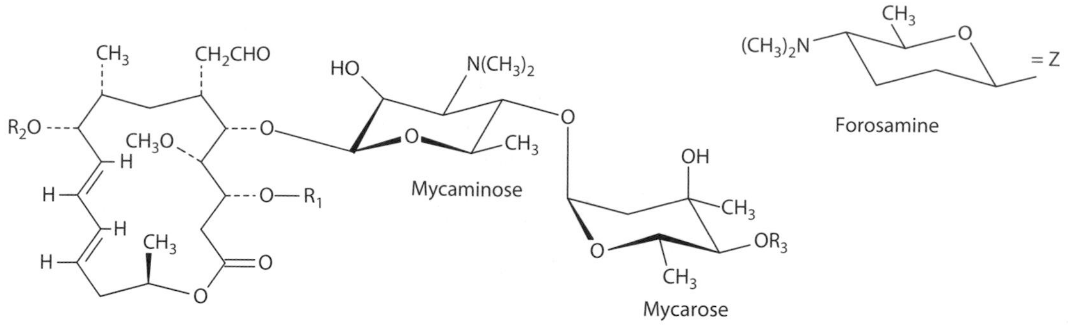

	R_1	R_2	R_3
Spiramycin I	—H	Z	H
Spiramycin II	—$COCH_3$	Z	H
Spiramycin III	—$COCH_2CH_3$	Z	H
Leucomycin A_1	H	H	—$COCH_2CH(CH_3)_2$

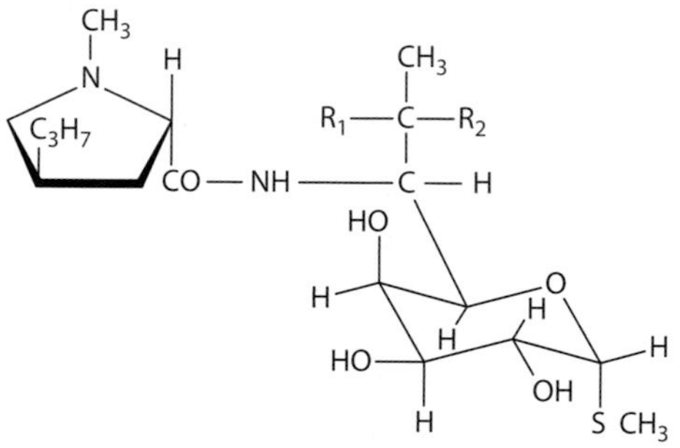

Lincomycin	$R_1 = OH$	$R_2 = H$
Clindamycin	$R_1 = H$	$R_2 = Cl$

FIGURE 25.17 Macrolides and lincomycin (derivatives).

(*N*-acetylglycosamine units). The carboxyl group of NAM is linked to a short peptide, which forms the bridge with another NAG–NAM strand.

One of the important new targets to develop antibiotics is the transglycosylase reaction, responsible for the polymerization of the disaccharide units. During the transglycosylase reaction, the configuration of C_1 is changed from α to β. Moenomycin A, active against Gram-positive bacteria, is a transglycosylase inhibitor (Figure 25.18).

The formation of the linkage between two peptide ends is catalyzed by a transpeptidase bound to the outside of the cytoplasmic membrane.

This reaction regulates the degree of peptidoglycan cross-linking. In *Staphylococcus aureus*, 90% of the peptide chains are involved in cross-links, whereas in Gram-positive and Gram-negative bacilli only 20%–50% of the strands are cross-linked.

Vancomycin kills bacteria by targeting lipid II. Lipid II is the target of several other classes of natural products like lantibiotics, ramoplanin, and mannopeptimycins. Nisin that acts through a lipid-II-dependent targeted pore formation mechanism is an example of a lantibiotic. Vancomycin and teicoplanin dissociate with the acyl-D-Ala-D-Ala terminus of the growing peptidoglycan chain. Both compounds are active against Gram-positive bacteria. Resistance against vancomycin is observed when the D-Ala-D-Ala terminus is changed to D-Ala-D-Lac. Nisin is active against vancomycin resistant bacteria because it does not interact with the amino acids of the pentapeptide chain (Figure 25.19).

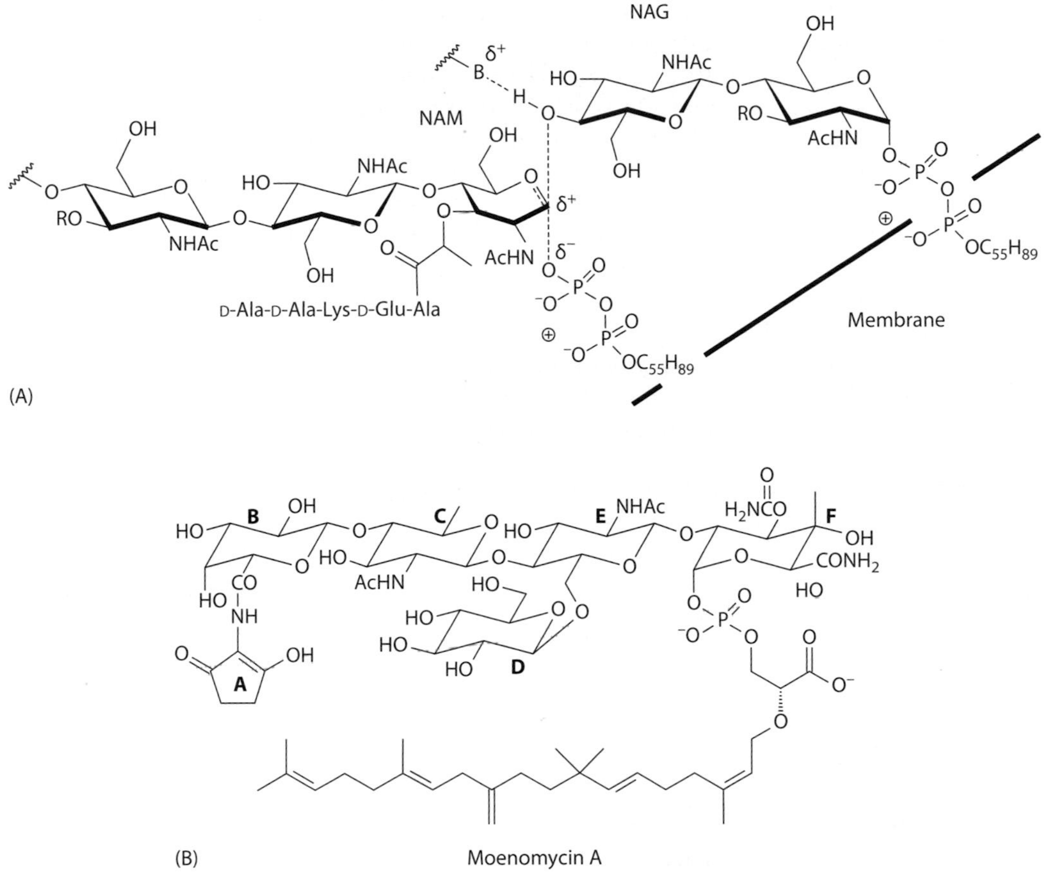

(A)

(B)   Moenomycin A

FIGURE 25.18   Transglycosylation step and structure of moenomycin A.

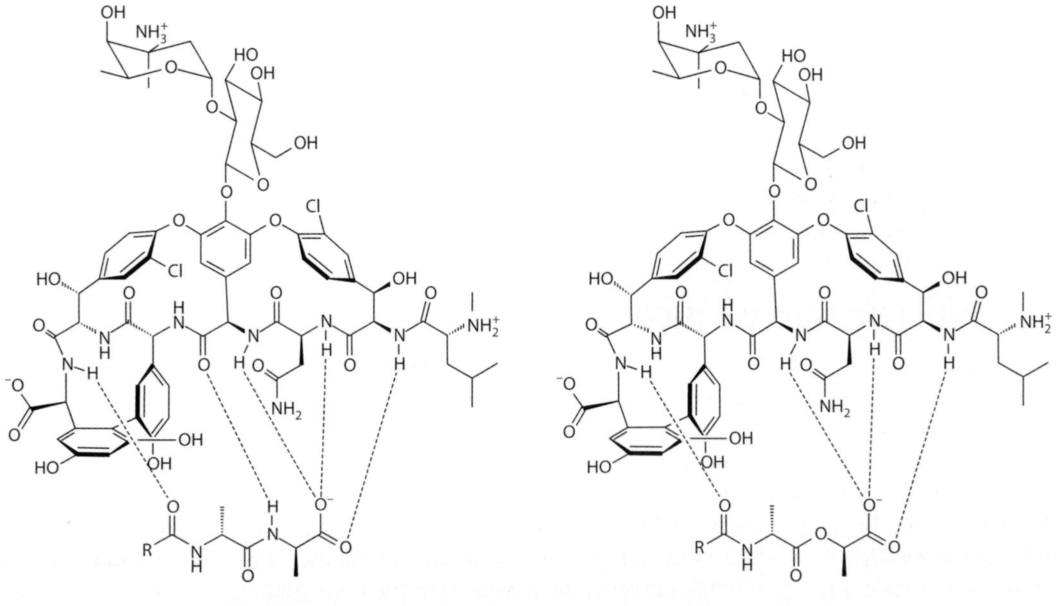

D-Ala-D-Ala   D-Ala-D-Lac

FIGURE 25.19   Binding of vancomycin at D-ALa-D-Ala and D-Ala-D-Lac.

β-Lactam antibiotics inhibit transpeptidase and carboxypeptidase. In this reaction, the β-lactam ring is opened and the serine of catalytic part of the enzyme is acylated. The inhibition of the cross-linking reaction yields a poorly structured cell wall and results in the lysis of the cell. This explains the early observation that penicillin affects only growing bacteria.

From 1972 onward, several investigators could show, using radioactive penicillin and separation techniques, that membranes of bacteria contained several penicillin-binding proteins (PBPs). Differences in susceptibility of bacteria to different β-lactams may be explained by the amounts of the different PBPs and their affinity for these antibiotics. The PBP of lower molecular mass are monofunctional carboxy- and endopeptidases, transpeptidases, and β-lactamase. The higher molecular mass PBPs are multimodular containing a transpeptidase and, for example, a transglycosylase. It seems that inhibition of at least two PBPs is required for efficient killing by β-lactams.

In the periplasmic space (between cytoplasmic membrane and outer membrane), several enzymes are present, e.g., β-lactamases (in all Gram-negative bacteria), enzymes that acetylate chloramphenicol, adenylate streptomycin, etc. and also proteins responsible for the transport of sugars and other nutrients.

In Gram-negative bacteria, antibiotics have to pass through the porins. This permeation is easiest for polar molecules. A second factor in their activity is the resistance to β-lactamase. A third factor, both in Gram-positive and Gram-negative bacteria is the affinity for PBP. There are marked differences in this affinity for different penicillins and cephalosporins. The minimum inhibitory concentration (MIC) that measures the *in vitro* activity of an antibiotic is the result of a series of different factors.

25.7   MEMBRANES

The cytoplasmic membrane of bacteria is also a lipoprotein structure. The major phospholipid is phosphatidylethanolamine. Several proteins (also enzymes) are located in and around this membrane. Polymyxin and colistin interact with the cell membrane. The binding of the drug involves the phosphate groups of the phospholipid. According to a mechanism similar to that of the quaternary ammonium detergents, the fatty acid tail penetrates into the hydrocarbon part of the phospholipid, while the cyclic peptide containing the free amino groups interacts with the phosphate groups. This disruption of the membrane structure brings about a loss of their permeability barrier property. Biochemical functions like respiration, nucleic acid and protein synthesis are perturbed. Tyrocidine that does not have a detergent-like structure nevertheless has a similar effect.

Ionophores cause the loss of essential monovalent cations (K^+) because of specific changes in the permeability of the membrane. The polyethers act as carriers, by providing lipid solubility of the transported cations. Gramicidin, which is a linear peptide, probably adopts a helical conformation and forms a hydrophobic channel of the ions.

25.8   NUCLEIC ACID SYNTHESIS

The planar phenoxazone ring of actinomycin intercalates at the level of GpC sequences in DNA. This intercalation partially unwinds the DNA helix and inhibits the use of DNA for replication and transcription. The anthracycline antibiotics have a more complex mode of action: they intercalate in DNA, they have alkylating properties, they inhibit the action of topoisomerase II and could give rise to hydroxyl radicals that damage DNA. Several antitumoral antibiotics like daunomycin and bleomycin may break strands of DNA and give rise to cross-linking. Rifampicin inhibits RNA polymerase by directly binding to the enzyme in a noncovalent manner. The drug does not inhibit transcription once it has begun, but prevents the initiation of the transcription.

A molecule of DNA consists of two linear strands intertwined to form a double helix. Those strands form often a ring. Such a relaxed bacterial DNA is too long to fit inside a bacterial cell.

Supercoiling is essential for housing inside the cell and strand unwinding is necessary for replication and transcription.

Topoisomerases are enzymes that convert DNA from one topological form to another. The enzyme hydrolyses the phosphodiester bond of DNA backbone, making use of tyrosine residues in the protein, allows the supercoiled DNA to pass into the relaxed form and reseals the strands.

Topoisomerase I cuts a single strand of the double helix, topoisomerase II cuts two strands simultaneously. DNA gyrase, discovered by Gellert in 1976, is a topoisomerase II, which occurs only in prokaryotes. It is able to supercoil a relaxed DNA ring (reaction A). ATP is required as an energy source in this process. Quinolones inhibit the catalytic activity of DNA gyrase and stabilize the DNA cut. This means that the higher the topoisomerase activity in a cell, the more active the quinolones will be. DNA gyrase consists of two A subunits and two B subunits and the A subunits are involved in cleavage and annealing of the DNA strands. The targeting of the enzymes (gyrase A, a subunit of topoisomerase IV) by fluoroquinolines is organism- and compound-specific.

25.9   PROTEIN SYNTHESIS

Ribosomes are cellular particles, 200–500 Å in diameter. About 1800 are present per bacterial cell, where they are bound to the cytoplasmic membrane. All ribosomes can dissociate into a small and a large subunit. In prokaryotes, the 70 S ribosome (sedimentation coefficient 70 S) consists of a 30 and a 50 S subunit. The 30 S subunit contains a 16 S RNA molecule, the 50 S subunit a 23 and a 5 S RNA molecule. The ribosomes of eukaryotic cells are larger (80 S) with 60 and 40 S subunits. Several ribosomes simultaneously translate mRNA. The complex of mRNA and several ribosomes is called as a polyribosome. Several antibiotics interfere with protein synthesis at the level of the ribosome. The macrolides (for example erythromycin) bind to the 50 S subunit and inhibit the translocation reaction. The aminoglycoside (for example streptomycin) interferes with the initiation step of protein synthesis and also induces miscoding during protein synthesis. They interact with the 30 S subunit. The binding mode of aminoglycosides to the A site of the 16 S ribosomal RNA has been determined using x-ray diffraction studies and is based on electrostatic interactions and direct and water-mediated hydrogen bonds. Aminoglycosides stabilize a conformation of the aminoacyl-tRNA decoding site that normally occurs when the tRNA–mRNA complex is bound. The ribosome now is unable to discriminate between cognate and noncognate tRNA–mRNA duplex.

Tetracyclines bind to the 30 S subunit and inhibit the binding of aminoacyl–tRNA at the A-site of the ribosome. Chloramfemicol is an inhibitor of the peptidyl transferase activity of the 50 S subunit and releases short oligopeptidyl–tRNA. Fusidic acid stabilizes the normally unstable ribosome-elongation factor G-GTP complex and inhibits translocation.

Puromycin is an antibiotic that has been isolated from *S. alboniger* in 1953 but that is not used in the clinic. It has been very useful in the study of the protein synthesis. It resembles the amino-acyl–adenosine part of aminoacyl-tRNA. It binds to the A-site in the ribosome in the place of an aminoacyl–tRNA. The amino group of puromycin forms a peptide bond with the peptide, a reaction catalyzed by peptidyl transferase. The peptidylpuromycin is then released from the ribosome and thus causes premature chain termination.

25.10   NEW DEVELOPMENTS IN ANTIBACTERIAL RESEARCH

Since the last new structural class of antibiotics has been discovered (quinolons), only few really new antibiotics have been approved, i.e., linezolid, daptomycin, tigecycline, and retapamulin. Tigecycline is a glycylcycline that is active against Gram-positive and Gram-negative bacteria. It inhibits protein synthesis. Daptomycin is used against Gram-positive infections, including resistant pathogens such as MRSA (methycillin-resistant *Staphylococcus aureus*), VRE (vancomycin-resistant *enterococci*), and PRSP (penicillin-resistant *Streptococcus pneumoniae*). It is a cyclic lipopeptide that originates from *Streptomyces roseosporus*, discovered at Eli Lilly in the early 1980s. It has a unique mode of action, as it disrupts several functions of the bacterial plasma membrane without penetrating into the cytoplasm. The insertion of daptomycin into the cytoplasmic membrane bilayers is calcium-dependent. In the presence of calcium, daptomycin readily forms aggregates. Its concentration in the membrane causes leakage leading to cell death. The oxazolidinones (i.e., linezolid) are active against Gram-positive pathogenic bacteria (including MRSA) and inhibit bacterial translation at the initiation phase of protein synthesis. It binds to the 50 S subunit and inhibits the interaction with the 30 S subunit. Retapamulin is a semisynthetic pleuromutilin, a natural product of the fungi *Pleurotus mutilis*. It is active against *S. aureus* and *S. pyogenes* and used for the treatment of impetigo. Retapamulin inhibits bacterial protein synthesis by binding to the bacterial ribosomal 50 S subunit. Its binding site is different from that of the classical ribosome-binding antibiotics (Figure 25.20).

During the last decennium, successful drug classes, like β-lactams (ceftazoline, ceftobiprole), tetracyclines (tigecycline), macrolides (cetromycin), and trimethoprim (iclaprim) have been modified by introduction of additional target binding sites, so that the compounds become active against resistant strains. Other possibilities to obtain improved antibiotics are the synthesis of hybrids of two antibiotic pharmacophores, the development of multitargeted antibiotics, and the combination therapy.

During the last two decennia, several discoveries and new technologies that could increase the likelihood of identifying new antibiotics or antibiotic targets became available. Examples are combinatorial library synthesis (i.e., the automatic synthesis of complex oligosaccharides), new screening methods, the availability of bacterial genome sequences, a better understanding of the host immune system, natural product screening, the discovery of riboswitches, and the availability of more x-ray structure of bacterial proteins. Despite this, however, the main target for antibiotic development is still not altered and includes DNA replication, cell-wall biosynthesis, ribosomal RNA function, and membrane functions. It seems that the use of other antibacterial targets mainly leads to antibiotics with a narrow spectrum and that targeting multiple enzymes will be needed together with the development of new chemistries (from natural or bioengineering origin). The de novo design of multiple-targeted antibiotics for monotherapy is a difficult process. An alternative is to target functions, essential for infection, such as bacterial virulence factors or disrupting the interactions between the host and the pathogen (with lesser risk to develop antibiotic resistance). Improvement of the treatment of bacterial infections might also be expected when the antibiotic is combined with a compound that inhibits the mechanisms of persistence.

Antibiotic resistance is presently a serious problem in the fight against bacterial infections. Three types of resistance mechanism can be distinguished: (a) natural or intrinsic resistance; (b) mutational

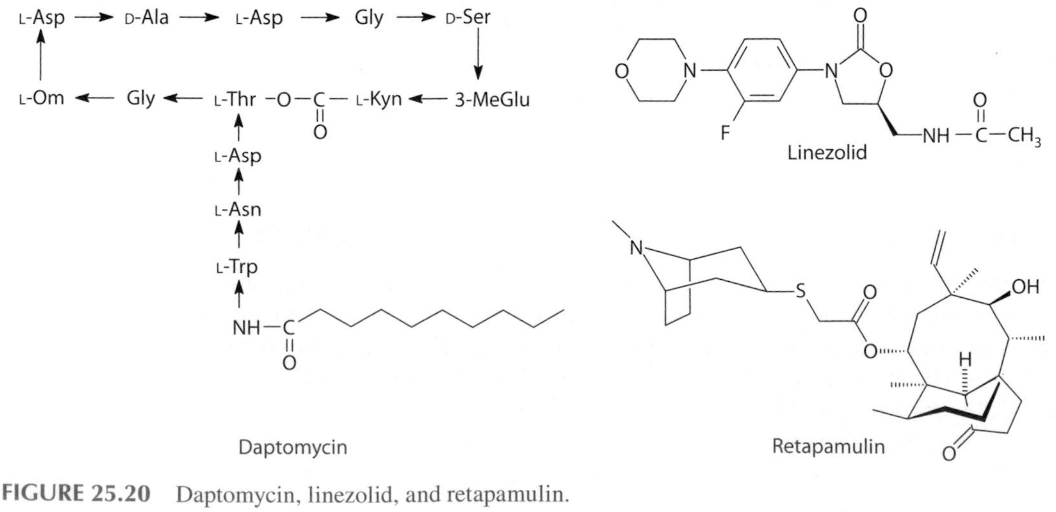

FIGURE 25.20   Daptomycin, linezolid, and retapamulin.

resistance; and (c) extrachromosomal acquired resistance, which is disseminated by plasmids or transposons. The first type of resistance can be due to the inaccessibility of the target, the presence of a multidrug efflux system, or due to inactivation of the antibiotic. The mutational resistance, likewise, may be of different origin. This resistance may be due to a reduced permeability or uptake, a metabolic bypass, a modification of the target site, or a repression of the multidrug efflux system. The third type of resistance may be caused by drug inactivation, target site modification, metabolic by-pass, or the efflux system. The existence of multidrug-resistance efflux pumps is the major mechanism of intrinsic and acquired resistance of bacteria against a variety of different antibiotics. The resistance-nodulation-cell-division (RND) family of transporters form a large multiprotein complex that transverse the inner and outer membrane of Gram-negative bacteria through the periplasmic region. Inhibition of the efflux pumps could be used in combination with antibiotics and a variety of such compounds have been identified. It has been observed that antibacterial targets, that give rise to low occurrence of resistance through single-step mutation, are of made of multiple genes or are structures that are synthesized by multiple genes. Antibacterials targeting single enzymes easily give rise to high-level resistance and are best used in combination therapy.

Another important observation is that most bacterial infections contain nonmultiplying bacteria that are resistant to treatment with antimicrobial drugs. They could exist as spores, in a dormant state, or in a clinically latent situation. The best-known example of a clinically latent bacterium is the *Mycobacterium tuberculosis*. The use of drugs that target nonmultiplying bacteria should result in shorter treatment periods and, hence, a lower level of resistance. Some of the known antibiotics (i.e., penems, nocardicin A, some quinolines, rifampicins) are able to kill some nonmultiplying bacteria.

25.11   CONCLUDING REMARKS

Development of antibiotics is one of the oldest fields in medicinal chemistry and antibiotics have saved uncountable lives of human beings during the last 60 years. It is a research domain that has been neglected during the last two decennia with serious consequences. The bugs are fighting back and they have learned how to defend themselves against their killers. There is an urgent need for new antibacterials with new mode of actions. Likewise, the way of discovery of new antibacterials should be altered. In the past, this was mainly focused on the discovery of inhibitors of bacterial growth. The example of tuberculosis shows that this is not sufficient. The main problem is that the development of new antibiotics is one of the difficult fields of medicinal chemistry. Using leads from nature combined with medicinal chemistry to influence the spectrum and the pharmacokinetics of the natural compounds will remain one of the most important ways to develop new antibiotics.

FURTHER READINGS

E. Breukink and B. de Kruiff. Lipid II as a target for antibiotics. *Nat Rev Drug Discov*. 5, 321–332, 2006.
A.E. Clatworthy, E. Pierson, and D.T. Hung. Targeting virulence: A new paradigm for antimicrobial therapy. *Nat Chem Biol*. 3, 541–548, 2007.
A. Coates, Y. Hu, R. Bax, and C. Page. The future challenges facing the development of new antimicrobial drugs. *Nat. Rev. Drug Discov*. 1, 895–910, 2002.
O. Lomovskaya, H.I. Zgurskayal, M. Totrov, and W.J. Watkins. Waltzing transporters and 'the dance macabre' between humans and bacteria. *Nat. Rev. Drug Discov*. 6, 56–65, 2007.
D.J. Payne, M.N. Gwynn, D.J. Holmes, and D.L. Pompliano. Drugs for bad bugs: Confronting the challenges of antibacterial discovery. *Nat. Rev. Drug Discov*. 6, 29–40, 2007.
L.L. Silver. Multi-targeting by monotherapeutic antibacterials. *Nat. Rev. Drug Discov*. 6, 41–55, 2007.
P.A. Smith and F.E. Romesberg. Combating bacteria and drug resistance by inhibiting mechanisms of persistence and adaptation. *Nat Chem Biol*. 3, 549–556, 2007.

Index